Intelligence Techniques Applied in Infrastructure, Engineering and Construction

Intelligence Techniques Applied in Infrastructure, Engineering and Construction

Guest Editors

Kaiwen Liu
Tengfei Wang
Xiaoning Zhang

MDPI

Basel • Beijing • Wuhan • Barcelona • Belgrade • Novi Sad • Cluj • Manchester

Guest Editors
Kaiwen Liu
School of Civil Engineering
Southwest Jiaotong
University
Chengdu
China

Tengfei Wang
School of Civil Engineering
Southwest Jiaotong
University
Chengdu
China

Xiaoning Zhang
School of Civil Engineering
Chongqing University
Chongqing
China

Editorial Office
MDPI AG
Grosspeteranlage 5
4052 Basel, Switzerland

This is a reprint of the Special Issue, published open access by the journal *Buildings* (ISSN 2075-5309), freely accessible at: https://www.mdpi.com/journal/buildings/special_issues/WUNJFZ3314.

For citation purposes, cite each article independently as indicated on the article page online and as indicated below:

Lastname, A.A.; Lastname, B.B. Article Title. *Journal Name* **Year**, *Volume Number*, Page Range.

ISBN 978-3-7258-4363-3 (Hbk)
ISBN 978-3-7258-4364-0 (PDF)
https://doi.org/10.3390/books978-3-7258-4364-0

Contents

About the Editors

Kaiwen Liu

Kaiwen Liu is a Professor and Doctoral Supervisor at Southwest Jiaotong University, recognized as a national high-level young talent. He has authored over 60 papers in leading SCI-indexed journals and holds more than 30 invention patents, including patents granted in the United States and Australia. Dr. Liu serves as the Associate Editor of *Geotextiles and Geomembranes*, Scientific Editor of the *Journal of Rock Mechanics and Geotechnical Engineering*, and Editorial Board Member of the *Canadian Geotechnical Journal*. His research achievements have earned him multiple awards, including the First Prize of the Science and Technology Progress Award from the China Transportation Association (1st place), Anhui Province (2nd place), and the Third Prize from Sichuan Province (3rd place).

Tengfei Wang

Tengfei Wang is an Associate Professor of Railway Engineering at Southwest Jiaotong University. He received his Ph.D. from Beijing Jiaotong University with joint training at the University of Wisconsin–Madison. His research focuses on the mechanics of transport earthworks, including pile behaviors, subgrade dynamics, and AI-assisted geotechnical analysis. Dr. Wang has led several national research projects and published over 50 SCI-indexed papers as first or corresponding author. He holds eleven Chinese invention patents and serves on technical committees of national engineering societies. He is also on the Youth Editorial Board of *Research in Cold and Arid Regions* and acts as a reviewer for more than twenty international journals, including *Géotechnique*. His honors include the ASCE Outstanding Reviewer Award and NSERC external assessment roles.

Xiaoning Zhang

Xiaoning Zhang is a Lecturer and Assistant Researcher at Chongqing University, and a recipient of the National Postdoctoral Innovation Talent Support Program. His research centers on intelligent, low-carbon geotechnical environments and sustainable construction technologies. Dr. Zhang has published 15 SCI/EI papers and holds 15 invention patents, including two in the U.S. He has contributed to multiple national and local engineering standards and received several top-tier science and technology awards, including the First Prize from the China Highway Society and China Transportation Association. He has led or participated in various national-level research projects, including those funded by the National Natural Science Foundation and major R&D programs.

Preface

The accelerating infusion of artificial intelligence, machine learning, and data-centric analytics into infrastructure and construction is redefining how engineers conceive, build, and maintain the built environment. Across their lifecycles, bridges, tunnels, railways, and buildings must withstand dynamic loading, harsh climates, and escalating performance expectations for sustainability and resilience. Intelligent techniques now make it possible to anticipate these demands, optimize structural responses in real time, and extend service life with a precision that traditional methods could scarcely imagine.

Building on this momentum, the present Special Issue, "Intelligent Techniques Applied in Infrastructure, Engineering and Construction," gathers twenty-three peer-reviewed studies that collectively chart the state of the art. The papers investigate topics ranging from smart excavation support and tunnel-boreability classification to UAV-assisted bridge-defect localisation, sensor-calibrated finite-element modelling for ballastless tracks, hybrid optimization of concrete thermal parameters, and explainable deep-learning frameworks for cost estimation. Materials explored include ultra-high-performance concrete, fiber-reinforced polymers, unsaturated clayey sands, and mechanically connected precast piles, while methodological advances span ensemble classifiers, Bayesian networks, transformer-enhanced computer vision, and physics-informed surrogate modelling. Together, these contributions showcase how interdisciplinary innovation can elevate safety, efficiency, and environmental stewardship across civil-engineering domains.

The Guest Editors extend their heartfelt gratitude to every author for entrusting their cutting-edge work to this collection and to the anonymous reviewers whose thoughtful evaluations ensured its scholarly rigor. Appreciation must also be extended to the *Buildings* editorial staff for their steadfast support throughout the publication process. It is our hope that the insights contained herein will guide practitioners, inspire researchers, and accelerate the responsible adoption of intelligent techniques in infrastructure worldwide.

Kaiwen Liu, Tengfei Wang, and Xiaoning Zhang
Guest Editors

buildings

MDPI

Editorial

Intelligent Techniques Applied in Infrastructure, Engineering, and Construction

Kaiwen Liu [1,*], Tengfei Wang [1,*] and Xiaoning Zhang [2]

1 MOE Key Laboratory of High-Speed Railway Engineering, School of Civil Engineering, Southwest Jiaotong University, Chengdu 610031, China
2 School of Civil Engineering, Chongqing University, Chongqing 400045, China; zhangxn@cqu.edu.cn
* Correspondence: kaiwen.liu@queensu.ca (K.L.); w@swjtu.edu.cn (T.W.)

1. Introduction

The accelerating convergence of artificial intelligence (AI) and smart construction [1] technologies is transforming every phase of infrastructure delivery—from concept [2] and design [3] to construction [4], operation, and renewal. In recent years, breakthroughs in predictive analytics [5], computer vision [6], and real-time sensing have begun to alter long-standing engineering workflows, enabling decisions [7,8] that are simultaneously faster, more accurate, and more sustainability-oriented than traditional heuristics permit [9–12]. Motivated by this profound shift, the present Special Issue of *Buildings* assembles 23 original papers that collectively illustrate how data-driven intelligence can unlock unprecedented levels of efficiency, resilience, and environmental stewardship across civil engineering domains. The contributions cover a range of topics, including underground construction, bridge and tunnel engineering, railway systems, concrete technology, cost optimization, risk governance, and Fourth Industrial Revolution (4IR) adoption, reflecting the wide thematic net cast by the guest editors. Beyond merely cataloging applications, the papers reveal emerging best practices for integrating domain knowledge with advanced algorithms, establishing robust data pipelines, and validating digital predictions against field measurements. Collectively, they confirm that intelligent techniques are no longer experimental add-ons but essential, increasingly standardized tools for twenty-first-century infrastructure planning and management.

2. Underground Construction and Geo-Infrastructure Intelligence

Snap-fit optimization for folding steel arches (Contribution 1) and stability evaluation of layered multi-stage fill slopes (Contribution 2) provide complementary advances in data-calibrated numerical modeling and field verification. Contribution 3 compares pile–anchor and double-row-pile supports for foundation pits adjacent to metro tunnels, while Contribution 4 integrates vehicle–track coupling dynamics with void detection in ballastless CA-mortar, illustrating how physics-based simulations can be fused with vibration signatures for rapid condition assessment. For TBM tunneling, Contribution 19 develops a boreability-aware rock-mass classification using ensemble learning. Contribution 14 predicts deformation when new tunnels overcross existing lines, and Contribution 15 introduces a self-developed drilling-test system linked to random-forest models for real-time rock-grade identification. These studies show that intelligent analytics can reduce uncertainty in subsurface construction, supporting safer and greener underground networks.

 https://doi.org/10.3390/buildings15111799

3. Smart Structures, Materials, and Performance Prediction

High-temperature debonding of FRP-strengthened beams is modeled analytically (Contribution 5), while Contribution 6 quantifies the dual effects of traffic-induced vibration on high-strength concrete curing and old-to-new interfaces—information critical to bridge widening without traffic closure. Contribution 7 uses refined finite elements to capture nonlinear stresses in spherical hinges during swivel-bridge construction, and Contribution 8 couples multi-field simulations with on-site crack-control trials for pier concrete in high-altitude climates. Contribution 9 enhances thermal-parameter inversion for mass concrete via a mixed-strategy sparrow-search algorithm, exemplifying how hybrid optimization accelerates calibration against sensor data. Contribution 10 blends laboratory and ABAQUS analysis to clarify load transfer in stepped DX piles, whereas Contribution 18 extends the classical method to mechanically connected precast piles under horizontal load. Contributions 21 and 22 explore seismic behavior of precast beam–column joints and fatigue of ultra-short studs in UHPC, both supported by detailed numerical or S–N modeling. Together, these 10 papers affirm that ML-enhanced analytics, surrogate modeling, and digital twins are maturing as reliable instruments for performance-critical decision-making.

4. Cost, Optimization, and Lifecycle Sustainability

Contribution 13 presents an adaptive self-explanatory CNN that predicts the cost of Huizhou replica vernacular dwellings to within 0.6% MAPE and makes feature importance fully transparent via SHAP values—evidence that trustworthy AI can penetrate cost-management workflows. Contribution 23 extends optimization to steel-plate fitting, reducing fabrication errors by up to 75% through a minimum-error interpolation method and thereby lowering both cost and embodied carbon.

From a materials perspective, Contribution 20 quantifies how wet–dry cycles alter shear strength and cohesion in unsaturated clayey sands, data essential for moisture-resilient roadway subgrades. Contributions 17 and 8 address plateau concrete cracking under harsh environmental gradients, offering deformation-compensated mix designs that double long-term compressive reserves. These results emphasize that intelligent techniques are key enablers of whole-life-cycle sustainability—from material selection to asset maintenance.

5. Risk Assessment, Digitalization, and 4IR Adoption

Contribution 12 couples triangular-fuzzy theory with Bayesian networks to rank construction-stage risks in cantilever-casting arch bridges, whereas Contribution 11 fuses UAV panoramas with a Swin Transformer–enhanced YOLOv8 to achieve 98.7% accuracy in bridge defect localization. Contribution 16 provides rare survey data on 4IR technology uptake in an emerging-economy construction sector, revealing virtualization as a gateway to broader AI adoption. Contributions 9 and 19 independently confirm that hybrid or ensemble algorithms outperform single-heuristic approaches for parameter inversion and classification, reinforcing the risk-management benefits of algorithmic diversity.

6. Outlook

Taken together, the articles in this Special Issue crystallize three interrelated trajectories. First, data-informed simulation loops are becoming the normative backbone of design: sensor-calibrated finite-element models, surrogate ML surrogates, and hybrid optimization routines now shorten iteration cycles while providing confidence bounds that were previously unattainable. Second, the rise of explainable intelligence is dismantling the "black-box" barrier; methods such as SHAP analyses, Bayesian inferences, and transformer-based visualizations are empowering practitioners to interrogate algorithmic reasoning, thereby enhancing adoption in safety-critical contexts. Third, an explicit focus on lifecycle resilience is redirecting innovation from short-term performance gains toward long-term durability and carbon reduction, as illustrated by studies on plateau-concrete cracking, UHPC fatigue, and AI-guided maintenance scheduling.

Looking ahead, further progress will depend on tighter coupling between domain-specific physics and general-purpose AI architectures, the establishment of standardized and interoperable data schemas that facilitate cross-project learning, and the development of governance frameworks that embed ethical considerations such as fairness, transparency, and data sovereignty. Equally important will be interdisciplinary collaboration—bringing together civil engineers, computer scientists, materials specialists, and policymakers—to translate laboratory advances into deployable tools that can scale across diverse regulatory and climatic contexts. With these foundations in place, intelligent techniques will not simply augment conventional practice but will redefine the baseline expectations for how infrastructure is conceived, built, and sustained.

Author Contributions: Writing—original draft preparation, K.L and T.W.; writing—review and editing, X.Z.; project administration, K.L. All authors have read and agreed to the published version of the manuscript.

Funding: This Special Issue is supported by the Major Program of the Natural Science Foundation of Sichuan Province of China (Grant No. 2024NSFSC0003) and the Overseas Expertise Introduction Project for Discipline Innovation ("111 Project", Grant No. B21011).

Institutional Review Board Statement: Not applicable.

Informed Consent Statement: Not applicable.

Data Availability Statement: Not applicable.

Acknowledgments: The Guest Editors gratefully acknowledge the authors for their high-quality submissions and the anonymous reviewers whose rigorous feedback enhanced every manuscript. We also thank the *Buildings* editorial team for their professional support.

Conflicts of Interest: The Guest Editors declare no conflicts of interest.

List of Contributions

1. Li, S.; Huang, C.; Yang, X.; Tao, Z.; Guo, J.; Li, H.; Yao, T.; Hu, J. Research on the Bearing Characteristics of Folding Steel Arch Frames with Different Snap-Fit Types Based on the Compensation Excavation Concept. *Buildings* **2025**, *15*, 1423. https://doi.org/10.3390/buildings15091423.
2. Li, X.; Ye, S.; Qiu, M.; Ye, W.; Li, J. Stability Analysis of Horizontal Layered Multi-Stage Fill Slope Based on Limit Equilibrium Method. *Buildings* **2025**, *15*, 1105. https://doi.org/10.3390/buildings15071105.
3. Mao, Z.; Ding, T.; Hu, F.; Ye, S.; Ding, L.; Zhang, X.; Li, P.; Li, N. The Impact of Different Excavation Support Structures on the Deformation and Stability of Adjacent Station and Tunnels. *Buildings* **2025**, *15*, 493. https://doi.org/10.3390/buildings15030493.
4. Chen, X.; Pei, Y.; Liu, K. Dynamic Response of Train–Ballastless Track Caused by Failure in Cement–Asphalt Mortar Layer. *Buildings* **2025**, *15*, 334. https://doi.org/10.3390/buildings15030334.
5. Zhang, X.; Hao, J.; Hou, W.; Yao, J.; Wang, Y.; Su, X.; Li, X. Debonding Analysis of FRP-Strengthened Concrete Beam in High-Temperature Environment: An Enhanced Understanding on Sustainable Structure. *Buildings* **2024**, *14*, 4079. https://doi.org/10.3390/buildings14124079.

6. Gu, P.; Wu, H.; Li, L.; Li, Z.; Hong, J.; Zhuang, M. Effect of Traffic Vibration on Compressive Strength of High-Strength Concrete and Tensile Strength of New-to-Old Concrete Interfaces. *Buildings* **2024**, *14*, 3765. https://doi.org/10.3390/buildings14123765.
7. Zhao, L.; Sun, X.; Wu, Z.; Chen, Y.; Liu, J.; Wang, Y. Nonlinear Static Analysis of Spherical Hinges in Horizontal Construction of Bridges. *Buildings* **2024**, *14*, 3726. https://doi.org/10.3390/buildings14123726.
8. Hu, X.; Liu, L.; Liao, M.; Li, M.; Lu, C.; Yao, Z.; Huang, Q.; Zhuang, M. Experimental Investigations on the On-Site Crack Control of Pier Concrete in High-Altitude Environments. *Buildings* **2024**, *14*, 3445. https://doi.org/10.3390/buildings14113445.
9. Wang, Y.; Gao, Y.; Zhang, K.; Zhuang, M.; Xu, R.; Yan, X.; Wang, Y. Inversion Analysis for Thermal Parameters of Mass Concrete Based on the Sparrow Search Algorithm Improved by Mixed Strategies. *Buildings* **2024**, *14*, 3273. https://doi.org/10.3390/buildings14103273.
10. Cheng, J.; Tong, L.; Sun, C.; Zhu, H.; Deng, J. Experimental and Numerical Simulation Investigations on the Bearing Capacity of Stepped Variable-Section DX Piles under Vertical Loading. *Buildings* **2024**, *14*, 3078. https://doi.org/10.3390/buildings14103078.
11. Yin, T.; Shen, G.; Yin, L.; Shi, G. Bridge Surface Defect Localization Based on Panoramic Image Generation and Deep Learning-Assisted Detection Method. *Buildings* **2024**, *14*, 2964. https://doi.org/10.3390/buildings14092964.
12. He, Z.; Xiang, Y.; Li, L.; Wei, M.; Liu, B.; Wu, S. Research on Construction Risk Assessment of Long-Span Cantilever Casting Concrete Arch Bridges Based on Triangular Fuzzy Theory and Bayesian Networks. *Buildings* **2024**, *14*, 2627. https://doi.org/10.3390/buildings14092627.
13. Huang, J.; Huang, W.; Quan, W.; Xing, Y. Hybrid Intelligent Model for Estimating the Cost of Huizhou Replica Traditional Vernacular Dwellings. *Buildings* **2024**, *14*, 2623. https://doi.org/10.3390/buildings14092623.
14. Xie, C.; Qu, Y.; Lu, H.; Song, S. Study on Deformation of New Tunnels Overcrossing Existing Tunnels Underneath Operating Railways. *Buildings* **2024**, *14*, 2420. https://doi.org/10.3390/buildings14082420.
15. Liu, Q.; Yan, J.; Li, H.; Zhang, P.; Liu, Y.; Liu, L.; Ye, S.; Liu, H. Intelligent Identification of Surrounding Rock Grades Based on a Self-Developed Rock Drilling Test System. *Buildings* **2024**, *14*, 2176. https://doi.org/10.3390/buildings14072176.
16. Lopes, J.M.; Pinto da Silva Filho, L.C. Adoption of Fourth Industrial Revolution Technologies in the Construction Sector: Evidence from a Questionnaire Survey. *Buildings* **2024**, *14*, 2132. https://doi.org/10.3390/buildings14072132.
17. Hu, X.; Liao, M.; Li, M.; Wang, F.; Lyu, X.; Zhuang, M. Investigations on the Environmental Characteristics and Cracking Control of Plateau Concrete. *Buildings* **2024**, *14*, 2104. https://doi.org/10.3390/buildings14072104.
18. Gao, L.; Zhuang, M.; Zhang, Q.; Bao, G.; Yu, X.; Du, J.; Zhou, S.; Wang, M. Displacement and Internal Force Response of Mechanically Connected Precast Piles Subjected to Horizontal Load Based on the *m-Method*. *Buildings* **2024**, *14*, 1943. https://doi.org/10.3390/buildings14071943.
19. Li, Z.; Tao, Y.; Du, Y.; Wang, X. Classification and Prediction of Rock Mass Boreability Based on Daily Advancement during TBM Tunneling. *Buildings* **2024**, *14*, 1893. https://doi.org/10.3390/buildings14071893.
20. Wang, C.; Yang, W.; Zhang, N.; Wang, S.; Ma, C.; Wang, M.; Zhang, Z. Effect of Moisture Content and Wet–Dry Cycles on the Strength Properties of Unsaturated Clayey Sand. *Buildings* **2024**, *14*, 1375. https://doi.org/10.3390/buildings14051375.

21. Zhuang, M.; Sun, C.; Yang, Z.; An, R.; Bai, L.; Han, Y.; Bao, G. Numerical Investigation on the Seismic Behavior of Novel Precast Beam–Column Joints with Mechanical Connections. *Buildings* **2024**, *14*, 1199. https://doi.org/10.3390/buildings14051199.
22. An, R.; Wang, Y.; Zhuang, M.; Yang, Z.; Tian, C.; Qiu, K.; Cheng, M.; Lv, Z. Study of Fatigue Performance of Ultra-Short Stud Connectors in Ultra-High Performance Concrete. *Buildings* **2024**, *14*, 1179. https://doi.org/10.3390/buildings14041179.
23. Huang, Z.; Ding, J. Minimum Error Interpolation Method for Quadrilateral Flat Plate Fitting in Steel Construction. *Buildings* **2025**, *15*, 1433. https://doi.org/10.3390/buildings15091433.

References

1. Xu, M.; Nie, X.; Li, H.; Cheng, J.C.P.; Mei, Z. Smart Construction Sites: A Promising Approach to Improving on-Site HSE Management Performance. *J. Build. Eng.* **2022**, *49*, 104007. [CrossRef]
2. Wang, T.; Luo, Q.; Li, Z.; Zhang, W.; Chen, W.; Wang, L. System Safety Assessment with Efficient Probabilistic Stability Analysis of Engineered Slopes along a New Rail Line. *Bull. Eng. Geol. Environ.* **2022**, *81*, 68. [CrossRef]
3. Wang, T.; Chen, W.; Li, T.; Connolly, D.P.; Luo, Q.; Liu, K.; Zhang, W. Surrogate-Assisted Uncertainty Modeling of Embankment Settlement. *Comput. Geotech.* **2023**, *159*, 105498. [CrossRef]
4. Hu, B.; Shan, Y.; Zhao, Y.; Wang, B.; Zhou, S.; Alberti, G.S.; Ma, W.; Detmann, B.; Briançon, L. Tunneling beneath the Pile-Raft Foundations of High-Speed Railways: Progressive Arching Deformation and Pile Settlement Behavior. *Undergr. Space* 2025, *in press*. [CrossRef]
5. Wang, T.; Ma, H.; Liu, J.; Luo, Q.; Wang, Q.; Zhan, Y. Assessing Frost Heave Susceptibility of Gravelly Soils Based on Multivariate Adaptive Regression Splines Model. *Cold Reg. Sci. Technol.* **2021**, *181*, 103182. [CrossRef]
6. Shao, Y.; Li, L.; Li, J.; Li, Q.; An, S.; Hao, H. Monocular Vision Based 3D Vibration Displacement Measurement for Civil Engineering Structures. *Eng. Struct.* **2023**, *293*, 116661. [CrossRef]
7. Chen, W.; Luo, Q.; Liu, J.; Wang, T.; Wang, L. Modeling of Frozen Soil-Structure Interface Shear Behavior by Supervised Deep Learning. *Cold Reg. Sci. Technol.* **2022**, *200*, 103589. [CrossRef]
8. Chen, W.; Qu, S.; Lin, L.; Luo, Q.; Wang, T. Ensemble Learning Methods for Shear Strength Prediction of Fly Ash-Amended Soils with Lignin Reinforcement. *J. Mater. Civil. Eng.* **2023**, *35*, 04023022. [CrossRef]
9. Liao, Y.; Liu, J.; Wang, W.; Huang, F.; Yang, W.; Peng, Y. Intelligent Virtual Trial Assembly for Large and Complex Box Girders Based on Point Cloud Data. *J. Transp. Sci. Eng.* **2023**, *39*, 33–39. [CrossRef]
10. Chen, Z.; Tan, S.; Li, C.; Peng, Y. Research on Ultrasonic Intelligent Recognition of CFRP-Steel Interface Defects. *J. Transp. Sci. Eng.* **2023**, *39*, 71–79. [CrossRef]
11. Yu, J.; Zhu, Y.; Chen, L. Prediction of Mud Fluid Properties in Rockfill Based on LS-GEP. *J. Transp. Sci. Eng.* **2024**, *40*, 28–35. [CrossRef]
12. Zhang, D.; Zhong, X.; Wang, X.; Wei, Y. Development Status and Trends of Intelligent Connected Transportation Technology Based on WOS. *J. Transp. Sci. Eng.* **2024**, *40*, 144–152. [CrossRef]

buildings

MDPI

Article

Research on the Bearing Characteristics of Folding Steel Arch Frames with Different Snap-Fit Types Based on the Compensation Excavation Concept

Shaohua Li [1], Changfu Huang [1,*], Xiaojie Yang [2], Zhigang Tao [2], Jiaqi Guo [3], Humin Li [1], Tiejun Yao [1] and Jie Hu [2]

[1] China Railway 15th Bureau Group Co., Ltd., Shanghai 200070, China; zz16lish@yeah.net (S.L.); huminli@outlook.com (H.L.); ytj1502@163.com (T.Y.)
[2] State Key Laboratory for Tunnel Engineering, China University of Mining and Technology-Beijing, Beijing 100083, China; yxjcumt@163.com (X.Y.); taozhigang1981@163.com (Z.T.); hujie@cumtb.edu.cn (J.H.)
[3] School of Civil Engineering, Henan Polytechnic University, Jiaozuo 454003, China; gjq519@163.com
* Correspondence: hcf1971@126.com

Abstract: As the core technology for mechanized installation of tunnel folding steel arch frames, snap-fit connection optimization proves critical in enhancing the load-bearing efficiency of support systems and addressing surrounding rock deformation and instability caused by excavation-induced stress redistribution. Addressing the theoretical gaps in existing research regarding snap-fit selection mechanisms and quantitative evaluation criteria, this study adopts a combined approach of numerical simulation and field monitoring verification based on the excavation compensation concept to systematically investigate the load-bearing characteristics of folding steel arch frames with different snap-fit configurations. Key findings include (1) identification of 20 mm as the optimal joint diameter, where the vertical displacements of Type A and B snap-fit connections reached their minimum values of 43.1 mm and 39.2 mm, respectively; (2) demonstration of significant geometric configuration effects on principal stress distribution, with Type B connections exhibiting 4.5% lower maximum principal stress compared to Type A, effectively mitigating stress concentration; and (3) field monitoring data verification, revealing that Type B connections achieved 15.8% lower stress values than Type A at critical crown sections, satisfying yield strength requirements while demonstrating enhanced resistance to surrounding rock deformation induced by excavation-induced geostress redistribution. These results confirm Type B snap-fit connections as superior structural solutions for folding steel arch frames, thereby facilitating the advancement of mechanized installation technology for tunnel steel arch frames.

Keywords: bearing characteristics; excavation compensation; folding steel arch frame; joint diameter; snap-fit connection

1. Introduction

Since the beginning of the 21st century, China has become the fastest-growing country in tunnel and underground engineering development due to continuous economic growth and major infrastructure programs [1–3]. By the end of 2022, China had 17,873 operational railway tunnels with a total length of 21,978 km, and 22,850 highway tunnels spanning 26,784 km [4]. The large-scale construction of transportation tunnels has driven advancements in construction technologies and mechanization, with innovative equipment such as three-boom hydraulic drills, arch frame assembly machines, and intelligent lining trolleys being successfully applied in projects like the Zhengzhou–Wanzhou Railway and the

Beijing–Zhangjiakou Railway [5–7]. Although mechanized construction has significantly improved, tunnel excavation disrupts the original stress equilibrium of surrounding rock masses, leading to stress redistribution and tensile stress concentration that frequently cause rock failure. The excavation compensation theory [8] emphasizes that implementing stress compensation support during the critical window period before tensile stress concentration can restore the pre-excavation three-dimensional stress state, thereby controlling rock deformation. The excavation compensation method [9] proposes that prestressed active support effectively limits deformation by rapidly compensating for stress loss. Academic He Manchao developed a large deformation control system using NPR materials, achieving remarkable success in tunnel stability [10]. Additionally, high-stiffness arch frame systems combined with rapid support technologies establish a "resist-then-yield" compensation mechanism to mitigate excavation-induced stress redistribution and associated hazards. Therefore, investigating the bearing performance of folding steel arch frames under excavation unloading conditions and their rapid installation methods is critical for enhancing the effectiveness of excavation compensation techniques, reducing the adverse effects of geostress, and advancing both the load-bearing capacity of primary support systems and the mechanization level of tunnel construction.

Regarding the erection equipment for tunnel steel arch frames, numerous scholars both domestically and internationally have conducted extensive research. Guo et al. [11] developed a novel mechanized supporting construction equipment integrating multiple functions for steel arch frame erection in tunnels. Jianglu Mechanical and Electrical Group Co., Ltd. Industrial Development Company designed the SGDC700 tunnel arch multi-function installing truck, which is capable of grasping and connecting arch frames in mid-air using flanges [12]. Wang et al. [13] independently developed an intelligent arch erection device, which is compatible with the supporting structure and employs an assembly approach to construction. Liu et al. [14] innovated the SCDZ133 tunnel arch multi-function installing truck, addressing issues such as high labor intensity and low efficiency in the erection of tunnel steel arch frames. Sun et al. [15] developed an intelligent arch frame erection machine along with assembly devices such as automatic assembly nodes and longitudinal positioning connections. He et al. [16] designed a grasping and docking mechanism for tunnel boring machine (TBM) steel arch splicing manipulators. Zhao et al. [17] analyzed the usage of the XZGMT411 multi-functional arch erection machine in the construction of the Yuelongmen Tunnel. Hudita Construction designed a remote and unmanned construction system for assembling steel arch frames [18]. Krauze et al. [19] developed a system for transporting and installing tunnel steel arch frames. To minimize manual operations in arch erection, Ni [20] developed a teleoperated assembly system for unmanned steel arch construction. The aforementioned literature primarily focuses on the research progress of steel arch frame erection equipment in the initial support phase of tunnels.

In terms of innovative steel arch frames, Sun et al. [21] introduced a snap-fit machining connection pre-stressed concrete solid square pile, providing a reliable basis for the improvement and application of steel arch frame connection methods. Wang [22] designed a retractable movable arch frame, analyzing its supportive effectiveness in tunnels with large deformations in layered soft rock beds. Li et al. [23] investigated the effects of multi-arch spatial combination and shotcrete on primary support, analyzing the influence of arch spacing and longitudinal connection spacing on the mechanical properties of the support system. Xu et al. [24] designed a novel grid steel frame core–tube bracing system and compared the strain characteristics of two arched supports under compression. Ma et al. [25] found that failures in tunnel steel arch supports are primarily induced by axial stress, and that incorporating longitudinal connections can significantly enhance their load-bearing capacity. Through experimental investigations on I-beam and hollow-pipe string joints under both static and cyclic loading conditions, Song et al. [26] systematically evaluated the structural behavior of connections, with particular emphasis on their failure mechanisms across different loading regimes. Yue et al. [27] conducted a comprehensive investigation into the critical challenges of segmented connections in steel arch frames for primary tunnel support systems, proposing targeted structural optimization strategies to improve their mechanical performance. Li et al. [28] developed an innovative longitudinal connection system utilizing steel plates. Through field-scale experimental studies, they quantitatively compared the structural efficacy of this plate-based solution against traditional reinforcing bar connections within steel arch support systems. He et al. [29] addressed the problem of large deformations in squeeze-type soft rock tunnels, researching an adaptive steel arch frame joint suitable for such tunnels. Zhang et al. [30,31] proposed design schemes for snap-fit, adhesive, and interference-fit assembly steel arch frame joints suitable for machining construction. Their research found that snap-fit steel arch frames exhibit the best load-bearing performance.

In conclusion, while significant progress has been made in tunnel steel arch frame installation equipment, traditional arch segments still require manual bolt connections even with mechanical assistance, resulting in high labor intensity, complex installation procedures, and low efficiency [32]. These limitations impede the timely stress compensation required by the excavation compensation method to control surrounding rock instability caused by geostress redistribution, thereby restricting deformation mitigation. Consequently, developing novel folding steel arch frames compatible with installation machinery is critical for achieving mechanized deployment and timely support. Although current research confirms that snap-fit folding steel arch frames offer promising solutions [33], systematic investigations remain lacking regarding the influence of snap-fit types on load-bearing characteristics, mechanisms governing stress states at joint sections, and the contact behavior of connecting plates. Additionally, selection criteria for snap-fit configurations to balance rapid mechanized erection with effective support remain unclear. To address these gaps, this study evaluates existing folding arch frame designs and systematically investigates the mechanical performance of snap-fit-type folding frames through displacement–stress characteristics, joint stress distribution, and plate contact dynamics. Field test comparisons further identify optimal snap-fit configurations, achieving mechanized rapid erection, prompt support, and effective deformation control. These findings provide critical insights for advancing mechanized installation technology in tunnel engineering.

2. Snap-Fit Folding Steel Arch Frames for Tunnels

2.1. Folding Steel Arch Frames for Machining Arch Frame Erection in Tunnels

The folding steel arch frame is a novel form suitable for machining assembly, comprising a minimum of two arch frame units. The adjacent arch frame units are joined using a piecing device. The piecing device facilitates the fixed connection between adjacent arch units, eliminating the need for manual bolt fixation at the connecting ends of the arch units. This enhances erection efficiency and quality, reducing labor intensity for operational personnel.

The folding steel arch frame, as depicted in Figure 1, encompasses various-sized segments and piecing devices. In accordance with tunnel cross-sectional dimensions and construction methods, the processed steel arch frame segments are initially assembled outside the tunnel opening. Using an arch erection machine, the entire steel arch frame is lifted in a single operation. One of the gripping arms is employed to secure the middle segment of the arch at the crown. Subsequently, the other two arms are used to unfold the steel arch frame segments on either side, and the connection devices facilitate their linkage. Once the steel arch frame is erected, at the tunnel construction site, a multi-functional operation trolley enables the machining erection of the tunnel steel arch frame. Prompt shotcrete operations follow to swiftly establish a collaborative structural load-bearing system between the steel arch frame and the sprayed concrete.

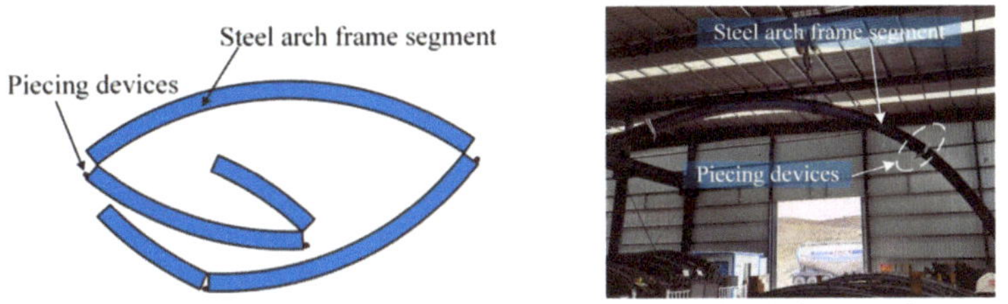

(**a**) Components of folding steel arch frame; (**b**) folding steel arch frame indoor assembly.

Figure 1. Folding steel arch frame.

2.2. Connection Methods for Folding Steel Arch Frames

Currently, there are three methods of piecing devices between segments of the folding steel arch frame, as illustrated in Table 1: snap-fit connections, adhesive connections, and interference-fit assembly connections. Snap-fit connections primarily rely on the tight clamping force between elastic joints and elastic snap-fit positions to secure the connection between arch segments. Elastic joints and elastic snap-fit positions are located on two connecting plates. When the elastic joint passes through the elastic snap-fit position, it causes the elastic ring to expand. After the elastic joint passes through the elastic ring, the elastic ring exerts clamping force on the joint through elastic reset.

Table 1. Connection types and characteristics of folding steel arch frames.

<table>
<tr><th>Snap-Fit Connection</th><th>Adhesive Connection</th><th>Interference-Fit Assembly Connection</th></tr>
<tr><td>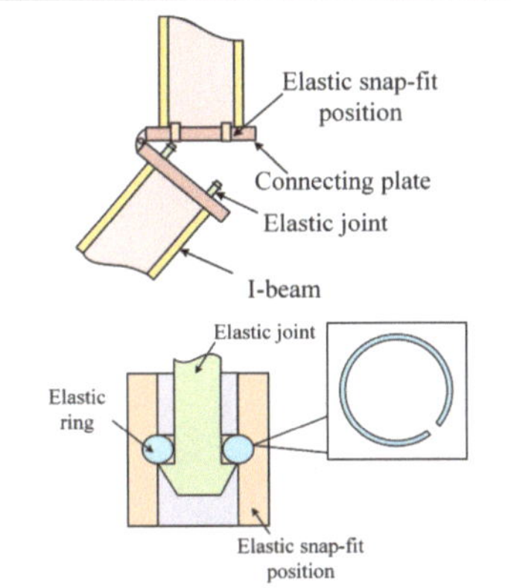
</td><td>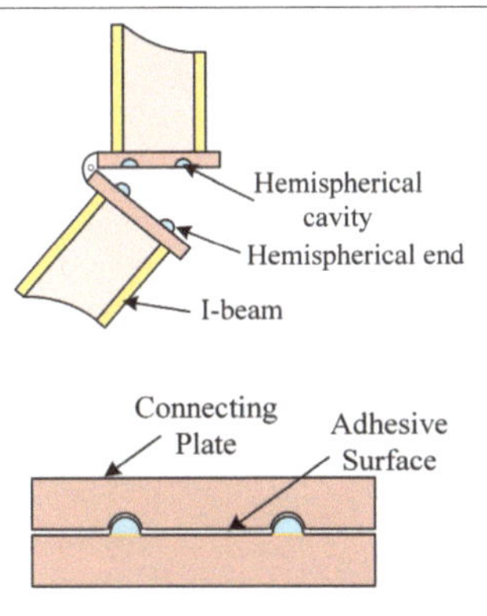
</td><td>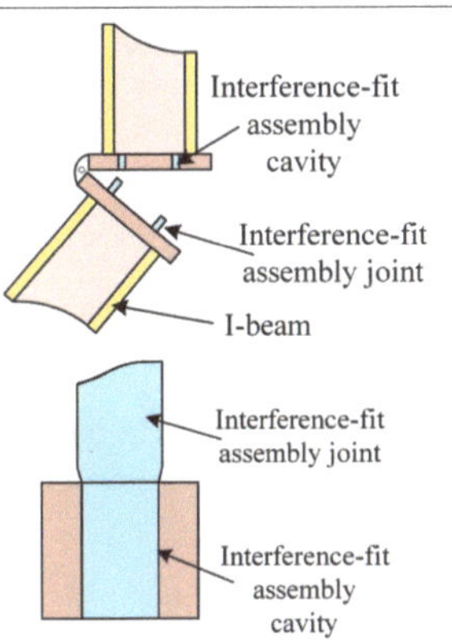
</td></tr>
<tr><td>Simple structure, easy to manufacture, and excellent mechanical performance</td><td>Relatively uniform stress distribution but requires waterproof treatment and has a shorter service life</td><td>Easy to manufacture but requires high structural dimensional accuracy</td></tr>
</table>

Adhesive connections consist of a semi-spherical end on one connecting plate and a cavity on another connecting plate, with the semi-spherical end and the cavity arranged in a floral pattern. This connection relies on pre-set adhesive within the cavity, which, when compressed, forms a stable adhesive surface between the connecting plates to achieve the purpose of connection. Interference-fit assembly connections mainly comprise interference-fit assembly joints and assembly cavities. This connection exploits the elastic deformation capability of materials. During the application of force, the hole diameter increases, and upon recovery, it generates a clamping force on the axis, facilitating the connection between joint components.

The snap-fit connection exhibits excellent assembly performance, making it easy for on-site processing and assembly and boasting favorable machining properties. While adhesive connections can uniformly transmit stress and withstand loads within the adhesive range, preventing stress concentration and potential damage to joint structures, they typically require waterproof treatment at the adhesive site. Moreover, adhesive connections possess a certain time-dependent strength, demanding high resistance to oxidation and considerable service life of the adhesive. Interference-fit assembly connections simplify joint structures, facilitating straightforward and convenient processing. However, this connection is suitable for erection under impact loads and demands high precision in the structural dimensions at the joint. Through a comparative analysis considering assembly, durability, and processability, and aligning with the on-site conditions necessary for tunnel steel arch frame erection, it can be concluded that snap-fit connections offer high practicality. They can substitute manual connections between steel arch frames, ensuring a tight connection among arch frame segments, enhancing overall structural stability, and minimizing safety risks during machining construction. This promotes the smooth progress of machining tunnel arch frame erection.

2.3. Types of Snap-Fit Connection

The local structure of the snap-fit connection, as illustrated in Figure 2a, is designed to facilitate the smooth passage of the elastic joint through the snap-fit position. To achieve it, the internal diameter of the cavity in the elastic snap-fit position is 1.05 to 1.1 times the diameter of the joint. The internal diameter of the elastic ring is 0.8 to 0.95 times the diameter of the joint. This configuration ensures that, as the elastic joint passes through the elastic ring, a self-locking force is provided. The elastic snap-fit position is equipped with an annular groove for installing the elastic ring, with the groove's specifications being 1.1 to 1.5 times the size of the elastic ring. This facilitates the erection of the elastic ring and provides expansion space for the elastic ring when the elastic joint passes through it. The elastic ring features notches, ensuring the passage of the elastic joint through the elastic ring while maintaining a self-locking fixation of the elastic joint by the elastic ring.

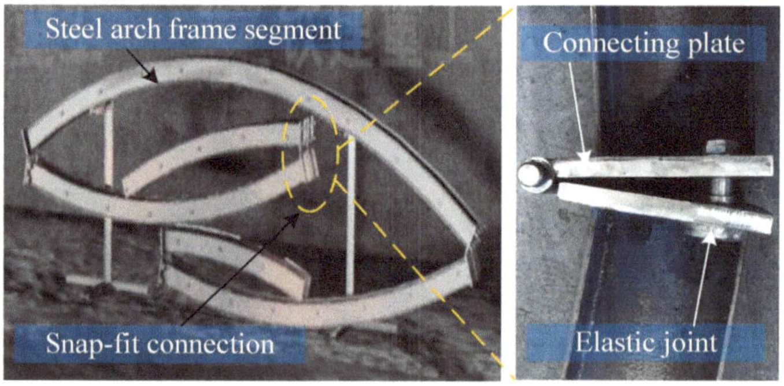

(**a**) Partial diagram of snap-fit connection.

Figure 2. *Cont.*

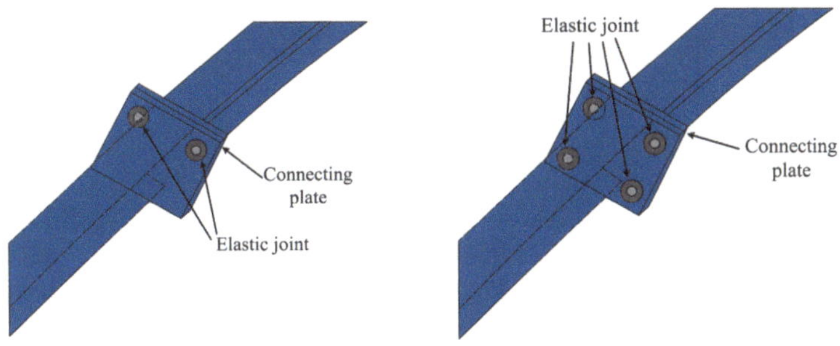

(**b**) A-type snap-fit connection. (**c**) B-type snap-fit connection.

Figure 2. Types of snap-fit connection.

The elastic ring is manufactured from high-quality spring steel, providing excellent elasticity and strength. The efficacy of folding steel arch frames in providing structural support is predominantly contingent on the design of snap-fit connections. Past research has indicated that when the thickness of the end plate exceeds 16 mm, it ceases to exert effective control over the nodal bearing capacity [34]. Consequently, this paper establishes the thickness of the snap-fit joint connecting plate at 16 mm, ensuring that the load-bearing capacity of the snap-fit joint is primarily determined by the number of enclosed elastic joints and the dimensions of the elastic joint. Two types of snap-fit connections are introduced, incorporating two elastic joints and four elastic joints, designated A-type and B-type, respectively (refer to Figure 2b,c).

3. Analysis of Mechanical Performance of Folding Steel Arch Frames with Different Snap-Fit Types

3.1. Numerical Calculation Model Building

The folding steel arch frame consists of I-beams, connection plates, and flexible joints. ABAQUS 6.14.4 enables independent material parameter definition for individual components with corresponding mesh generation, demonstrating robust stability in handling highly nonlinear problems. Static analysis of the folding steel arch frame was performed using ABAQUS [35]. A three-dimensional model was developed on a Windows 10 system (Intel® Core™ i7-12700F CPU @2.10 GHz, 16GB RAM) utilizing hexahedral elements (HEX), achieving high computational accuracy with relatively low resource consumption. The model mesh was partitioned using C3D8R elements (8-node linear brick reduced integration elements) for more accurate displacement results.

In the model development process, interactions among various components of the steel arch frame joints are considered. The interaction function module is utilized for definition, and the model is appropriately simplified to obtain reliable and accurate computational results. For the numerical simulation of the snap-fit joints, the following assumptions are made: (1) the self-locking effect of the elastic ring on the elastic joint is realized through the Tie constraint; (2) mutual contact between the connecting plates of the snap-fit joint is considered, neglecting tangential friction and only accounting for normal contact through "hard" contact, allowing for separation after contact; and (3) the hinged structure between the connecting plates of the snap-fit connection primarily facilitated the rotation between steel arch frame segments, simulated through the HINGE combination property. In practical engineering, the presence of longitudinal reinforcement limits the out-of-plane buckling deformation of the steel arch frame to some extent. Therefore, this paper only considers the in-plane deformation of the folding steel arch frame. Constraints are applied in the Z-direction in the model, and due to the action of the lock foot anchor rod, the foot position of the folding steel arch frame is set as completely fixed. The model is illustrated in Figure 3. The typical grid size of a steel frame is 50 × 50 × 22 mm. The typical grid size at the joint is about 3 × 3 × 3 mm, and the grid is encrypted at the connections.

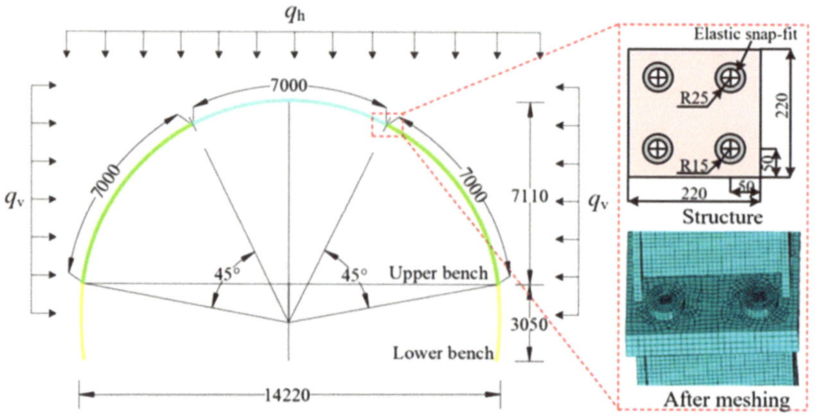

Figure 3. Numerical model of folding steel arch frame (unit: mm).

3.2. Numerical Model Parameters

The A- and B-type snap-fit connections are both positioned at a 45° angle to the arch crown. The folding steel arch frame is constructed using No. 16 I-beams, and the material parameters for the folding steel arch frame are presented in Table 2. The initial support steel arch frame of the tunnel is subjected to surrounding rock pressure, with its load decomposed into horizontal pressure q_v and vertical pressure q_h acting on the steel arch frame. Therefore, the effect of the surrounding rock on the folding steel arch frame can be equivalently simulated by applying horizontal and vertical pressures. According to the Code for the Design of Railway Tunnels (2016) [36] regarding the calculation method for deep-buried tunnel loads, the vertical pressure q_h is set to 259.2 kPa, and the horizontal pressure q_v is set to 103.7 kPa.

Table 2. Material parameters of the folding steel arch frame.

Component	Material	*E*/GPa	μ	Sectional Area/cm^2
Steel arch frame segments	Q235	200	0.3	26.1
Connecting plate	Q215	187	0.41	58.3
Elastic joint	Q215	187	0.41	-

In summary, the calculation scenarios for folding steel arch frames with different specifications under surrounding rock pressure are outlined in Table 3. To investigate the impact of the elastic joint diameter on the folding steel arch frame, in this study, we design elastic joints with diameters of d = 16 mm, d = 20 mm, and d = 24 mm, analyzing the influence of different diameter snap-fit types on the load-bearing performance of the folding steel arch frame.

Table 3. Different calculation conditions of folding steel frames.

	Vertical Surrounding Rock Pressure/kPa	Horizontal Surrounding Rock Pressure/kPa	Snap-Fit Connection	
			Type	Cone Head Diameter/mm
1	259.2	103.7	A	16
2	259.2	103.7	B	16
3	259.2	103.7	A	20
4	259.2	103.7	B	20
5	259.2	103.7	A	24
6	259.2	103.7	B	24

3.3. Simulation Results Analysis

3.3.1. Displacement Characteristics of the Folding Steel Arch Frame

Using various joint diameters, the vertical displacements of A- and B-type snap-fit connected folding steel arch frames are illustrated in Figure 4. It can be observed that the deformation settlement of folding steel arch frames under surrounding rock pressure is not influenced by the type of snap-fit, exhibiting a symmetrical distribution with maximum vertical displacement at the arch crown, gradually decreasing towards both sides of the arch feet. The maximum vertical displacements for folding steel arch frames with A- and B-type snap-fit connections under different joint diameters are 70.6 mm, 43.1 mm, and 43.1 mm, and 43.1 mm, 39.2 mm, and 87.1 mm, respectively. The maximum vertical displacement of folding steel arch frames is significantly influenced by the snap-fit type and joint diameter. When the joint diameter does not exceed 20 mm, the maximum vertical displacement of both A- and B-type snap-fit steel arch frames decreases with increasing joint diameter. Moreover, the maximum displacement of B-type folding steel arch frames is noticeably smaller than that of A-type folding steel arch frames. This is attributed to the increase in stiffness at the snap-fit joint due to the larger elastic joint diameter and the greater number of joints, resulting in a gradual reduction in the arch crown settlement of folding steel arch frames. Therefore, within the joint diameter range of no more than 20 mm, B-type snap-fit effectively enhances the load-bearing capacity of the folding steel arch frame, demonstrating advantages in controlling arch deformation and reducing surrounding rock settlement. However, when the joint diameter exceeds 20 mm, A- and B-type snap-fit folding steel arch frames exhibit varying degrees of increased maximum vertical displacement with larger joint diameters. Additionally, the maximum displacement of B-type folding steel arch frames is significantly greater than that of A-type folding steel arch frames. This analysis suggests that when the joint diameter and quantity surpass certain values, they may weaken the strength and stiffness of the snap-fit connecting plates, leading to an increase in folding steel arch frame settlement, which is unfavorable for surrounding rock stability.

As shown in Figure 5, the vault settlement of Type A and Type B snap-fit connected folding steel arch frames under different joint diameters exhibits the following patterns. For Type A, the vault settlement decreases with increasing joint diameter, but when the diameter exceeds 20 mm, further diameter variations show negligible effects on settlement. This occurs because the enlarged flexible joints enhance the stiffness at snap-fit connections, thereby reducing settlement, yet beyond 20 mm diameter, additional increments no longer significantly affect joint stiffness. For Type B, when the joint diameter is below 20 mm, the vault settlement decreases with increasing diameter due to improved strength and stiffness at snap-fit connections; however, diameters exceeding 20 mm conversely increase settlement, primarily because Type B contains more flexible joints than Type A, and oversized diameters (>20 mm) weaken the strength and stiffness of snap-fit connection plates.

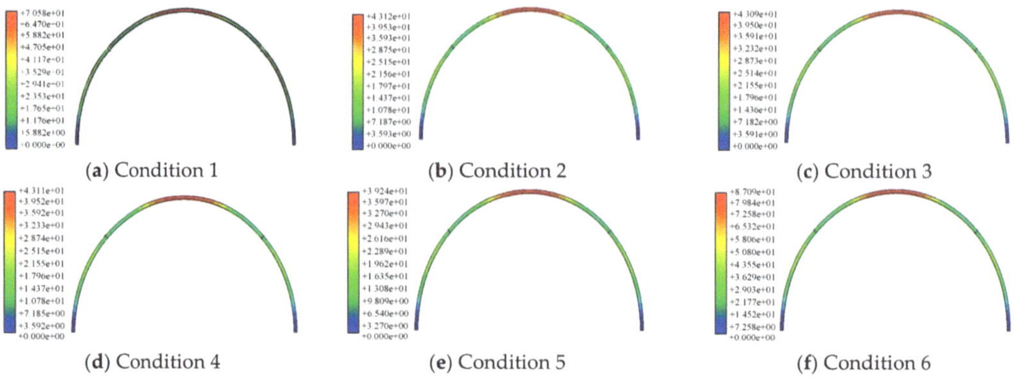

Figure 4. Deformation of A- and B-type snap-fit folding steel arch frames.

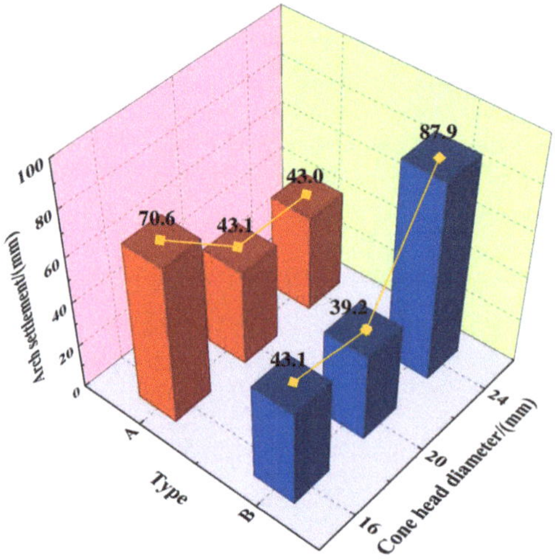

Figure 5. Arch settlement of A- and B-type snap-fit folding steel arch frames.

3.3.2. Stress Characteristics of the Folding Steel Arch Frame

The stress distribution of folding steel arch frame structures with A- and B-type connectors under different elastic joint diameters is illustrated in Figures 6 and 7, respectively. As depicted in the figures, it is evident that the maximum stresses in folding steel arch frames with various elastic joint diameters and quantities exhibit a symmetrical distribution, with compression being the primary stress. The highest compressive stress occurs at the crown, gradually decreasing towards the left and right arch feet. Notably, stress concentrations are predominantly localized at the arch feet positions. Therefore, reinforcement at the arch feet positions is essential during the erection process of folding steel arch frames to prevent structural damage at these locations, ensuring the overall load-bearing performance of the arch. When the type of snap-fit is constant, the maximum principal stresses in A-type folding steel arch frames show a negative correlation with the increase in elastic joint diameter.

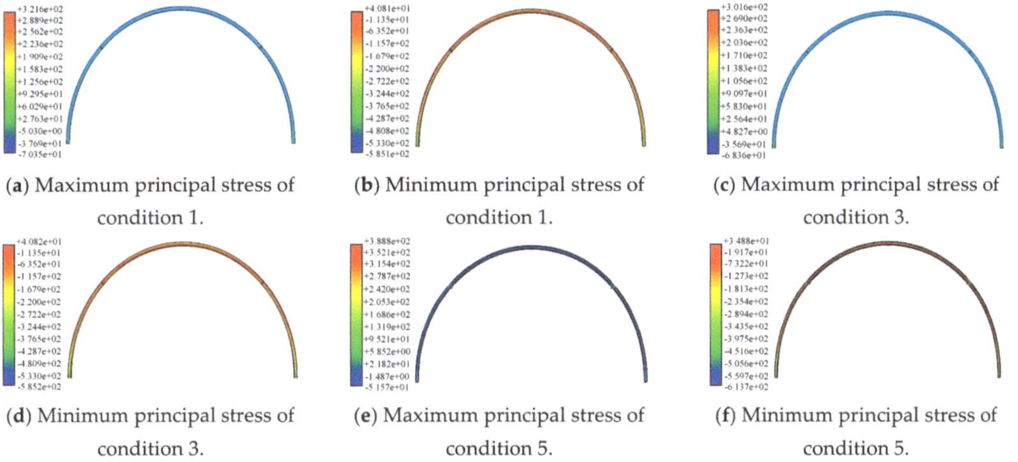

(**a**) Maximum principal stress of condition 1. (**b**) Minimum principal stress of condition 1. (**c**) Maximum principal stress of condition 3. (**d**) Minimum principal stress of condition 3. (**e**) Maximum principal stress of condition 5. (**f**) Minimum principal stress of condition 5.

Figure 6. Principal stresses of the A-type snap-fit folding steel arch frame.

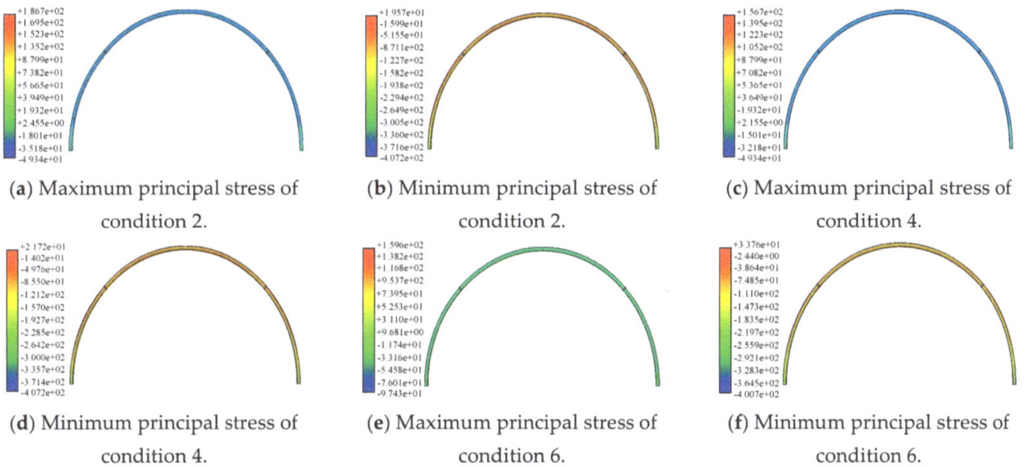

(**a**) Maximum principal stress of condition 2. (**b**) Minimum principal stress of condition 2. (**c**) Maximum principal stress of condition 4. (**d**) Minimum principal stress of condition 4. (**e**) Maximum principal stress of condition 6. (**f**) Minimum principal stress of condition 6.

Figure 7. Principal stresses of the B-type snap-fit folding steel arch frame.

The values are 70.4 MPa, 68.4 MPa, and 51.6 MPa, indicating the enhanced load-bearing capacity of A-type steel arch frames with increasing joint diameter. For B-type folding steel arch frames, when the joint diameter is below 20 mm, the maximum principal stress remains relatively stable at 49.3 MPa. However, when the joint diameter exceeds 20 mm, a noticeable increase in the maximum principal stress is observed, reaching 97.4 MPa. This increase may be attributed to the adverse effects of excessively large joint diameters on the strength of the connecting plates, resulting in a decline in the load-bearing capacity of the folding steel arch frame.

Furthermore, with equivalent elastic joint diameters, the B-type snap-fit folding steel arch frames consistently exhibit significantly lower maximum principal stresses compared to A-type snap-fit folding steel arch frames, showcasing superior resistance to deformation and load-bearing capacity. Consequently, the safety of the support structure is enhanced. In conclusion, the B-type snap-fit with a joint diameter of 20 mm demonstrates outstanding mechanical performance, positioning it as an ideal form of connection for folding steel arch frames.

3.3.3. Stress Characteristics of Joints

The principal stress of the A-type elastic joint with different joint diameters is shown in Figure 8. As illustrated in the figure, the principal stress concentration occurs mainly in the root area where the elastic joint contacts the connecting plate, predominantly experiencing tensile stress. Among the different joint diameters, the elastic joint with a diameter of 16 mm exhibits the highest maximum principal stress, reaching 139.3 MPa, while the elastic joint with a diameter of 24 mm shows the lowest maximum principal stress at 92.2 MPa.

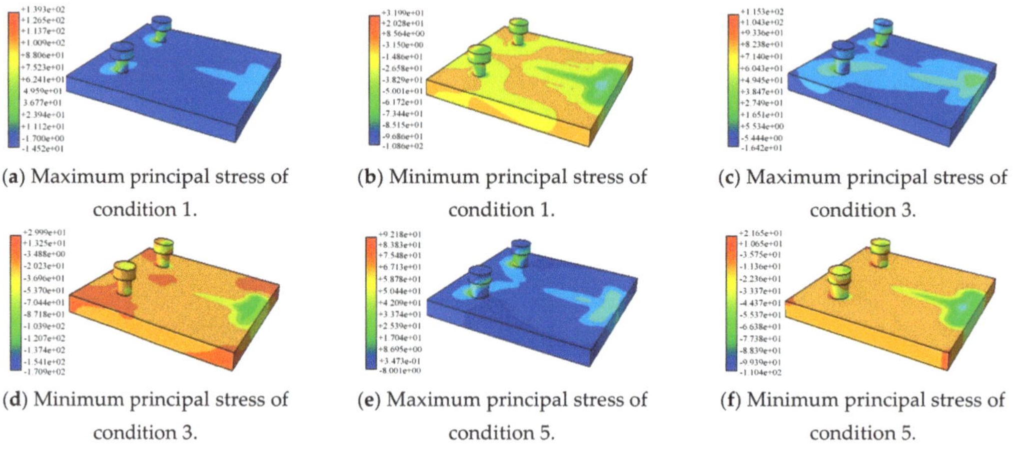

(**a**) Maximum principal stress of condition 1. (**b**) Minimum principal stress of condition 1. (**c**) Maximum principal stress of condition 3.
(**d**) Minimum principal stress of condition 3. (**e**) Maximum principal stress of condition 5. (**f**) Minimum principal stress of condition 5.

Figure 8. Principal stress of the A-type elastic joint.

The principal stress of the B-type elastic joint is shown in Figure 9. Similar to the A-type snap-fit joints, stress concentration is mainly observed at the connection point between the elastic joint and the connecting plate.

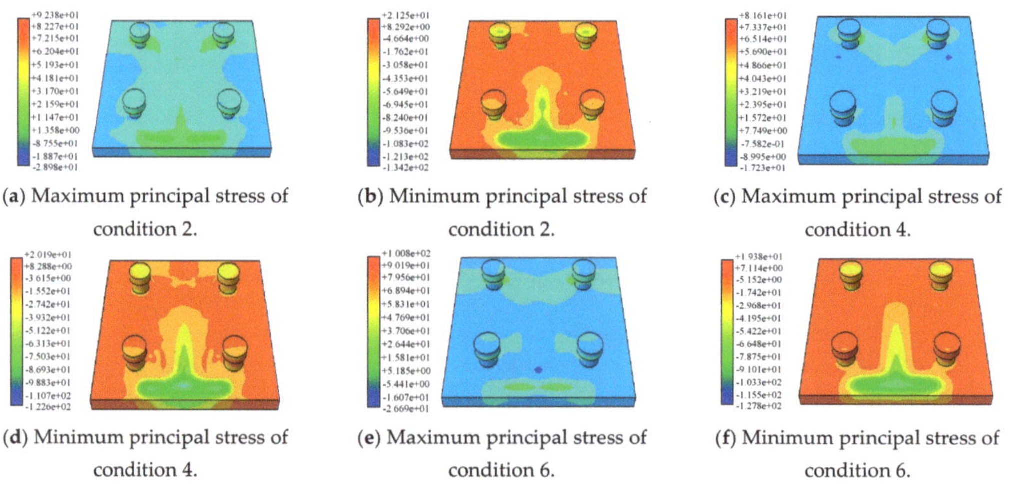

(**a**) Maximum principal stress of condition 2. (**b**) Minimum principal stress of condition 2. (**c**) Maximum principal stress of condition 4.
(**d**) Minimum principal stress of condition 4. (**e**) Maximum principal stress of condition 6. (**f**) Minimum principal stress of condition 6.

Figure 9. Principal stress of the B-type elastic joint.

However, with an incremental increase in elastic joint diameter, the stress in the elastic joint decreases initially and then increases. The elastic joint with a diameter of 24 mm exhibits the highest maximum principal stress at 100.8 MPa, while the one with a 20 mm diameter corresponds to the lowest maximum principal stress at 81.6 MPa. In comparison to A-type snap-fit connections, tensile stress remains predominant in elastic joints; however, the maximum principal stress values at the stress concentration sites in the elastic joints experience a certain degree of reduction.

By extracting the maximum and minimum principal stresses of A-type and B-type snap-fit flexible joints under different working conditions, variation patterns can be obtained, as shown in Figure 10. As depicted in Figure 10a, the maximum principal stress of A-type snap-fit connections decreases gradually as the joint diameter increases. Furthermore, from a force perspective of flexible joints, increasing the joint diameter can effectively enhance the safety and resistance to damage of the snap-fit joint. Regarding the minimum principal stress, the change in joint diameter has a more significant impact on its maximum principal stress. When the joint diameter increases from 16 mm to 24 mm, the maximum principal stress decreases by 47.2 MPa, representing a reduction of approximately 33.8%. However, the rate of decrease in the maximum principal stress diminishes when the joint diameter exceeds 20 mm. By extracting the maximum and minimum principal stresses of B-type snap-fit flexible joints and analyzing the influence of different joint diameters on the principal stresses, we can observe a trend in their variation, as shown in Figure 10b. The maximum principal stress of B-type snap-fit connections exhibits a trend of initially decreasing and then increasing with the increase in joint diameter. This phenomenon is attributed to the fact that an excessively large joint diameter can alter the strength and stiffness of the connecting plate where the joint is located, thereby weakening the overall connection effect of the snap-fit assembly. When the joint diameter is 20 mm, the maximum principal stress of B-type snap-fit connections reaches its minimum value of 81.6 MPa.

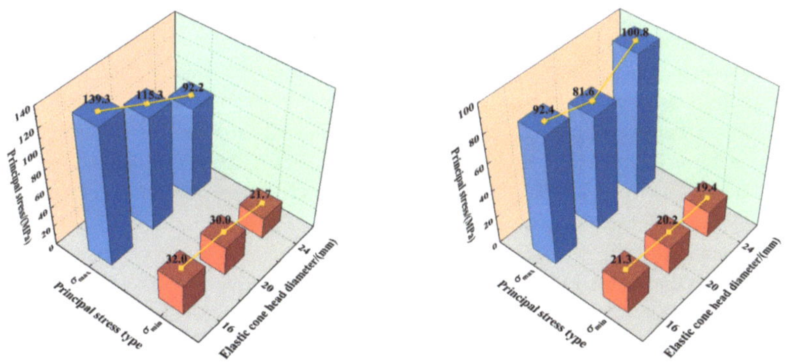

(**a**) A-type (**b**) B-type

Figure 10. Principal stress of the A-type and B-type elastic joints.

3.3.4. Contact Status Analysis of Connecting Plates

The displacement of the connecting plates can serve as an indicator of the contact state between the two plates, as illustrated in Figures 11 and 12. Figure 11 reveals that in the A-type snap-fit connection, there is mutual compression between the connecting plates near the inner arc side of the tunnel, with no apparent trend of detachment. The separation zone of the connecting plates is primarily concentrated on the outer arc side closer to the surrounding rock. With the increase in the joint diameter, the displacement of the separation site of the connecting plates gradually decreases. This indicates that the spacing between the connecting plates gradually reduces as well. This indicates a gradual reduction in the detachment between the connecting plates and a decreased likelihood of compression failure. When the elastic joint diameter exceeds 20 mm, there is a certain alteration in the contact area of the connecting plates, but the impact on the detachment distance between the plates becomes less pronounced.

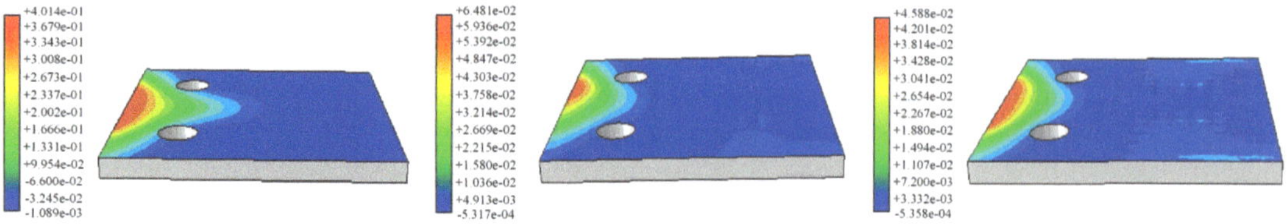

(**a**) Condition 1. (**b**) Condition 3. (**c**) Condition 5.

Figure 11. Contact state between the snap-fit connecting plates of Type A.

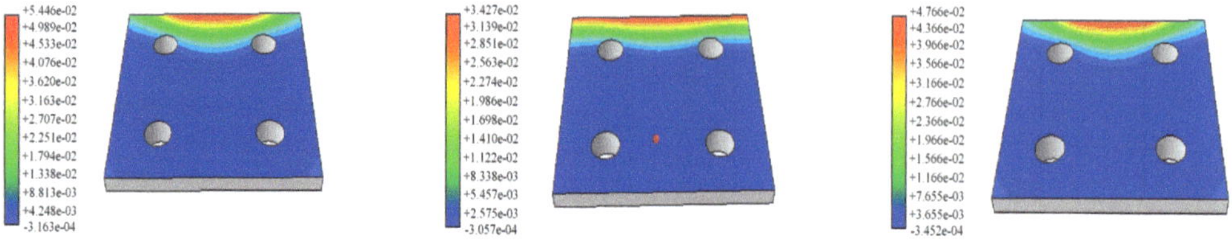

(**a**) Condition 2. (**b**) Condition 4. (**c**) Condition 6.

Figure 12. Contact state between the snap-fit connection plates of Type B.

The contact status between the connecting plates of the B-type snap connection device under different joint diameters is shown in Figure 12. The extrusion area between the connecting plates is mainly concentrated on the inner arc side close to the tunnel, and the disengagement area between the connecting plates is mainly concentrated on the outer arc side close to the surrounding rocks. However, in comparison to A-type snap-fit connection assemblies, the detachment distance between the connecting plates is significantly reduced. For B-type snap-fit connection assemblies, with the increase in elastic joint diameter, the detachment distance between connecting plates decreases initially and then increases. When the joint diameter is 20 mm, the minimum detachment distance between the connecting plates is observed. However, as the joint diameter increases to 24 mm, both the detachment area and distance increase. This suggests that exceeding a specific value of elastic joint diameter weakens the stiffness and strength of the connecting plates.

The above analysis concludes that B-type snap-fit connections exhibit better control over surrounding rock deformation while experiencing lower stress levels. The stress concentration at the connection site and the detachment distance between connecting plates are both minimized, indicating that the comprehensive performance of B-type snap-fit connections in folding steel arch frames is superior to that of A-type snap-fit connections. Furthermore, the load-bearing capacity of the B-type snap-fit connection, particularly that with an elastic joint diameter of 20 mm, surpasses that of other sizes of connectors, making it more suitable for use in machining folding steel arch frame structures with snap-fit connections.

4. Engineering Application and Verification

4.1. Project Overview

The New Wushaoling tunnel is a pivotal project in the construction of the Lanzhou to Zhangye No. 3 and No. 4 Line Railway, specifically the Chuan Airport to Wuwei section. Positioned to the east of the existing Lanzhou–Wuwei No. 2 Line's Wushaoling Extra-Long Tunnel, the tunnel is approximately 210 m and 571 m away from the existing right-line tunnels at the Zhangye and Lanzhou ends, respectively. The rail's surface elevation is approximately 36 m higher at the Zhangye end and 110 m higher at the Lanzhou end compared to the Zhangye end of the existing Wushaoling Extra-Long Tunnel. This newly designed tunnel is a double-track railway tunnel with a design speed of 250 km/h (see Figure 13). Situated in Dachaigou Town, Tianzhu County, at the entrance and in Anyuan Town at the exit, the tunnel traverses the Wushaoling Mountain range. It spans from DK160+920 to DK178+045, with a maximum burial depth of 952 m and a total length of 17,125 m. The geological conditions comprise 3275 m of Class III surrounding rock, accounting for 19.12%; 6680 m of Class IV surrounding rock, constituting 39.01%; and 7275 m of Class V surrounding rock, making up 41.87%. To address challenges such as high erection risks, intensive labor requirements, and low erection precision associated with the erection of folding steel arch frames in the tunnel, from October 2020 to March 2021, the China Railway 15th Bureau Group Fifth Engineering Co., Ltd. conducted a research trial for tunnel machining arch frame erection in the entrance section of the New Wushaoling tunnel, spanning DK160+920 to DK168+941. The average burial depth of the trial section is 450 m, and the geological conditions are complex, featuring exposures of sedimentary, igneous, and metamorphic rocks, with sedimentary rocks being predominant.

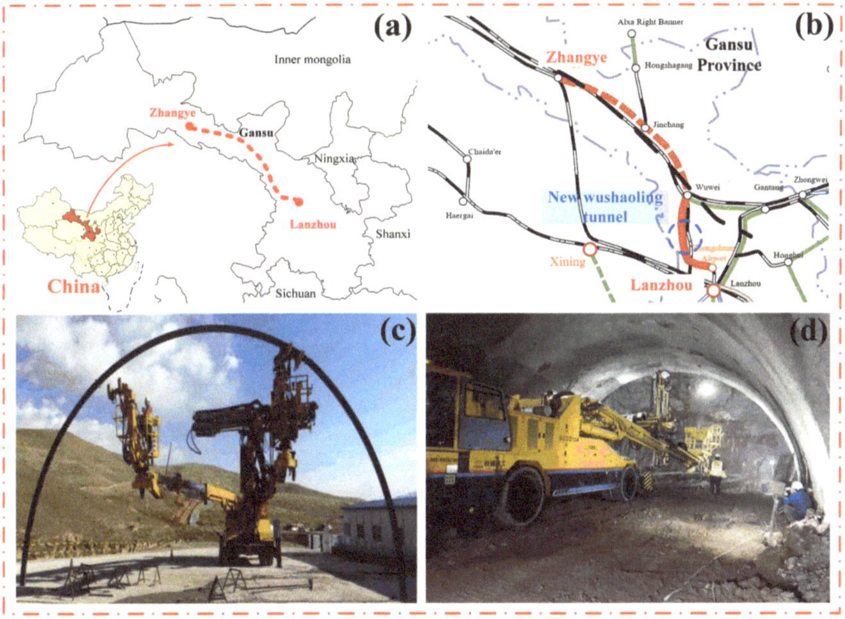

Figure 13. Folding steel arch frame machining erection test in New Wushaoling tunnel: (**a**) Lanzhou to Zhangye No. 3 and No. 4 Line Railway; (**b**) New Wushaoling tunnel; (**c**) splicing outside the tunnel; (**d**) erection inside the tunnel.

4.2. Result Analysis

To avoid the influence of irrelevant factors, two representative sections with favorable subsurface conditions and similar characteristics were selected for the mechanical arch construction test at the import site. These sections are located at the monitoring faces DK162+100 and DK162+105, with a burial depth of 260 m and IV surrounding rock. The arch construction was carried by opening excavation at the monitoring faces. Two arch frames were constructed using the mechanical arch method at the monitoring faces, namely an A-type snap-fit connection folding steel arch frame (hereinafter referred to as the A-type steel arch) at the DK162+100 monitoring face and a B-type snap-fit connection folding steel arch frame (hereinafter referred to as the B-type steel arch) at the DK162+105 monitoring face. Monitoring points were set at key positions along the excavation profile and installed at seven measurement points on the monitoring faces, as shown in Figure 14. From top to bottom, these points include the arch crown, left and right spandrels, left and right sidewalls, and left and right arch feet. The installation of the field measurement instruments is shown in Figure 14, with TZX-R-type tension steel strain gauges installed on both sides of the steel arch rib plate. The changes in the steel arch's vibrational frequency after loading are recorded daily to measure the dynamic rules of steel arch axial loading. Due to the axial symmetry of steel arch loading, only the mechanical responses of the two steel arches at the arch crown, left spandrels, left sidewalls, and left arch feet are compared in this section for analysis.

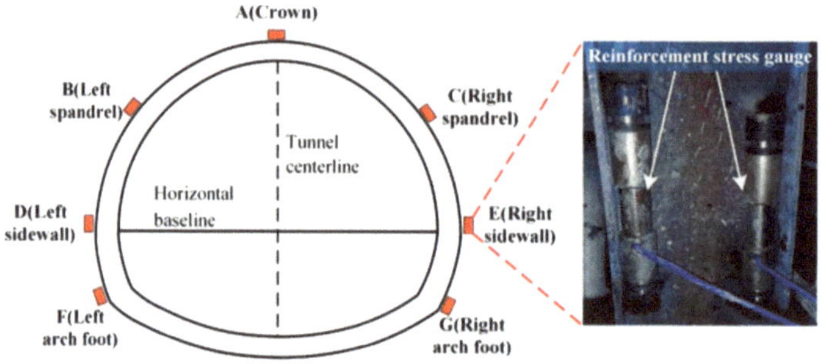

Figure 14. Measurement point layout of the tunnel monitoring section.

Figure 15 shows stress monitoring curves for the steel arch frames. The two types of steel arch frames exhibited the maximum stress values at the crown of the arch due to the key node effect, as the crown is the primary load-bearing point. Both A-type and B-type steel arch frames showed significant differences in axial stress at the crown, arch soffit, sidewall, and arch springing; the A-type steel arch frames exhibited higher stress levels compared to the corresponding components of the B-type steel arch frames, indicating that under similar conditions, the A-type steel arch frames experience greater axial stress, while the B-type steel arch frames demonstrated better support performance with lower overall stress levels. The axial stress changes in both steel arch frames exhibited similar trends over time, but the A-type steel arch frames showed a higher rate of stress increase. The stress at the crown of the steel arch frames gradually reached values close to 25 MPa before stabilizing briefly, followed by a rapid increase from the 10th to the 20th day. This trend was closely associated with the progress of the lower bench excavation. Before the 10th day, the stress changes at the crown of the steel arch frames were relatively small, primarily due to the diminishing influence of the creation of the construction face. However, after the 10th day, the stress increase rate in the steel arch frames became significantly larger, indicating that the lower bench excavation had a greater influence on the axial stress of the steel arch frames at the crown. Additionally, the A-type steel arch frames showed higher rates of stress increase compared to the B-type steel arch frames at the arch soffit and sidewall, while the B-type steel arch frames exhibited a higher rate of stress increase at the sidewall compared to the A-type steel arch frames. However, after the stress stabilized, the A-type steel arch frames showed higher stress levels in all components compared to the corresponding components of the B-type steel arch frames. These differences further indicate that under similar conditions, the A-type steel arch frames experience greater axial stress, while the B-type steel arch frames demonstrate better support performance with more reasonable axial stress distribution, reduced concentration of stresses on the connection plates, and improved overall load-bearing capacity and safety. This result further validates the stability and reliability of the B-type snap-fit connection folding steel arch frames under large loads and complex construction conditions, providing effective support for the redistribution of stresses caused by removal of load during bench excavation, thereby effectively controlling rock deformation. This makes the B-type snap-fit connection folding steel arch frame an ideal choice for optimizing the structure of folding steel arch frames.

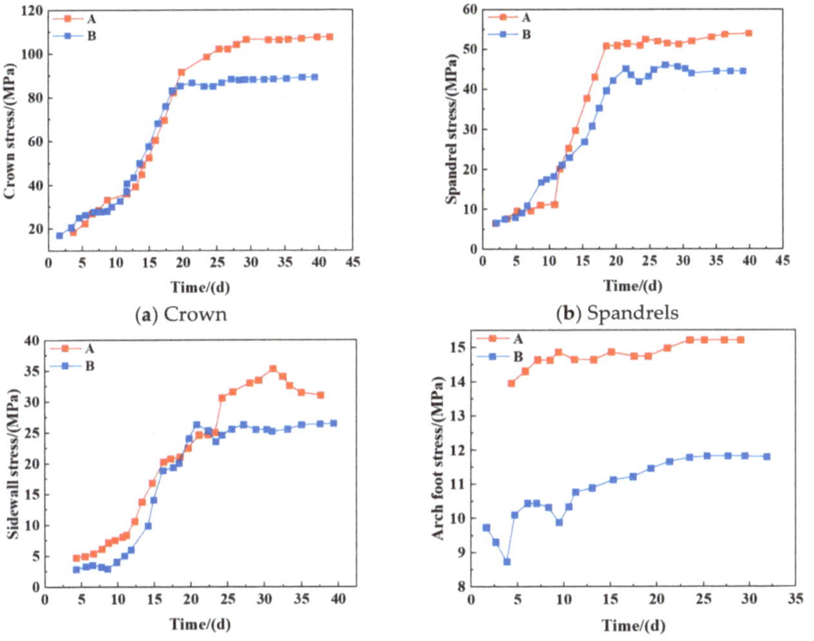

(**a**) Crown (**b**) Spandrels (**c**) Sidewalls (**d**) Arch feet

Figure 15. Stress–time history curve of steel arch at various measuring points.

Based on experimental data, a comparative analysis was conducted between the A-type and B-type steel arch frames regarding the stress values measured at the arch crown, spandrel, sidewall, and arch foot. The results (Figure 16) indicate that regardless of the type of connection, the numerical simulation results consistently exceeded the measured values, likely due to delays in steel arch installation after excavation, allowing the surrounding rock to release some energy. However, both the numerical simulation and field monitoring results showed similar trends in terms of stress changes. Additionally, the B-type steel arch frames exhibited lower stress levels at all components compared to the corresponding components of the A-type steel arch frames. This result is primarily attributed to the B-type snap-fit connection, which results in more uniform stress distribution and reduces the concentration of stresses on the connection plates, thereby improving the overall load-bearing capacity and safety of the steel arch frames. These findings further validate the stability and reliability of the B-type snap-fit connection folding steel arch frames under large loads and complex construction conditions. The B-type steel arch frames are capable of providing strong support for the redistribution of stresses caused by the removal of loads during bench excavation, effectively controlling rock deformation. As a result, the B-type snap-fit connection folding steel arch frame represents an optimal choice for optimizing the structure of folding steel arch frames.

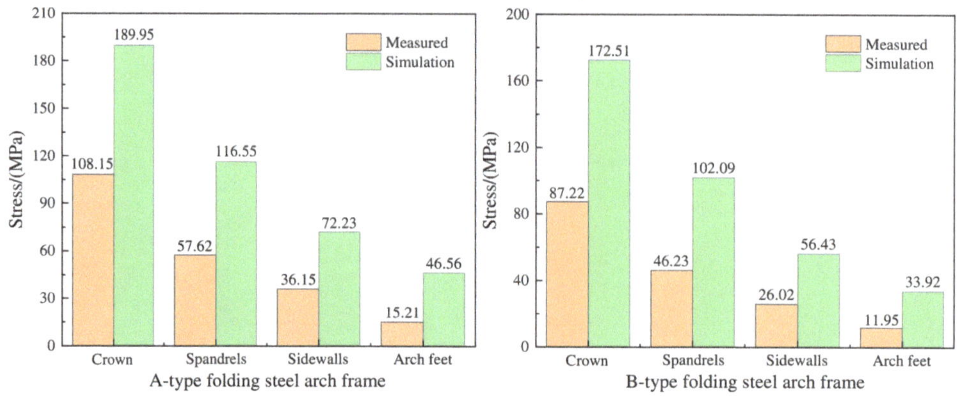

Figure 16. Comparison of field test and simulation results.

5. Recommendations

In order to meet the requirements for "unmanned" construction of arch spans in tunnels, this study investigated the load characteristics of folding steel arch frames with different snap-fit connection types. Although certain achievements have been made, there are still some limitations and shortcomings, detailed as follows.

(1) The study of the load characteristics of the folding steel arch frame did not consider the planar bending and twisting deformation of the steel arch frame.

(2) When establishing the finite element model, due to the greater burial depth of the tunnel, the load equivalent of the upper rock mass was used without considering the effects of construction stresses, dynamic loads, and fatigue loads, leading to some discrepancies compared to the actual situation.

(3) Due to limited monitoring points and environmental factors under actual construction conditions, it is difficult to comprehensively simulate long-term performance. Additionally, the simulation model's complexity in material models and joint failure mechanisms led to certain simplifications, which may not fully reflect the actual complex conditions.

It is recommended that in future research, experimental conditions are expanded to study the load stability of folding steel arch frames under different working conditions; the finite element model be improved by developing more precise material models and joint failure mechanism models. Long-term testing and actual application monitoring could be conducted to evaluate long-term performance and efforts made to promote the development and application of mechanized tunnel installation technologies.

6. Conclusions

This study is based on the excavation compensation concept and combines numerical simulation with field testing to investigate the effects of different snap-fit connections on the load characteristics of folding steel arch frames in tunnels. The main conclusions are as follows.

1. The maximum vertical displacements of A- and B-type snap-fit connections decrease and then increase with increasing joint diameter. With a joint diameter of 20 mm, the minimum displacements are 43.1 mm and 39.2 mm, respectively, for A- and B-type snap-fit connections. The vertical displacement of the B-type snap-fit is reduced by 9.1% compared to the A-type snap-fit, indicating that the B-type snap-fit has stronger anti-deformation capacity and greater load-bearing performance.
2. The maximum principal stress of the B-type folding steel arch frames is 49.3 MPa, which is 4.5% lower than that of the A-type folding steel arch frames (51.6 MPa).

3. The field test results indicate that for the same joint diameter, the maximum stress of the B-type folding steel arch frames is reduced by 15.8% compared to the A-type folding steel arch frames, significantly improving their ability to compensate for rock deformation. This makes the B-type snap-fit a more suitable form of connection for folding steel arch frames in tunnels and facilitates the development and application of mechanized installation technologies for steel arch frames in tunnels.

Author Contributions: Conceptualization, S.L.; Methodology, S.L., C.H., X.Y., Z.T. and J.G.; Software, S.L., X.Y. and Z.T.; Formal analysis, J.G.; Investigation, C.H.; Writing—original draft, S.L. and C.H.; Writing—review & editing, S.L., C.H., X.Y., Z.T., J.G., H.L., T.Y. and J.H. All authors have read and agreed to the published version of the manuscript.

Funding: This research was funded by the National Natural Science Foundation of China—Railway Fundamental Research Joint Fund Project (U2468219); the Research and Development Projects of Science and Technology of China Railway Construction Co., Ltd. (Project No. N2020G040); the Science and Technology Research and Development Project of China Railway Construction Co., Ltd. (Project No. 2024-W25); and the China Association of Construction Enterprise Management (Grant No. 2023-B-028).

Data Availability Statement: Data will be made available on request.

Acknowledgments: The authors greatly appreciate financial support from various funding bodies and are grateful to the reviewers for their valuable comments and suggestions, which improved the quality of the paper.

Conflicts of Interest: Authors Shaohua Li, Humin Li and Tiejun Yao were employed by the company China Railway 15th Bureau Group Co., Ltd. The remaining authors declare that the research was conducted in the absence of any commercial or financial relationships that could be construed as a potential conflict of interest. The authors declare that this study received funding from China Railway Construction Co., Ltd. The funder was not involved in the study design, collection, analysis, interpretation of data, the writing of this article or the decision to submit it for publication.

References

1. China Civil Engineering Society. *Report on Advance in Tunnel and Underground Engineering*; Science and Technology of China Press: Beijing, China, 2020.
2. Qian, Q.H.; Lin, P. Safety risk management of underground engineering in China: Progress, challenges and strategies. *J. Rock Mech. Geotech. Eng.* **2016**, *8*, 423–442. [CrossRef]
3. Guo, D.S.; Song, Z.P.; Xu, T.; Zhang, Y.W.; Ding, L.B. Coupling analysis of tunnel construction risk in complex geology and construction factors. *J. Constr. Eng. Manag.* **2022**, *148*, 04022097. [CrossRef]
4. Gong, J.F.; Wang, W.; Li, X.; Zhu, Y. Statistics of railway tunnels in China by the end of 2022 and overview of key tunnels of projects newly put into operation in 2022. *Tunn. Constr.* **2023**, *43*, 721.
5. Wang, Q.S. Some thoughts on development of drilling-and-blasting construction technology for mountain tunnels in China. *Tunn. Constr.* **2023**, *43*, 37–45.
6. Wang, Q.; Qin, Q.; Jiang, B.; Xu, S.; Zeng, Z.N.; Luan, Y.C.; Liu, B.H.; Zhang, H.J. Mechanized construction of fabricated arches for large-diameter tunnels. *Autom. Constr.* **2021**, *124*, 103583. [CrossRef]
7. He, Y.F.; Yang, M.; Xu, Z.; Li, S.Y.; Zhang, B.W. Design of tunnel steel arch looping manipulator with multiple actuators in limited space. *Sci. Prog.* **2023**, *106*, 368504231180025. [CrossRef]
8. Tao, Z.G.; Xie, D.; Sui, Q.R.; Sun, J.H.; He, M.C. Study on active support method and control effect of NPR anchor cables for large deformation of tunnel surrounding rock under complex geological conditions. *Chin. J. Rock Mech. Eng.* **2024**, *43*, 275–286.
9. Sui, M.C.H.Q.R.; Tao, Z.G. Excavation compensation theory and supplementary technology system for large deformation disasters. *Deep. Undergr. Sci. Eng.* **2023**, *2*, 105–128. [CrossRef]
10. Guo, L.J.; Tao, Z.G.; He, M.C.; Coli, M. Excavation compensation and bolt support for a deep mine drift. *J. Rock Mech. Geotech. Eng.* **2024**, *16*, 3206–3220. [CrossRef]
11. Guo, J.B.; Chen, J.; Liu, J.Z.; Niu, J.C. Multifunctional integration equipment design and construction technology research. *Adv. Mater. Res.* **2013**, *650*, 359–363. [CrossRef]

12. Cheng, Y. The Research of Trajectory Planning for Manipulator of SGDC700 Tunnel Arch Multi-Function Installing Truck. Master's Thesis, Xiangtan University, Xiangtan, China, 2015.
13. Wang, Q. *Control Theory and Engineering Practice of Confined Concrete in Underground Engineering*; Science Press: Beijing, China, 2019.
14. Liu, F.X. SCDZ133 intelligent multi-function trolley and its application in tunneling. *Mod. Tunn. Technol.* **2019**, *56*, 1–7.
15. Sun, H.L. Study on Stability Bearing Mechanism and Key Technologies of Assembly Confined Concrete Support for Large Section Tunnel. Master's Thesis, Shandong University, Jinan, China, 2020.
16. He, Y.F.; Xia, Y.M.; Long, B.; Deng, Z.H.; Lei, M.L.; Yao, J. Grasping docking mechanism of TBM steel arch splicing robot. *J. Zhejiang Univ. (Eng. Sci.)* **2021**, *54*, 2204–2213.
17. Zhao, Y. Subsidiary technology for rapid construction of multifunctional steel arch erection machine in Yuelongmen tunnel. *Railw. Constr. Technol.* **2020**, *2*, 96–98+132.
18. Yan, Q. A remote unmanned construction system for steel arch erection developed by Japan Toda Construction Corporation. *Tunn. Constr.* **2021**, *41*, 387.
19. Krauze, K.; Boloz, L.; Mucha, K.; Wydro, T. The mechanized supporting system in tunnelling operations. *Tunn. Undergr. Space* **2021**, *113*, 103929. [CrossRef]
20. Ni, P.L. Research and application of new type of tunnel supporting equipment, steel arch installation trolley. *China Plant Eng.* **2020**, *18*, 12–13. (In Chinese)
21. Sun, C.Z.; Miao, C.Q.; Li, A.Q.; Qian, Y. Finite element analysis of mechanical properties of pile cap connections for prestressed concrete solid pile using snap-in mechanical connection. *J. Disaster Prev. Mitig. Eng.* **2019**, *39*, 37–44.
22. Wang, J.B. Numerical Simulation of Bending and Large Deformation Retractable Steel Arch Frame Support in Layered Soft Rock Tunnel. Master's Thesis, Chengdu University of Technology, Chengdu, China, 2020.
23. Li, W.T.; Yang, N.; Mei, Y.C.; Zhang, Y.H.; Wang, L.; Ma, H.Y. Experimental investigation of the compression-bending property of the casing joints in a concrete filled steel tubular supporting arch for tunnel engineering. *Tunn. Undergr. Space Technol.* **2020**, *96*, 103184. [CrossRef]
24. Xu, F.; Li, S.C.; Zhang, Q.Q.; Li, L.P.; Shi, S.S.; Zhang, Q. A new type support structure in-troduction and its contrast study with traditional support structure used in tunnel construction. *Tunn. Undergr. Space Technol.* **2017**, *63*, 171–182. [CrossRef]
25. Ma, K.M.; Zhang, J.C.; Zhang, J.R.; Wang, Z.Y.; Feng, J.M. Longitudinal connection effect on initial support steel frames in tunnels—Take the traffic tunnels as examples. *Undergr. Space* **2022**, *7*, 608–622. [CrossRef]
26. Song, Q.Y.; Heidarpour, A.; Zhao, X.L.; Han, L.H. Performance of flange-welded/web-bolted steel I-beam to hollow tubular column connections under seismic load. *Thin Walled Struct.* **2017**, *116*, 250–264. [CrossRef]
27. Yue, G.Z. Some suggestions on the connection of steel arch frame in tunnel construction. *J. Shandong Ind. Technol.* **2015**, *23*, 86–87. (In Chinese) [CrossRef]
28. Li, G.A.; Wang, Y.D.; Nie, Y.W.; Wan, S.T.; Wang, C.Y. Study on longitudinal connection support structure of steel plate between steel arches of soft rock tunnel. *Highway* **2022**, *67*, 411–418. (In Chinese)
29. He, M.C.; Wang, B.; Tao, Z.G.; Qiao, Y.F.; Xiao, Y.M. Axial compression behavior of adaptive steel arch joint for large deformation tunnels. *China J. Highw. Transp.* **2021**, *34*, 1–10.
30. Zhang, G.W.; Li, D.W.; Lei, X.T. Folding steel arch frame applicable to mechanized construction of tunnels and its key techniques. *Tunn. Constr.* **2022**, *42*, 445–450.
31. Zhang, G.W. Study on Mechanical Effect of Connection Mode Between Segments of Steel Arch Frame Suitable for Unmanned Installation. Master's Thesis, Lanzhou Jiaotong University, Lanzhou, China, 2022.
32. Wu, Y.M.; Tian, C.Q.; Xu, P.; Zhao, Z.Z.; Zhang, J.W.; Wang, S.J. Design optimization method of feet-lock steel pipe for soft-rock tunnel based on load-deformation coordination. *Appl. Sci.* **2022**, *12*, 3866. [CrossRef]
33. Huang, C.F.; Li, S.H.; Li, D.W.; Li, W.B.; Yao, T.J.; Xiao, Y. Mechanical properties of folding arch frame joints for unmanned arch erection. *Buildings* **2024**, *14*, 1480. [CrossRef]
34. Liu, Z.Y.; Tang, Q.S.; Dong, X.Y.; Weng, W.S.; Zhao, J.X.; Zhang, J.M. Experimental study on performance of the connection joints of cold-formed square steel tubular columns and H-shaped steel beams extension endplate using blind bolts. *Build. Struct.* **2021**, *51*, 57–63.
35. Wang, H.; Zheng, P.Q.; Zhao, W.J.; Tian, H.M. Application of a combined supporting technology with U-shaped steel support and anchor-grouting to surrounding soft rock reinforcement in roadway. *J. Cent. South Univ.* **2018**, *25*, 1240–1250. [CrossRef]
36. *TB 10003-2016*; Code for Design of Railway Tunnel. Ministry of Railways: Beijing, China, 2016.

buildings

MDPI

Article

Stability Analysis of Horizontal Layered Multi-Stage Fill Slope Based on Limit Equilibrium Method

Xiaohui Li [1], Shuaihua Ye [1,*], Manman Qiu [2], Weina Ye [3] and Jingbang Li [3]

1 School of Civil Engineering, Lanzhou University of Technology, Lanzhou 730050, China; lxh121366@163.com
2 School of Civil Engineering and Water Resources, Qinghai University, Xining 810016, China; xrwmmz@163.com
3 School of Civil Engineering, Lanzhou Institute of Technology, Lanzhou 730050, China; yewn@lzit.edu.cn (W.Y.); lijingbang@lzit.edu.cn (J.L.)
* Correspondence: yesh@lut.edu.cn

Abstract: The stability of a multi-stage fill slope composed of horizontal layered soil is particularly prominent due to its complex structural characteristics, the variability of filler properties, and the combined effects of external environmental factors. Therefore, it is of great significance to clarify whether such slopes are in a safe state. Based on the limit equilibrium method, this paper divides the soil horizontally and obliquely and analyzes the stress and establishes the potential failure mechanism (such as slope toe circle, midpoint circle, etc.) of the multi-stage fill slope in the overall failure mode and local failure mode. The analytical expressions of slope safety factors corresponding to various failure mechanisms are further derived, and the stability analysis process of multi-stage fill slope and the determination method of the most dangerous slip surface are proposed. Through the verification of two examples, the results show that the safety factor obtained by this method is similar to the minimum safety factor obtained by the traditional slice method, and the error is small. At the same time, the most dangerous slip surface and slip range are basically consistent with the traditional method. The research results can provide a theoretical basis and practical reference for stability analysis, filler selection, and engineering design of the fill slope.

Keywords: multistage fill slope; limit equilibrium; horizontal stratification; stability; failure mechanism

1. Introduction

In recent years, with the rapid development of infrastructure construction in western China, large-scale fill projects such as highways, railway subgrades, reservoir dams, and industrial site leveling have been increasing, resulting in a large number of fill slopes. The stability of these slopes is directly related to the safety and durability of the project. However, due to the frequent occurrence of landslides and collapses caused by slope instability, it has a serious impact on regional economic development and people's lives and health. In particular, the multi-level fill slope is composed of a horizontal layered soil layer; because of its complex structural characteristics, the variability of filler properties, and the combined effect of external environmental factors (such as rainfall, earthquake, etc.), its stability problem is particularly prominent [1–6]. Therefore, it is of great significance to clarify whether such slopes are in a safe state for ensuring engineering safety and promoting regional sustainable development.

Many experts and scholars have studied the stability of fill slope by combining theoretical calculation, model test, numerical analysis, and other methods [7–12]. In terms of numerical simulation and test, Amena [13] analyzed the applicability of plastic waste-treated clay as embankment filler and used PLAXIS 2D software to analyze the slope stability by the finite element method. Gong et al. [14] studied the disaster-hidden danger of the high and steep original slope-fill slope interface through the indoor model test and compared and analyzed the influence of working conditions such as no weak zone, different thickness of the weak zone, and weakening coefficient with numerical simulation. Taking the loess high fill slope in Lanzhou as an example, Yong et al. [15] evaluated the influence of water content change on slope stability during rainfall through indoor and outdoor tests. The limitation of experiment and numerical simulation lies in the specificity of the results and the dependence on the initial conditions. The theoretical analysis method can reveal the internal mechanism of slope instability and provide a universal theoretical framework and design basis. In this regard, based on the upper bound theorem of limit analysis, Yan et al. [16] constructed an extended three-dimensional horn-shaped failure mechanism for the three-dimensional stability analysis of high slopes with an oblique intersection between the excavation-filling interface and the strike line of the slope surface. By introducing the inclination angle parameter of the excavation-filling interface, the functional equilibrium equation was established, and the sequential quadratic programming optimization algorithm was used to solve the upper bound solution of the slope safety factor after the strength reduction. Mostafaei et al. [17] studied the seismic safety of the abutment of the Bakhtiari double-curvature arch dam, using time-history analysis under DBE and MCE hazard levels. The stability of the wedge is calculated by the Londe limit equilibrium method and MATLAB (R2022b), and the thrust is obtained by ABAQUS. The effects of foundation flexibility, grouting curtain performance, seismic vertical component, material, and geometric nonlinearity on the safety factor were investigated. In addition, Chen et al. [18] combined the limit equilibrium method and the logarithmic spiral curve model to evaluate the seismic stability of the reinforced soil slope and revealed the influence of the uneven distribution of tensile strength on the seismic reinforcement effect. Based on the quasi-static method, Fatehi et al. [19] carried out the limit equilibrium analysis of the reinforced slope under seismic load and discussed the influence of different parameters on the design of the reinforcement layer. Zhang et al. [20] proposed a stability evaluation method of a V-shaped fill slope based on the combination of the limit equilibrium method and quasi-static method and analyzed the influence of seismic load and geometric parameters on the three-dimensional stability and the most dangerous slip surface of the slope. Wang et al. [21] analyzed the probability model of shear strength distribution of typical loess Q2 and Q3 for a high fill project in northern Shaanxi and applied it to the reliability analysis of loess high slope. Huang et al. [22] used the grey correlation analysis method to identify the sensitive factors of high-fill slope stability, which provides a basis for the selection of design parameters. Although the above theoretical analysis of slope stability has been fully carried out, compared with the traditional limit equilibrium method (such as the Bishop method and Morgenstern-Price method), the proposed method has obvious advantages in the calculation of safety factor and the identification of the most dangerous sliding surface by considering the non-uniformity of soil parameters and accurate sliding surface search, and taking into account the overall and local instability. It improves the accuracy and computational efficiency of the analysis and has good engineering practicability.

A large number of calculations and analyses have been carried out on the failure mechanism and stability of the fill slope. However, in the existing research, the fill slope is usually simplified as a homogeneous body, which fails to fully consider the physical characteristics of its layered filling. In fact, the filling part of the fill slope is usually prepared according to the design requirements, and the required degree of compaction is achieved by compaction. However, due to the influence of filling construction technology, there are significant differences in the compaction coefficient of filling soil in different regions, resulting in obvious spatial distribution uncertainty of soil parameters such as unit weight, internal friction angle, and cohesion. In addition, most of the existing research methods on the stability of fill slopes are aimed at single-stage slopes, while there are relatively few studies on the stability of multi-stage slopes. Therefore, it is of great theoretical significance and engineering value to explore the influence of the layered characteristics of the fill slope on the stability, especially the stability of the multi-stage slope.

In summary, aiming at the stability problem of horizontal layered multi-level fill slope, based on the limit equilibrium method, the horizontal slice method and the inclined slice method are used to systematically analyze the failure mechanism that can occur under the overall failure mode and local failure mode of the fill slope. The two typical examples are verified. The calculation results of this method are close to the minimum safety factor obtained by the traditional slice method, the error is small, and the most dangerous slip surface position and slip range determined are basically consistent with the traditional method.

2. Basic Assumption of Sliding Soil and Stress Condition of Soil Strip

2.1. Basic Assumptions

In this paper, the limit equilibrium method is used to analyze the stability of horizontal layered multi-stage filled slope, and the following assumptions are applied:

(1) The slope soil is an ideal rigid-plastic body;
(2) The shear strength of slope soil complies with the Mohr–Coulomb criterion;
(3) Ignore the interaction force between the soil strips taken;
(4) The slope is a heterogeneous slope and the slip surface is an arc slip surface.

For the circular slip surface, it is assumed that the failure mechanism can be divided into three cases: ① The circular slip surface passes through the slope toe, which is called the slope toe circle; ② The circular slip surface passes through a point outside the toe of the slope, which is called the midpoint circle; ③ The circular slip surface passes through a point on the slope, which is called the slope circle.

2.2. Stress Condition of Soil Strip

The N-grade slope is taken as an example for analysis, as shown in Figure 1, and the cdef division of the horizontal strip and agh division of the inclined strip are shown in Figure 2. This study takes the heterogeneity of soil slope into consideration. The potential slip surface is circular, the center of the slip surface is $O\ (a,b)$, and the radius is r. Slope height is H, soil weight is γ, cohesion is c, and internal friction angle is φ.

In the sliding body, a horizontal soil strip i and an inclined soil strip j are selected, and the stress conditions of the soil strip are as follows: the normal stress N_i and the shear stress T_i on the sliding surface of the horizontal soil strip and the weight of the soil strip W_i; the normal stress N_j and shear stress T_j on the sliding surface of the sloping soil strip and the dead weight of soil strip W_j;

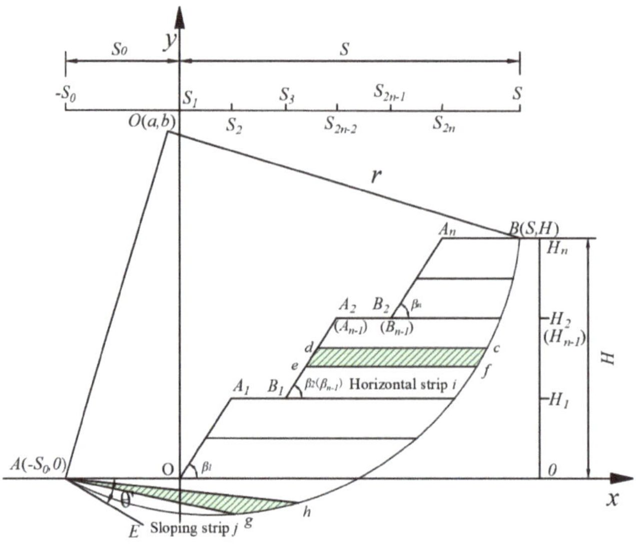

Figure 1. Calculation model of multi-stage fill slope.

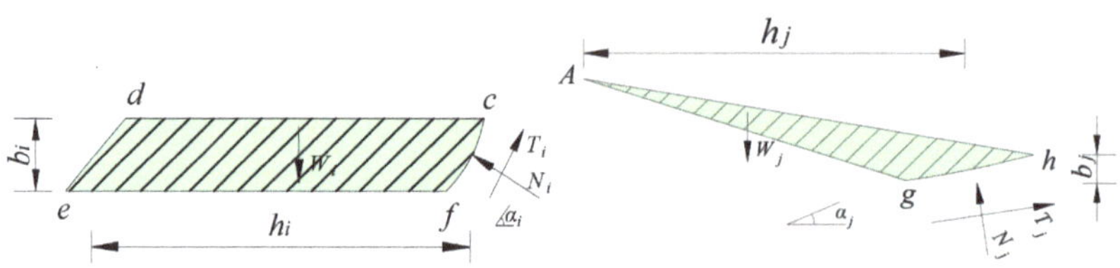

Figure 2. Stress analysis of horizontal and sloping soil strips.

3. Stability Analysis of Multi-Stage Fill Slope Under Overall Failure Mechanism

3.1. Stability Analysis Under Circular Failure Mechanism of Slope Toe

3.1.1. Damage Mechanism

Considering the specific geometric structure of the multi-fill slope, take the N-grade slope $AA_1B_1A_{n-1}B_{n-1}A_nD$ in Figure 3 as an example, where $n = 1, 2, 3, \ldots$ is the slope series, the slope surface is AA_1, $B1A_{n-1}$, $B_{n-1}A_n$, the platform width is A_1B_1, $A_{n-1}B_{n-1}$, the slope angle is β_1, β_{n-1}, β_n, the A_{xy} plane rectangular coordinate system is established. The projections of points A, A_1, B_1, A_{n-1}, B_{n-1}, A_n, B on the X-axis is successively S_1, S_2, S_3, S_{2n-2}, S_{2n-1}, S_{2n}, S, and the projections of points A, $A_1(B_1)$, $A_{n-1}(B_{n-1})$, and $A_n(B)$ on the Y-axis is successively 0, H_1, H_{n-1}, H_n. The center of the circle is O (a,b), the radius is r, and the arc slip surface is assumed to be AB. The angle between the arc tangent AE and the X-axis through point A is θ, then the slope $k = \tan(\theta)$ of AE, the coordinate of point A is (0,0), and the coordinate of point B is (S,H), where S is the abscissa of the intersection point between the critical slip surface and the slope top plane.

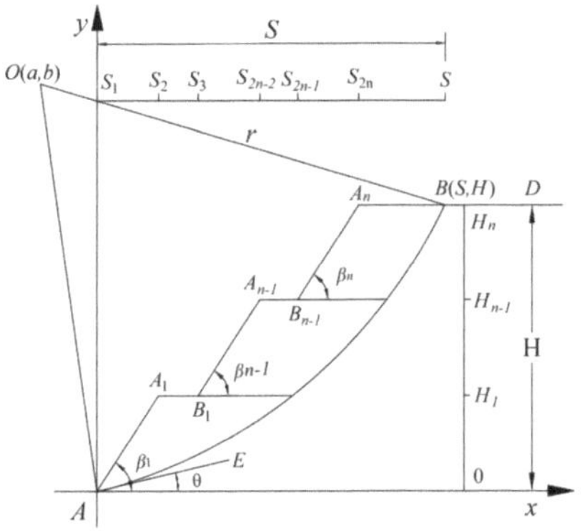

Figure 3. Multi-stage filling edge slope angle circle failure mechanism.

Based on the failure mechanism in Figure 3, the arc slip surface equation *AB* and the geometric parameters of the equation center *O* and radius *r* are given. The equation of the circular slip surface *AB* of the multi-stage filled slope can be expressed as follows:

$$(x-a)^2+(y-b)^2=r^2 \tag{1}$$

And satisfy:

$$\left.\begin{array}{l} a^2+b^2=r^2 \\ (a-x)^2+(b-y)^2=r^2 \\ a=-kb \end{array}\right\} \tag{2}$$

According to Equation (2), it can be obtained:

$$\left.\begin{array}{l} a=\frac{-k(H^2+S^2)}{2H-2kS} \\ b=\frac{H^2+S^2}{2H-2kS} \\ r=\frac{\sqrt{1+k^2}(H^2+S^2)}{2H-2kS} \end{array}\right\} \tag{3}$$

Among them:

$$\left.\begin{array}{ll} S_{2n}\leq S & k\leq 0 \\ S_{2n}\leq S<\frac{H}{k} & k>0 \end{array}\right\} \tag{4}$$

At this point, the relationship between *S* and *k* can be established. This equation represents the slip surface control condition of the circular failure mechanism of the slope toe. The slip plane of failure is determined by controlling *S* and *k* variables.

3.1.2. Stability Analysis

For the calculation and analysis of slope stability under the slope angle circular failure mechanism, the multi-stage fill slope stability analysis and calculation model as shown in Figure 4 is established. The specific derivation process is as follows:

Based on the traditional limit equilibrium method, the sliding soil $AA_1B_1A_{n-1}B_{n-1}A_nB$ is divided into soil strips with a height of b_i, and the whole sliding soil is divided into *n* soil strips. The height of each soil strip can be approximately infinitely small, and the calculation process is simply summed. The soil strip element cdef is arbitrarily taken out for force analysis and calculation. The force diagram of cdef of soil strips is shown in Figure 5. Based on the equilibrium condition of soil strips, it can be obtained as follows:

$$\left.\begin{array}{ll} \sum F_x = 0 & -N_i \sin\alpha_i + T_i \cos\alpha_i = 0 \\ \sum F_y = 0 & W_i - N_i \cos\alpha_i - T_i \sin\alpha_i = 0 \end{array}\right\} \tag{5}$$

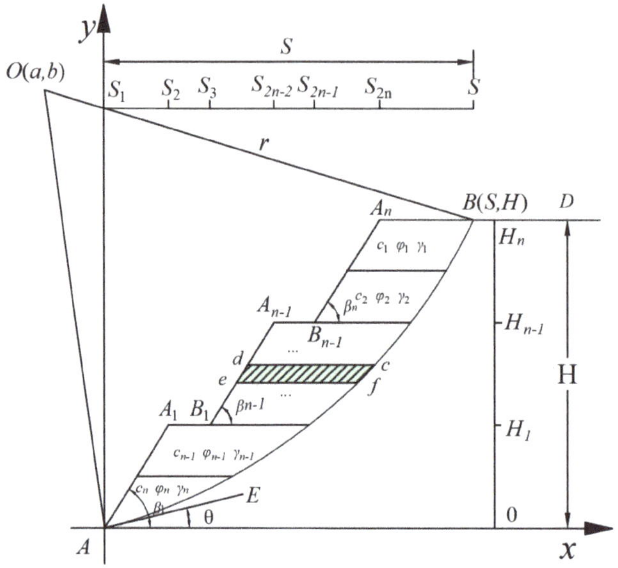

Figure 4. Calculation model of the stability analysis of the angle circle of the multi-stage filling slope.

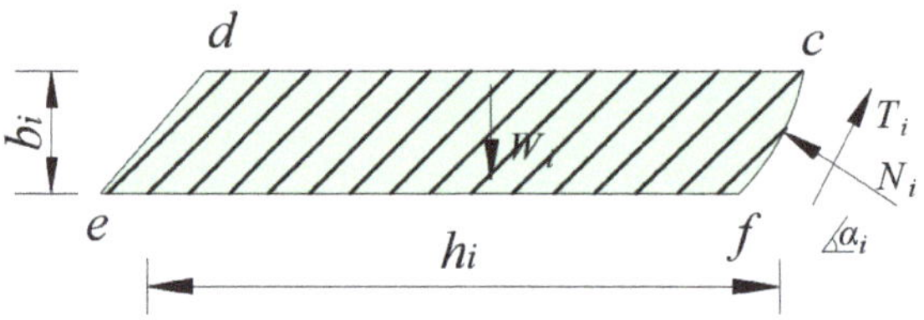

Figure 5. Schematic diagram of soil strip element cdef force.

In the formula: W_i is the gravity of the i-th soil strip; $W_i = \gamma h_i b_i$; γ is the unit soil weight; h_i is the length of the i-th soil strip; N_i is the normal stress of the i-th soil strip; T_i is the shear stress of the i-th soil strip; α_i is the angle between the normal line and the vertical line of the middle point of the bottom edge of the i-soil strip.

For the convenience of calculation, let the equations of line segment AA_1, B_1A_{n-1}, $B_{n-1}A_n$ and arc AB be x_1, x_2, x_3, x_4, respectively, then the stability analysis model of slope angle circle of multi-stage filling edge slope and the governing equation of slip surface can be obtained, as follows:

$$\left.\begin{array}{l} x_1 = \frac{y}{\tan\beta_1} \\ x_2 = \frac{y-H_2}{\tan\beta_2} + S_2 \\ x_3 = \frac{y-H_3}{\tan\beta_3} + S_3 \\ x_4 = a - \sqrt{r^2 - (y-b)^2} \end{array}\right\} \tag{6}$$

From Equation (6) and Figure 4, we can obtain the following:

$$h_i = \begin{cases} x_1 - x_4 & (0 \leq y < H_1) \\ x_2 - x_4 & (H_1 \leq y < H_{n-1}) \\ x_3 - x_4 & (H_2 \leq y < H_n) \end{cases} \quad (7)$$

The moment balance equation for the center of the circle is established from the soil strip i:

$$\sum M_O = 0 \quad W_i d_i - T_i r = 0 \quad (8)$$

In the formula: d is the distance between the center of gravity of the i-th soil strip and the center of the sliding circle, $d_i = r \sin \alpha_i - h_i/2$.

Since the soil on the sliding surface of the soil strip is in the limit equilibrium state, according to the Mohr–Coulomb criterion:

$$\frac{c_i l_i + N_i \tan \varphi}{F_{s-t}} = T_i \quad (9)$$

In the formula: c is the cohesion force of soil on the sliding surface of strip i; φ is the internal friction angle of soil on the sliding surface of strip i; l_i is the length of the sliding surface of strip i; F_{s-t} is the safety factor of slope.

According to Equations (5) and (8), the stability calculation formula of the multi-stage loess fill slope can be obtained as follows:

$$F_{s-t} = \frac{\sum_{i=1}^{n} (c_i l_i + N_i \tan \varphi_i) r}{\sum_{i=1}^{n} W_i d_i} \quad (10)$$

It can be seen from the above formula that the established objective function is the minimum safety factor F_{s-t} of the slope angle circle of the multi-stage loess fill side slope. When a specific multi-stage slope is given, the objective function is essentially a binary function about S and k, and the corresponding objective function can be described as follows:

$$F_{s-t,\min} = \min F(S,k) \quad (11)$$

When S and k satisfy the Equation (12), $F_{s-t,\min}$ can obtain the minimum value, and the minimum value of the objective function can be obtained by taking different values of S and k within a reasonable range. According to the obtained S and k, the slip surface corresponding to the minimum value of the safety factor can be determined by substituting it into the Equation (3). At this point, the calculation $F_{s-t,\min}$ can be converted into a mathematical optimization problem, and the minimum value of the stability coefficient is optimized and solved by MATLAB.

$$\frac{\partial F_{s-t}}{\partial S} = \frac{\partial F_{s-t}}{\partial k} = 0 \quad (12)$$

3.2. *Stability Analysis Under the Mid-Point Circle Failure Mechanism*

3.2.1. Failure Mechanism

In addition to the circular failure mechanism of slope toe mentioned above, the whole failure mechanism of multi-stage loess fill slope may also be a mid-point circle. Similarly, take n-level $AA_1B_1A_{n-1}B_{n-1}A_nB$ as an example, as shown in Figure 6, where $n = 1, 2, 3, \ldots$ is the slope progression. According to the actual engineering situation, θ' is always negative.

In addition, the meaning of the remaining symbols is the same as before. The specific process is as follows:

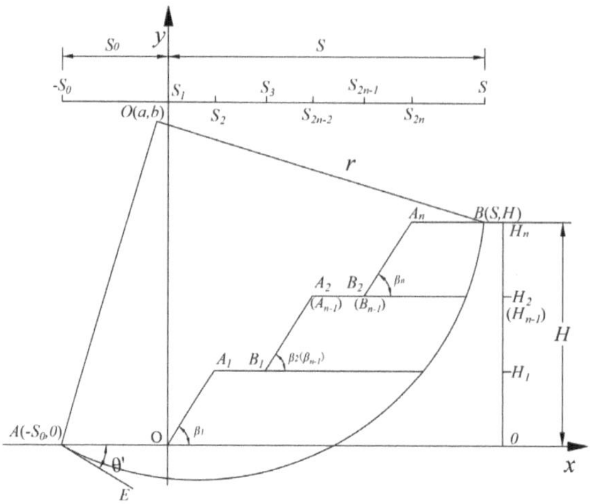

Figure 6. Mid-point circular failure mechanism of multi-stage filled slope.

In the plane cartesian coordinate system, the center point O is (a,b), point B is (S,H), and S is the abscissa of the intersection point between the critical glide surface and the slope top plane. H is the total slope height. The point F is $(-S_0,0)$, where the intersection of the arc BF with the negative half-axis of the X-axis is F, the intersection with the positive half-axis is E, and the projection of the point F on the X-axis is $-S_0$ and the projection on the Y-axis is 0, and its angle with the X-axis is θ'. The slope of the line FE is k', and $k' = -\tan\theta$ is known by θ'.

$$(x-a)^2 + (y-b)^2 = r^2 \tag{13}$$

And satisfy:

$$\left.\begin{array}{l} (-S_0 - a)^2 + b^2 = r^2 \\ (S-a)^2 + (H-b)^2 = r^2 \\ a = -(k'b + S_0) \end{array}\right\} \tag{14}$$

Among them:

$$\left.\begin{array}{l} a = -S_0 - \frac{k'[(S+S_0)^2+H^2]}{2H-2k'(S+S_0)} \\ b = \frac{(S+S_0)^2+H^2}{2H-2k'(S+S_0)} \\ r = \frac{\sqrt{1+k'^2}[(S+S_0)^2+H^2]}{2H-2k'(S+S_0)} \end{array}\right\} \tag{15}$$

At this point, it can be seen from Equation (15) that a relational equation about S, S_0, k' can be established. This equation represents the control condition of the slip surface of the mid-point circular failure mechanism. By controlling S, S_0, k' three variables, the slip surface of failure is determined.

3.2.2. Stability Analysis

For the calculation and analysis of slope stability under the slope angle circular failure mechanism, a multi-stage fill slope stability analysis and calculation model was established, as shown in Figure 7. The specific derivation process is as follows:

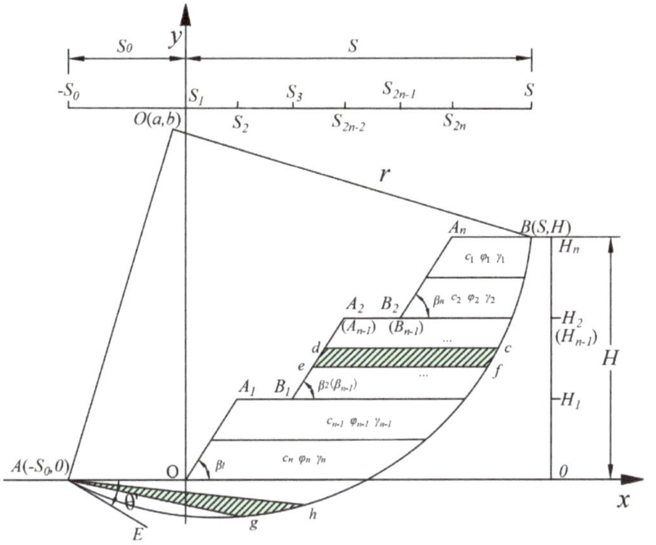

Figure 7. Stability analysis and calculation model of the mid-point circle of multi-stage fill slope.

For the slope with a general sliding surface, the soil above the toe of the slope is treated by horizontal strips, while the soil below the toe is divided by inclined strips. The soil above the toe of the slope is divided into horizontal soil strips with a height of b_i, and the soil below the toe is divided into sloping soil strips with a height of b_j. Above the toe of the slope, the calculation process of the slope toe circle is followed; below the toe of the slope, the soil strip element agh is taken for force analysis and calculation. The force diagram of soil strip agh is shown in Figure 8. From the equilibrium condition of the soil strip, the following is obtained:

$$\left.\begin{array}{ll}\sum F_x = 0 & -N_j \sin\alpha_j + T_j \cos\alpha_j = 0 \\ \sum F_y = 0 & W_j - N_j \cos\alpha_j - T_j \sin\alpha_j = 0\end{array}\right\} \tag{16}$$

where: W_j is the gravity of the soil strip; $W_j = \gamma h_j b_j$; γ is the unit weight of soil; h_j is the j soil strip length; N_j is the normal stress of the j soil strip. T_j is the shear stress of the j soil strip; α_j is the angle between the normal line and the vertical line at the midpoint of the bottom edge of the soil strip.

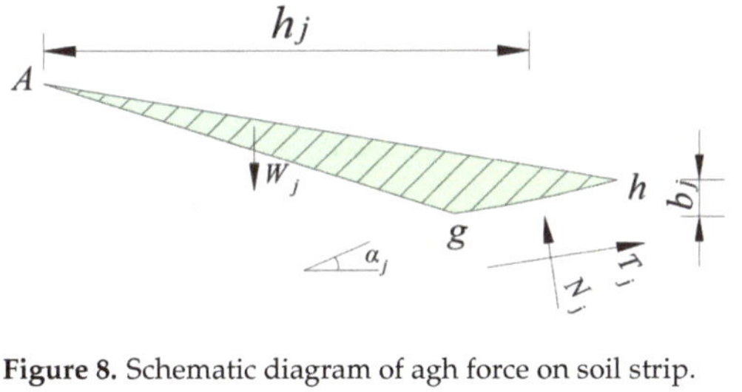

Figure 8. Schematic diagram of agh force on soil strip.

Among them:

$$h_j = 0.5b_j + \sum_{x=1}^{j-1} b_x \tag{17}$$

Soil bar j establishes a moment balance equation for the center of a circle:

$$\sum M_O = 0 \quad W_j d_j - T_j r = 0 \tag{18}$$

In the formula: d is the distance between the center of gravity of the j soil strip and the center of the sliding circle, $d_j = r\sin\alpha_j - h_j/2$.

From Equations (16) and (18), the formula for calculating the stability of sloping soil strips can be obtained:

$$F_{sx-m} = \frac{\sum\limits_{j=1}^{n} (c_j l_j + N_j \tan\varphi_j) r}{\sum\limits_{j=1}^{n} W_j d_j} \tag{19}$$

The soil above the toe of the slope is divided into horizontal strips, and the soil below the toe is divided into oblique slices. According to the above Formulas (10) and (19), the F_{s-m} final expression of the safety factor of the mid-point circular failure mechanism under the overall failure mode of multi-stage fill slope can be obtained, namely:

$$F_{s-m} = \frac{\sum\limits_{i=1}^{n} (c_i l_i + N_i \tan\varphi_i) r}{\sum\limits_{i=1}^{n} W_i d_i} + \frac{\sum\limits_{j=1}^{n} (c_j l_j + N_j \tan\varphi_j) r}{\sum\limits_{j=1}^{n} W_j d_j} \tag{20}$$

Similarly, it can be seen from the above equation that the objective function of the minimum safety factor F_{s-m} of the midpoint circle of a multi-stage filled slope has been established. When a specific multi-stage slope is given, the objective function is essentially a ternary function about S, S_0, k', and the corresponding objective function can be described as follows:

$$F_{s-m,\min} = \min F(S, S_0, k') \tag{21}$$

When S, S_0, k' satisfies Formula (22), $F_{s-m,\min}$ can obtain the minimum value, and the minimum value of the objective function can be obtained by taking different values of S, S_0, k' within a reasonable range, and the slip surface corresponding to the minimum value of the safety factor can be determined by substituting the obtained S, S_0, k' into Formula (15). Similarly, the minimum value of the stability coefficient is optimized by MATLAB.

$$\frac{\partial F_{s-m}}{\partial S} = \frac{\partial F_{s-m}}{\partial S_0} = \frac{\partial F_{s-m}}{\partial k} = 0 \tag{22}$$

For the above two kinds of slope failure, when the midpoint circle is $S_0 = 0$, the damage will also become the slope toe circle failure. However, through comprehensive analysis, the specific process of the analysis of the two is different, and the conditions to be met are different, so they cannot be classified into one type for discussion and analysis. Therefore, the situation of the midpoint circle and the slope toe circle is necessary to be classified and discussed.

3.3. The Minimum Safety Factor of Multi-Stage Fill Slope Under the Overall Failure Mode

In this paper, two types of sliding surface failure modes of horizontal layered multi-fill slopes are proposed, and the minimum value of slope safety factor $F_{s-t,\min}$ and $F_{s-m,\min}$ under the two types of failure modes can be obtained respectively. Therefore, the minimum value of the two results can be obtained by Equation (23), and then the minimum safety factor $F_{gs,\min}$ under the overall failure mode of the multi-fill slope can be determined.

$$F_{gs,\min} = \min\{F_{s-t,\min}, F_{s-m,\min}\} \tag{23}$$

4. Stability Analysis of Multi-Stage Fill Slope Under Local Failure Mechanism

In the multi-stage fill slope, the failure of the slope is not a one-time occurrence, and there will be local damage or deformation in some areas. Due to the uneven thickness of the filling soil layer, the local stress distribution may be uneven, the slope is steep and slow, the slope body is irregular, and there may be local sliding before the overall sliding along the slip zone. In this paper, the local failure modes are divided into two categories: one is the single-stage slope as the local single-stage failure mode; secondly, the multi-stage slope is the local multi-stage (m is the series, $2 \leq m \leq n - 1$) failure mode. The stability of these two types of local failure modes of multi-stage fill slope will be analyzed in detail.

4.1. Slope Stability Analysis Under Local Single-Stage Failure Mode

Under the local single-stage failure mode, it can be concluded that the local single-stage failure mode can be regarded as the failure mechanism of a single-stage slope. There may be three failure mechanisms of the single-stage slope, as shown in Figure 9: Toe circle (glide surface a), midpoint circle (glide surface b), and slope circle (glide surface c). In the stability analysis of a multi-stage fill slope, the slope toe of each single-stage slope is taken as the origin, the local rectangular coordinate system is established, the possible failure mechanism of each single-stage slope is analyzed, and the minimum value is taken as the minimum safety factor of the single-stage slope. Finally, the minimum safety factor of the multi-stage loess slope under the local single-stage failure mode can be determined by finding the minimum value from the minimum safety factor set of each single-stage slope. This method helps to consider the effects of multiple levels and multiple failure mechanisms on the overall stability.

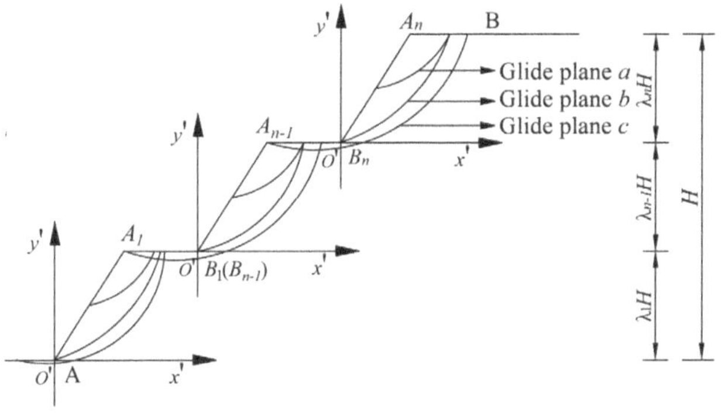

Figure 9. Stability analysis model of multi-stage fill slope under local single-stage failure mode.

4.1.1. Damage Mechanism

Taking the *i*-grade slope as an example, the schematic diagram of the failure mechanism shown in Figure 10 is established, and the arc slip surface equations of different control failure mechanisms and the center and radius of the geometric parameters of the equation are given.

(1) Toe circle of slope

Based on the failure mechanism in Figure 10a, a local plane rectangular coordinate system $O'x'y'$ is established, whose center O' is (a', b'), arc radius is r_i', the point C_i is $(S_i, \lambda_i H)$, the point B_i is (0,0), and the slope of the line B_iE is k_i.

Then, the equation of arc C_iB_i is as follows:

$$(x' - a_i')^2 + (y' - b_i')^2 = r'^2 \tag{24}$$

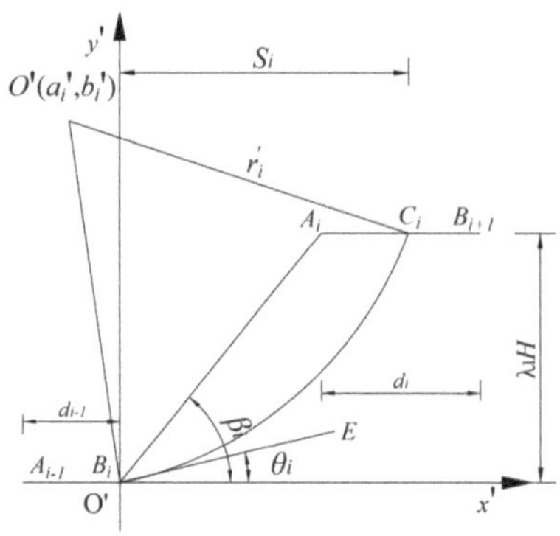

(**a**) Schematic diagram of the toe circle failure of the *i*-th grade slope

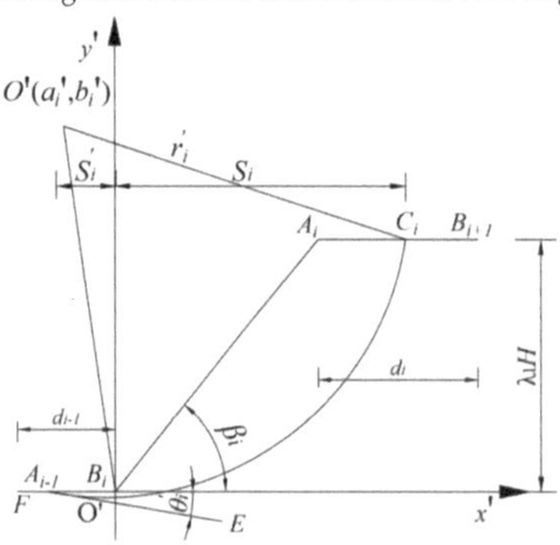

(**b**) Schematic diagram of the midpoint circle failure of the *i*-th grade slope

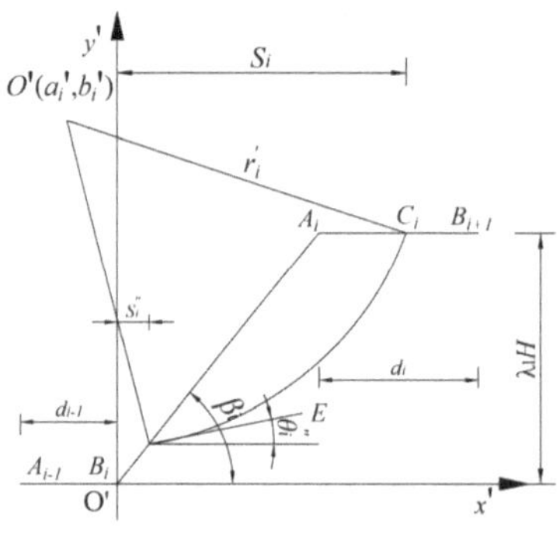

(**c**) Schematic diagram of the slope circle failure of the *i*-th grade slope

Figure 10. Schematic diagram of the failure mechanism of the *i*-th grade slope.

Among them:

$$\left.\begin{aligned} a_i' &= \frac{k_i(S_i^2+\lambda_i^2H^2)}{2k_iS_i-2\lambda_iH} \\ b_i' &= \frac{S_i^2+\lambda_i^2H^2}{2\lambda_iH-2k_iS_i} \\ r_i' &= \frac{\sqrt{1+k_i^2}(S_i^2+\lambda_i^2H^2)}{2\lambda_iH-2k_iS_i} \end{aligned}\right\} \tag{25}$$

(2) Midpoint circle

Based on the failure mechanism in Figure 10b, a local plane rectangular coordinate system $O'x'y'$ is established, point F is $(-S_i', 0)$, line FE slope is k_i', and $k_i' < 0$.

Then, the equation of arc C_iF is as follows:

$$(x' - a_i')^2 + (y' - b_i')^2 = r'^2 \tag{26}$$

Among them:

$$\left.\begin{array}{l} a_i' = -S_i' - \frac{k_i'[(S_i+S_i')^2+\lambda_i^2H^2]}{2\lambda_iH-2k_i'(S_i+S_i')} \\ b_i' = \frac{(S_i+S_i')^2+H^2}{2\lambda_iH-2k_i'(S_i+S_i')} \\ r_i' = \frac{\sqrt{1+k_i'^2}[(S_i+S_i')^2+\lambda_i^2H^2]}{2\lambda_iH-2k_i'(S_i+S_i')} \end{array}\right\} \tag{27}$$

(3) Arc-shaped sliding surface on a slope

Based on the failure mechanism in Figure 10c, a local plane rectangular coordinate system $O'x'y'$ is established, point G is $(S_i'', S_i'' \tan\beta_i)$, and the slope of line GE is k_i''.

Then the equation of the arc G_iG is as follows:

$$(x' - a_i')^2 + (y' - b_i')^2 = r_i'^2 \tag{28}$$

Among them:

$$\left.\begin{array}{l} a_i' = S'' - \frac{k_i''(S_i-S_i'')^2+k_i''(\lambda_iH-S_i''\tan\beta_i)^2]}{2(\lambda_iH-S_i''\tan\beta_i)-2k_i''(S_i-S_i'')} \\ b_i' = \frac{(S_i-S_i'')^2+\lambda_i^2H^2-S_i''\tan\beta_i[S_i''\tan\beta_i+2k_i''(S_i-S_i'')]}{2(\lambda_iH-S_i''\tan\beta_i)-2k_i''(S_i-S_i'')} \\ r_i' = \frac{\sqrt{1+k_i''^2}[(S_i-S_i'')^2+(\lambda_iH-S_i''\tan\beta_i)^2]}{2(\lambda_iH-S_i''\tan\beta_i)-2k_i''(S_i-S_i'')} \end{array}\right\} \tag{29}$$

4.1.2. Stability Analysis

(1) Toe circle of slope

$$F_{1s-t} = \frac{\sum\limits_{i=1}^{n}(c_il_i + N_i\tan\varphi_i)r}{\sum\limits_{i=1}^{n}W_id_i} \tag{30}$$

where a_i', b_i', r_i' expressions are brought in by Equation (25).

Assuming that the minimum safety factor F_{1s-t} for the circular failure mechanism at the toe of the i-th slope under the local single-stage failure mode has already been established, the objective function is essentially a binary function related to S_i and k_i. In that context, the corresponding objective function can be described as follows:

$$F_{1s-t,\min} = \min F_{1s-t}(S_i, k_i) \tag{31}$$

When S_i and k_i satisfy Equation (32), F_{1s-t} can achieve its minimum value. By varying S_i and k_i within a reasonable range, the minimum value of the objective function can be obtained.

$$\frac{\partial F_{1s-t}}{\partial S_i} = \frac{\partial F_{1s-t}}{\partial k_i} = 0 \tag{32}$$

(2) Midpoint circle

$$F_{1s-m} = \frac{\sum\limits_{i=1}^{n}(c_il_i + N_i\tan\varphi_i)r}{\sum\limits_{i=1}^{n}W_id_i} + \frac{\sum\limits_{j=1}^{n}(c_jl_j + N_j\tan\varphi_j)r}{\sum\limits_{j=1}^{n}W_jd_j} \tag{33}$$

where a_i', b_i', r_i' expressions are brought in by Equation (27).

Similarly, the minimum value F_{1s-m} of the safety factor corresponding to the mid-point circle failure mechanism of the *i*-grade slope in the local single-stage failure mode has been established, and the objective function is essentially a ternary function about S_i, S_i', k_i', so the corresponding objective function can be described as follows:

$$F_{1s-m,\min} = \min F_{1s-m}(S_i, S_i{}', k_i') \tag{34}$$

When S_i, S_i', k_i' satisfies Equation (35), F_{1s-m} can obtain the minimum value, and S_i, S_i', k_i' can obtain the minimum value of the objective function by taking different values within a reasonable range.

$$\frac{\partial F_{1s-m}}{\partial S_i} = \frac{\partial F_{1s-m}}{\partial S_i'} = \frac{\partial F_{1s-m}}{\partial k_i'} = 0 \tag{35}$$

(3) Arc-shaped sliding surface on a slope

$$F_{1s-s} = \frac{\sum\limits_{i=1}^{n} (c_i l_i + N_i \tan \varphi_i) r}{\sum\limits_{i=1}^{n} W_i d_i} \tag{36}$$

where a_i', b_i', r_i' expressions are brought in by Equation (29).

Similarly, the minimum value of the safety factor F_{1s-s} corresponding to the circular failure mechanism of the *i*-grade slope in the local single-stage failure mode has been established, and the objective function is essentially a ternary function about S_i, S_i'', k_i'', so the corresponding objective function can be described as follows:

$$F_{1s-s,\min} = \min F_{1s-m}(S_i, S_i'', k_i'') \tag{37}$$

When S_i, S_i'', k_i'' satisfies Equation (38), F_{1s-s} can obtain the minimum value, and the minimum value of the objective function can be obtained by taking different values within a reasonable range of S_i, S_i'', k_i''.

$$\frac{\partial F_{1s-s}}{\partial S_i} = \frac{\partial F_{1s-s}}{\partial S_i''} = \frac{\partial F_{1s-s}}{\partial k_i''} = 0 \tag{38}$$

Through the calculation of the above three failure mechanisms, the minimum safety factor $F_{1s-i,\min}$ of the *i*-grade slope under the local single-stage failure mode can be determined by Equation (39).

$$F_{1s-i,\min} = \min\{F_{1s-t,\min}, F_{1s-t,\min}, F_{1s-s,\min}\} \tag{39}$$

Through the above method, $F_{1s-t,\min}$ is calculated for each level of slope in turn. Finally, the minimum safety factor $F_{1s,\min}$ of the multi-stage fill slope under the local single-stage failure mode can be determined by Equation (40):

$$F_{1s,\min} = \min\{F_{1s-1,\min}, F_{1s-2,\min}, \ldots, F_{1s-n,\min}\} \tag{40}$$

4.2. Slope Stability Analysis Under Local Multi-Stage Failure Mode

The stability analysis model under the local multi-stage failure mode is shown in Figure 11. Using the above method, m = 2, 3, ... When $n - 1$, the local minimum safety factor $F_{1s,\min}, F_{2s,\min}, \ldots, F_{(n-1)s,\min}$ of the multi-stage fill slope is obtained.

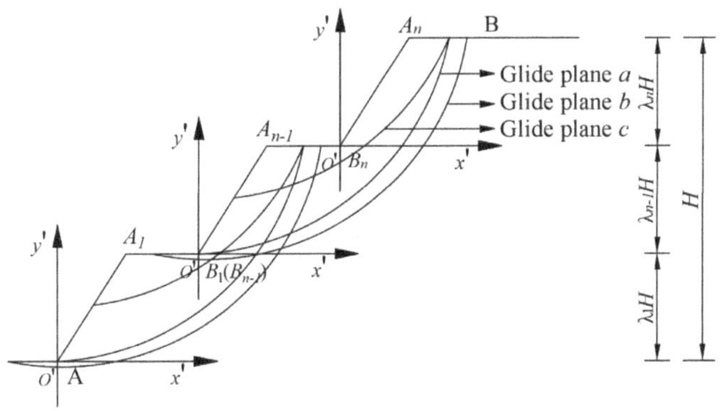

Figure 11. Stability analysis model of multi-stage fill slope under local multi-stage failure mode.

5. Multi-Stage Fill Slope Stability Analysis Process

In order to more intuitively represent the slope stability analysis process in this paper, the multi-stage fill slope stability analysis flow chart as shown in Figure 12 is drawn, and the analysis process is carried out during the stability analysis.

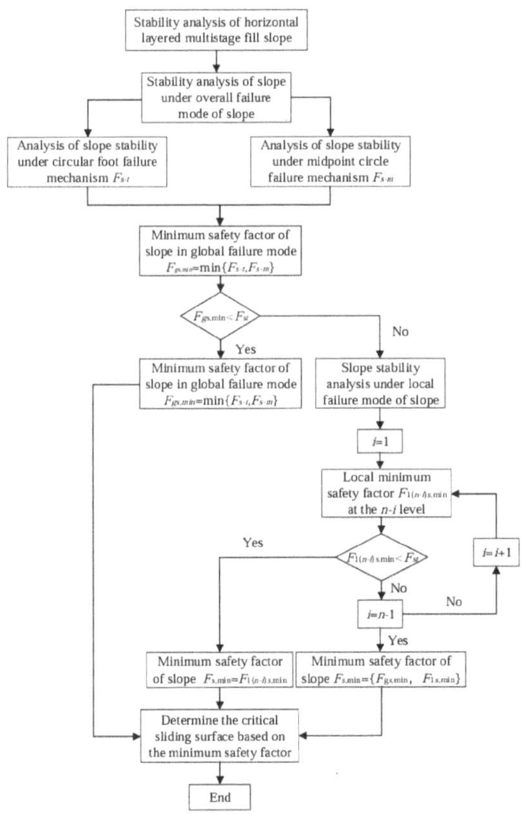

Figure 12. Stability analysis flow chart of horizontal layered multi-stage fill slope.

6. Example Verification

6.1. Example 1

In order to verify the correctness of the derived formula, the calculated results in this paper are compared with those calculated by Deng Dongping, Gao Liansheng, and the traditional strip method, as shown in Table 1. The ratio of slope potential slip face corresponding to the slope safety factor *Fs* is shown in Figure 13.

Table 1. Comparison of stability analysis results by different methods.

Calculation Method	Center of a Circle (a,b)	Radius r/m	Safety Factor $F_{s,min}$
Bishop method	(22.564, 37.182)	25.972	2.288
Janbu method	(22.319, 37.408)	25.884	2.107
The algorithm in this paper	(22.385, 37.047)	25.817	2.124
Deng Dongping	-	-	2.197
Gao Liansheng	-	-	2.072

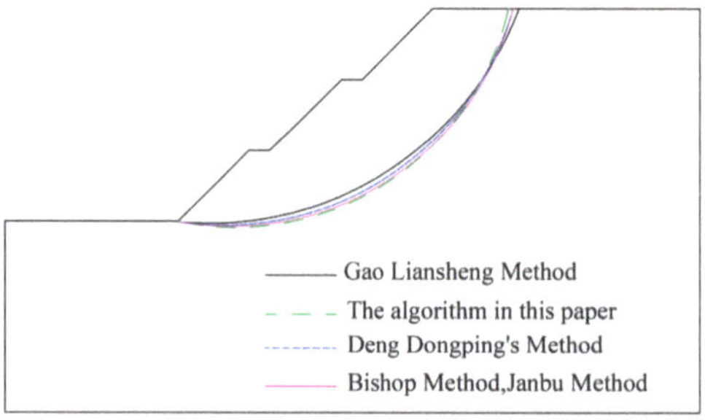

Figure 13. Schematic diagram of the critical slip surface determined by different methods.

Comparing the calculation methods of the safety factor when local instability occurs in the three-level slope, the total height of the slope if H = 60 m, the height of the first-, second-, and third-grade slopes are 20 m, the platform width d_1 = 4 m, d_2 = 2 m. Side slope toe $\beta_1 = \beta_2 = 45°$, $\beta_3 = 50°$. Cohesion c = 32 kPa, internal friction angle $\varphi = 28°$, unit weight of soil $\gamma = 18.7\text{kN/m}^3$. As shown in Figure 14, the calculated minimum safety factor of the slope is 1.091. The potential slip-face ratio of the slope is shown in Figure 15.

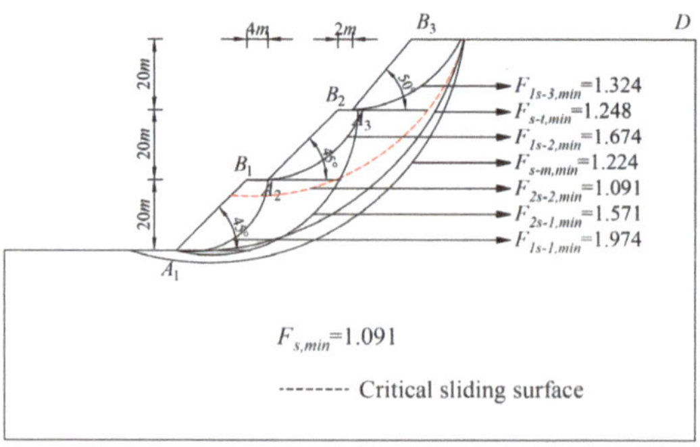

Figure 14. Slope stability analysis results of the proposed algorithm.

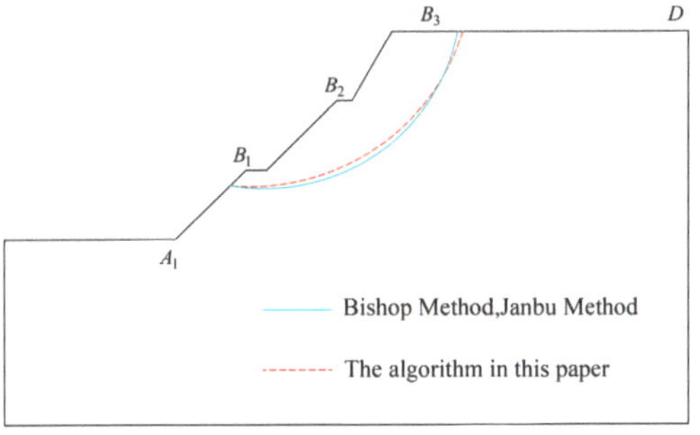

Figure 15. Schematic diagram of the critical slip surface determined by different methods.

6.2. Example 2

It is known that a third-stage heterogeneous fill slope project is a third-stage heterogeneous fill slope composed of 5 layers of soil, and the material parameters are shown in Table 2. The total height of the slope is H = 30 m, the height of the first, second, and third slopes is 10 m, the width of the platform is d_1 = 3 m, d_2 = 3 m, and the slope toe of the slope surface $\beta_1 = \beta_2 = \beta_3 = 45°$.

Table 2. Soil material parameters.

Soil Layer Numbering	$c/(kPa)$	$\varphi/(°)$	$\gamma/(kN/m^{-3})$
① Crushed stone	27.6	38.5	20.2
② Sandy soil I	27.0	28.0	19.0
③ Sandy soil II	19.3	25.0	19.0
④ Silty clay I	15.3	33.0	18.5
⑤ Silty clay II	17.2	25.0	19.5

6.2.1. Algorithm in This Study

The limit equilibrium method proposed in this paper is used to analyze the stability of a horizontal layered multi-stage filled slope, and the specific calculation results are shown in Figure 16.

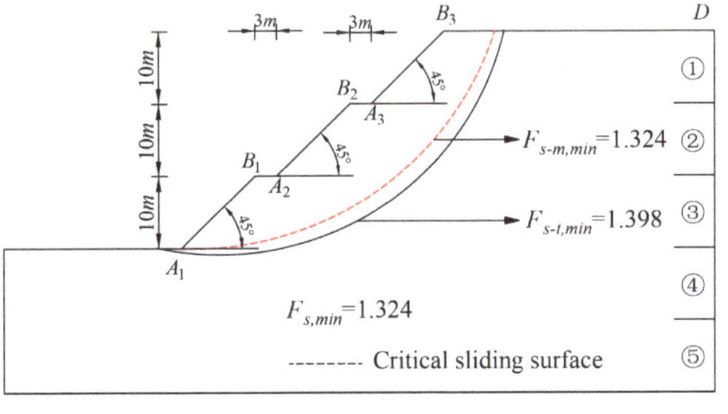

Figure 16. Stability results of the third-grade fill slope calculated by the method presented in this paper.

6.2.2. Traditional Strip Method

For this calculation example, the traditional strip method is used for stability analysis and calculation, and the SLOPE/W module is used for slope stability analysis. In Geo

Studio, the M-P method, Bishop method, Janbu method, and Spencer method (collectively referred to as the traditional strip method) were used to solve the minimum safety factor of the multi-stage fill slope and the coordinate radius of the circle center, as shown in Table 3.

Table 3. Stability analysis results of the slope with the traditional strip method.

Calculation Method	Center of a Circle (a,b)	Radius r/m	Safety Factor $F_{s,min}$
Bishop method	(89.28, 80.125)	61.04	1.384
Janbu method	(89.28, 80.125)	61.04	1.297
M-P method	(89.28, 80.125)	61.04	1.395
Spencer method	(89.28, 80.125)	61.04	1.395

6.2.3. Comparative Analysis

The above two methods were used to calculate the stability of example 2. The specific comparison results are shown in Table 4 and Figure 17. It can be seen from Table 4 that the relative deviations of the minimum safety factor obtained by the proposed algorithm and the traditional slice methods in Geo Studio are −0.4%, 3.5%, −3.8%, and −3.8%, respectively. It can be seen that the calculation and analysis methods in this chapter have little difference compared with the results of the traditional slice method and all meet the requirements of the specification. It can be shown that the method in this chapter has a certain rationality.

Table 4. Comparison of different stability analysis results.

Calculation Method	Safety Factor $F_{s,min}$	Minimum Relative Deviation of Safety Factor
The algorithm in this paper	1.342	-
Bishop method	1.348	−0.4%
Janbu method	1.297	3.5%
M-P method	1.395	−3.8%
Spencer method	1.395	−3.8%

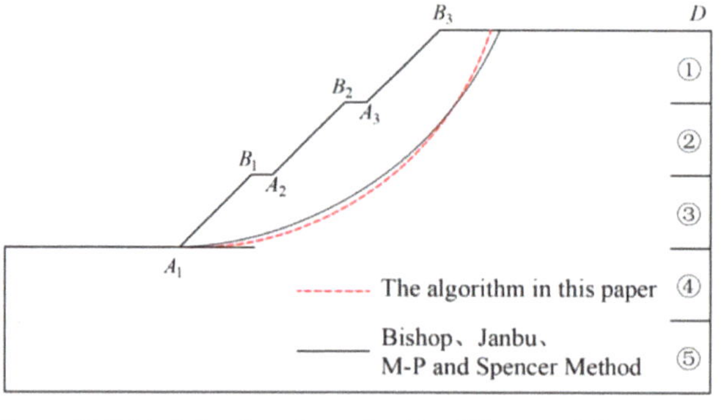

Figure 17. Critical slip plane determined by different stability analysis results.

7. Conclusions

This paper conducts a horizontal and inclined slice analysis of soil based on the limit equilibrium method and performs a stress analysis. It establishes potential failure mechanisms, such as toe circles and midpoint circles, for multi-stage fill slopes under both global and local failure modes. The main conclusions are as follows:

(1) The method in this paper not only considers the overall instability failure of multi-stage fill slope but also considers the local instability failure and provides an analysis process that can comprehensively analyze the stability of multi-stage fill slope.

(2) The comparative analysis of the calculation results obtained by the horizontal minimum safety factor between the two is not more than 5%, and the obtained most dangerous sliding surface is basically the same, which can prove the rationality of the proposed method.

(3) The method presented in this paper can be applied to the stability calculation of all kinds of heterogeneous and homogeneous single-stage and multi-stage slopes, and the influence of geometric parameters and soil parameter values on slope stability can be considered, providing a certain reference for the stability research of multi-stage fill slopes.

Author Contributions: Conceptualization, X.L.; Methodology, X.L. and S.Y.; Validation, S.Y. and W.Y.; Investigation, M.Q.; Data curation, M.Q. and J.L.; Resources, S.Y.; Writing—original draft, X.L. and M.Q.; Writing—review and editing, S.Y., W.Y. and J.L. All authors have read and agreed to the published version of the manuscript.

Funding: This research was funded by the National Natural Science Foundation of China (Grant No. 52168050), the Gansu Province Higher Education Teacher Innovation Fund Project (Grant No. 2025A-231), the Gansu Province Science and Technology Plan Project (Grant No. 24JRRA292), the Gansu Provincial Department of Education Young Doctor Support Project (Grant No. 2023QB-048).

Data Availability Statement: Some or all data, models, or codes that support the findings of this study are available from the corresponding author upon reasonable request.

Conflicts of Interest: The authors declare that they have no known competing financial interests or personal relationships that could have appeared to influence the work reported in this paper.

References

1. Yang, Z.; Cheng, X. High Fill Slope Collapse: Stability Evaluation Based on Finite Element Limit Analyses. *Transp. Geotech.* **2024**, *44*, 101156.
2. Mei, Y.; Zhou, D.; Hu, C.; Wang, X.; Zhang, Y.; Xiao, N.; Shi, W. Study on Deformation Characteristics of Loess Ultrahigh-Fill Slope Based on Large-Scale Undisturbed Soil Centrifugal Model Tests. *Front. Earth Sci.* **2022**, *10*, 848542.
3. Jin, L.; Liao, Y. Upper Bound Analysis of Two-Layered Slopes Subjected to Seismic Excitations Using the Layer-Wise Summation Method. *Buildings* **2024**, *14*, 1990. [CrossRef]
4. Zhang, F.; Jia, S.; Gao, Y. Stability Analysis of Cut-Fill and Deep Patch Unsaturated Slopes Under Steady Infiltration Conditions. *Sci. China Technol. Sci.* **2025**, *68*, 1120701.
5. Wu, Y.; Zhou, X.; Liu, J. Long-Term Performance of the Water Infiltration and Stability of Fill Side Slope against Wetting in Expressways. *Appl. Sci.* **2022**, *12*, 5809. [CrossRef]
6. Yao, Y.; Zhang, J.; Li, X.; Tu, Y.; Zhong, Z. The Stability of Slopes and Building Structures Using an Energy Visualization Procedure. *Buildings* **2024**, *14*, 3705. [CrossRef]
7. Ren, Y.; Li, T.; Yang, L.; Wei, D.; Tang, J. Stability Analysis of Ultra-High-Steep Reinforced Soil-Filled Slopes Based on Centrifugal Model Tests and Numerical Calculation. *Chin. J. Geotech. Eng.* **2022**, *44*, 836–844.
8. Ye, S.; Zhuang, Y.; Fang, G. Stability Factors and Deformation Law of Loess High Fill Slope. *J. Lanzhou Univ. Technol.* **2021**, *47*, 120–126.
9. Liu, X.; Song, B.; Sun, Z.; Jiao, W. Critical Filling Height of Embankment over Soft Soil: A Three-Dimensional Upper-Bound Limit Analysis. *Buildings* **2025**, *15*, 395. [CrossRef]
10. Renaud, V.; Al Heib, M.; Muller, R.; Burda, J. Long-Term Slope Stability of Abandoned Mine Lake—Numerical Modelling and Risk Assessment. *Mater. Proc.* **2021**, *5*, 88. [CrossRef]
11. Zhang, H.; Zhang, C.; Zheng, W.; Wang, X.; Zhang, J. Analysis of the Stability of a High Fill Slope under Different Gradients and Precipitation Conditions. *Appl Sci.* **2024**, *14*, 7590. [CrossRef]
12. Li, Z.; Sun, L.; Liu, L.; Guan, C. Research on Safety Evaluation Method of High-Fill Slope Considering Heterogeneity and Anisotropy. *Bull. Eng. Geol. Environ.* **2022**, *81*, 275.
13. Amena, S. Analysis of the Stability of Reinforced Plastic Waste Treated Clay as Embankment Fill on Soft Soils. *Adv. Civ. Eng.* **2022**, *2022*, 1831970.

14. Gong, W.; Zhang, X.; Pei, X.; Chen, L.; Jiang, R.; Sun, H.; Zhong, Y. Instability Modelling of Loess Fill Slope with Influence of Weak Zone of High and Steep Filling Interface. *J. Eng. Geol.* **2023**, *31*, 288–298.
15. Yong, X.; Zhang, Y.; Hou, Y.; Han, B.; An, N.; Zhang, H.; Ma, Y. Stability of Loess High-Fill Slope Based on Monitored Soil Moisture Changes. *Res. Cold Arid Reg.* **2023**, *15*, 191–201.
16. Yan, C.; Wang, H. Three-Dimensional Stability of Slopes with Cut-Fill Interface Based on Upper-Bound Limit Analysis. *Chin. J. Geotech. Eng.* **2024**, *46*, 174–181.
17. Mostafaei, H.; Behnamfar, F.; Alembagheri, M. Nonlinear analysis of stability of rock wedges in the abutments of an arch dam due to seismic loading. *Struct. Monit. Maint.* **2020**, *7*, 295–317.
18. Chen, X.; Zhou, Z.; Zhu, Y.; Pu, S.; Zhang, D.; Li, E. Evaluating Seismic Stability of Reinforced Soil Slopes Using the Limit Equilibrium Method: Influence of Geosynthetic Strength Indicators. *Front. Phys.* **2025**, *12*, 1526223.
19. Fatehi, M.; Hosseinpour, I.; Jamshidi Chenari, R.; Payan, M.; Javankhoshdel, S. Deterministic Seismic Stability Analysis of Reinforced Slopes Using Pseudo-Static Approach. *Iran. J. Sci. Technol. Trans. Civ. Eng.* **2023**, *47*, 1025–1040.
20. Zhang, F.; Jia, S.; Shu, S. Effects of Seismic Amplification on the 3D Stability of Fill Slopes in V-Shaped Valleys. *Acta Geotech.* **2024**, *19*, 3241–3255.
21. Wang, X.; Yuan, Y.; Zhang, S. Reliability Analysis of Natural High Loess Slope and Backfill Slope in Northern Shanxi area. *J. Xi'an Univ. Archit. Technol.* **2021**, *53*, 665–672.
22. Huang, A.; Ye, S. Sensitivity of High Fill Slope Stability Factors Under Seismic Conditions. *Soil Mech. Found. Eng.* **2020**, *57*, 356–363.

buildings

MDPI

Article

The Impact of Different Excavation Support Structures on the Deformation and Stability of Adjacent Station and Tunnels

Zhitong Mao [1], Tian Ding [1], Fengchao Hu [1], Shuaihua Ye [2,*], Linzhao Ding [3], Xiaoning Zhang [4], Peiqiang Li [2] and Nianxiang Li [2]

1 Gansu Construction Investment (Holdings) Group Co., Ltd., Lanzhou 730050, China; maozhitong1819@163.com (Z.M.); dingtian1819@163.com (T.D.); 13893497029@163.com (F.H.)
2 School of Civil Engineering, Lanzhou University of Technology, Lanzhou 730050, China; 13689442127@163.com (P.L.); linianxiang55@163.com (N.L.)
3 School of Civil Engineering, Shandong University, Jinan 250061, China; 202335074@mail.sdu.edu.cn
4 School of Civil Engineering, Chongqing University, Chongqing 400045, China; zhangxn@cqu.edu.cn
* Correspondence: yeshuaihua@163.com

Abstract: This study uses the finite element software Midas GTS NX (2019), combined with actual engineering projects, to establish numerical models and analyze the impact of different support types (pile-anchor support and double-row pile support) on the excavation of foundation pits near metro station tunnels. The results indicate that under both support methods, the vertical displacement of the tunnel is the greatest at the interface between the station and the tunnel, with greater vertical displacement occurring under double-row pile support. Under pile-anchor support, the horizontal displacement of the tunnel reaches its maximum value during the sixth excavation stage, while under double-row pile support, the horizontal displacement increases steadily, and the overall displacement is small. The horizontal displacement under pile-anchor support is significantly greater than that under double-row pile support. For the station, the maximum vertical displacement under pile-anchor support is smaller than that under double-row pile support. The horizontal displacement under pile-anchor support exhibits a linear change, while under double-row pile support, the displacement continuously increases from the end of the foundation pit farther from the excavation to the end closer to it. The model tests are consistent with the numerical simulation results, verifying the correctness of the numerical simulation. This study can provide references for relevant engineering projects to ensure the safety and stability of metro structures.

Keywords: foundation pit excavation; metro station tunnel; support structure; Midas GTS NX; scaled model test

1. Introduction

With the continuous advancement of urbanization, the population in cities is steadily increasing, placing significant pressure on urban housing, transportation, and other infrastructure. The subway, as one of the city's key public transportation systems, can effectively alleviate the traffic congestion caused by population density and thus has been widely promoted in many cities. However, a notable feature of subway construction is that it must pass through densely populated areas, as only in such locations can the advantages of rail transit be fully realized. Densely populated areas are typically accompanied by high-density buildings and are often the core zones of economic activity. At the same time, land resources are becoming increasingly scarce, making excavation for foundation pits near

 https://doi.org/10.3390/buildings15030493

subway structures nearly unavoidable. Therefore, ensuring the safety of the surrounding environment and the existing subway structures during foundation pit excavation, as well as guaranteeing the continuous safe operation of the subway, has become an urgent issue that needs to be addressed. Foundation pit excavation is a process of unloading, which inevitably induces additional stresses on existing metro structures (stations, tunnels) [1], causing deformation and displacement that can damage the metro structure and affect its safe operation [2–5]. The selection of the support structure is a critical step in foundation pit engineering. The correct choice of support structure directly impacts the safety of construction. Analyzing the impact of foundation pit excavation with different support types on the adjacent metro station tunnels is of great significance for the safe operation of the metro and the safe construction of the foundation pit.

In recent years, foundation pit construction adjacent to metro structures has received increasing attention. Many scholars conducted extensive research on this topic. Some used theoretical analysis to study the excavation of foundation pits near existing structures [6–11], while others employed model tests (scale model tests and centrifuge model tests) to study field construction patterns [12,13]. Some scholars also used numerical simulations for their research [14–16]. Zheng Gang et al. [17–19] applied a combination of field measurements, finite element modeling, and detailed analysis to investigate the effects of different support structures (such as concave, cantilever, composite, and kickboard types) on deep soil, existing tunnels, and neighboring buildings during foundation pit excavation. They analyzed the changes in and ranges of vertical and horizontal displacement fields of the soil under different deformation modes and explored the mechanisms by which tunnel location, support structure deformation characteristics, and maximum displacement affect tunnel deformation. Additionally, they examined the relationship between building settlement, deflection, and wall strain with factors such as the distance from the foundation pit, support structure deformation characteristics, and building parameters, emphasizing the importance of these factors in ensuring foundation pit safety and controlling environmental impacts. These findings provided critical guidance and theoretical support for related engineering practice and future research, reminding engineers of the importance of these factors to ensure the stability of both the project and the environment. Ye S. et al. [20] developed a numerical model to study the impact of foundation pit excavation on adjacent metro tunnel deformation and additional stress in the lining segment. They analyzed the influence of factors such as the distance between the metro tunnel and foundation pit, single excavation volume, and geological conditions on tunnel deformation characteristics and the internal forces in the tunnel lining. Wang Y. et al. [21] used SEG to monitor soil deformation, establishing a new type of monitoring and early warning system to ensure the accuracy and reliability of soil monitoring and providing an effective method for field monitoring. Liu B. et al. [22] studied the impact of excavation and support methods on metro tunnels under different spatial configurations of foundation pits and tunnels. Ding Z. et al. [23] conducted a study based on full-process monitoring data of foundation pit excavation, investigating the relationship between deep soil lateral displacement and adjacent metro tunnel deformation throughout the excavation stages. They found that early-stage support structure construction and dewatering had significant impacts on the foundation pit, and excavation exhibited spatial effects. Zhou Z. et al. [24] developed a numerical model to analyze the impact of overlapping foundation pit excavation on adjacent tunnel deformation characteristics. They analyzed key factors such as tunnel depth, horizontal distance between the tunnel and foundation pit support structure, and tunnel burial depth on tunnel displacement variations. Liu J. et al. [25] used numerical simulations to study the impact of distance, geological conditions, and hydrogeological factors on nearby tunnels. However, most current studies have issues such as a single research object, excessive model simplification, and a lack of comprehensive analysis of multiple factors. Research considering the simultaneous presence of station tunnels is extremely rare, so this study is very necessary and can provide valuable reference for subsequent related engineering projects.

This study, based on a practical engineering project, utilizes finite element software Midas GTS NX (2019) to establish and analyze numerical models, investigating the impact of different support types on the excavation of foundation pits near metro station tunnels. This research provides valuable references for related engineering projects.

2. Engineering Situation

This project is located in Gansu, with a metro station to the east, residential complexes to the north and west, and a commercial street to the south. The upper edge of the foundation pit on the eastern side is 16.2 m from the metro station. The excavation depth of the foundation pit is 16 m, and the excavation area (fan-shaped) is approximately 50,188 m^2. According to the geological survey report, the stratum within the excavation range consists of mixed fill soil and three layers of gravel, with increasing density from top to bottom. The groundwater table is located between 6.9 m and 12.6 m deep, and the permeability coefficient of the gravel layer is 50 m/d. Since the foundation slab is below the groundwater table, dewatering is required before excavation, the dewatering plan adopts well-point dewatering outside the foundation pit plus secondary dewatering inside the foundation pit. The well-points are arranged along the outer perimeter of the foundation pit support piles. During the earthwork excavation of the foundation pit, simple open drainage ditches and sump wells should be set at the bottom of the excavation pit. These should be deepened as the foundation pit is dug deeper to maintain smooth water flow. The planar relationship between the foundation pit and the existing metro is shown in Figure 1. The depth of the metro station roof is 3.5 m, the depth of the station bottom is 14 m, the depth of the tunnel top is 7.9 m, the depth of the tunnel bottom is 13.4 m, and the diameter of the tunnel is 5.5 m. The closest distance between the left side of the station and the foundation pit is 16.2 m, the width of the station is 17.5 m, the distance from the center line of the left tunnel to the edge of the foundation pit is 19.5 m, and the distance from the center line of the right tunnel to the edge of the foundation pit is 30 m. The spatial relationship between the foundation pit and the existing metro is shown in Figure 1.

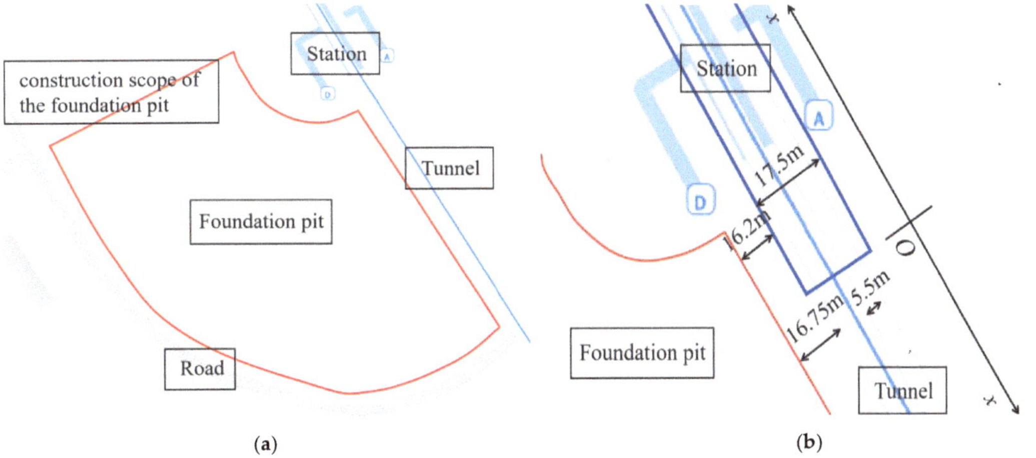

Figure 1. Planar relationship between the foundation pit and the existing metro: (**a**) overall diagram; (**b**) detailed diagram.

Based on the actual situation of the project and in combination with commonly used support structures, the pile-row + anchor cable support and double-row pile support are selected as the supporting structures. Using Midas GTS NX (2019) software, excavation models for the foundation pit are established under both support structures. Scaled model experiments are conducted based on the corresponding dimensions, and a comparative analysis is performed to assess the impact of deep foundation pit excavation on the adjacent subway station and tunnel under different support structures.

In this study, a series of key assumptions were made for establishing the numerical model and conducting the model test: The Mohr–Coulomb criterion was modified to simulate the soil, which was regarded as a homogeneous and isotropic material, ignoring dynamic characteristics such as non-uniformity. The concrete support piles and metro structures were assumed to be linear elastic materials, and the anchor cables were simulated using elastic truss elements, without considering some complex effects. The foundation boundary was set with fixed constraints, assuming that the boundary was far enough to avoid boundary effects. A partial section was intercepted for research, believing that it could reflect the overall deformation law. The simulated construction conditions were consistent with the actual sequence, and there was no initial displacement in the stratum before construction. Only the soil unloading caused by the foundation pit excavation was considered, while external factors, such as earthquakes and wind loads, were ignored. In the model test, a geometric similarity ratio of 1/50 was used, and the materials were simplified based on the equivalence of flexural stiffness, assuming that the deformation laws of the model and the prototype were the same. Local sand was used to simulate the soil, and the reduction in cohesion might lead to deviations in strength similarity. The slip effects between the pile and the soil, and between the anchor cable and the soil were not simulated, and co-deformation was assumed. The connection nodes between the station and the tunnel were assumed to be rigid, ignoring the problems at the joints.

3. Establishing the Numerical Model

Due to the large scale of the foundation pit in this project, a partial section analysis is conducted for the establishment of the numerical model. The research scope is selected with the section at the junction of the station and tunnel as the center, extending 65 m in both directions along the longitudinal axis of the station and tunnel. This scope includes both the station and the tunnel, and the edge of the foundation pit within this range is straight and parallel to the longitudinal direction of the station and tunnel, making it an ideal research area.

3.1. Calculation Model

Considering the relative spatial relationship between the foundation pit and the existing metro structure and taking into account the foundation pit support design scheme, excavation construction plan, and geological survey report, a finite element model was constructed using Midas GTS NX (2019) software. In order to account for the impact of boundary conditions on the model calculation, a model with dimensions of 130 m $\times$ 90 m $\times$ 60 m was established. The final model results are shown in Figure 2a,b.

3.2. Parameters and Boundary Conditions

The soil material is modeled using the modified Mohr–Coulomb criterion, with the physical and mechanical properties of the soil layers specified in Table 1. Concrete and anchor cable materials are simulated using an elastic constitutive model. The supporting piles are represented using one-dimensional elements, while the soil and subway station tunnel are modeled with three-dimensional solid elements. The anchor cables are simulated using embedded truss elements. In terms of boundary conditions, the Midas GTS NX (2019) software is used to automatically constrain and generate the foundation boundaries. Specifically, the model is constrained for displacement in the x-direction on the left and right, in the y-direction at the front and back, and in the z-direction at the top and bottom.

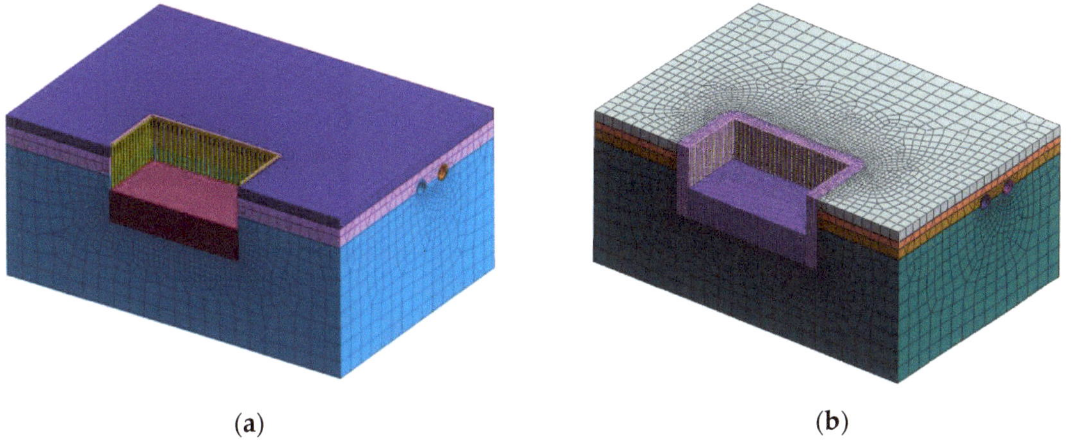

Figure 2. Excavation with different support: (**a**) pile-row and anchor cable; (**b**) double-row pile.

Table 1. Indices of physical and mechanical properties of soil layers.

Layer Number	Soil Categories	Layer Thickness (m)	γ (kN/m^3)	c (kPa)	φ (°)	E (GPa)	μ
1	Mixed fill	4.00	16.0	5.00	26.00	22	0.2
2	Gravel Layer 1	3.00	20.0	2.00	35.00	43	0.2
3	Gravel Layer 2	4.00	20.5	3.00	38.00	43	0.2
4	Gravel Layer 3	49.00	21.0	5.00	40.00	60	0.25

3.3. Working Condition Settings

In order to ensure the model has practical significance, not only is the relative spatial relationship between the foundation pit and the metro station tunnel considered but also the construction conditions, which are made consistent with the actual construction site. The construction stages set in the numerical model are aligned with those at the site, and the displacement is reset to zero in the initial state, avoiding simulation deviations caused by inaccurate construction processes and conforming more to the actual deformation law. The construction conditions are shown in Table 2. The construction conditions for the double-row pile support structure are the same as those for the pile-anchor support structure, except that no anchor cables and waist beams are installed.

Table 2. Construction conditions.

Construction Conditions	Construction Scope	Excavation Depth (m)
1	Establishment of initial stress field, displacements set to zero	0
2	Pile driving	0
3	First excavation	2.5
4	Second excavation	2.5
5	Third excavation, construction of the first layer of anchor cables and the first diaphragm beam	3

Table 2. *Cont.*

Construction Conditions	Construction Scope	Excavation Depth (m)
6	Fourth excavation, construction of the second layer of anchor cables and the second diaphragm beam	3
7	Fifth excavation	2.5
8	Sixth excavation, construction of the third layer of anchor cables and the third diaphragm beam	2.5

4. Analysis of Model Results

Analyzing the magnitude and direction of the displacement of tunnels and stations can provide a better understanding of their deformation trends. This is beneficial for analyzing patterns and drawing conclusions that are advantageous for the engineering project. Determining the direction of displacement allows for a better analysis of the stress conditions, enabling targeted reinforcement measures, which will provide valuable support for the subsequent stages of the project. The displacement sampling points are all taken longitudinally along the station and the tunnel. Specifically, the vertical displacement of the station is sampled along the center line on the station roof, the horizontal displacement of the station is sampled along the middle height line of the wall on the left side of the station close to the foundation pit, the vertical displacement of the tunnel is sampled longitudinally along the crown position of the left tunnel, and the horizontal displacement of the tunnel is sampled longitudinally along the left springline position of the left tunnel. Horizontal displacement is considered positive when moving away from the pit, and vertical displacement is considered positive when moving upwards.

A one-dimensional coordinate axis is established along the longitudinal direction of the metro tunnel to represent the relative positions between the foundation pit, the tunnel, and the station. The origin of the coordinate system is located at the intersection section of the station and the tunnel, and the axis extends in both directions, in which x is the coordinate established along the longitudinal direction of the station and the tunnel. Because the excavation of the foundation pit only affects the interval metro, it is necessary to compare and analyze the position of the foundation pit, the station, and the tunnel in the longitudinal direction, as shown in Figure 3. In Figures 4–8, the values of x are the same as those in Figure 3. All of them represent the corresponding positions perpendicular to the x-axis, and the corresponding data collection positions all have the same perpendicular distance from the x-axis. R represents the number of excavation layers of the foundation pit.

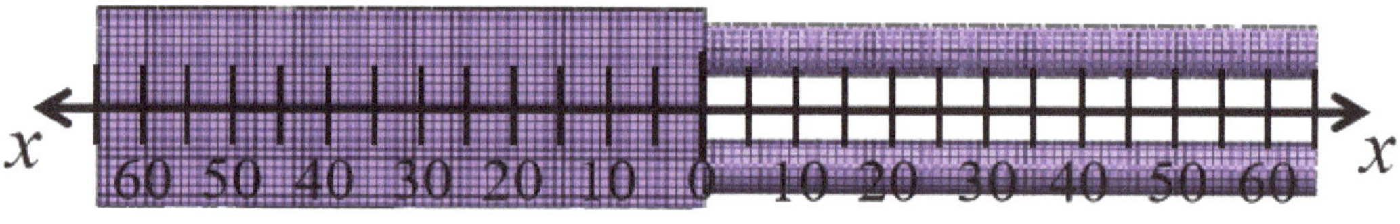

Figure 3. Schematic diagram of the x-coordinate.

4.1. *Analysis of Vertical Displacement of Metro Structures*

Due to the consolidation and compression of the soil under gravitational forces, during the foundation pit excavation process, as the soil is removed, the originally consolidated soil may experience rebound due to unloading. This causes upward pressure on the metro station tunnel, leading to vertical heave. The displacement and deformation of the soil layers generated during the excavation may potentially cause structural damage to the metro station tunnel. The excavation and unloading on one side may also lead to the supporting structure exerting compressive forces on the soil at the bottom of the subway station and tunnel, which can result in the uplift deformation of the subway station and tunnel.

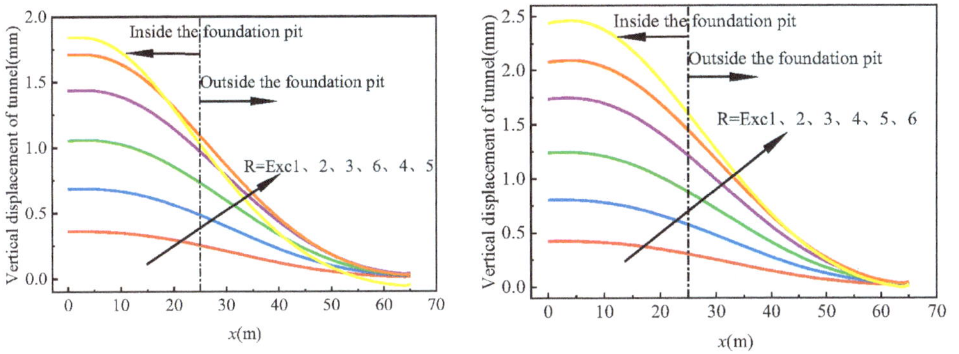

(a) (b)

Figure 4. Vertical displacement of tunnel: (**a**) pile-anchor; (**b**) double-row pile.

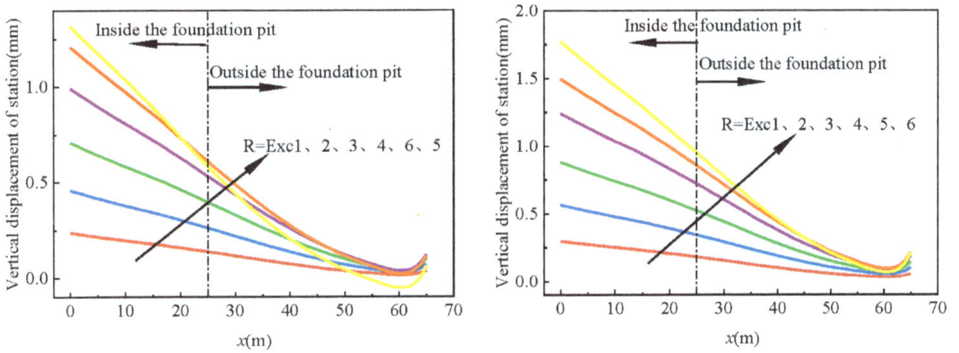

(a) (b)

Figure 5. Vertical displacement of station: (**a**) pile-anchor; (**b**) double-row pile.

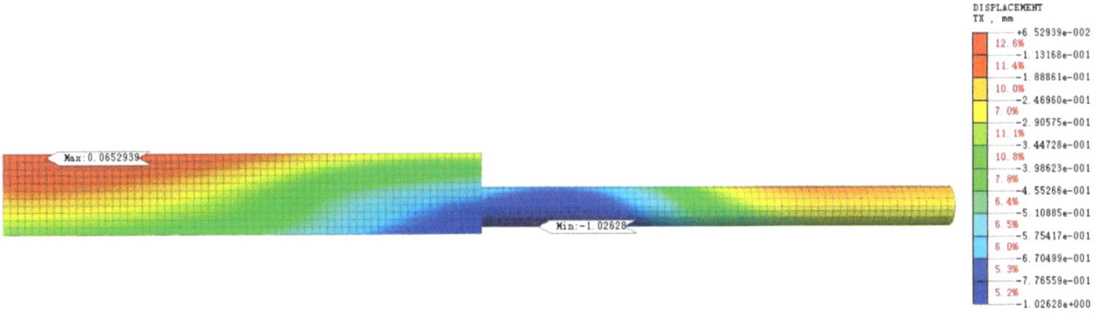

(a)

Figure 6. *Cont.*

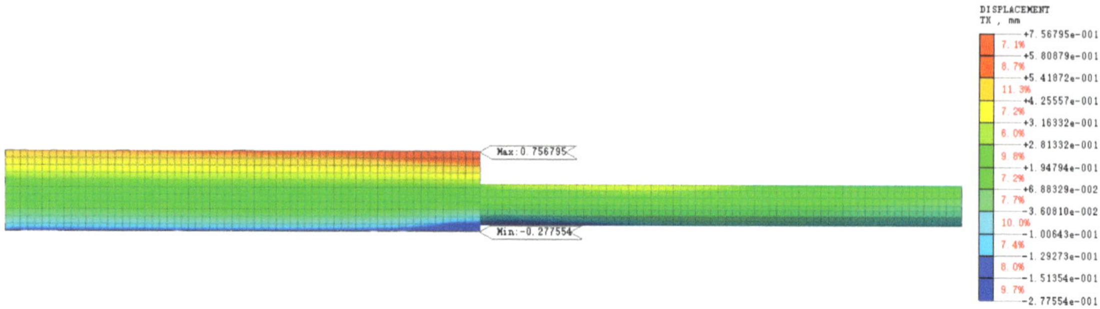

(b)

Figure 6. Vertical displacement cloud map of the fifth excavation layer: (**a**) pile-anchor; (**b**) double-row pile.

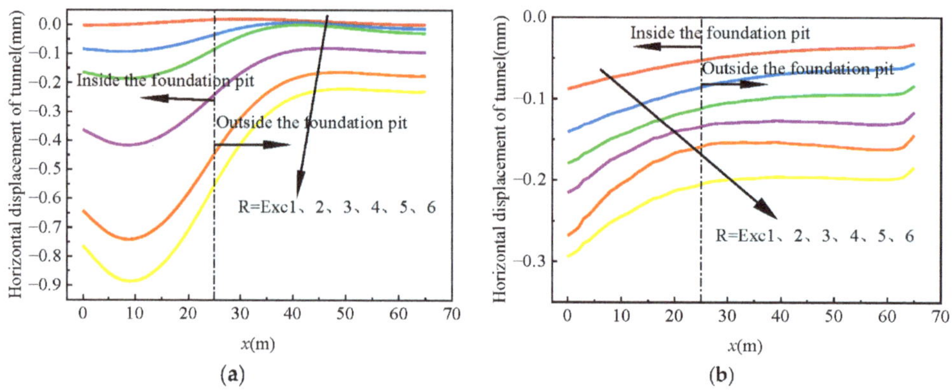

Figure 7. Horizontal displacement of tunnel: (**a**) pile-anchor; (**b**) double-row pile.

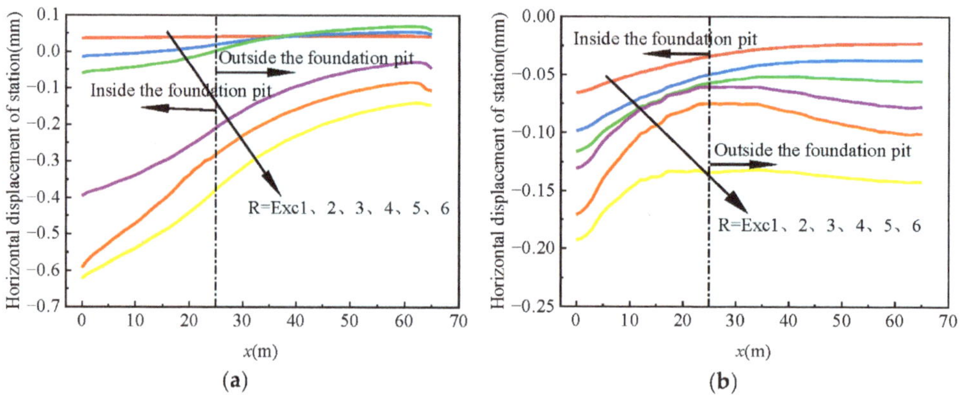

Figure 8. Horizontal displacement of station: (**a**) pile-anchor; (**b**) double-row pile.

4.1.1. Analysis of Tunnel Vertical Displacement

As shown in Figure 4a, the areas inside and outside the foundation pit are defined as follows: the extension of the edge plane of the foundation pit perpendicular to the *x*-axis. The tunnels and stations within the two extended planes are considered to be inside the foundation pit, while the remaining parts of the tunnels and stations are outside the foundation pit. All the displacements are measured along the displacement measurement lines described above, under the pile-anchor support structure, the vertical displacement of the tunnel increases continuously towards the direction near the foundation pit center, reaching its maximum at the center of the foundation pit. As the excavation depth increases, the upward displacement of the tunnel also increases, and it reaches its maximum value when the excavation depth reaches 16 m. At the end far from the foundation pit, there is almost no effect, with the displacement approaching zero.

As shown in Figure 4b, the vertical displacement trend of the tunnel under the double-row pile support structure is generally consistent with that under the pile-anchor support structure. The vertical displacement increases with the excavation depth, reaching its maximum at the position near the foundation pit center, while at the end far from the pit, the displacement is negligible. The reason for this behavior is similar to that observed under the pile-anchor support structure, where vertical heave of the tunnel occurs.

A comparison reveals that the vertical displacement of the tunnel under both support structures increases at a similar rate, and the overall trends are almost identical. However, the maximum vertical displacement under the double-row pile support structure is 0.614 mm larger than that under the pile-anchor support structure, representing an increase of 33.2%. The reason for the analysis is that the stiffness of the double-row pile support structure is greater than that of the pile-anchor support structure. Therefore, during the unloading process of the upper soil layer, the double-row pile support structure has a greater squeezing effect on the soil under the tunnel than the pile-anchor support structure. Additionally, in both support structures, the highest rate of vertical displacement growth occurs during the excavation of the third and fourth layers, indicating that excavation at the tunnel burial depth causes the greatest disturbance to the tunnel.

4.1.2. Analysis of Station Vertical Displacement

Referring to Figure 5, it is clear that under both the pile-anchor support structure and the double-row pile support structure, the variation trend of the vertical displacement of the station shows high consistency. Specifically, in both cases, the displacement reaches its maximum at the position corresponding to the foundation pit center, while in areas further from the pit, the displacement approaches zero, exhibiting very slight variations. Under the pile-anchor support structure, the maximum vertical displacement of the station is 1.320 mm, whereas under the double-row pile support structure, the corresponding vertical displacement reaches 1.774 mm, indicating a more significant displacement in the latter case.

Further analysis reveals that the variation trend of vertical displacement of the station under both support structures follows a generally linear growth pattern. The root cause of this behavior lies in the station's inherent structural stiffness. During the foundation pit excavation process, although there is an unloading effect, the disturbance caused is limited and insufficient to generate significant additional deformation in the station, resulting only in overall rotation.

The displacement nephogram of the subway station tunnel is selected after the excavation of the fifth layer. The areas selected are the surfaces of the station and the tunnel. The station and the tunnel are observed from the position of the foundation pit to visually obtain the displacement and deformation of the subway structure. Figure 6 shows the vertical displacement contour map of the metro structure after the fifth layer of excavation. Figure 6a displays the horizontal displacement of the metro station tunnel under the pile-anchor support structure, while Figure 6b shows the vertical displacement of the metro station tunnel under the double-row pile support structure. Among them, the station is on the left and the tunnel is on the right.

4.2. Analysis of Horizontal Displacement of Metro Structures

During the foundation pit excavation process near the metro station tunnel, horizontal displacement primarily results from the redistribution of soil stresses and the lateral movement of the soil. As the foundation pit is excavated, the soil, which was originally in a state of equilibrium, is disturbed, causing a change in its internal stress state. Due to the space created by the excavation, the surrounding soil tends to move inward toward the pit. This movement can cause the metro station tunnel to experience compression or tension, leading to horizontal displacement.

Moreover, if the excavation depth is large or the geological conditions are poor—such as having soft soil or high water content—the lateral movement of the soil becomes more pronounced, further exacerbating the horizontal displacement of the metro station tunnel. Since the station and tunnel are integrated, they influence each other, leading to different types of deformation. Similarly, due to the significant difference in stiffness between the station and tunnel, they will experience different deformations and effects as a result of the disturbances caused by foundation pit excavation.

4.2.1. Analysis of Tunnel Horizontal Displacement

As shown in Figure 7a, under the pile-anchor support system, the horizontal displacement of the tunnel continuously increases with the excavation depth of the foundation pit and reaches its peak during the sixth excavation (with a depth of 16 m). The maximum displacement near the foundation pit is 0.76 mm, while at the far end, it is 0.23 mm. The displacement at a distance of 10 m from the pit center reaches its maximum value of 0.88 mm. The deformation of the tunnel intensifies as it approaches the foundation pit, with the fastest increase near the pit's edge. Overall, the displacement trend follows an increase and then a decrease. After analysis, the maximum displacement at 10 m is likely related to the close connection between the tunnel and the station, as the station has significant longitudinal stiffness and is relatively resistant to deformation. At the foundation pit center, the tunnel and station are at the boundary section, where the deformations of both are continuous, resulting in the maximum horizontal displacement not occurring exactly at the pit center. After the fourth excavation, the displacement increase rate becomes more gradual, and the curve trends during the third to sixth excavations remain similar. This may be because the excavation has reached the depth of the metro structure, significantly enhancing the impact on the metro structure. In particular, the excavation range from the fourth to fifth layers covers the height of the tunnel, leading to the most pronounced impact on its deformation.

From Figure 7b, it can be seen that under the double-row pile support structure, the horizontal displacement of the tunnel also increases with the excavation depth and reaches its maximum during the sixth excavation at 16 m. The deformation rate continues to increase as it approaches the pit edge and moves toward the center, showing a generally monotonic increasing trend, with the maximum horizontal displacement occurring at the pit center. The maximum horizontal displacement at the far end from the foundation pit is 0.198 mm, while at the near end, it is 0.293 mm. Throughout the excavation process, the maximum horizontal displacement of the tunnel consistently occurs at the pit center. Additionally, the displacement increase shows no significant fluctuations, and the rate of increase is almost the same on both the far side and the side near the pit center. The reason for this may be the overall small displacement, indicating that whether the excavation reaches below the metro roof or within the tunnel burial depth, the impact on the tunnel is relatively limited. This metro tunnel is located near the uplift zone and the deformation transition area in the research results of Zheng Gang et al. [19]. The occurrence of the uplift phenomenon is consistent with their research findings.

Comparing the horizontal displacement of the metro tunnel under both support structures, both structures show that the displacement increases as the tunnel approaches the foundation pit, with the far side from the pit being less affected. Moreover, the horizontal displacement of the tunnel is significantly affected by the station, especially when the displacement is relatively large. Since the longitudinal stiffness of the station is greater than that of the tunnel, the station is less likely to deform than the tunnel. Therefore, being close to the station will instead suppress the horizontal deformation of the tunnel. However, the horizontal displacement of the tunnel under the double-row pile support structure does not show this characteristic because the overall displacement is relatively small. It is worth noting that the horizontal displacement under the pile-anchor support structure is significantly greater than that under the double-row pile support structure. During the first two layers of excavation, the horizontal displacement under the pile-anchor support structure is slightly smaller than that under the double-row pile support structure. However, from the third layer of excavation onwards, the displacement under the pile-anchor support structure begins to exceed that under the double-row pile support structure. After the sixth excavation, the maximum horizontal displacement under the pile-anchor support structure is 0.591 mm larger than that under the double-row pile support structure. This difference indicates a significant variation in the impact of the two support structures on tunnel horizontal displacement. In practical engineering, the appropriate support system should be selected based on specific conditions to effectively control tunnel deformation and ensure the safety and stability of the metro structure.

4.2.2. Analysis of Station Horizontal Displacement

From Figure 8a, it can be seen that under the pile-anchor support structure, the horizontal displacement of the station exhibits a more pronounced "linear" characteristic compared to the tunnel. The reason for this is that the station has higher longitudinal stiffness than the tunnel, so under slight disturbances, it experiences overall rotation first. During the excavation of the first three layers, the entire station undergoes a certain degree of rotation, with the part near the pit center shifting inward, and the far end of the station rotating outward. As excavation reaches the fourth layer and subsequent stages, the station begins to move inward toward the foundation pit, and this displacement increases with the excavation depth. This is because only half of the station is located on the side corresponding to the excavation area of the foundation pit, and the other half is outside the excavation area of the foundation pit. Therefore, during the excavation of the first three layers of the foundation pit, there is less unloading. Due to the high overall stiffness of the station, it will experience overall rotational deformation. When the excavation reaches a certain depth and the unloading reaches a certain level, the station will be pushed by the soil on the side away from the foundation pit and move as a whole.

As shown in Figure 8b, under the double-row pile support structure, the station displacement follows a trend of increasing deformation from the edge of the foundation pit toward the center. However, it is noteworthy that there is virtually no deformation occurring outside the pit area, and the station as a whole shows an inward displacement toward the pit.

When comparing the horizontal displacement of the station under both the pile-anchor support structure and the double-row pile support structure, it is evident that the displacement under the pile-anchor support structure is significantly larger than under the double-row pile support structure. In the pile-anchor support structure, the displacement exhibits a linear change, whereas in the double-row pile structure, such linear displacement is not observed. The likely reason is that the horizontal displacement of the station under the double-row pile support structure is too small. Only the part of the station within the excavation range of the foundation pit is slightly affected, resulting in inward deformation toward the pit.

Still taking the displacement nephogram after the fifth-layer excavation as an example, the displacement and deformation of the surfaces of the station and the tunnel are selected. We observed the process from the foundation pit towards the station and the tunnel. The direction perpendicular to the x-axis and away from the foundation pit is defined as the positive direction. Figure 9 shows the horizontal displacement cloud map of the metro structure after the excavation of the fifth layer. Figure 9a shows the horizontal displacement of the metro station tunnel under the pile-anchor support structure, while Figure 9b shows the horizontal displacement of the metro station tunnel under the double-row pile support structure.

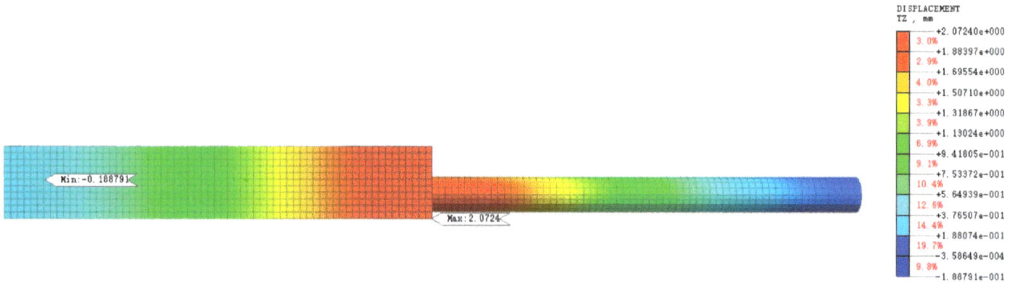

(**a**)

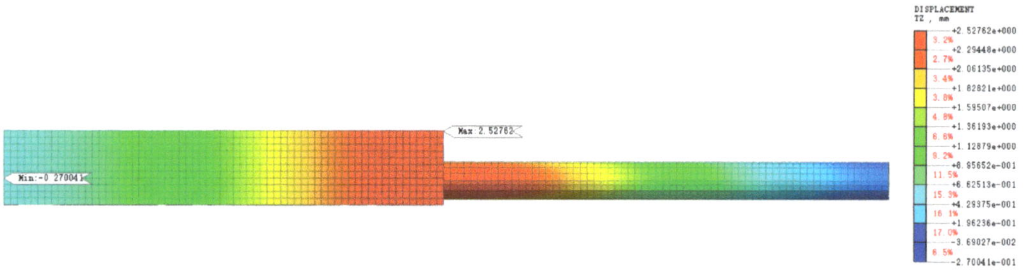

(**b**)

Figure 9. Horizontal displacement cloud map of the fifth excavation layer: (**a**) pile-anchor; (**b**) double-row pile.

5. Comparison Between Numerical Model and Model Test Results

In order to make the numerical simulation results more convincing, scaled model tests were conducted. According to the similarity principle, a geometric similarity ratio of 1/50 was set, the method of bending stiffness equivalence and dimensional analysis is adopted, with different materials selected as model test materials. The soil materials should ensure consistent density and internal friction angle, with the cohesion reduced to 1/50. The materials for piles, tunnels, stations, and other components are all considered with bending stiffness equivalence. And a comparison analysis was conducted between the model test results and the numerical model results. Lanzhou local sand was selected as the soil for the model test. PP pipes were used to simulate the piles, PVC pipes simulated the tunnel, and acrylic sheets were used to assemble the station. Iron wires were used to simulate the anchor cables. In the tunnel, soil pressure gauges (YITUO SENSING TECHNOLOGY, Changsha, China) were arranged every 60° along the circumferential direction, and strain gauges (Chengdu Electrical Measurement & Sensing Technology Co., Ltd., Chengdu, China) were attached every 90° to measure the tunnel's strain and additional loads. The excavation conditions in the model test were set to be consistent with those in the numerical model. The variation in the metro structure under the pile-anchor support structure was analyzed. The model test setup is shown in Figure 10. The figure shows the excavation of a foundation pit with a pile-anchor support structure, and one layer was excavated.

Figure 10. Conducting model tests.

By analyzing the bending moment of the tunnel, the deformation of the tunnel can be clearly observed. Typically, the maximum bending moment occurs after the final excavation. A comparative analysis was conducted on the impact of foundation pit excavation on the existing metro structure after the final excavation. The selected tunnel planes are all perpendicular to the x-axis, and the observation is carried out with the foundation pit on the right and the tunnel on the left. From Figure 11, it can be seen that a tunnel cross-section near the junction with the foundation pit was selected for analysis. The 0° position corresponds to the horizontal crown waist of the tunnel near the foundation pit, and the 90° position corresponds to the tunnel crown.

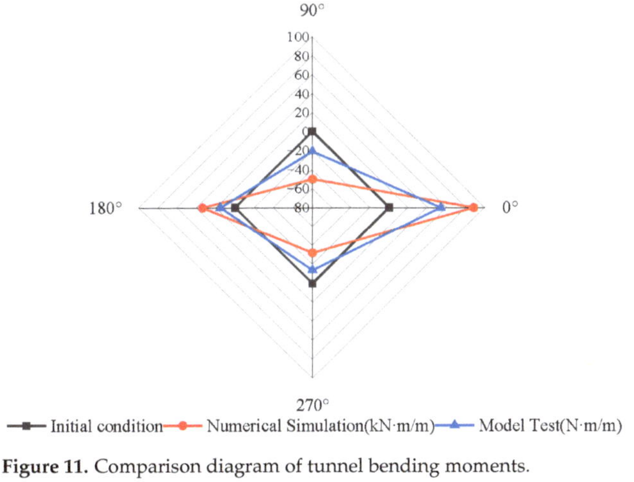

Figure 11. Comparison diagram of tunnel bending moments.

A comparison of the model test and numerical simulation reveals that both show the same trend in the variation in the tunnel's bending moment. Specifically, at both sides of the crown waist, the bending moment in the tunnel is positive. At the crown and invert positions, the bending moment is negative. The absolute value of the bending moment at the right crown waist is greater than that at the crown and invert, indicating that the tunnel's horizontal diameter decreases while the vertical diameter increases. This results in a "vertical duck-egg" deformation of the tunnel, where it experiences vertical stretching and horizontal flattening.

From Figure 12, the horizontal springline of the tunnel on the side close to the foundation pit is at 0°, and the tunnel crown is at 90°, it can be observed that in the numerical model, a tunnel cross-section corresponding to the edge of the foundation pit was selected for analysis. This result is obtained from numerical simulation. Under the pile-anchor support structure, the maximum displacement of the tunnel segment occurs at 20°, with a value of 1.537 mm. Under the double-row pile support structure, the maximum displacement of the tunnel segment occurs at 0°, with a value of 1.126 mm. In comparison, the tunnel's circumferential displacement under the pile-anchor support structure is more regular, and the displacement value is much smaller than that under the double-row pile support structure.

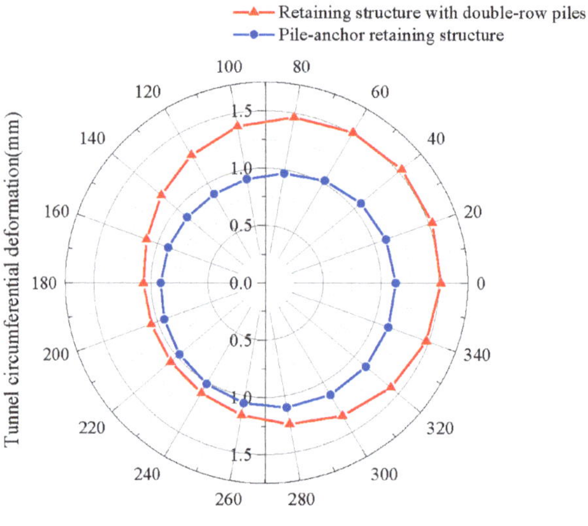

Figure 12. Circumferential displacement of the tunnel.

As shown in Figure 13, the horizontal springline of the tunnel on the side close to the foundation pit is at 0°, and the tunnel crown is at 90°, the variation in additional load is also consistent. Specifically, at 60°, the additional load is either zero or negative, indicating that significant unloading occurs on the side of the tunnel near the foundation pit during the excavation process. Additionally, smaller load values are observed both above the tunnel and on the side near the foundation pit, suggesting that the additional load exerted by the foundation pit excavation is mainly applied to the side and below the tunnel that is farthest from the foundation pit. This also confirms the phenomenon of tunnel uplift and lateral movement towards the side near the foundation pit observed in the numerical simulation.

By selecting the cross-section at the center of the foundation pit for analysis, Figure 14 shows that the surface settlement trends in both the model test and numerical simulation are essentially consistent. A slight settlement occurs in the pile body, with the maximum settlement occurring between the pile and the metro structure. However, above the station, a clear uplift is observed, which gradually diminishes as the distance from the foundation pit increases. This is basically consistent with the research results of Zheng Gang et al. [18]. The uplift of the surface soil corresponds to the observed uplift of the tunnel at the location directly above the center of the foundation pit, thereby confirming that both the numerical simulation and model test results are correct.

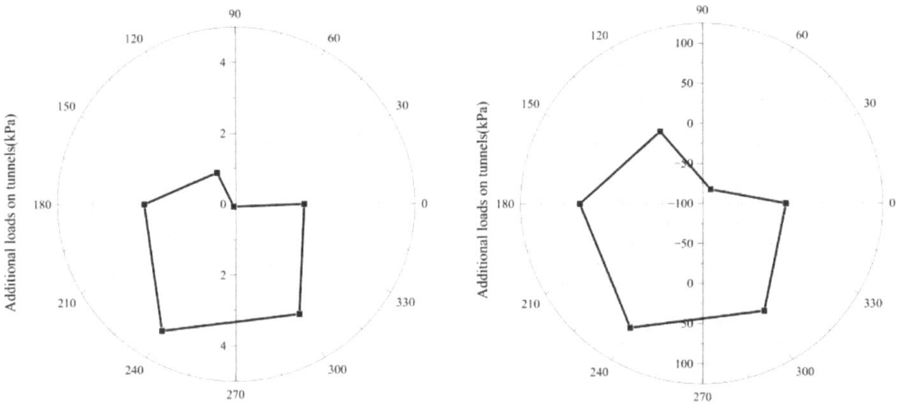

(a) (b)

Figure 13. Comparison of additional loads on the tunnel: (**a**) model test; (**b**) numerical simulation.

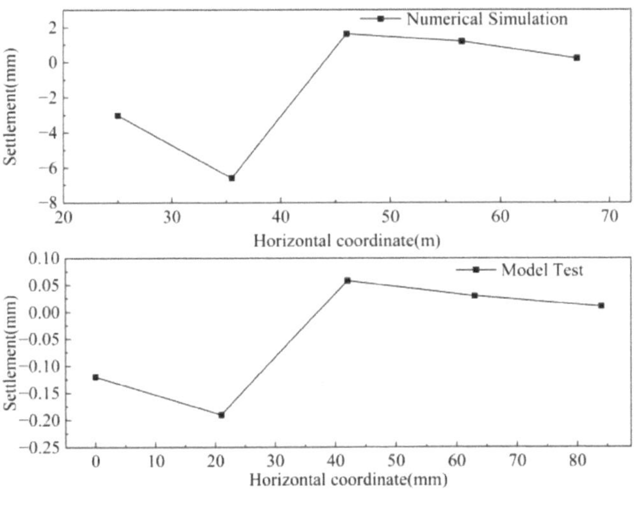

Figure 14. Variation in surface settlement.

A comprehensive comparative analysis shows that the displacement of the tunnel and station, as well as the settlement of the ground surface around the foundation pit and the bending moment of the tunnel, all demonstrate a high degree of consistency between the numerical simulation results and the model test results, thereby confirming the accuracy of the results.

6. Conclusions

(1) This study investigates the impact of foundation pit excavation on adjacent metro station tunnels. A numerical model was established using Midas GTS NX (2019)

software, and model tests were conducted to thoroughly analyze the deformation characteristics of the metro structure under different support conditions. In terms of vertical displacement, significant changes were observed in both the tunnel and station near the center of the foundation pit. Under pile-anchor and double-row pile support, the tunnel's vertical displacement increased towards the center of the pit, with the maximum displacement under double-row pile support being 33.2% greater than that under pile-anchor support. The station's vertical displacement also reached its maximum at the foundation pit's center, and the displacement variation followed a generally linear trend. These findings emphasize the critical role of the support structure in controlling vertical displacement.

(2) Regarding horizontal displacement, under pile-anchor support, the tunnel's horizontal displacement increased with excavation depth, peaking at the sixth excavation stage, showing an initial increase followed by a decrease. In contrast, under double-row pile support, the displacement steadily increased, reaching its maximum at the foundation pit's center. For the station, under pile-anchor support, horizontal displacement initially caused a rotational shift before moving inward, while under double-row pile support, the displacement increased from the pit's edge towards its center. The differences in displacement trends and magnitudes between the two support systems highlight the importance of selecting an appropriate support structure to effectively control horizontal displacement.

(3) The strong consistency between the model test and numerical simulation results validates the reliability of the research methodology. This demonstrates that the numerical model can accurately simulate real-world engineering conditions, providing a solid foundation for practical applications and confirming the accuracy of the observed deformation patterns and displacement trends of the metro structure.

(4) In summary, this study offers theoretical and practical guidance for metro construction in addressing the impact of foundation pit excavation on adjacent structures. It contributes to optimizing support design schemes, ensuring the safety and stability of metro structures in complex construction environments.

(5) Most current studies involve model simplification and idealization of construction conditions. They do not take into account the non-uniformity, stratification, and dynamic mechanical behavior of soil or the impact of construction errors or unexpected interferences on deformation. The substitution of materials in model tests may lead to local deviations in strength and deformation characteristics.

(6) Future research can further consider multi-field coupling, conduct refined modeling, and take into account various construction scenarios. For scaled-down model tests, further optimization is needed, such as optimizing the materials used in the tests to ensure the accuracy of the model tests as much as possible.

Author Contributions: Conceptualization, Z.M.; Methodology, T.D.; Validation, F.H.; Resources, S.Y., L.D. and X.Z.; Data curation, P.L. and N.L.; Writing—review and editing, L.D.; Visualization, L.D.; Supervision, S.Y. All authors have read and agreed to the published version of the manuscript.

Funding: The National Natural Science Foundation of China (52168050 and 52408465); China Postdoctoral Science Foundation (BX20240451 and 2024M753850); Major Science and Technology Special Project Plan (24ZDFA010).

Data Availability Statement: The data that support the findings of this study are available from the corresponding author upon reasonable request.

Acknowledgments: The authors acknowledge the support from Gansu Construction (Investment) Holding Group Co., Ltd., and Lanzhou University of Technology.

Conflicts of Interest: Authors Zhitong Mao, Tian Ding and Fengchao Hu were employed by the company Gansu Construction (Investment) Holding Group Co., Ltd. The remaining authors declare that the research was conducted in the absence of any commercial or financial relationships that could be construed as a potential conflict of interest.

References

1. Zhang, Z.G.; Zhang, M.X.; Wang, W.D. Two-stage method for analyzing effects on adjacent metro tunnels due to foundation pit excavation. *Rock Soil Mech.* **2011**, *32*, 2085–2092.
2. Chang, C.T.; Sun, C.W.; Duann, S.W.; Hwang, R.N. Response of a Taipei Rapid Transit System (TRTS) tunnel to adjacent excavation. *Tunn. Undergr. Space Technol.* **2001**, *16*, 151–158. [CrossRef]
3. Chen, R.; Meng, F.; Li, Z.; Ye, Y.; Ye, J. Investigation of response of metro tunnels due to adjacent large excavation and protective measures in soft soils. *Tunn. Undergr. Space Technol.* **2016**, *58*, 224–235. [CrossRef]
4. Zhu, M.; Gong, X.; Gao, X.; Liu, S.; Yan, J. Remediation of damaged shield tunnel using grouting technique: Serviceability improvements and prevention of potential risks. *J. Perform. Constr. Facil.* **2019**, *33*, 04019062. [CrossRef]
5. Sharma, J.S.; Hefny, A.M.; Zhao, J.; Chan, C.W. Effect of large excavation on deformation of adjacent MRT tunnels. *Tunn. Undergr. Space Technol.* **2001**, *16*, 93–98. [CrossRef]
6. Zhang, D.M.; Zong, X.; Huang, H.W. Longitudinal deformation of existing tunnel due to underlying shield tunneling. *Rock Soil Mech.* **2014**, *35*, 2659–2666.
7. Xu, R.Q.; Cheng, K.; Ying, H.W.; Lin, C.G.; Liang, R.Z.; Feng, S.Y. Deformation response of a tunnel under foundation pit unloading considering buried depth and shearing effect. *Rock Soil Mech.* **2020**, *41* (Suppl. S1), 195–207.
8. Attewell, P.B.; Woodman, J.P. Predicting the dynamics of ground settlement and its derivitives caused by tunnelling in soil. *Ground Eng.* **1982**, *15*, 13–22.
9. Zhang, Z.G.; Lu, M.H.; Gong, J.F. Time-domain solution for influence of excavation on adjacent pile deformation in visco-elastic foundation. *Rock Soil Mech.* **2017**, *38*, 3017–3028.
10. Cheng, K.; Xu, R.; Ying, H.; Liang, R.Z.; Lin, C.G.; Gan, X.L. Simplified method for evaluating deformation responses of existing tunnels due to overlying basement excavation. *Chin. J. Rock Mech. Eng.* **2020**, *39*, 637–648.
11. Zhou, Z.L.; Chen, S.G.; Chen, L.; Tu, P. Analysis of uplift deflection of metro tunnel due to adjacent pit excavation. *Chin. J. Geotech. Eng.* **2015**, *37*, 2224–2234.
12. Huang, X.; Huang, H.; Zhang, D. Centrifuge modelling of deep excavation over existing tunnels. *Proc. Inst. Civ. Eng.-Geotech. Eng.* **2014**, *167*, 3–18. [CrossRef]
13. Liang, F.Y.; Chu, F.; Song, Z.; Li, Y.S. Centrifugal model test research on deformation behaviors of deep foundation pit adjacent to metro stations. *Rock Soil Mech.* **2012**, *33*, 657–664.
14. Li, M.G.; Chen, J.J.; Wang, J.H.; Zhu, Y.F. Comparative study of construction methods for deep excavations above shield tunnels. *Tunn. Undergr. Space Technol.* **2018**, *71*, 329–339. [CrossRef]
15. Huang, K.; Yang, W.; Ma, Q.; An, Y.; Li, Y.; Zhou, J.; Qiu, L. Influence of foundation excavation pit on adjacent metro tunnel using fluid-solid mechanics theory. *J. Cent. South Univ.* **2019**, *50*, 198–205.
16. Zhang, D.M.; Gao, C.P.; Yin, Z.Y. CFD-DEM modeling of seepage erosion around shield tunnels. *Tunn. Undergr. Space Technol.* **2019**, *83*, 60–72. [CrossRef]
17. Zheng, G.; Li, Z.-W. Comparative analysis of responses of buildings adjacent to excavations with different deformation modes of retaining walls. *Chin. J. Geotech. Eng.* **2012**, *34*, 969–977.
18. Zheng, G.; Deng, X.; Liu, C.; Liu, Q.-C. Comparative analysis of influences of different deformation modes of retaining structures on displacement field of deep soils outside excavations. *Chin. J. Geotech. Eng.* **2014**, *36*, 273–285.
19. Zheng, G.; Wang, Q.; Deng, X.; Du, Y.-M. Comparative analysis of influences of different deformation modes of retaining structures on deformation of existing tunnels outside excavations. *Chin. J. Geotech. Eng.* **2015**, *37*, 1181–1194.
20. Ye, S.; Zhao, Z.; Wang, D. Deformation analysis and safety assessment of existing metro tunnels affected by excavation of a foundation pit. *Undergr. Space* **2021**, *6*, 421–431. [CrossRef]
21. Wang, Y.; Cui, X.; Liu, K.; Jiang, P. Manufacture, development, and application of sensor-enabled geosynthetics: State-of-the-art review. *Intell. Transp. Infrastruct.* **2022**, *1*, liac012. [CrossRef]
22. Liu, B.; Lin, H.; Chen, Y.; Liu, J.; Guo, C. Deformation stability response of adjacent metro tunnels considering excavation and support of foundation pit. *Lithosphere* **2022**, *2022*, 7227330. [CrossRef]
23. Ding, Z.; Zhang, X.; Jin, J.K.; Wang, L. Measurement analysis on whole excavation of foundation pit and deformation of adjacent metro tunnel. *Rock Soil Mech.* **2019**, *40* (Suppl. S1), 415–423.

24. Zhou, Z.; Zhou, Y.; Zhang, H.; Chen, S.; Xiang, L.; Wang, L. Effects of Foundation Excavation on Metro Tunnels at Different Locations and Performance of Corresponding Reinforcement Measures: A Case of Shenzhen Metro Line 11, China. *Symmetry* **2022**, *14*, 2561. [CrossRef]
25. Liu, J.; Xue, B.; Wang, H.; Zhang, X.; Zhang, Y. Numerical study on the behavior of an existing tunnel during excavating adjacent deep foundation pit. *Sustainability* **2023**, *15*, 9740. [CrossRef]

buildings

MDPI

Article

Dynamic Response of Train–Ballastless Track Caused by Failure in Cement–Asphalt Mortar Layer

Xicheng Chen [1], Yanfei Pei [2,3,*] and Kaiwen Liu [2,3]

[1] Sichuan Provincial Engineering Research Center of Rail Transit Lines Smart Operation and Maintenance, Chengdu Vocational & Technical College of Industry, Chengdu 610218, China; earth_structures@outlook.com
[2] School of Civil Engineering, Southwest Jiaotong University, Chengdu 610031, China
[3] Key Laboratory of High-Speed Railway Engineering of Ministry of Education, Southwest Jiaotong University, Chengdu 610031, China
* Correspondence: yanfeipei@my.swjtu.edu.cn

Abstract: Cement–asphalt (CA) mortar voids in earth's structure are prone to inducing abnormal vibrations in vehicle and track systems and are more difficult to recognize. In this paper, a vehicle–ballastless track coupling model considering cement–asphalt mortar voids is established and the accuracy of the model is verified. There are two main novel results: (1) The displacement of the track slab in the ballastless track structure is more sensitive to the void length. Voids can lead to blocked vibration transmission between the ballastless track slab and concrete base. (2) The wheel–rail vibration acceleration is particularly sensitive to voids in cement–asphalt mortar, making the bogie pendant acceleration a key indicator for detecting such voids through amplitude changes. Additionally, the train body pendant acceleration provides valuable feedback on the cyclic characteristics associated with single-point damage in the cement–asphalt mortar, thereby enhancing the accuracy of dynamic inspections for vehicles. In the sensitivity ordering of the identification indexes of voids, the bogie's vertical acceleration in high-speed trains > the nodding acceleration of the bogie > the vehicle's vertical acceleration. Adaptive suspension parameters can be designed to accommodate changes in track stiffness.

Keywords: high-speed railway; ballastless track; cement–asphalt (CA) mortar; void defects; dynamic response; health monitoring system (HMS)

1. Introduction

The cement–asphalt (CA) mortar layer is a key structural element in the ballastless track systems of high-speed railways. This structural layer is used in railway tracks in China, Japan, Italy, and other countries. As the duration of the use of the CA mortar layer has increased and the base material itself has deteriorated [1,2], a state of void failure has been observed in ballastless tracks in various countries [3–9]. In China, the cement–asphalt (CA) mortar layer is widely used in the railway track system, specifically in the China Railway Track System II (CRTS-II) structure. The CRTS-II structure consists of three components: the track slab, the CA mortar layer, and the concrete base. However, over time, numerous voids have been observed in the CA mortar layer at project sites, as shown in Figure 1. This layer, located between the track slab and the concrete base, plays a crucial role in vibration damping and energy absorption. Consequently, the presence of voids in the CA mortar layer directly compromises the vibration-damping function of the ballastless track structure. Additionally, the lack of elasticity causes inconsistencies in the stiffness of the track structure along the railway's direction. This inconsistency leads to a stronger impact from the operation of high-speed trains, posing safety risks to passengers. Furthermore, the voids exacerbate the dynamic effects of the high-speed trains, increasing the likelihood of damage to other components of the ballastless track.

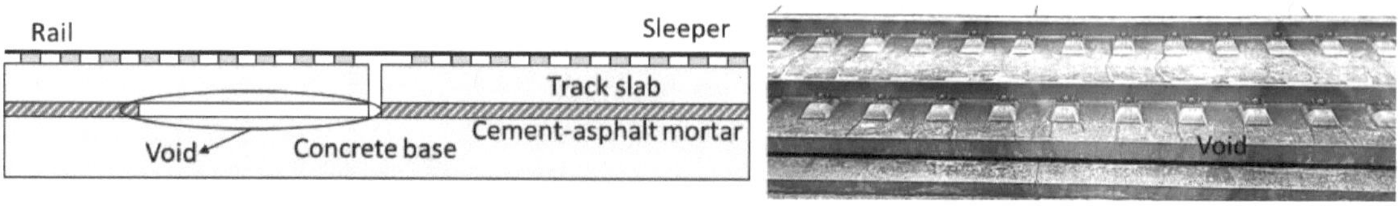

Figure 1. Cement–asphalt (CA) mortar layer void of CRTS-II ballastless track.

The cement–asphalt mortar layer functions as an elastic adjustment layer, playing key roles such as buffering and vibration reduction [8,10]. However, with the prolonged operation of high-speed railway ballastless tracks, the CA mortar layer experiences various forms of damage, including gaps, mud pumping, spalling, and voids [11–13], due to long-term train vibrations, temperature fluctuations, humidity, and other external factors. These operational defects ultimately lead to the formation of voids in the CA mortar layer. The vehicle–ballastless track system operates as an integrated unit, coupled through wheel–rail interactions [14]. As a crucial elastic adjustment layer, any weakening of the CA mortar's vibration-reducing function can cause abnormal dynamic interactions within the system, posing a long-term threat to both traffic safety and the service life of the structure [15–19].

Extensive research has been conducted on the damage and vibration behavior of ballastless track composite structures [20]. The High-Speed Railway Design Code [21] specifies that the design life of ballastless track structures is 60 years, but due to long-term cyclic loading from trains and the compounded effects of damage, such as from the CA mortar layer, the actual service life of ballastless tracks is significantly shortened. Firstly, for the failure problem of the CRTS-II ballastless track structure, research on the damage mechanism of the concrete [22,23] and CA mortar of CRTS-II ballastless tracks is focused on environmental temperature [15], trains' dynamic loads [24,25], track irregularity [26], material fatigue and impact damage [27,28], foundation deformation [29], and other aspects.

Currently, the main research is focused on the damage mechanisms of ballastless tracks' concrete and CA mortar layers, as well as the application of some non-destructive testing means to identify CA mortar void lesions [30]. But, the working efficiency is very low. However, in recent years, line infrastructure damage detection based on moving vehicles has gradually become a research hotspot. Kordestani et al. [31] used the acceleration response law to identify bridge damage. There are fewer studies on the comparison of vibration characteristics of vehicle–ballastless track systems caused by different failure modes in CA mortar [32].

In this paper, to address the abnormal vibration problem induced by varying CA mortar cavity lengths, a vehicle–ballastless track–earth structure model considering CA mortar failure is developed based on multi-body dynamics and the finite element method. The effects of four types of short-shaped void distributions on the dynamic response of the rail and ballastless track are investigated. Subsequently, the internal vibration transfer state of the ballastless track structure under CA mortar void conditions is analyzed using an innovative transfer function. Finally, a comparative analysis is performed to identify differences in the wheel–rail force, train body vibration, and bogie acceleration using high-speed trains, and a method is proposed to adapt the suspension parameters of high-speed trains to changes in track stiffness.

2. Numerical Modeling and Methods

The research approach and analysis process of this thesis is shown in Figure 2a. The model established in this paper contains the vehicle system, track system, and subgrade; the vehicle system is considered modeled as a pendant model with 10 degrees of freedom (DOF), the rails are modeled using a bending beam model, the track system is modeled using a solid model, and the subgrade is represented as springs and damping to achieve the support of the subgrade [33,34]. A combination of a vehicle and ballastless track is produced through wheel–rail contact. After forming the overall stiffness, damping, and mass matrices again, the solution is performed using the backward difference method. The accuracy of the model is verified. On this basis, parametric analysis of CA mortar void length is performed to compare the dynamic response of the vehicle and ballastless track and to propose a more sensitive index for CA mortar failure. The system dynamics model is shown in Figure 2b.

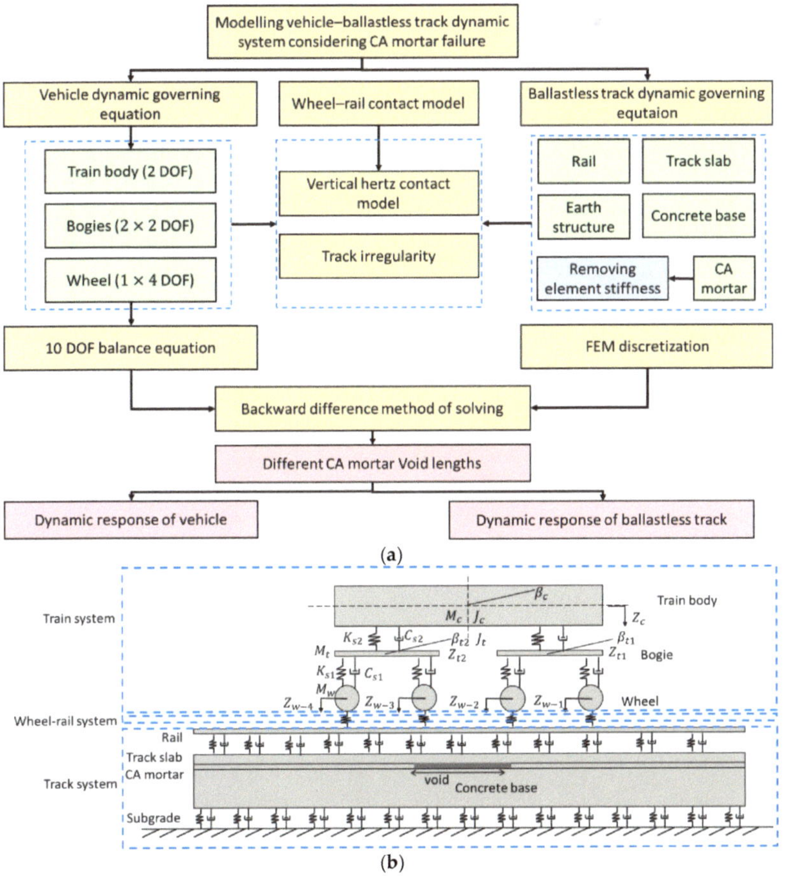

Figure 2. Research method and analysis process: (**a**) research flowchart; (**b**) physical model of problem.

(1) Train system

Since damage to the CA mortar layer leads to the deterioration of the stiffness and damping of this layer, which mainly affects the vertical dynamic behavior, the vehicle system was modeled using a 10 DOF multi-rigid-body dynamics model, taking into account the solution efficiency and the properties of the analytical problem. The solution variables of the vehicle system dynamics are the nodding and sinking of the vehicle body, the nodding and sinking of the front and rear bogies, and the nodding and sinking of the four wheel pairs. The vibration equation of the system is as follows (1):

$$[\mathbf{M}][\ddot{\mathbf{z}}] + [\mathbf{C}][\dot{\mathbf{z}}] + [\mathbf{K}][\mathbf{z}] = [\mathbf{F}] \tag{1}$$

where $[\mathbf{M}]$ represents the mass matrix of the train body, bogie, and wheelset; $[\mathbf{K}]$ is the stiffness matrix of the vehicle assembly; $[\mathbf{C}]$ is the damping matrix of the vehicle; $[\mathbf{z}]$, $[\dot{\mathbf{z}}]$, $[\ddot{\mathbf{z}}]$ represent the displacement matrix, velocity matrix, and acceleration matrix of the vehicle system, respectively; $[\mathbf{F}]$ is the load vector of the vehicle; and the model parameters for the numerical calculation of the vehicle are shown in the table below. The computational parameters of the system dynamics model are shown in Table 1. The matrix of the train system as follows:

$$\mathbf{M} = diag\{\mathbf{J_c M_c M_t J_t M_{w1} M_{w2} M_{w3} M_{w4}}\}$$

$$\mathbf{K_u} = \begin{bmatrix} 2K_{s2} & 0 & -K_{s2} & -K_{s2} & 0 & 0 & 0 & 0 & 0 & 0 \\ & 2K_{s2}l_2^2 & K_{s2}l_2 & -K_{s2}l_2 & 0 & 0 & 0 & 0 & 0 & 0 \\ & & 2K_{s1}+K_{s2} & 0 & 0 & 0 & -K_{s1} & -K_{s1} & 0 & 0 \\ & & & 2K_{s1}+K_{s2} & 0 & 0 & 0 & 0 & -K_{s1} & -K_{s1} \\ & & & & 2K_{s1}l_1^2 & 0 & K_{s1}l_1 & -K_{s1}l_1 & 0 & 0 \\ & & & & & 2K_{s1}l_1^2 & 0 & 0 & K_{s1}l_1 & -K_{s1}l_1 \\ & & & & & & K_{s1} & 0 & 0 & 0 \\ & & & & & & K_{s1} & 0 & 0 & \\ & & & & & & & K_{s1} & 0 & \\ & & & & & & & & K_{s1} & \end{bmatrix}$$

$$\mathbf{C_u} = \begin{bmatrix} 2C_{s2} & 0 & -C_{s2} & -C_{s2} & 0 & 0 & 0 & 0 & 0 & 0 \\ & 2C_{s2}l_2^2 & C_{s2}l_2 & -C_{s2}l_2 & 0 & 0 & 0 & 0 & 0 & 0 \\ & & 2C_{s1}+C_{s2} & 0 & 0 & 0 & -C_{s1} & -C_{s1} & 0 & 0 \\ & & & 2C_{s1}+C_{s2} & 0 & 0 & 0 & 0 & -C_{s1} & -C_{s1} \\ & & & & 2C_{s1}l_1^2 & 0 & C_{s1}l_1 & -C_{s1}l_1 & 0 & 0 \\ & & & & & 2C_{s1}l_1^2 & 0 & 0 & C_{s1}l_1 & -C_{s1}l_1 \\ & & & & & & C_{s1} & 0 & 0 & 0 \\ & & & & & & & C_{s1} & 0 & 0 \\ & & & & & & & & C_{s1} & 0 \\ & & & & & & & & C_{s1} & \end{bmatrix}$$

$$\mathbf{Z} = \left\{ \begin{matrix} v_c & \beta_c & v_{t1} & v_{t2} & \beta_{t1} & \beta_{t2} & v_{w1} & v_{w2} & v_{w3} & v_{w4} \end{matrix} \right\}^T$$

$$\mathbf{F} = \left\{ \begin{matrix} -M_c g & 0 & -M_t g & -M_t g & 0 & 0 & P_1 & P_2 & P_3 & P_4 \end{matrix} \right\}^T$$

M_c and J_c represent the mass and rolling moment of the inertia of the rigid body of the car, respectively; M_t and J_t refer to the mass and rolling moment of the inertia of the bogies; and $\mathrm{M_{wi}}$ denotes the mass of the ith wheel. ($\mathrm{v_c}$) and (β_c) represent the vertical displacement and pitch angle of the car body, respectively; ((v_{t1}, β_{t1})) and ((v_{t1}, β_{t2})) indicate the vertical displacement and pitch angle of the front and rear bogies, respectively; and (v_{wi}) denotes the vertical displacement of the ith wheel. Detailed equations for the damping matrix [C] and the stiffness matrix [K] can be found in Lei [35].

Table 1. Calculation parameters of China railway high-speed train-3 (CRH3) (Lei [35]).

Parameters	Value
Half-car body mass, M_c (kg)	20,000
Bogie mass, M_t (kg)	3200
Wheelset mass, M_w (kg)	2400
Primary suspension stiffness, k_{s1} (MN/m)	2.08
Primary suspension damping, c_{s1} (kN·s/m)	100
Secondary suspension stiffness, k_{s2} (MN/m)	0.8
Secondary suspension damping, c_{s2} (kN·s/m)	120
Train body pitch moment of inertia, J_c $\left(\text{kg}\cdot\text{m}^2\right)$	5.47×10^5
Bogie pitch moment of inertia, J_t $\left(\text{kg}\cdot\text{m}^2\right)$	6800
Wheel base, (m)	2.5
Bogie center distance, (m)	17.375
Axle weight, (t)	14
Wheel–rail contact ratio, G $(\text{m}/\text{N}^{\frac{2}{3}})$	$3.86R^{-0.115} \times 10^{-8}$
Wheel rolling radius, $R(\text{m})$	0.457

(2) Track system

Computational simulations of rails are generally simplified using the Euler–Bernoulli beam model, with Euler beam governing equations [35]:

$$E_r I_r \frac{\partial^4 w_r}{\partial x^4} + m_r \frac{\partial^2 w_r}{\partial t^2} + c_r \left(\frac{\partial w_r}{\partial t} - \frac{\partial w_s}{\partial t} \right) + k_r (w_r - w_s) = -F(t)\delta(x - vt) \tag{2}$$

where w_r and w_s represent the vertical deformation of the rail and track slab, respectively; $E_r I_r$ represents the bending stiffness of the rail; m_r represents the mass per unit length of the rail; k_r and c_r represent the fastener stiffness coefficient and damping coefficient; $F(t)$ is the wheel–rail contact force; V is the speed of the train; and δ is the Dirac function. The calculated parameters of the ballastless track model are shown in Table 2. A beam element model was used for discretization to obtain the mass, stiffness, and damping matrices.

Table 2. Calculation parameters of track structure (Lei [35]).

Parameters	Value
Mass of rail (kg/m)	60
Flexural stiffness of rail beam (kN·m^2)	6756
Fastener stiffness (kN/m)	60
Fastener damping (MN·s/m)	47.7
Density of track slab (kg/m^3)	2500
Elastic modulus of track slab (MPa)	36,000
Poisson's ratio of track slab	0.2
CA mortar stiffness (MN/m)	900
CA mortar damping (kN·s/m)	83
Density of concrete base (kg/m^3)	2500
Elastic modulus of concrete base (MPa)	34,000
Poisson's ratio of concrete base	0.2

The fastener system and the CA mortar layer mainly provide the vertical elasticity of the superstructure, which was modeled by spring-damping. The track slab and concrete base were modeled as if they were continuous elastic bodies, and the track slab and concrete base are described by Navier's elastodynamic equations, both of which are expressed in a unified expression in tensor form [36]:

$$\sigma_{ij,j} - \rho \ddot{u}_i = 0 \quad (3)$$

where σ represents the stress tensor, ρ is the density, and u is the displacement.

As shown in Figure 3, CA mortar voids were modeled by putting the stiffness of the CA mortar at 0 in this range (k = 0, c = 0), considering that the CA mortar in the void area could not provide support.

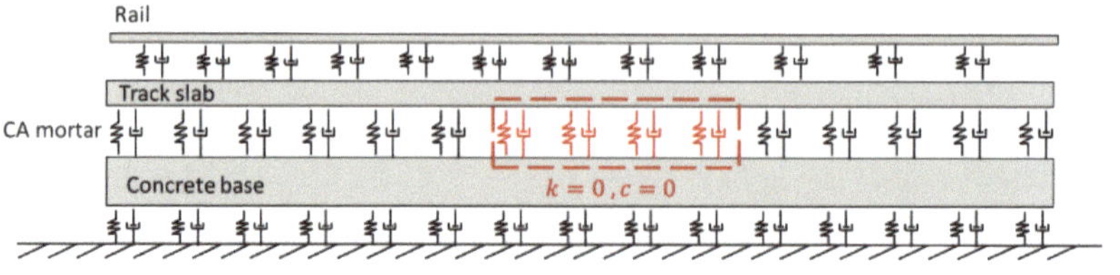

Figure 3. CA mortar modeling.

(3) Wheel–rail contact behavior

In the vertical vibration analysis, the current mainstream wheel–rail Hertzian contact theory was used, and the Hertzian contact theory was chosen because it allows wheel–rail separation and the occurrence of rail hopping. Hertz contact is established under the assumption that under the condition of small deformation, the wheel and the rail are regarded as isotropic elastic cylinders. According to the Hertz contact theory to achieve the coupling of the vehicle system and the rail system, the wheel–rail contact force can be expressed as follows [35]:

$$F_{W-R} = \begin{cases} G^{\frac{2}{3}}[|z_{Wi} - z_{Ri} - \eta|]^{\frac{3}{2}}, & z_{Wi} - z_{Ri} \leq 0 \\ 0 & , z_{Wi} - z_{Ri} > 0 \end{cases} \quad (4)$$

where z represents the displacements of wheels and rails at the coordinate i, respectively, and G is the contact deflection coefficient. η is the value of the irregularity of the rail surface at different locations on the track.

The track irregularity used the ballastless track spectrum from China's high-speed rail network and applied the triangular gradient method to generate track height variations. The resulting track height deviations, generated by the program, are shown in Figure 4 [35].

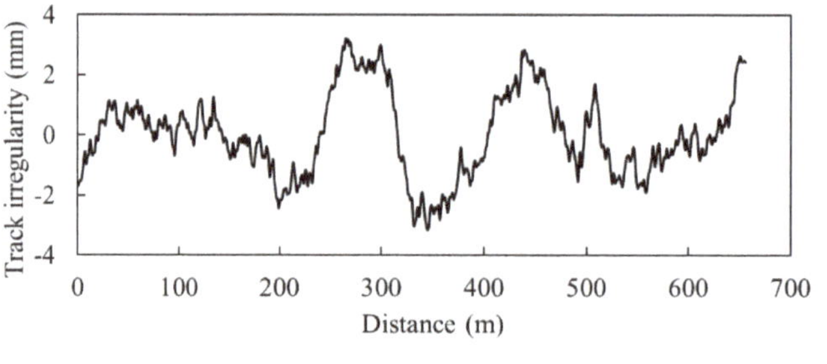

Figure 4. Distribution of track irregularities [35].

(4) Model solver

The total mass matrix, damping matrix, and stiffness matrix were constructed using the conventional finite element method. The time integration of the governing equations was performed using the implicit backward difference method. Specifically, the second-order backward difference method was employed for time integration. Further details of the backward difference method can be found in [37,38]. A time step of 0.0001 s was used for solving the model. For nonlinear analysis, the Newton–Raphson method was applied. For the ballastless track slabs and concrete bases, the mesh size was set to 0.1 m. The length of the numerical model was 150 m. The solid structure was discretized using first-order Lagrangian elements. The observation location of the dynamic response of the ballastless track structure was placed in the middle of the CA mortar void.

3. Model Validation

The simulated train speed was 72 km/h, and the rail surface was smooth. The initial conditions for displacement, velocity, and acceleration were all set to zero. A comparison of the dynamic response of high-speed trains under gravity was conducted. The model's calculation parameters were consistent with those in the literature. The vehicle used was of the CRH3 type, and the ballastless track followed the CRTS-II type. All computational parameters in the validation model were taken from the literature [39]. The vehicle–track coupling dynamics calculation procedure employed the backward difference method to solve the problem, with a time step of 0.0001 s. The physical system used for model validation is shown in Figure 5.

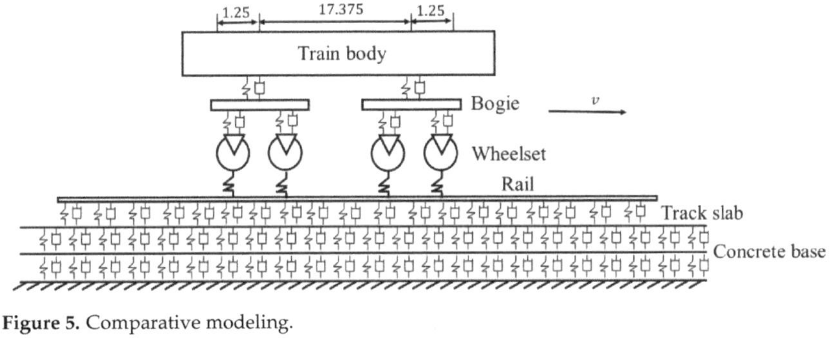

Figure 5. Comparative modeling.

The calculation procedure developed in this paper was compared with the results from the literature [39]. The comparison, shown in Figure 6, indicates that the results of the numerical method are in good agreement, demonstrating the accuracy of the proposed calculation approach.

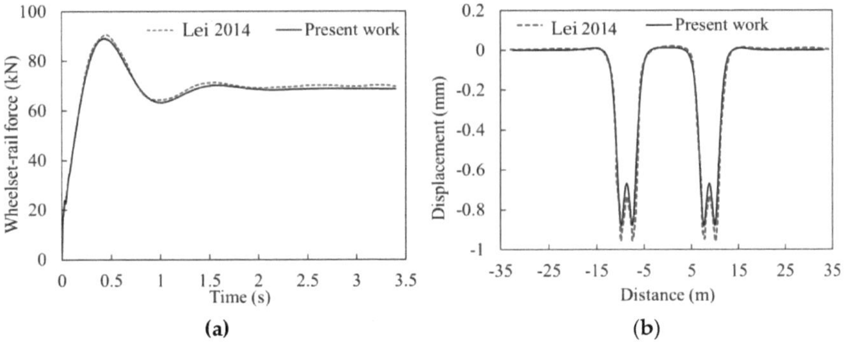

Figure 6. Comparison of calculation results: (**a**) dynamic wheelset–rail force; (**b**) displacement of rail Lei et al. [39].

In addition, in order to fully verify the accuracy of the model, different control parameters were selected and the calculated parameters were consistent with the original literature. From the comparison in Table 3, we find that the calculated results are very close to each other, indicating the accuracy of the model in this paper.

Table 3. Comparison of the results of some calculations.

Parameter	Ref. [40]	Ref. [41]	Ref. [39]	Present Work
Speed (km/h)	300	300	300	300
Rail displacement (mm)	1.3	1.34	\	1.23
Dynamic stress of subgrade surface (kPa)	14.6	14.03	\	13.81
Displacement of track slab (mm)	/	/	0.69	0.61

4. Results and Discussion

In the following analysis, the vehicle, ballastless track, and subgrade calculation parameters are all from the literature [35]. A CA mortar void is a kind of imperceptible disease; therefore, analyzing the dynamic response of the vehicle–track system and its degree of change is of great significance in exploring the identification of this kind of disease. However, in the operation of high-speed railways, the problem of the small length of voids is generally hidden, and this potential damage has not been noticed so far; therefore, the extent of its impact on the dynamic performance of high-speed trains and ballastless track systems needs to be investigated in order to better identify its characteristics. Therefore, a total of four analytical conditions were set up using this model, namely the following: (1) CA mortar non-void + random irregularities; (2) CA mortar void 100 cm + random irregularities; (3) CA mortar void 200 cm + random irregularities; and (4) CA mortar void 300 cm + random irregularities.

4.1. Dynamic Response of Rail Track

(1) Wheelset–rail system

Figure 7a shows the dynamic response behavior of the wheel–rail system of the vehicle under different line foundation conditions. From the analysis of the figure, it is clear that the value of the wheel–rail force was 73.56 kN when there were no voids in the CA mortar, which was small compared to the different lengths of voids, and the values of the wheel–rail force were 74.61 kN, 80.83 kN, and 88.92 kN for the change in the length of voids from 1 m to 3 m. The percentage of the increase in the value of the wheel–rail force compared to the absence of voids was 1.43%, 9.88%, and 20.88%.

Figure 7b shows the amplitude of the wheelset vibration acceleration as the development of the void was 1.62 m/s^2, 5.19 m/s^2, 9.81 m/s^2, and 17.54 m/s^2. With the development of the length of the void, the wheelset vibration acceleration increased by 220.37%, 505.55%, and 982.82%, respectively. The vertical vibration acceleration of the wheelset exhibited an obvious amplification law for the void in the CA mortar, and this feature was very significant for the response to void defects in the CA mortar.

Figure 8 shows the rail vibration acceleration and the time curve of rail dynamic displacement. From the perspective of rail vibration acceleration, the peak acceleration reached 60 m/s^2. When comparing the three working conditions (non-void and void), the acceleration of the rail did not exhibit a significant difference. The maximum values of rail vibration acceleration were 63.6 m/s^2, 55.1 m/s^2, 53.81 m/s^2, and 42.01 m/s^2, with corresponding percentage changes of 13.36%, 15.39%, and 33.95%.

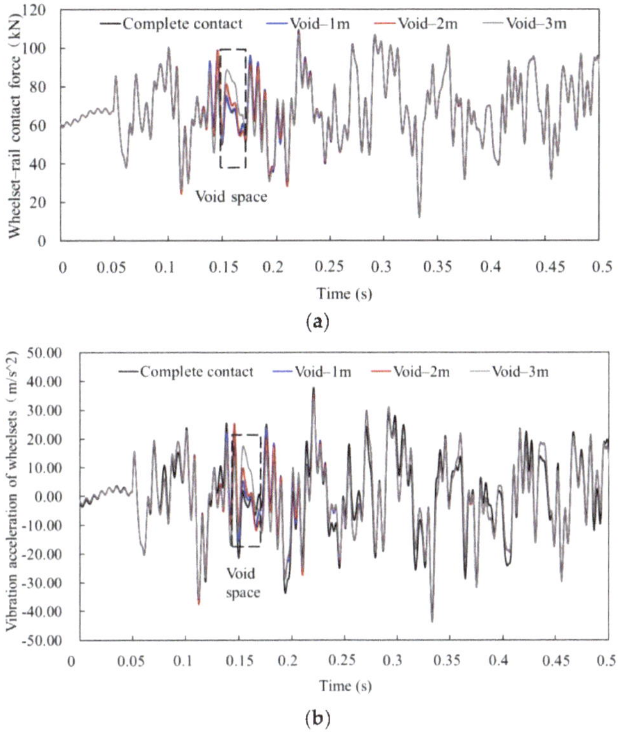

Figure 7. Dynamic response of wheelset: (**a**) wheelset-rail contact force; (**b**) vibration acceleration of wheelsets.

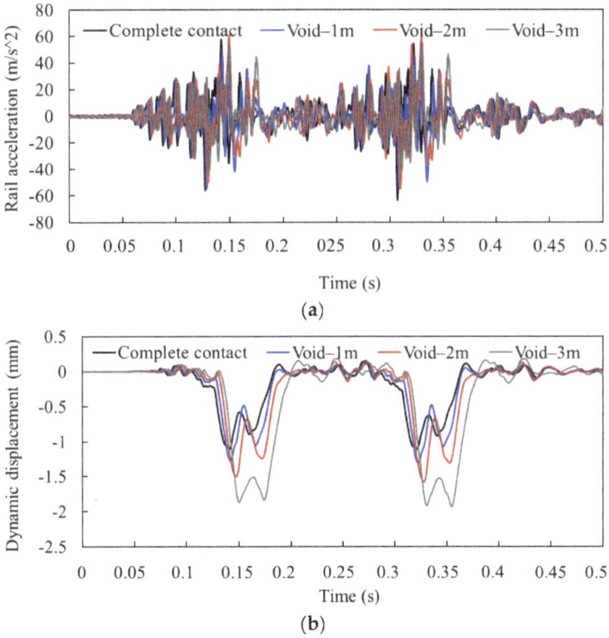

Figure 8. Dynamic response of rail: (**a**) rail vertical acceleration; (**b**) rail vertical displacement.

Regarding the rail vibration displacement, this index clearly reflected the presence of voids. The dynamic displacement of the rail was 1.11 mm in the non-void condition. However, as the void length increased to 1 m, 2 m, and 3 m, the vertical displacement of the rail increased to 1.27 mm, 1.5 mm, and 1.83 mm, respectively. The percentage increases in displacement were 14.4%, 34.14%, and 64.86%, respectively.

(2) Ballastless track system

Figure 9 illustrates the dynamic response of the track slab. First, the vibration acceleration of the track slab is analyzed. The maximum acceleration at the surface of the track slab was 11.77 m/s^2 when there was no void. As the void length increased, the vibration acceleration of the track slab reached 14.90 m/s^2, 27.43 m/s^2, and 17.80 m/s^2, respectively. This was due to the void length reaching up to 3 m. When the distance between the adjacent axles of the same bogie was 2.5 m, interference between the front and rear wheels occurred, resulting in an increase in dynamic displacement while causing acceleration to decrease.

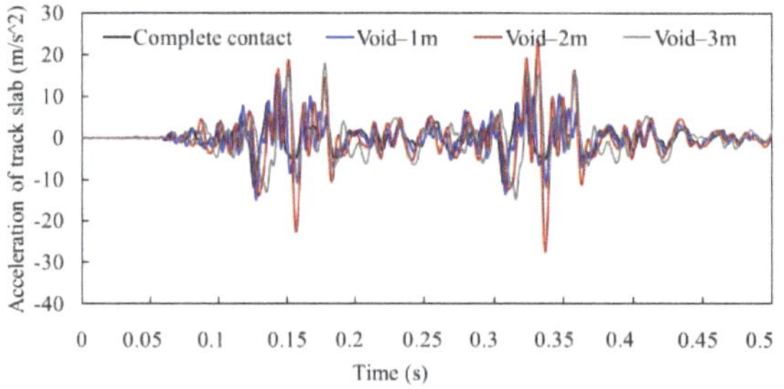

(a)

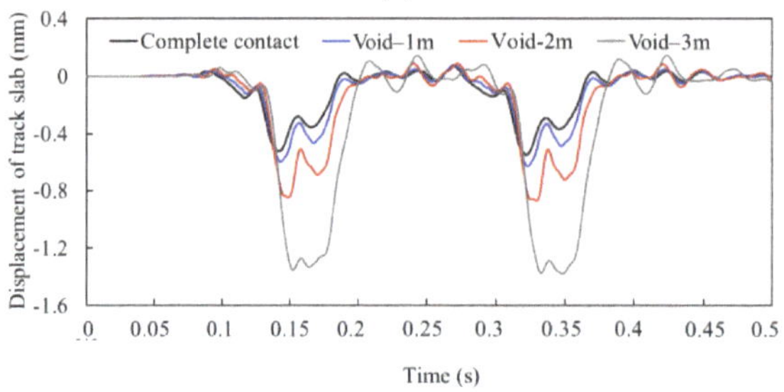

(b)

Figure 9. Dynamic response of track slab: (**a**) track slab vertical acceleration; (**b**) track slab vertical displacement.

Regarding the dynamic displacement of the track slab, Figure 9 clearly shows that the change in void length significantly impacted the deformation of the track slab. As the void length in the CA mortar increased, the bottom face-to-face support became localized. With an increasing void span, the displacement of the track slab increased substantially. The dynamic settlement displacement of the track slab was 0.52 mm, 0.60 mm, 0.84 mm, and 1.33 mm, corresponding to increases of 15.38%, 61.54%, and 155.77%, respectively. The dynamic displacement of the track slab strongly reflected the deformation response due to the CA mortar void. Therefore, the dynamic displacement of the track slab could be recorded by the dynamic acquisition device of the high-speed train, which could be used to better identify the CA mortar layer void and assess its damage range.

Figure 10 presents the dynamic response of the concrete base. The vibration acceleration on the surface of the concrete base was 9.77 m/s^2 when there was complete contact. As the void length increased from 1 m to 3 m, the acceleration decreased to 7.45 m/s^2,

7.08 m/s^2, and 6.46 m/s^2, respectively. The reduction percentages were 23.75%, 27.53%, and 33.88%. This decrease in peak acceleration was attributed to the reduced effective energy transfer downward due to the expansion of the void area. Regarding the displacement of the concrete base, the dynamic displacement was 0.48 mm, 0.46 mm, 0.38 mm, and 0.33 mm for the condition of no void and void lengths of 1 m, 2 m, and 3 m, respectively. The reduction percentages were 4.17%, 20.83%, and 31.25%.

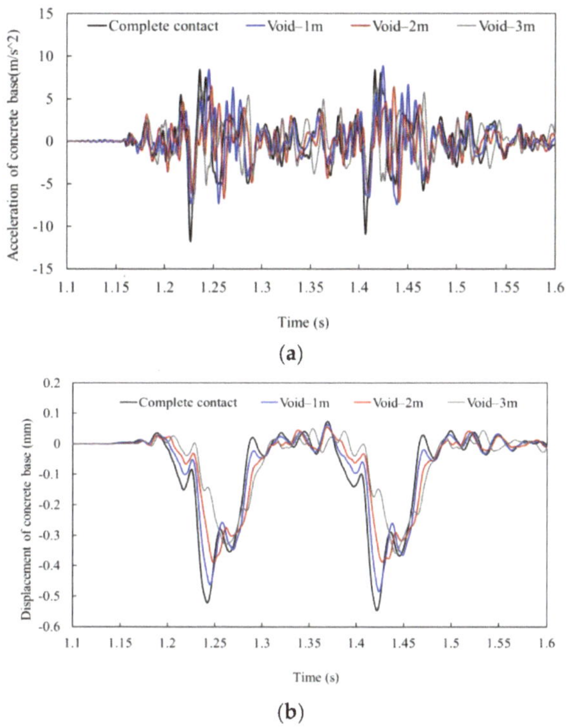

Figure 10. Dynamic response of concrete base: (**a**) concrete base's vertical acceleration; (**b**) concrete base's vertical displacement.

4.2. *Transfer Function of Ballastless Track*

The effect of the CA mortar void damage on the system's dynamic response was influenced by several factors. The vibration transfer function was based on the frequency response characteristics between the system's substructures, which describe the intrinsic properties of vibration transfer between different components within the system in the frequency domain. This understanding is crucial for analyzing the role of various parameters within the system, which can be derived directly from the input–output frequency response relationships of the computational mode.

$$H(\omega) = \frac{\frac{1}{2\pi}\int_{-\infty}^{+\infty} Y(t)e^{-j2\pi\omega t}dt}{\frac{1}{2\pi}\int_{-\infty}^{+\infty} X(t)e^{-j2\pi\omega t}dt} = \frac{Y(\omega)}{X(\omega)} \tag{5}$$

where H(w) is the vibration transfer function; Y(t) and X(t) are the input and output time-domain signals of the system; and, correspondingly, Y(w) and X(w) are the input and output Fourier transform frequency-domain signals of the system. Therefore, in this paper, the transfer function calculation program was implemented using Python 3.7, and the transfer function was directly output through the results of the vehicle–rail coupling calculation model.

Figure 11 shows the amplitude–frequency response curves of the transfer function between the vibration displacements of the track slab and the concrete base. From the transfer function characterization model, the values of the transfer function were generally in the range of 0.87–1 under the condition of no void, indicating that the vibration distribution of the ballasted track system under optimal contact conditions was excellent. As the length of the void increased, the transfer function values varied within the following intervals: 0.43–0.98, 0.27–0.74, 0.08–0.32, and 0.08–0.32, respectively.

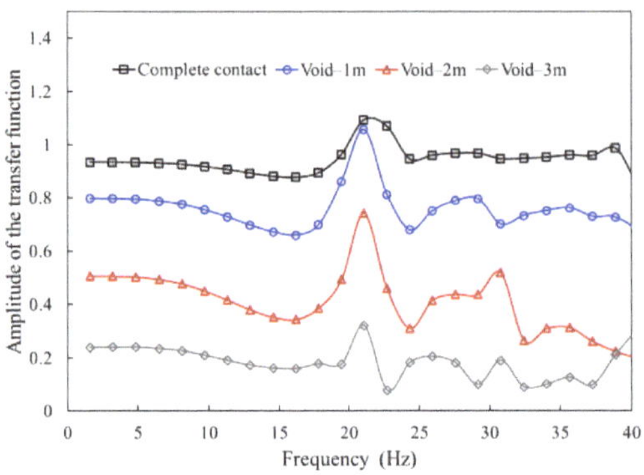

Figure 11. Transfer function of displacement between track slab and concrete base.

Figure 12 shows the transfer function of acceleration between the track slab and the concrete base. The transfer function ranged from 0.81 to 1 when there was no void. When the void was 1 m, the transfer function ranged from 0.66 to 0.98. For a void of 2 m, the transfer function ranged from 0.34 to 0.75, and for a void of 3 m, it ranged from 0.07 to 0.30. Overall, the degradation of the transfer function due to the removal of CA mortar was significant and clearly observed across all frequency-domain stages.

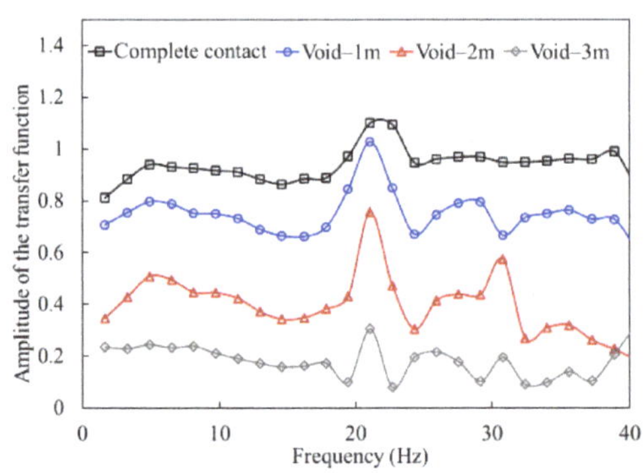

Figure 12. Transfer function of acceleration between track slab and concrete base.

4.3. Dynamic Response of Vehicle

Figure 13a shows the vertical vibration acceleration of the bogie. The acceleration exhibited a significant increasing trend at the same position. When the deflection grew, the acceleration increased by 1.592 m/s^2, 1.597 m/s^2, 2.104 m/s^2, and 3.281 m/s^2, respectively. The vertical acceleration of the bogie increased by 106.09% when the deflection length reached 3 m.

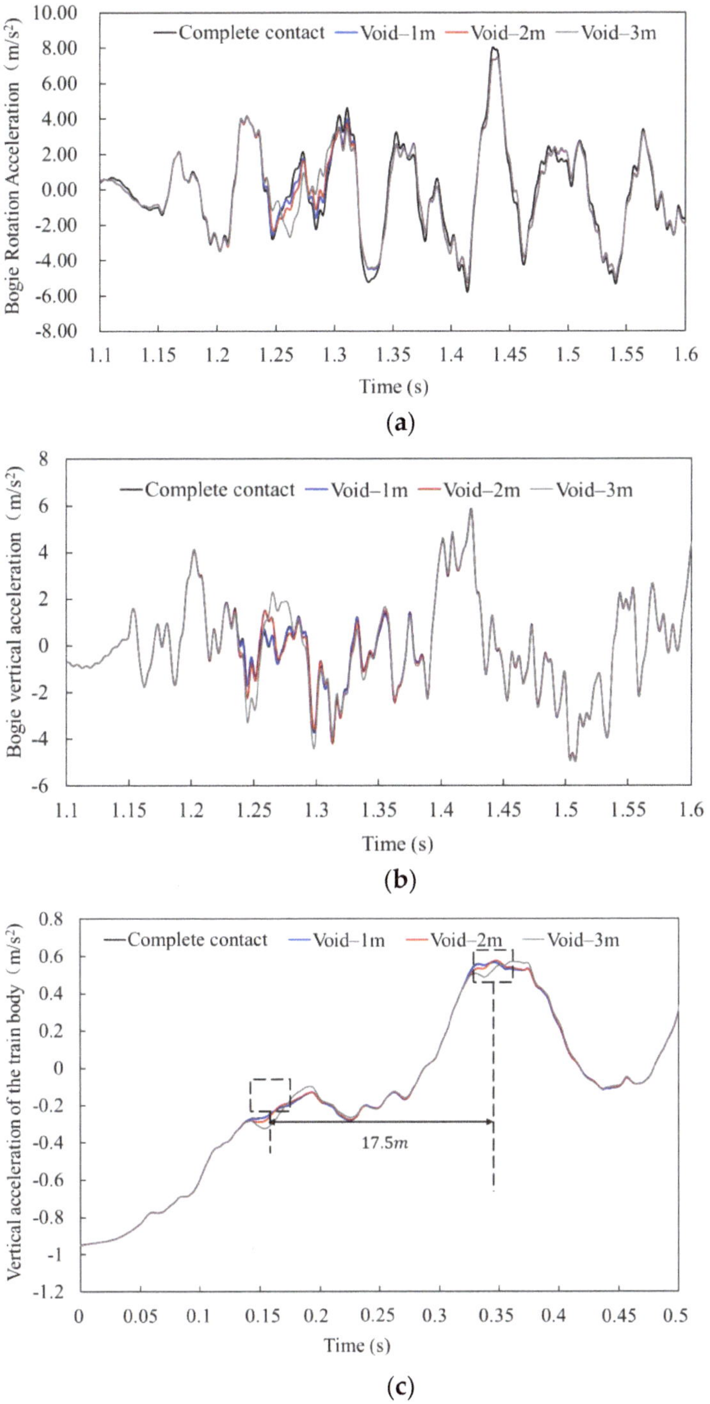

Figure 13. Dynamic response of train body: (**a**) bogie rotation acceleration; (**b**) bogie vertical acceleration; (**c**) vertical acceleration of train body.

Figure 13b illustrates the nodding acceleration of the bogie. The analysis indicates that the nodding acceleration exhibited more pronounced abnormal characteristics when the void length was 3 m. Conversely, the nodding acceleration showed some memory characteristics in the normal section. As the bogie passed through the void region, the nodding acceleration behaved similarly to that observed in the normal section without a void. However, in the void area, the nodding acceleration exhibited a different response amplitude compared to the normal section, which aided in identifying the CA mortar damage.

Figure 13c presents the body's vertical acceleration. The time-domain characteristics of the body's vertical acceleration reveal clear periodic behavior. When the first wheelset of the body passed through the de-embedded area, it triggered an increase in the body's vibration acceleration. As different bogies of the same body passed through the same area, the body experienced vertical swaying at the same time. The periodic swaying occurred over a distance of 17.5 m, which corresponded exactly to the distance between the front and rear bogie centers. This characteristic is useful for high-speed dynamic inspection vehicles to identify CA mortar damage, although it may cause a bumpy experience for passenger vehicle operation.

Figure 14 presents an analysis of the damping parameters for the primary and secondary suspensions of the vehicle. Although identifying damage to the substructure beneath the rails through the dynamic response of inspection vehicles is desirable, locomotive vehicles can adaptively enhance stiffness compensation during operation to meet ride comfort requirements. From the perspective of general passenger vehicles, computational analysis showed that increasing the damping parameters of the primary suspension system did not significantly reduce the vertical acceleration of the vehicle body. However, a 10% increase in the damping of the secondary suspension significantly decreased vertical acceleration. When the secondary suspension was increased by 20%, the reduction in acceleration compensated for the stiffness irregularities of the substructure. Therefore, designing an adaptive suspension system that automatically senses changes in substructure stiffness and makes real-time compensations holds promise for improving the operational comfort of passenger vehicles.

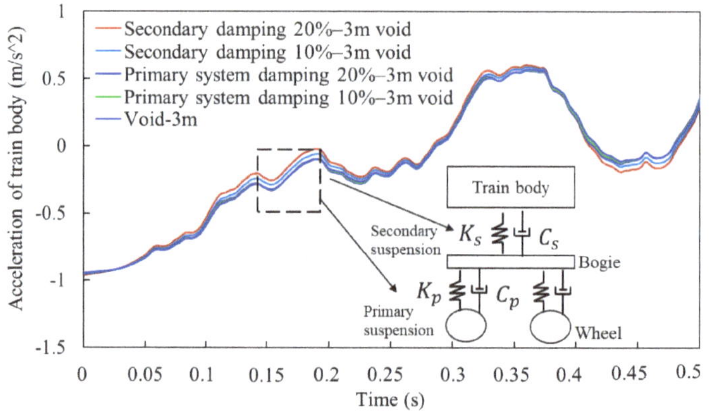

Figure 14. Parameter analysis of vehicle suspension system.

By summarizing the vibration indicators of each subsystem in Table 4, statistical analysis revealed that the wheelset acceleration was most sensitive to voids in the CA mortar layer, followed by the dynamic displacement of the track slab. Additionally, the vertical acceleration of the high-speed train bogie showed significant variations. These indicators were useful for detecting hidden voids in the CA mortar layer.

Table 4. Statistical analysis of system dynamic response.

Index	Complete Contact (0 m)	Void—1 m	Void—2 m	Void—3 m	Statistically Significant (0 m → 3 m)
Wheel–rail force (kN)	73.56	74.61	80.83	88.92	20.88%
Acceleration of wheelset (m/s^2)	1.62	5.19	9.81	17.54	982.71%
Rail displacement (mm)	1.11	1.27	1.50	1.83	64.86%
Rail acceleration (m/s^2)	63.6	55.10	53.81	42.01	33.94%
Track slab displacement (mm)	0.52	0.60	0.84	1.33	155.76%
Track slab acceleration (m/s^2)	11.77	14.90	27.43	17.80	51.23%
Concrete base displacement (mm)	0.48	0.46	0.38	0.33	31.25%
Concrete base acceleration (m/s^2)	9.77	7.45	7.08	6.46	33.87%
Bogie rotation acceleration (m/s^2)	2.75	2.54	2.31	1.12	59.27%
Bogie vertical acceleration (m/s^2)	1.592	1.597	2.104	3.281	106.09%
Train body vertical acceleration (m/s^2)	0.26	0.27	0.29	0.32	23.07%

5. Conclusions

This paper establishes a vehicle–ballastless track dynamic coupling model and uses an FEM solution and compares the accuracy of the calculation results. Different void lengths in CA mortar are compared and the effects of the railway track's dynamic response and transfer function between the ballastless track layers are analyzed. The main conclusions are as follows:

(1) The vibration acceleration of the high-speed train wheelsets was sensitive to the change in the CA mortar void length. It could increase from 1.62 m/s^2 under complete contact to 17.54 m/s^2. The displacement of the track slab in the ballastless track structure was more sensitive to the void length.

(2) In the sensitivity ordering of the identification indexes of the voids, the bogie's vertical acceleration in the high-speed train > the nodding acceleration of the bogie > the vehicle's vertical acceleration. When the length of the void was 3 m, the bogie's vertical acceleration increased by 106.09%.

(3) Under the complete-contact condition, the value of the dynamic displacement transfer function between the track slab and concrete base was basically between 0.87 and 1. When the void length increased, the transfer function varies in the intervals of (0.43–0.98), (0.27–0.74), and (0.08–0.32), respectively. The values of the acceleration transfer function between the track slab and the concrete base varied between 0.81 and 1, and when the length of the void increased, it varied between 0.66 and 0.98, 0.34 and 0.75, and 0.07 and 0.30.

Although this paper numerically analyzes the CA mortar layer voids in ballastless track structures, there are still some limitations. At present, there are many ballastless track structure types in the world, among which CRTS-I, CRTS-II, and CRTS-III exist in China, and this paper only analyzes the CRTS-II type. In addition, because in real engineering high-speed trains and ballastless track structures are three-dimensional and the geometry of ballastless track structures is also very complicated, in future research, the three-dimensional characteristics of ballastless track structures can be further considered, and the damage model of concrete can be embedded in the numerical model so as to analyze it at a deeper level.

Author Contributions: X.C. designed the research. Y.P. processed the corresponding data. X.C. wrote the first draft of the manuscript. K.L. helped to organize the manuscript. Y.P. and K.L. revised and edited the final version. All authors have read and agreed to the published version of the manuscript.

Funding: This work was supported by the Foundation of Sichuan Provincial Engineering Research Center of Rail Transit Lines Smart Operation and Maintenance, Chengdu Vocational and Technical

College of Industry (2024GD-Y15). The research was supported by the National Natural Science Foundation of China (Grant No. U24A20173 and 51978588) and Joint Fund for Basic Railway Research (Grant No. U2268213).

Data Availability Statement: The original contributions presented in this study are included in the article. Further inquiries can be directed to the corresponding author.

Acknowledgments: The support of the National Natural Science Foundation of China is gratefully acknowledged.

Conflicts of Interest: No conflicts of interest.

References

1. Qian, Y.; Liu, C.; Yuan, Y.; Xu, J.; Wang, P.; Wang, K. Numerical characterization and formation process study of rail light bands in high-speed turnout areas. *Eng. Fail. Anal.* **2024**, *168*, 109083. [CrossRef]
2. Kong, G.; Sun, G.; Liu, H.; Li, J. Dynamic response of ballastless track XCC pile-raft foundation under train axle loads. *J. Test. Eval.* **2021**, *49*, 1691–1704. [CrossRef]
3. Matias, S.R.; Ferreira, P.A. Railway slab track systems: Review and research potentials. *Struct. Infrastruct. Eng.* **2020**, *16*, 1635–1653. [CrossRef]
4. Jadidi, K.; Esmaeili, M.; Kalantari, M.; Khalili, M.; Karakouzian, M. A review of different aspects of applying asphalt and bituminous mixes under a railway track. *Materials* **2020**, *14*, 169. [CrossRef] [PubMed]
5. Inaba, K.; Takahashi, T.; Fuchigami, S.; Momoya, Y.; Naito, H. Study on detecting voids within a concrete slab track of shinkansen with non-destructive testing by impact acoustics. *J. Jpn. Soc. Civ. Eng. Ser. E1 (Pavement Eng.)* **2022**, *78*, 1–11. [CrossRef] [PubMed]
6. Inaba, K.; Tanigawa, H.; Naito, H. Improvement of impact acoustic inspection for high-speed railway track slabs using time–frequency analysis and non-defective machine learning. *NDT E Int.* **2024**, *145*, 103125. [CrossRef]
7. Auersch, L.; Said, S. Track-soil dynamics—Calculation and measurement of damaged and repaired slab tracks. *Transp. Geotech.* **2017**, *12*, 1–14. [CrossRef]
8. Dong, W.; Wang, X.; He, Q.; Ren, J. A survey of Ballastless track defects in China's high-speed railway after ten years of service. *Intell. Transp. Infrastruct.* **2022**, *1*, liac023. [CrossRef]
9. Xie, Y.-j.; Fu, Q.; Zheng, K.-r.; Yuan, Q.; Song, H. Dynamic mechanical properties of cement and asphalt mortar based on SHPB test. *Constr. Build. Mater.* **2014**, *70*, 217–225. [CrossRef]
10. Cui, X.; Du, B.; Xiao, H.; Zhou, R.; Guo, G.; Liu, H. Interface damage and arching mechanism of CRTS II slab track under temperature load. *Constr. Build. Mater.* **2021**, *291*, 123258. [CrossRef]
11. Feng, Y.; Li, J.; He, B.; Zhou, W.; Jiang, L.; Wu, B. Experimental Study on Seismic Performance of CRTS II Slab-Type Ballastless Track Lateral Block of HSRs. *J. Earthq. Eng.* **2024**, *28*, 4003–4024. [CrossRef]
12. Guo, G.; Wang, J.; Du, B.; Du, Y. Application study on fiber optic monitoring and identification of CRTS-II-Slab ballastless track debonding on viaduct. *Appl. Sci.* **2021**, *11*, 6239. [CrossRef]
13. Guo, W.; Hu, Y.; Gou, H.; Du, Q.; Fang, W.; Jiang, L.; Yu, Z. Simplified seismic model of CRTS II ballastless track structure on high-speed railway bridges in China. *Eng. Struct.* **2020**, *211*, 110453. [CrossRef]
14. Liu, K.; Shao, K.; El Naggar, M.H.; Su, Q.; Wang, T.; Qiu, R. Dynamic Performance of a Railway Subgrade Reinforced by Battered Grouted Helical Piles. *J. Geotech. Geoenviron. Eng.* **2024**, *150*, 05024009. [CrossRef]
15. Li, Y.; Li, H.; Zhou, R.; Chen, J.; Song, Y.; Cao, K. Mechanical performance of CRTS II slab tracks in reinforced-unreinforced transition zone in extreme heat events. *Transp. Geotech.* **2024**, *45*, 101201. [CrossRef]
16. Song, L.; Liu, H.; Cui, C.; Yu, Z.; Li, Z. Thermal deformation and interfacial separation of a CRTS II slab ballastless track multilayer structure used in high-speed railways based on meteorological data. *Constr. Build. Mater.* **2020**, *237*, 117528. [CrossRef]
17. Zeng, Z.; Ye, M.; Wang, W.; Liu, J.; Shen, S.; Qahtan, A.A.S. Analysis on mechanical characteristics of CRTSII slab ballastless track structures in rectification considering material brittleness. *Constr. Build. Mater.* **2022**, *319*, 126058. [CrossRef]
18. Zhao, L.; Zhou, L.; Yu, Z.; Mahunon, A.D.; Peng, X.; Zhang, Y. Experimental study on CRTS II ballastless track-bridge structural system mechanical fatigue performance. *Eng. Struct.* **2021**, *244*, 112784. [CrossRef]
19. Zhou, L.-Y.; Zhao, L.; Mahunon, A.D.; Zhang, Y.-Y.; Li, H.-Y.; Zou, L.-F.; Yuan, Y.-H. Experimental study on stiffness degradation of Crts II ballastless track-bridge structural system under fatigue train load. *Constr. Build. Mater.* **2021**, *283*, 122794. [CrossRef]
20. Xin, X.; ZunSong, R.; Xue, W. Effects of Mortar Void of CRTS I Slab Track on the DynamicPerformances of Vehicle-track System. *Chin. J. Mech. Eng.* **2022**, *58*, 177–183+193. [CrossRef]
21. *TB 10621-2014*; Code for Design of High Speed. National Railway Administration of Science and Technology and Legal System: Beijing, China, 2015.

22. Yang, Y.; Zhang, G.; Wu, G.; Cao, D. Study on fatigue damage laws and life prediction of CRTS-II ballastless track slab. *Eng. Struct.* **2022**, *252*, 113659. [CrossRef]
23. Chen, R.; Yang, K.; Qiu, X.; Zeng, X.; Wang, P.; Xu, J.; Chen, J. Degradation mechanism of CA mortar in CRTS I slab ballastless railway track in the Southwest acid rain region of China–Materials analysis. *Constr. Build. Mater.* **2017**, *149*, 921–933. [CrossRef]
24. Wang, Y.; Zhao, L.; Cai, X.; Liu, Y.; Wang, T. Dynamic damage evolution of double-block ballastless track structure under train derailment impact. *Eng. Fail. Anal.* **2024**, *162*, 108347. [CrossRef]
25. Qingsong, F.; Sihao, Z.; Kui, S.; Wei, W. Analysis of the effects of CA mortar gap on the CRTS I slab ballastless track on bridges. *J. Railw. Sci. Eng.* **2020**, *17*, 1061–1069. [CrossRef]
26. Xin, X.; Zhou, S.; Ren, Z.; Yang, C. Influence and sensitivity analysis of mortar void in CRTS II ballastless slab track on the vehicle-track dynamics. *J. Mech. Sci. Technol.* **2024**, *38*, 557–568. [CrossRef]
27. Huang, Y.; Gao, L.; Zhong, Y.; Zhou, C. Study on the damage evolution of the joint and the arching deformation of CRTS-II ballastless slab track under complex temperature loading. *Constr. Build. Mater.* **2021**, *309*, 125083. [CrossRef]
28. Deng, S.; Ren, J.; Wei, K.; Ye, W.; Du, W.; Zhang, K. Fatigue damage evolution analysis of the CA mortar of ballastless tracks via damage mechanics-finite element full-couple method. *Constr. Build. Mater.* **2021**, *295*, 123679. [CrossRef]
29. Cui, X.; Du, X.; Du, B.; Zhao, M.; Xiong, C.; Xu, F.; Xu, H. Effect of Tunnel Floor Heave on the Deformation and Damage Behavior of Ballastless Track Structures in High-Speed Railways. *Transp. Geotech.* **2024**, *49*, 101375. [CrossRef]
30. Hu, M.; Xu, Y.; Xue, Z.; Wang, S. An inversion method for evaluating ballastless track degradation based on multi-channel analysis of surface wave. *Mech. Syst. Sig. Process.* **2023**, *200*, 110572. [CrossRef]
31. Kordestani, H.; Zhang, C.; Arab, A. An Investigation into the Application of Acceleration Responses' Trendline for Bridge Damage Detection Using Quadratic Regression. *Sensors* **2024**, *24*, 410. [CrossRef]
32. Feng, Q.; Chao, H.; Lei, X. Influence of the seam between slab and CA mortar of CRTSII ballastless track on vibration characteristics of vehicle-track system. *Procedia Eng.* **2017**, *199*, 2543–2548. [CrossRef]
33. Liu, K.; Pei, Y.; Zhou, P.; Liu, B.; Chen, Y.; Wang, T. Numerical Methods of Hydro-Dynamic Coupling in Saturated Soil and Its Application to Railway Engineering. *Int. J. Multiscale Comput. Eng.* **2024**, *22*, 27–47. [CrossRef]
34. Liu, K.; Su, Q.; Yue, F.; Liu, B.; Qiu, R.; Liu, T. Effects of suffosion-induced contact variation on dynamic responses of saturated roadbed considering hydro-mechanical coupling under high-speed train loading. *Comput. Geotech.* **2019**, *113*, 103095. [CrossRef]
35. Lei, X. *High Speed Railway Track Dynamics*; Springer: Berlin/Heidelberg, Germany, 2017. [CrossRef]
36. Pei, Y.; Su, Q.; Liu, K.; Liu, B.; Wang, T.; Zhao, C. Numerical investigation into hydrodynamic behavior of graded aggregates of high-speed ballastless track considering the degree of saturation. *Transp. Geotech.* **2023**, *42*, 101055. [CrossRef]
37. Wanner, G.; Hairer, E. *Solving Ordinary Differential Equations II*; Springer: Berlin/Heidelberg, Germany; New York, NY, USA, 1996; Volume 375. [CrossRef]
38. Cash, J. Second derivative extended backward differentiation formulas for the numerical integration of stiff systems. *SIAM J. Numer. Anal.* **1981**, *18*, 21–36. [CrossRef]
39. Lei, X.; Wang, J. Dynamic analysis of the train and slab track coupling system with finite elements in a moving frame of reference. *J. Vib. Control* **2014**, *20*, 1301–1317. [CrossRef]
40. Chen, M.; Sun, Y.; Zhu, S.; Zhai, W. Dynamic performance comparison of different types of ballastless tracks using vehicle-track-subgrade coupled dynamics model. *Eng. Struct.* **2021**, *249*, 113390. [CrossRef]
41. Xie, H.; Luo, Q.; Wang, T.; Jiang, L.; Connolly, D.P. Stochastic analysis of dynamic stress amplification factors for slab track foundations. *Int. J. Rail Transp.* **2024**, *12*, 281–303. [CrossRef]

buildings

MDPI

Article

Debonding Analysis of FRP-Strengthened Concrete Beam in High-Temperature Environment: An Enhanced Understanding on Sustainable Structure

Xiaoning Zhang [1], Jianwen Hao [2,*], Wei Hou [2], Jiancheng Yao [2], Yazhuo Wang [2], Xiaojian Su [2] and Xiangyang Li [3,*]

1 School of Civil Engineering, Chongqing University, Chongqing 400045, China; sdjtuzxn@163.com
2 Shandong High-Speed Infrastructure Construction Co., Ltd., Jinan 250061, China
3 Department of Global Smart City, Sungkyunkwan University, Suwon 16419, Republic of Korea
* Correspondence: hjwww1717@163.com (J.H.); lixy@g.skku.edu (X.L.)

Abstract: FRP (fiber-reinforced composite) is generally regarded as the repair and enhancement material for existing concrete structures in extreme service environments such as high temperatures or fire exposure. In order to reveal the effect of high temperatures (i.e., thermal load) on the interfacial debonding behavior of a FRP-strengthened concrete beam, the novel closed-form analytical model was established and validated while considering the interfacial bond-slip constitutive. Based on the analytical model, solutions to the distributions of interfacial slip, interfacial shear stress, and debonding load were derived. Moreover, the effects of temperature variations and the FRP's bonded thickness and length on interfacial bond behavior were also evaluated. The results indicated that the increase in temperature variations accelerated the development trends of interfacial slip and shear stress, where the affected range was mainly concentrated in the bonded plate end. The relationship between temperature variations and debonding loads presented a changing linear trend, and a prediction model for the debonding load was also proposed. Meanwhile, the increase in the FRP's bonded thickness decreased the bond performance and accelerated the degradation trend of the debonding load. However, the increase in FRP's bonded length improved the bearing capacity of the FRP-strengthened concrete beam. This paper provides meaningful guidelines for the sustainable design and construction of FRP-strengthened concrete structures in high-temperature environments.

Keywords: FRP; debonding analysis; prediction model; concrete beam; high temperature

1. Introduction

The application of externally bonded (EB) fiber-reinforced polymer (FRP) is widely used for strengthening or retrofitting existing concrete structures, particularly concrete beams [1–4]. For strengthened structures, the interfacial bond behavior is a critical factor affecting the strengthening effect induced by bonded FRP plates [5]. However, due to the development of stress concentration on the end of the bonded interface, the end debonding is a critical and common failure mode for the EB–FRP system [6–11]. Nevertheless, during the service period, FRP-strengthened concrete beams usually face significant temperature variations [12–15], which means that the FRP-strengthened concrete beam is subject to the coupled effect induced by mechanical and thermal loads [16–18]. Under this coupled effect, the interfacial bonded performance would be degraded, which would threaten the safety of the strengthened beam [18–23]. Therefore, a better understanding of the debonding behavior of FRP-strengthened concrete beams subjected to the coupled effect of mechanical and thermal loads can provide strength design engineers with some meaningful reminders to prevent the bonded interface failure of FRP-strengthened concrete beams in their service period.

In existing research, numerous experimental investigations were carried out to explore the interfacial bond behavior of FRP-to-substrate bonded joints combining mechanical and thermal loads, including pull-out tests, pull-off tests, and fire tests [15–17,19,21,24–32]. It has been found that the effect of temperature variations (i.e., thermal load) on strengthened structures is reflected and concluded in the following two perspectives: (1) thermal stress induced by temperature variations exists at the bonded interface due to the different thermal expansion coefficients of FRP and concrete materials; (2) the degradation of the bonded interface is due to the degradation of mechanical properties of the adhesive due to the elevated temperature, such as a service temperature reaching the glass transition temperature of adhesive. In order to predict and further analyze the bond strength or debonding behavior of FRP-strengthened structures combined with mechanical and thermal loads, numerous analytical models were developed. Based on the existing experimental results for a FRP-to-concrete bonded interface, Dai et al. [33] proposed the exponential bond-slip model while considering temperature variations, which were determined by two parameters: the brittleness index B and fracture energy G_f. Then, Dong et al. established a prediction model on bond strength [34] and a nonlinear analytical model [35] on thermal stress for FRP-to-concrete bonded joints at high temperatures by adopting the above exponential bond-slip model. Moreover, the bi-linear bond-slip model is simply expressed and solved, which is applied widely to obtain the closed-form solutions to the full-range bond behavior of the FRP-to-substrate interface [36–39]. Subsequently, Gao et al. [34] first established an analytical model of full-range debonding behavior of FRP-to-concrete bonded joints combined with a pull-out load and thermal load; then, the solutions showed how the temperature variations affected the distributions of interfacial shear stress and slip and load-slip curves. Biscaia et al. [25] developed the determination of a temperature-dependent bi-linear model considering the glass transition temperature of adhesives. The closed-form solution to complete the debonding progress of FRP-to-substrate was derived subjected to the only thermal load [40]. Moreover, Guo et al. [31] established a FEM numerical model simulating the pull-out tests of FRP-to-steel bonded joints at elevated temperatures to explore the effect of temperature variations on the interfacial shear stress, debonding load, and CFRP strain, where the interfacial relationship was presented by the bi-linear bond-slip model.

The mentioned research proved that there is lacking analyses on the debonding behavior of FRP-strengthened concrete beams combined with mechanical and thermal loads from a theoretical perspective, where the research almost concentrated on the mechanical behavior of bonded joints. Of course, several analytical models were developed to describe the end debonding and intermediate debonding behaviors of FRP-strengthened concrete beams. In contrast, these models ignored the coupled effect of mechanical and thermal loads [7,8,41–45].

Therefore, this paper presents a novel closed-form analytical model to explore the effect of temperature variations on the bond behavior of a simple FRP-strengthened supported concrete beam, which employed the bi-linear model to reflect the bond-slip relationship between the FRP plate and concrete substrate. Based on the analytical model validated by the FEM approach, solutions to the distributions of interfacial shear stress and slip during complete debonding progress while the debonding loads are obtained. Moreover, the evolutions of bond behavior under different temperature variations and FRP bonded sizes are studied, which supplies useful information on the application of FRP-strengthened structures' combined mechanical and thermal loads in their service period.

2. Closed-Form Analytical Model

2.1. Basic Assumption

The closed-form analytical model takes a three-point bending concrete beam strengthened with FRP as the research object, and the FRP-strengthened concrete beam is also subjected to a uniform temperature field, as shown in Figure 1. Under the load condition of three-point bending and temperature variations, the mechanical response of the FRP-strengthened concrete beam is symmetrically distributed along the mid-span position (midpoint of the bond length), where the analytical model is carried out on the semi-symmetric structure of the reinforced concrete beam. Meanwhile, Firmo et al. [18] proved that the debonding failures initiated at the FRP's bonded end when the FRP-strengthened concrete beam was subjected to single-point load and temperature variations without end anchorage measures. Therefore, the debonding mode of the subsequent analytical solution is end debonding. Furthermore, the following main hypotheses are assumed:

The FRP and concrete are considered ideal elastic materials, and the elastic modulus of these materials remains constant when subjected to temperature variations.

The creep deformation of all the materials is assumed to not happen during the temperature variations.

The bond interface only undertakes the interfacial shear stress, and the normal stress is ignored, which means the interfacial failure mode is mode II.

FRP plates do not bear the shear force, and the flexural stiffness of FRP plates can be ignored compared to the flexural stiffness of concrete beams.

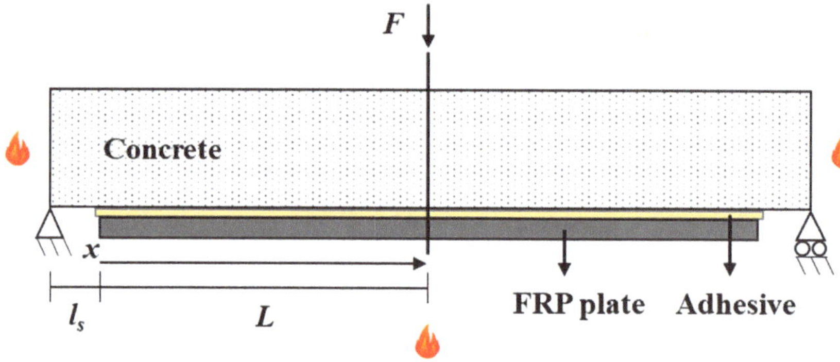

Figure 1. Schematic illustration of FRP-strengthened concrete beam under mechanical load and temperature variations.

2.2. Equilibrium Equations

According to the assumption mentioned above, it can be concluded that the microelement state of the FRP-strengthened concrete beam is as shown in Figure 2, and the equilibrium equations are established, as shown in Equations (1)–(3).

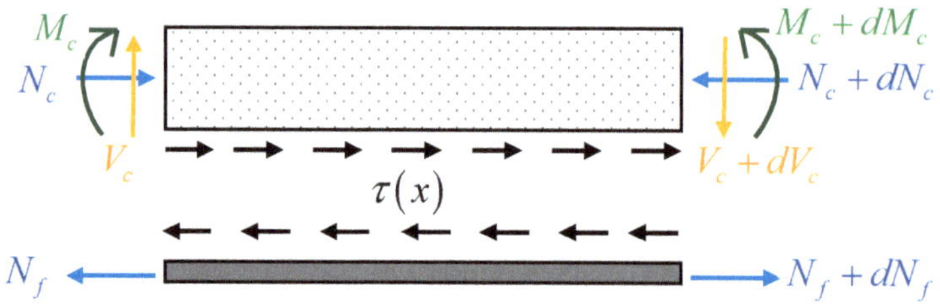

Figure 2. Mechanical analysis of the microelement state of the strengthened beam.

$$\frac{dM_c(x)}{dx} = V_c(x) - \frac{1}{2}h_c\tau(x)b_f = \frac{F}{2} - \frac{1}{2}h_c\tau(x)b_f \quad (1)$$

$$\frac{dN_c(x)}{dx} = \tau(x)b_f \quad (2)$$

$$\frac{dN_f(x)}{dx} = \tau(x)b_f \quad (3)$$

where in Figure 2, the N_f is the axial force of the FRP plate, the N_c, V_c, and M_c are the axial force, shear force, and bending moment of the concrete beam, and τ is the interfacial shear stress, respectively.

Meanwhile, the constitutive equations of all the materials subjected to a certain temperature variation are described as follows:

$$\varepsilon_c(x) = \frac{du_c(x)}{dx} - \alpha_c \Delta T = \frac{6M_c(x)}{E_c b_c {h_c}^2} - \frac{N_c(x)}{E_c b_c h_c} \tag{4}$$

$$\varepsilon_f(x) = \frac{du_f(x)}{dx} - \alpha_f \Delta T = \frac{N_f(x)}{E_f A_f} \tag{5}$$

where ε_c and ε_f are the values of the axial strain at the bottom of the concrete beam and the upper section of the FRP plate, respectively; E_c and E_f are the values of the elastic modulus of concrete and FRP plate, respectively; b_c and b_f are the values of the width of concrete and FRP plate, respectively; h_c and h_f are the values of the thickness of the concrete and FRP plate, respectively; α_c and α_f are the linear thermal expansion coefficients of the concrete and FRP plate, respectively; and ΔT is the temperature variation.

The interfacial slip is defined as the relative displacement between the FRP plate and concrete beam, which is noted as s and described by Equation (6). Moreover, the interfacial shear stress is expressed as a function of the interfacial slip, which is also named the bond-slip model, as shown in Equation (7).

$$s = u_f - u_c \tag{6}$$

$$\tau = f(s) \tag{7}$$

where u_c and u_f are the values of the axial deformation of the concrete and FRP plate, respectively.

Further, differential equations of the bond interface can be obtained and described as follows, which is through the derivation of Equations (4)–(6) and then the combination of Equations (1)–(3).

$$\frac{d^2u_c(x)}{dx^2} = \frac{6}{E_c b_c {h_c}^2}\frac{dM_c(x)}{dx} - \frac{1}{E_c b_c h_c}\frac{dN_c(x)}{dx} = \frac{6}{E_c b_c {h_c}^2}\left(\frac{F}{2} - \frac{1}{2}h_c \tau(x) b_f\right) - \frac{\tau(x) b_f}{E_c b_c h_c} \tag{8}$$

$$\frac{d^2u_f(x)}{dx^2} = \frac{1}{E_f b_f h_f}\frac{dN_f(x)}{dx} = \frac{\tau(x) b_f}{E_f b_f h_f} \tag{9}$$

$$\frac{ds}{dx} = \frac{N_f(x)}{E_f b_f h_f} + \alpha_f \Delta T - \frac{6M_c(x)}{E_c b_c {h_c}^2} + \frac{N_c(x)}{E_c b_c h_c} - \alpha_c \Delta T \tag{10}$$

2.3. Bond-Slip Model

The interfacial bond-slip model is critical and fundamental to the analysis of the debonding behavior of the FRP-to-concrete bonded interface [46–48]. The bi-linear model can reflect the relationship between the interfacial shear stress and slip well, which is widely applied for evaluating the bond behavior of the FRP-to-substrate bonded interface [8,22,38–40,49]. As shown in Figure 3, this constitutive model presents a linear ascending trend and a decreased trend after the interfacial shear stress reaches peak values; then, the interfacial shear stress drops to zero, and the initiation of debonding occurs along with the development of interfacial slip. Moreover, the mathematical equations of the bi-linear model are described as Equation (11).

$$\tau = \frac{\tau_f}{s_1} s \quad 0 < s \leq s_1 \tag{11a}$$

$$\tau = \frac{\tau_f}{s_f - s_1}(s_f - s) \quad s_1 < s \le s_f \tag{11b}$$

$$\tau = 0 \quad s > s_f \tag{11c}$$

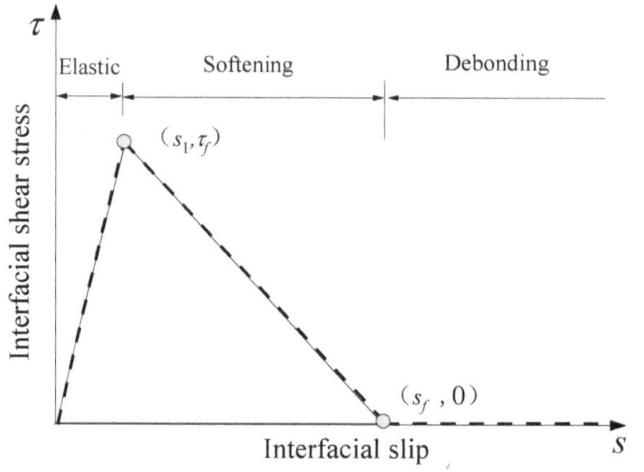

Figure 3. Bi-linear bond-slip model.

2.4. Analytical Solution

During the interfacial debonding of the FRP-strengthened concrete beam under the coupled load of mechanical and thermal, the debonding progress is divided into three stages based on the bond-slip model: elastic stage, elastic-soften stage, and elastic-soften-debonding stage, as shown in Figure 4.

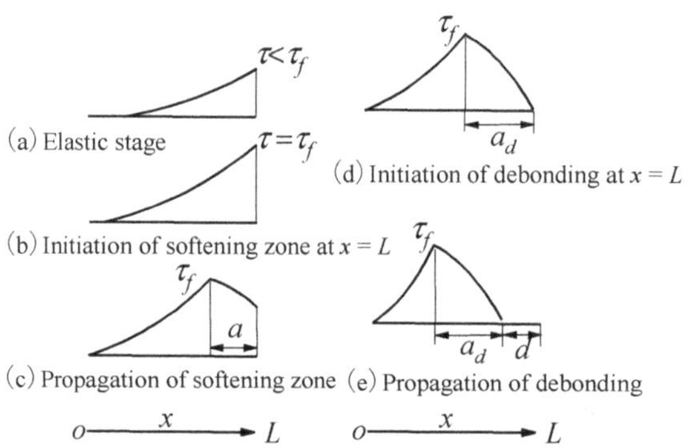

Figure 4. Interfacial shear stress distribution during the debonding progress.

2.4.1. Elastic Stage

When the coupled load is small, the full range of the bonded interface of the FRP-strengthened concrete beam is in a pure elastic stage, where there is no softening or debonding region. Deriving Equation (10) and combining Equation (11a), the following differential equation is calculated:

$$\frac{d^2s}{dx^2} - {\lambda_1}^2 s + BF = 0 \tag{12}$$

where ${\lambda_1}^2 = \frac{\tau_f}{s_1}\left(\frac{1}{E_f h_f} + \frac{4b_f}{E_c b_c h_c}\right)$ and $B = \frac{3}{E_c b_c {h_c}^2}$.

And the boundary conditions are obtained as follows:

$$\begin{cases} x = 0 & \frac{ds}{dx} = -BFl_s + \left(\alpha_f - \alpha_c\right)\Delta T \\ x = L & s = 0 \end{cases} \quad (13)$$

where L is the FRP's bonded length, and l_s is the distance between the end of the bonded interface and the concrete beam support, respectively, as shown in Figure 1.

Substituting the boundary conditions of Equation (13) into the differential equation of Equation (12), the expressions for the distributions of interfacial shear stress and slip are obtained:

$$s = C_1 \cosh(\lambda_1 x) + C_2 \sinh(\lambda_1 x) + \frac{BF}{{\lambda_1}^2} \quad (14)$$

$$\tau = \frac{\tau_f}{s_1}\left[C_1 \cosh(\lambda_1 x) + C_2 \sinh(\lambda_1 x) + \frac{BF}{{\lambda_1}^2}\right] \quad (15)$$

$$C_1 = \left[\frac{l_s}{\lambda_1}\cdot BF - \left(\alpha_f - \alpha_c\right)\Delta T\frac{1}{\lambda_1}\right]\cdot\tanh(\lambda_1 L) - \frac{BF}{{\lambda_1}^2\cosh(\lambda_1 L)} \quad (16)$$

$$C_2 = -\frac{l_s}{\lambda_1}\cdot BF + \left(\alpha_f - \alpha_c\right)\Delta T\frac{1}{\lambda_1} \quad (17)$$

2.4.2. Elastic-Softening Stage

Once the interfacial shear stress reaches its peak value, the softening regions occur at the end of the bond interface with the increase in coupled load. In this state, some regions of the bond interface present an elastic stage, and the other regions present a softening stage near the end of the bond interface, where we note the length of softening region a and the length of elastic region L-a.

For the softening region ($0 \le x \le a$), deriving Equation (10) and combining Equation (11b), the following differential equation is given by:

$$\frac{d^2 s_s}{dx^2} + {\lambda_2}^2 s_s + BF - {\lambda_2}^2 s_f = 0 \quad (18)$$

where ${\lambda_2}^2 = \frac{\tau_f}{s_f - s_1}\lambda^2$.

The solution to Equation (18) is similar to Equation (12) and can be obtained through the following boundary conditions:

$$\begin{cases} x = 0 & \frac{ds}{dx} = -BFl_s + \left(\alpha_f - \alpha_c\right)\Delta T \\ x = a & s = s_1 \end{cases} \quad (19)$$

The expressions for the softening region of the bond interface can be calculated by:

$$s = C_3 \cos(\lambda_2 x) + C_4 \sin(\lambda_2 x) + s_f - \frac{BF}{{\lambda_2}^2} \quad (20)$$

$$\tau = \frac{\tau_f}{s_f - s_1}(s_f - s) \quad (21)$$

$$C_3 = \left(s_1 - s_f + \frac{BF}{{\lambda_2}^2}\right)\frac{1}{\cos(\lambda_2 a)} + \left[\frac{l_s}{\lambda_2}\cdot BF - \left(\alpha_f - \alpha_c\right)\Delta T\frac{1}{\lambda_2}\right]\cdot\tan(\lambda_2 a) \quad (22)$$

$$C_4 = -\frac{l_s}{\lambda_2}\cdot BF + \left(\alpha_f - \alpha_c\right)\Delta T\frac{1}{\lambda_2} \quad (23)$$

For the elastic region ($a \le x \le L$), the differential equation is derived by the similar method of Equation (18) combining Equations (10) and (11).

$$\frac{d^2s}{dx^2} - {\lambda_1}^2 s + BF = 0 \tag{24}$$

And the boundary conditions for this state are:

$$\begin{cases} x = a & s = s_1 \\ x = L & s = 0 \end{cases} \tag{25}$$

Similarly, by substituting the boundary conditions of Equation (25) to the differential equation of Equation (24), the solutions for interfacial shear stress and slip in the elastic region are given by:

$$s = C_5 \cosh(\lambda_1 x) + C_6 \sinh(\lambda_1 x) + \frac{BF}{{\lambda_1}^2} \tag{26}$$

$$\tau = \left[C_5 \cosh(\lambda_1 x) + C_6 \sinh(\lambda_1 x) + \frac{BF}{{\lambda_1}^2}\right] \tag{27}$$

$$C_5 = \frac{1}{\sinh[\lambda_1(L-a)]}\left\{\sinh(\lambda_1 L)s_1 + \frac{BF}{{\lambda_1}^2}[\sinh(\lambda_1 a) - \sinh(\lambda_1 L)]\right\} \tag{28}$$

$$C_6 = -\frac{1}{\sinh[\lambda_1(L-a)]}\left\{\cosh(\lambda_1 L)s_1 + \frac{BF}{{\lambda_1}^2}[\cosh(\lambda_1 a) - \cosh(\lambda_1 L)]\right\} \tag{29}$$

Moreover, at $x = a$, the first derivative of Equations (20) and (26) is equal, which concludes the calculation of the softening length related to mechanical load F and temperature variations ΔT.

$$F = \frac{1}{B} f_1(a) \tag{30}$$

$$\begin{aligned} f_1(a) = {} & \left\{\lambda_1 s_1 \coth[\lambda_1(L-a)] + \left(s_f - s_1\right)\lambda_2 \tan(\lambda_2 a) + \left(\alpha_f - \alpha_c\right)\Delta T \cdot [\sin(\lambda_2 a) \cdot \tan(\lambda_2 a) + \cos(\lambda_2 a)]\right\} \cdot \\ & \left\{\frac{\cosh[\lambda_1(L-a)]-1}{\lambda_1 \sinh[\lambda_1(L-a)]} + \frac{\tan(\lambda_2 a)}{\lambda_2} + l_s[\sin(\lambda_2 a) \cdot \tan(\lambda_2 a) + \cos(\lambda_2 a)]\right\}^{-1} \end{aligned} \tag{31}$$

2.4.3. Elastic-Softening-Debonding Stage

When the interfacial slip (s) is up to the maximum slip (s_f), the initiation of debonding occurs with the increase in the coupled load, which means that s ($x = 0$) is equal to s_f at the initiation of debonding. The length of the softening region is defined as $a = a_d$.

Substituting the boundary condition ($x = 0$, $s = s_f$) to Equation (26), the debonding load (F_{deb}) is derived.

$$F_{deb} = \frac{1}{B} f_2(a_d) \tag{32}$$

$$f_2(a_d) = \left[\frac{s_1 - s_f}{\cos(\lambda_2 a_d)} - \frac{\left(\alpha_f - \alpha_c\right)\Delta T \tan(\lambda_2 a_d)}{\lambda_2}\right] \cdot \left[\frac{1}{{\lambda_2}^2} - \frac{l_s \tan(\lambda_2 a_d)}{\lambda_2} - \frac{1}{{\lambda_2}^2 \cos(\lambda_2 a_d)}\right]^{-1} \tag{33}$$

And the equation can be obtained by combining Equations (31) and (32) at the condition of $a = a_d$.

$$f_2(a_d) = f_1(a_d) \tag{34}$$

The value of a_d can be calculated by solving Equation (34) and then substituting the value of a_d with Equation (32). Thus, the debonding load of the FRP-strengthened concrete beam subjected to the coupled mechanical and thermal load can be obtained.

For the bond interface at the elastic-softening-debonding stage, the debonding length is introduced, noting it as d. The solutions of Equations (20)–(23) and Equations (26)–(29) are validated if the L is replaced by (L-d), which can be used to calculate the distribution of interfacial shear stress and slip of any region at the elastic-softening-debonding stage. Similarly, the length of the softening region is calculated by substituting the boundary condition ($x = d$, $\tau = 0$) into Equation (27).

2.5. Validation of Analytical Model

In the existing literature [12,50], the experimental investigations on the FRP-strengthened concrete beam were conducted at an elevated temperature environment, but the critical parameters of the linear thermal expansion coefficients of concrete and FRP materials were not mentioned, which cannot be used for the validation of the established closed-form analytical model. The good performance of finite element (FE) analysis has been proven in simulating the multi-physical analysis of structural response, such as the coupled effects of mechanical and thermal loads [24,35,51]. Therefore, the numerical simulation based on ABAQUS 6.14 software was used to perform the validation. As shown in Figure 5, the 2D FE model demonstrates the three-point bending concrete beam at an elevated temperature, where the concrete beam and FRP plate are established as a 2-node truss element (T2D2) and a 4-node plane stress element (CPS4R). And the 2-node connector element (CONN2D2) was set to simulate the relationship between the interfacial shear stress and slip.

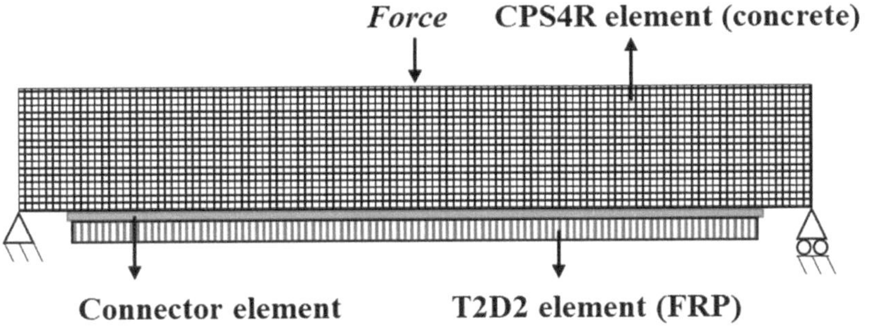

Figure 5. Typical meshed FE model.

The geometry parameters of the FE model are tabulated in Table 1. The properties of the materials are E_c = 34,000 MPa, E_f = 372,000 MPa, $\alpha_c = 8 \times 10^{-6}$/°C, and $\alpha_f = -1.1 \times 10^{-6}$/°C. The FR-E3PSK adhesive was adopted to characterize the bond-slip relationship because this adhesive has a high glass transition temperature (T_g = 180 °C). The parameters of the bi-linear model under different temperature variations are also tabulated in Table 2, and the constitutive models are shown in Figure 6, which are calculated by the determination proposed by Dong et al. [34], Gao et al. [49], and Dai et al. [33].

Table 1. Geometry parameters of FE model.

L/mm	l_s/mm	h_c/mm	h_f/mm	b_c/mm	b_f/mm
1600	200	200	0.167	100	100

Table 2. Parameters of bi-linear model.

Temperature Variations/°C	20	40	80	140	160	170	175	180
τ_f	6.334	6.334	6.334	6.334	2.4724	1.2057	0.7387	0.3819
s_1	0.0666	0.0666	0.0666	0.0666	0.0723	0.0794	0.0842	0.1005
s_f	0.3846	0.3846	0.3846	0.3846	0.4173	0.4528	0.4857	0.5802

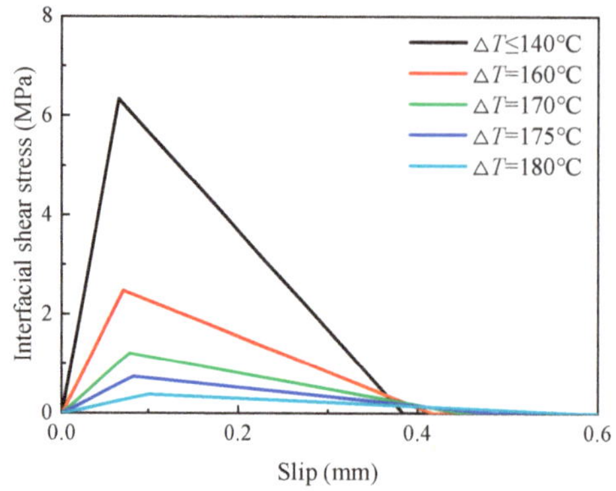

Figure 6. Bi-linear model under different temperature variations.

Figures 7 and 8 show the comparisons of the analytical and numerical solutions to the interfacial shear stress and slip at the elastic stage and elastic-softening stage. The comparisons confirm that the proposed closed-form analytical model can describe the interfacial bond behavior of FRP-strengthened concrete beams subjected to temperature variations well.

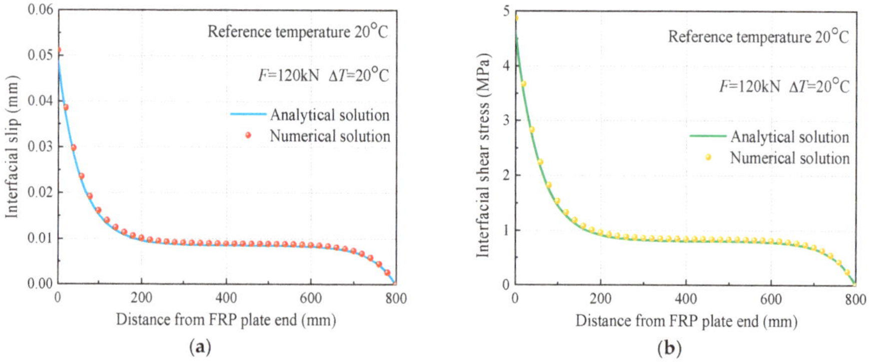

Figure 7. Comparisons of different solutions at the elastic stage. (**a**) Interfacial slip; (**b**) Interfacial shear stress.

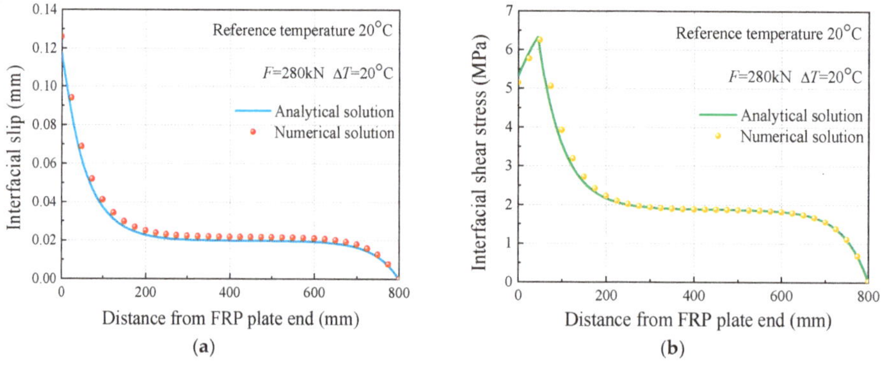

Figure 8. Comparisons of different solutions at the elastic-softening stage. (**a**) Interfacial slip; (**b**) Interfacial shear stress.

3. Effect of Temperature Variations on Bond Behavior and Debonding Load

Figure 9 shows the comparisons of the distributions of interfacial slip and shear stress under different temperature variations. From Figure 9a, the interfacial slip is mainly concentrated in a certain region near the FRP plate's end and presents a rapid nonlinear elastic growth subjected to the coupled mechanical and thermal loads. And in the region near the midpoint of the bonded length, the interfacial slip presents a nonlinear increased trend until a plateau is reached; the interfacial slip increases no longer significantly, where the length of this platform is about 50% of the total bonded length. Then, it can be seen that with the increase in temperature variations, the interfacial slip increases, while the effect of the temperature variations on the interfacial slip is mainly reflected in the regions near the FRP's plate end and is not obvious in the other bonded region. As shown in Figure 9b, the effect of an increase in temperature variations on the interfacial shear stress is also reflected in the region near the FRP's plate end. In addition, when the mechanical load does not change and only temperature variations increase, the softening region occurs at the FRP's plate end, which means end debonding failure can be produced due to the interfacial temperature stress.

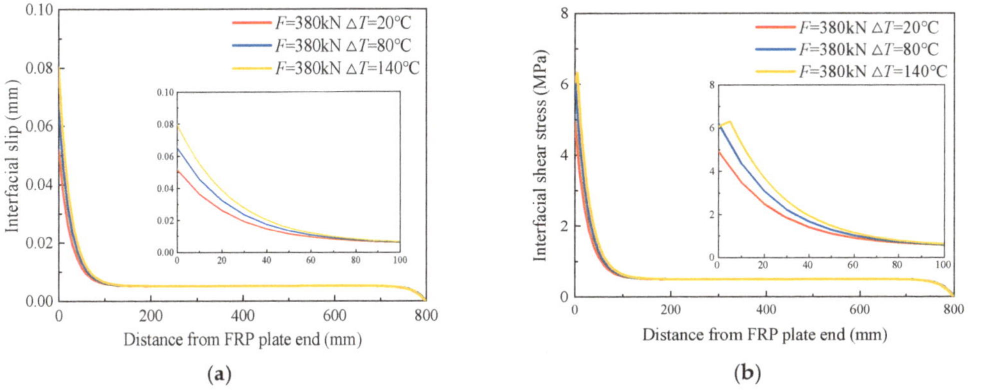

Figure 9. Bond behavior under different temperature variations. (**a**) Interfacial slip; (**b**) Interfacial shear stress.

In Figure 10, the vertical coordinate is the normalized debonding load, which is the ratio of the debonding loads under different temperatures to the debonding load under the reference temperature (20 °C), and the horizontal coordinate is the ratio of the temperature to the glass transition temperature. It can be concluded that the debonding load is negatively related to the temperature variations. And the debonding load decreases to 80% of the debonding load at the reference temperature before the temperature reaches the glass transition temperature range, where this temperature range is usually defined as the glass transition temperature plus or minus 20 °C, i.e., $T_g \pm 20$ °C. However, once the temperature reaches the glass transition temperature range, the debonding load presents a rapid decline. When the temperature reaches the glass transition temperature, the debonding load is zero. Because, at this stage, the decomposition and softening of the adhesive intensify dramatically, and the cohesive strength of the bonded interface almost disappears, which means that there is almost no bond force between the FRP plate and concrete substrate, resulting in a dangerous state regarding the FRP-strengthened concrete beam.

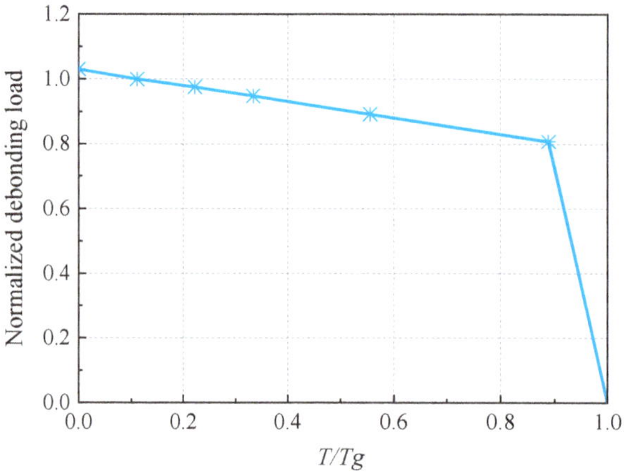

Figure 10. Effect of temperature variations on debonding load.

Replacing the horizontal coordinate T/T_g by the $T/(T_g - 20\ °C)$, the prediction on the degradation of the debonding load can be obtained, as shown in Equation (35), which is fitted by the values of the debonding load before the temperature reaches the glass transition temperature range. Meanwhile, this fitted result is acceptable due to the $R^2 = 0.99$, as shown in Figure 11.

$$\frac{F_{deb,T}}{F_{deb,20°C}} = a + b\left(\frac{T}{T_g - 20°C}\right) \tag{35}$$

where the $a = 1.03$, $b = -0.22$, and $0 \le T \le T_g - 20\ °C$.

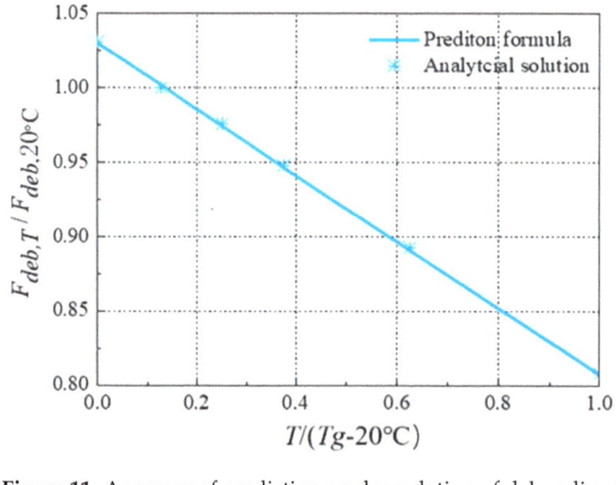

Figure 11. Accuracy of prediction on degradation of debonding load.

4. Effect of FRP Sizes on Bond Behavior and Debonding Load

Based on the analysis of the effect of the temperature variations on the bond behavior of FRP-strengthened concrete beams, the evolution of bond behavior with different FRP bonded sizes of strengthened concrete beams combined with mechanical and thermal loads are explored to provide better guidelines for strengthened designs.

Figure 12 shows the effect of temperature variations on the debonding load with different FRP thicknesses and lengths. Figure 13 shows the distribution of interfacial shear stress of debonding initiation with different FRP thicknesses and lengths at the 80 °C temperature variation. Figure 14 shows the effects of the FRP thicknesses and lengths on the debonding load under different temperature variations, where the debonding load is normalized by the debonding load at the reference FRP thickness and bonded length, i.e., one layer and 1200 mm, respectively.

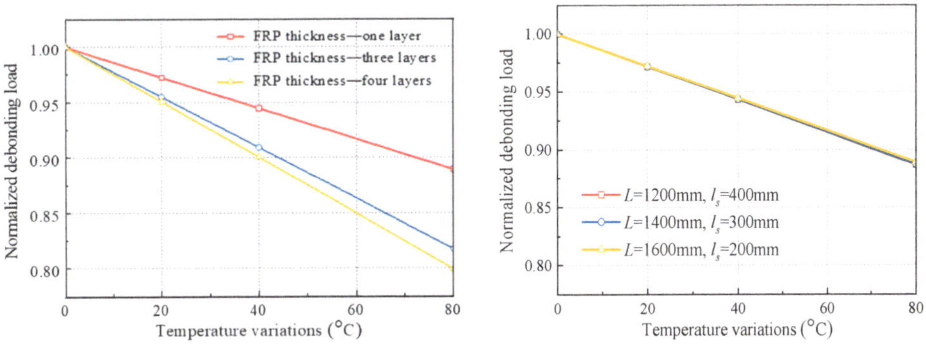

(a) (b)

Figure 12. Effect of temperature variations on the debonding load under different FRP sizes. (**a**) FRP thickness; (**b**) Bonded length.

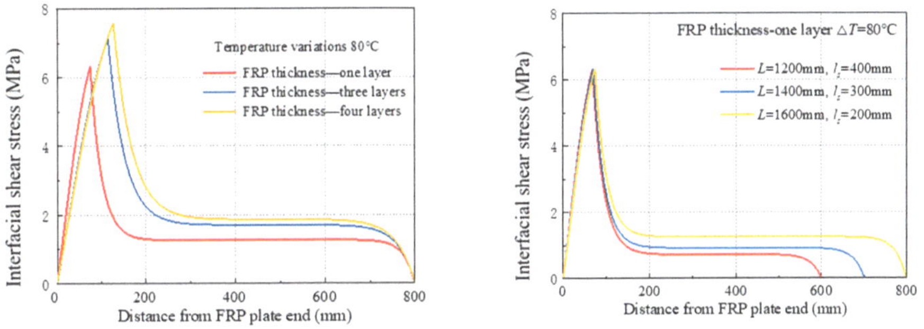

(a) (b)

Figure 13. Distribution of interfacial shear stress at the initiation of debonding. (**a**) FRP thickness; (**b**) Bonded length.

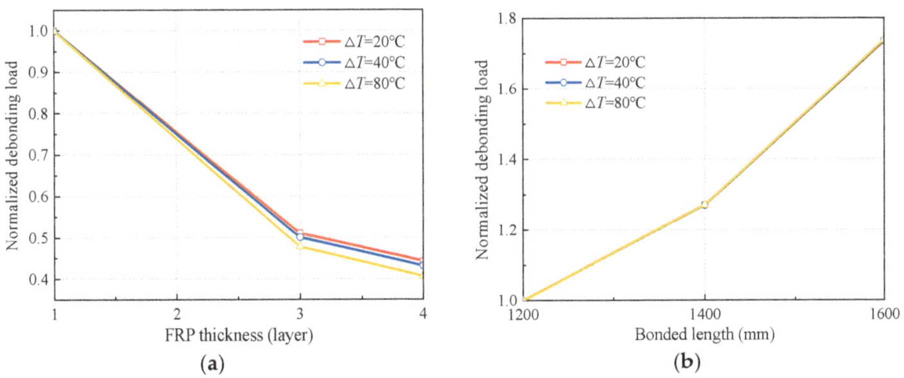

(a) (b)

Figure 14. Effect of FRP sizes on the debonding load. (**a**) FRP thickness; (**b**) Bonded length.

From Figure 12a, it can be seen that the sensitivity of temperature variations is more significant with the bigger FRP thickness, where the degradation trend of the debonding load accelerates with the increase in temperature variations. From Figure 13a, the distribution of interfacial shear stress changes along the total bonded length with a change in FRP thickness, and the softening length at the debonding initiation increases with the increase in temperature variations. From Figure 14a, at the same temperature variation, the increase in FRP thickness decreases the debonding load.

Combining Figures 12b and 13b, it can be concluded that the degradation of debonding load, distribution of interfacial shear stress, and the softening length at the initiation of debonding change little under the different bonded lengths, whereas the interfacial shear stress is higher in the elastic region along the bonded interface with the increase in bonded length. From Figure 14b, the debonding load is positively related to the bonded length, where this positive correlation between the debonding load and bonded length is not changed by the temperature variations.

5. Conclusions

In this paper, the evolution of the bond behavior of FRP-strengthened concrete beams subjected to temperature variations is studied through analytical approaches, and the main conclusions obtained are as follows:

(1) A new analytical model is established to predict the bond behavior and debonding progress of FRP-strengthened concrete beams under coupled mechanical and thermal loads. Given the clarity and simplicity of the bi-linear model in characterizing bond-slip relationships, here, the bi-linear model is introduced to obtain closed-form solutions for bond behavior.

(2) Based on the FE approach to model the FRP-strengthened concrete beam under coupled mechanical and thermal loads, the proposed analytical model is validated by comparisons of the distributions of bond behavior obtained from analytical and numerical models.

(3) The increase in temperature variation accelerates the development of interfacial slip and shear stress, which is mainly reflected near the FRP's end region. The debonding load presents a degradation with an increase in the temperature variations, and the debonding load decreases rapidly once the temperature reaches the glass transition temperature range. The prediction formula for the debonding load is proposed with high accuracy.

(4) An increase in FRP thickness decreases the interfacial bond behavior at elevated temperatures, while an increase in bonded length can promote interfacial bond behavior. This conclusion suggests that longer bonded lengths and thicker FRP plates should be designed to inform the design of strengthened concrete structures.

In the future, the intermediate crack-induced debonding and the behavior of interfacial debonding between two adjacent cracks of FRP-strengthened concrete beams under mechanical and thermal loads should be explored using experimental, numerical, or analytical methods.

Author Contributions: Conceptualization, X.Z.; Methodology, X.L.; Software, W.H.; Formal analysis, J.H.; Investigation, W.H. and X.S.; Data curation, J.Y.; Writing—original draft, J.H.; Writing—review & editing, X.L.; Visualization, Y.W.; Supervision, X.Z. and X.L. All authors have read and agreed to the published version of the manuscript.

Funding: This research was funded by National Natural Science Foundation of China (No. 52408465), China Postdoctoral Science Foundation (No. BX20240451, 2024M753850).

Data Availability Statement: Data are contained within the article.

Conflicts of Interest: Authors Jianwen Hao, Wei Hou, Jiancheng Yao, Yazhuo Wang and Xiaojian Su were employed by the company Shandong High-Speed Infrastructure Construction Co., Ltd. The remaining authors declare that the research was conducted in the absence of any commercial or financial relationships that could be construed as a potential conflict of interest.

References

1. Teng, J.G.; Chen, J.F.; Smith, S.T.; Lam, L. Behaviour and strength of FRP-strengthened RC structures: A state-of-the-art review. *Proc. Inst. Civ. Eng. Struct. Build.* **2003**, *156*, 51–62. [CrossRef]
2. Hollaway, L.C. A review of the present and future utilisation of FRP composites in the civil infrastructure with reference to their important in-service properties. *Constr. Build. Mater.* **2010**, *24*, 2419–2445. [CrossRef]
3. Li, D.; Zhou, J.; Ou, J. Damage, nondestructive evaluation and rehabilitation of FRP composite-RC structure: A review. *Constr. Build. Mater.* **2021**, *271*, 121551. [CrossRef]
4. Mugahed Amran, Y.H.; Alyousef, R.; Rashid, R.S.M.; Alabduljabbar, H.; Hung, C.C. Properties and applications of FRP in strengthening RC structures: A review. *Structures* **2018**, *16*, 208–238. [CrossRef]
5. Teng, J.G.; Chen, J.F.; Smith, S.T. Debonding failures in FRP-strengthened RC beams: Failure modes, existing research and future challenges. In *Composites in Construction: A Reality*; American Society of Civil Engineers: Reston, VA, USA, 2001; pp. 139–148.
6. Smith, S.T.; Teng, J.G. Interfacial stresses in plated beams. *Eng. Struct.* **2001**, *23*, 857–871. [CrossRef]
7. De Lorenzis, L.; Fernando, D.; Teng, J.-G. Coupled mixed-mode cohesive zone modeling of interfacial debonding in simply supported plated beams. *Int. J. Solids Struct.* **2013**, *50*, 2477–2494. [CrossRef]
8. De Lorenzis, L.; Zavarise, G. Cohesive zone modeling of interfacial stresses in plated beams. *Int. J. Solids Struct.* **2009**, *46*, 4181–4191. [CrossRef]
9. Silva, M.A.G.; Biscaia, H. Degradation of bond between FRP and RC beams. *Compos. Struct.* **2008**, *85*, 164–174. [CrossRef]
10. Kang, T.H.K.; Howell, J.; Kim, S.; Lee, D.J. A State-of-the-Art Review on Debonding Failures of FRP Laminates Externally Adhered to Concrete. *Int. J. Concr. Struct. Mater.* **2012**, *6*, 123–134. [CrossRef]
11. Cabral-Fonseca, S.; Correia, J.R.; Custódio, J.; Silva, H.M.; Machado, A.M.; Sousa, J. Durability of FRP—Concrete bonded joints in structural rehabilitation: A review. *Int. J. Adhes. Adhes.* **2018**, *83*, 153–167. [CrossRef]
12. Krzywoń, R. Behavior of EBR FRP Strengthened Beams Exposed to Elevated Temperature. *Procedia Eng.* **2017**, *193*, 297–304. [CrossRef]
13. Heshmati, M.; Haghani, R.; Al-Emrani, M. Environmental durability of adhesively bonded FRP/steel joints in civil engineering applications: State of the art. *Compos. Part B Eng.* **2015**, *81*, 259–275. [CrossRef]
14. Silva, M.A.G.; Biscaia, H.; Ribeiro, P. On factors affecting CFRP-steel bonded joints. *Constr. Build. Mater.* **2019**, *226*, 360–375. [CrossRef]
15. Firmo, J.P.; Correia, J.R.; Bisby, L.A. Fire behaviour of FRP-strengthened reinforced concrete structural elements: A state-of-the-art review. *Compos. Part B Eng.* **2015**, *80*, 198–216. [CrossRef]
16. Ferrier, E.; Agbossou, A. Temperature effects on the shear capacity of external bonded fiber reinforced polymer on concrete. *Constr. Build. Mater.* **2017**, *152*, 333–344. [CrossRef]
17. Ferrier, E.; Rabinovitch, O.; Michel, L. Mechanical behavior of concrete–resin/adhesive–FRP structural assemblies under low and high temperatures. *Constr. Build. Mater.* **2016**, *127*, 1017–1028. [CrossRef]
18. Firmo, J.P.; Arruda, M.R.T.; Correia, J.R. Numerical simulation of the fire behaviour of thermally insulated reinforced concrete beams strengthened with EBR-CFRP strips. *Compos. Struct.* **2015**, *126*, 360–370. [CrossRef]
19. Firmo, J.P.; Correia, J.R.; Pitta, D.; Tiago, C.; Arruda, M.R.T. Experimental characterization of the bond between externally bonded reinforcement (EBR) CFRP strips and concrete at elevated temperatures. *Cem. Concr. Compos.* **2015**, *60*, 44–54. [CrossRef]
20. Gamage, J.C.P.H.; Al-Mahaidi, R.; Wong, M.B. Bond characteristics of CFRP plated concrete members under elevated temperatures. *Compos. Struct.* **2006**, *75*, 199–205. [CrossRef]
21. Leone, M.; Matthys, S.; Aiello, M.A. Effect of elevated service temperature on bond between FRP EBR systems and concrete. *Compos. Part B Eng.* **2009**, *40*, 85–93. [CrossRef]
22. Zhou, H.; Torres, J.P.; Fernando, D.; Law, A.; Emberley, R. The bond behaviour of CFRP-to-steel bonded joints with varying bond properties at elevated temperatures. *Eng. Struct.* **2019**, *183*, 1121–1133. [CrossRef]
23. Biscaia, H.C.; Chastre, C.; Viegas, A.; Franco, N. Numerical modelling of the effects of elevated service temperatures on the debonding process of FRP-to-concrete bonded joints. *Compos. Part B Eng.* **2015**, *70*, 64–79. [CrossRef]
24. Dong, K.; Hu, K.X.; Gao, W.Y.; Yang, S.T. Fire endurance tests of CFRP-strengthened RC beams with different insulation schemes. *Structures* **2023**, *56*, 104887. [CrossRef]
25. Biscaia, H.C.; Ribeiro, P. A temperature-dependent bond-slip model for CFRP-to-steel joints. *Compos. Struct.* **2019**, *217*, 186–205. [CrossRef]
26. Klamer, E. Influence of temperature on concrete beams strengthened in flexure with CFRP. Ph.D. Thesis, Eindhoven University of Technology, Eindhoven, The Netherlands, 2009.
27. Kodur, V.; Solhmirzaei, R.; Agrawal, A.; Aziz, E.M.; Soroushian, P. Analysis of flexural and shear resistance of ultra high performance fiber reinforced concrete beams without stirrups. *Eng. Struct.* **2018**, *174*, 873–884. [CrossRef]
28. Kodur, V.K.R.; Bhatt, P.P.; Naser, M.Z. High temperature properties of fiber reinforced polymers and fire insulation for fire resistance modeling of strengthened concrete structures. *Compos. Part B Eng.* **2019**, *175*, 107104. [CrossRef]

29. Altunişik, A.C.; Akbulut, Y.E.; Adanur, S.; Kaya, A.; Günaydin, M.; Mostofi, S.; Mosallam, A. Evaluating the high-temperature endurance of FRP-strengthened concrete using an innovative insulation system: Experimental investigation. *J. Build. Eng.* **2023**, *73*, 106444. [CrossRef]
30. Guo, D.; Gao, W.Y.; Liu, Y.L.; Dai, J.G. Intermediate crack-induced debonding in CFRP-retrofitted notched steel beams at different service temperatures: Experimental test and finite element modeling. *Compos. Struct.* **2023**, *304*, 116388. [CrossRef]
31. Guo, D.; Liu, Y.L.; Gao, W.Y.; Dai, J.G. Bond behavior of CFRP-to-steel bonded joints at different service temperatures: Experimental study and FE modeling. *Constr. Build. Mater.* **2023**, *362*, 129836. [CrossRef]
32. Guo, D.; Wang, H.P.; Liu, Y.L.; Gao, W.Y.; Dai, J.G. Structural behavior of CFRP-strengthened steel beams at different service temperatures: Experimental study and FE modeling. *Eng. Struct.* **2023**, *293*, 116646. [CrossRef]
33. Dai, J.-G.; Gao, W.Y.; Teng, J.G. Bond-Slip Model for FRP Laminates Externally Bonded to Concrete at Elevated Temperature. *J. Compos. Constr.* **2013**, *17*, 217–228. [CrossRef]
34. Dong, K.; Hu, K. Development of bond strength model for CFRP-to-concrete joints at high temperatures. *Compos. Part B Eng.* **2016**, *95*, 264–271. [CrossRef]
35. Dong, K.; Hao, J.; Li, P.; Zhong, C. A nonlinear analytical model for predicting bond behavior of FRP-to-concrete/steel substrate joints subjected to temperature variations. *Constr. Build. Mater.* **2022**, *320*, 126225. [CrossRef]
36. Caggiano, A.; Martinelli, E.; Faella, C. A fully-analytical approach for modelling the response of FRP plates bonded to a brittle substrate. *Int. J. Solids Struct.* **2012**, *49*, 2291–2300. [CrossRef]
37. Liu, S.; Yuan, H.; Wu, J. Full-range mechanical behavior study of FRP-to-concrete interface for pull-pull bonded joints. *Compos. Part B Eng.* **2019**, *164*, 333–344. [CrossRef]
38. Teng, J.G.; Yuan, H.; Chen, J.F. FRP-to-concrete interfaces between two adjacent cracks: Theoretical model for debonding failure. *Int. J. Solids Struct.* **2006**, *43*, 5750–5778. [CrossRef]
39. Yuan, H.; Teng, J.G.; Seracino, R.; Wu, Z.S.; Yao, J. Full-range behavior of FRP-to-concrete bonded joints. *Eng. Struct.* **2004**, *26*, 553–565. [CrossRef]
40. Biscaia, H.C. The influence of temperature variations on adhesively bonded structures: A non-linear theoretical perspective. *Int. J. Non-Linear Mech.* **2019**, *113*, 67–85. [CrossRef]
41. Bennegadi, M.L.; Hadjazi, K.; Sereir, Z.; Amziane, S.; El Mahi, B. General cohesive zone model for prediction of interfacial stresses induced by intermediate flexural crack of FRP-plated RC beams. *Eng. Struct.* **2016**, *126*, 147–157. [CrossRef]
42. Wang, J. Cohesive zone model of intermediate crack-induced debonding of FRP-plated concrete beam. *Int. J. Solids Struct.* **2006**, *43*, 6630–6648. [CrossRef]
43. Wang, J. Debonding of FRP-plated reinforced concrete beam, a bond-slip analysis. I. Theoretical formulation. *Int. J. Solids Struct.* **2006**, *43*, 6649–6664. [CrossRef]
44. Tounsi, A.; Hassaine Daouadji, T.; Benyoucef, S.; Adda bedia, E.A. Interfacial stresses in FRP-plated RC beams: Effect of adherend shear deformations. *Int. J. Adhes. Adhes.* **2009**, *29*, 343–351. [CrossRef]
45. Guenaneche, B.; Krour, B.; Tounsi, A.; Fekrar, A.; Benyoucef, S.; Adda Bedia, E.A. Elastic analysis of interfacial stresses for the design of a strengthened FRP plate bonded to an RC beam. *Int. J. Adhes. Adhes.* **2010**, *30*, 636–642. [CrossRef]
46. Lu, X.Z.; Teng, J.G.; Ye, L.P.; Jiang, J.J. Bond–slip models for FRP sheets/plates bonded to concrete. *Eng. Struct.* **2005**, *27*, 920–937. [CrossRef]
47. Biscaia, H.C.; Chastre, C.; Silva, M.A.G. Linear and nonlinear analysis of bond-slip models for interfaces between FRP composites and concrete. *Compos. Part B Eng.* **2013**, *45*, 1554–1568. [CrossRef]
48. Biscaia, H.C.; Chastre, C.; Silva, M.A.G. Bond-slip model for FRP-to-concrete bonded joints under external compression. *Compos. Part B Eng.* **2015**, *80*, 246–259. [CrossRef]
49. Gao, W.Y.; Teng, J.G.; Dai, J.-G. Effect of Temperature Variation on the Full-Range Behavior of FRP-to-Concrete Bonded Joints. *J. Compos. Constr.* **2012**, *16*, 671–683. [CrossRef]
50. Shier, G.W.R.; Green, M.F. Performance of postcured CFRP-Strengthened reinforced concrete beams at elevated temperatures. *J. Compos. Constr.* **2017**, *21*, 04017008. [CrossRef]
51. Jawdhari, A.; Fam, A. Thermal-Structural Analysis and Thermal Bowing of Double Wythe UHPC Insulated Walls. *Energy Build.* **2020**, *223*, 110012. [CrossRef]

buildings

MDPI

Article

Effect of Traffic Vibration on Compressive Strength of High-Strength Concrete and Tensile Strength of New-to-Old Concrete Interfaces

Pingping Gu [1], Hao Wu [2], Luchang Li [1], Zhanghao Li [1], Jingyi Hong [1] and Mei-Ling Zhuang [1,3,*]

1 School of Transportation and Civil Engineering, Nantong University, Nantong 226019, China; gupingping@ntu.edu.cn (P.G.); 2233110324@stmail.ntu.edu.cn (L.L.); 2233110339@stmail.ntu.edu.cn (Z.L.); 2233110322@stmail.ntu.edu.cn (J.H.)
2 College of Architectural Science and Engineering, Yangzhou University, Yangzhou 225100, China; mz120200976@stu.yzu.edu.cn
3 Water Resources Research Institute of Shangdong Province, Jinan 250013, China
* Correspondence: ml_zhuang99@163.com

Abstract: Widening existing bridges is an important way to meet the surge in traffic demand, which is often carried out in a way that does not interrupt traffic. To investigate the effect of traffic vibration on the compressive strength of high-strength concrete and the splitting strength of new-to-old concrete interfaces, the initial to final set time of high-strength concrete C60 was first investigated in this article. Then, the traffic disturbance parameters were determined. Later, the compressive strength of C60 concrete at different stages under traffic disturbance parameters was carried out. Finally, the splitting tensile strength of new-to-old concrete specimens at different stages with different loading modes was tested. The test results indicated that the compressive strength of the specimens vibrated for 3 h and cured for 3, 7, and 28 days was increased by 4.3%, 5.7%, and 11.9%, respectively; those of the specimens vibrated for 7 h and cured for 3, 7, and 28 days was decreased by 13.7%, 20.4%, and 19.9%, respectively; the effect traffic vibration on the compressive strength of the specimens vibrated for 5 h was not obvious. When loaded along the old and new concrete joint, the specimens cracked along the joint; the splitting tensile strengths of the specimen at different disturbed stages were significantly decreased. When loaded perpendicular to the joint, the specimens cured for 3 and 7 days still cracked along the joint, and the splitting tensile strengths of the specimen at different disturbed stages were significantly decreased; while the specimens cured for 28 days cracked in the direction perpendicular to the joint, the tensile strengths of the specimens at different disturbed stages were significantly decreased. This study can promote the widening and improvement of existing concrete highways and bridges, which can save resources and improve land use.

Keywords: traffic vibration; high-strength concrete; compressive strength; splitting tensile strength

1. Introduction

With the development of science and technology, concrete has been widely used in many engineering fields [1–5]. Many investigations were carried out into the mechanical properties of concrete structures [6–8]. Widening existing bridges is an important way to meet the surge in traffic demand, which is often carried out in a way that does not interrupt traffic. In the open traffic situation for the widening of the old bridge, traffic vibration will inevitably have an impact on the mechanical properties and durability of the newly poured concrete [9]. Newly poured concrete is often referred to as fresh concrete. Old concrete is concrete that has been in use for some time. The interface between old and new concrete is naturally weak, which will affect the reliability and durability of the strengthened structure [10]. The interface between the old and new concrete is naturally weak. Therefore, the effect of traffic vibration on the compressive strength of high-strength concrete and the splitting strength of new-to-old concrete interfaces in bridge widening is of great significance.

In the last two decades, different vibration effects on the mechanical properties of concrete have been investigated by indoor experiments and on-site tests. Harsh et al. [11,12] studied the effect of traveling vibration on the compressive strength of newly poured concrete for repairing bridges, and the bond between reinforcement and concrete. The results showed that continuous vibration at the beginning of pouring reduced the compressive strength of the plastic concrete by 10% and the bond strength by 5.9%, which had a negative effect on the mechanical properties of the concrete. Issa et al. [13] simulated the deformation of concrete bridges under the vibration of traffic load. They concluded that the vibration of traffic load had a significant effect on the early strength of concrete. The early strength of concrete was affected by the vibration of traffic load when the concrete was poured for 7~8 h, and the vibration of traffic load affected the early strength of concrete. Dunham et al. [14] studied the two peak vibration speeds and five starting times of the mechanical properties of concrete and found that the vibration on the compressive strength of the impact was not significant, while the flexural strength had an adverse effect, the maximum loss of flexural strength of about 8%. Jiang et al. [15] found that an increase in the frequency of axle-coupled vibration exacerbated the degree of early segregation of concrete, reduced the uniformity and compactness of concrete, in the early and late stages of concrete setting and hardening, and the impact of traveling vibration on the performance of concrete was relatively small; traveling vibration occurred in the middle stage of setting and hardening, the impact of the mechanical properties of concrete was greater. Zhang and Xu [16] investigated the changes in compressive strength of concrete from a microscopic point of view by applying indoor perturbations of various parameters in combination with an acoustic emission device during the initial and initial-final setting stages of concrete. It was found that the compressive strength of concrete was improved when the disturbance was applied at the initial setting stage and decreased when the disturbance was applied at the initial–final setting stage. Li et al. [17] measured the frequency and amplitude of vibration in different directions during concrete placement and simulated the effect of three-dimensional disturbance on the mechanical properties of concrete. It was found that the compressive strength of the concrete specimens decreased by about 15% one hour before and after the initial setting of the concrete. Dai and Li [18] improved the effect of disturbance on the mechanical properties of concrete by mixing basalt fibers in concrete. It was found that when the volume fraction of basalt fibers is 0.1%, it can effectively improve the phenomenon of concrete delamination caused by disturbance. However, the concrete strength grade of C20 used in the experiments is more limited and a lot of subsequent experimental studies are needed. Pan et al. [19] applied different durations of disturbance during the initial and final setting stages of concrete and found that the increase in disturbance duration did not change after the compressive strength of the concrete decreased to a certain value, while the flexural strength showed greater damage as the disturbance duration increased.

At present, existing research has conducted in-depth research on settlement, shrinkage, and creep of new bridges. During the construction process of new bridges, due to the requirement of uninterrupted traffic, the vibration caused by vehicle loads throughout the entire process of concrete solidification, and hardening of the bridge affects the strength growth of newly poured concrete at the wet joint. The performance of the interface between new and old concrete has received little attention. Research on the compressive strength of concrete under static loading has been very mature, and the current stage of research is to develop high-strength, high-performance concrete. However, there is little experimental research on the compressive strength of concrete under traffic loads, and a large amount of experimental exploration is needed to study the impact of traffic disturbances on the compressive strength of concrete. In this current project, the combination of new and old concrete components in the same stress conditions as the whole specimen is more likely to be damaged. The reason is that the bond strength of the interface between the new and old concrete combination is not up to the strength of the whole specimen. In the study of the interface between new and old concrete, Marceau [20] found that the water-cement ratio at the interface of the new concrete was high, and a layer of water film was formed in the interface. The hydration product crystallization generated by the hydration of new concrete at the interface prevented good contact between the cementitious material and the aggregate, thereby reducing the bonding strength of the interface area. Wall and Shrive [21] regarded old concrete as a whole block of aggregate, but because of the different pore characteristics and surface roughness forms of the old concrete, the new-to-old concrete bond can only be similarly viewed as a cement-aggregate bond, with significant limitations. Wei et al. [22] tested the mechanical properties of molded concrete specimens after continuous vibration for 30 min using different frequencies and amplitudes. It was found that the splitting tensile strength was more sensitive to the change in amplitude when the applied vibration frequency was small in the concrete setting and hardening period, and the splitting tensile strength started to decrease when the amplitude was greater than 5 mm. Pan et al. [23] investigated the effect of the maximum coarse aggregate particle size on the mechanical properties of concrete after early disturbance. It was found that the compressive strength and split tensile strength damage of concrete were the largest when the maximum coarse aggregate particle size was 16 mm, and they were the smallest when the coarse aggregate particle size was 20 mm. Zhu et al. [10] studied the splitting tensile strength of the interface between the full lightweight ceramsite concrete and ordinary concrete. The results indicated that the tensile strength of the interface first grew and then decreased with the increase in roughness.

Therefore, it is important to investigate the effect of traffic vibration on the compressive strength of high-strength concrete and the splitting tensile strength of new-to-old concrete interfaces.

This study took the widening project of two existing bridges in the Lin'an Jiande section of the Linjin Expressway in China as the background. Before widening, the old concrete guardrail at the widening edge of the old bridge was removed, and the gap between the new and old bridges was filled and spliced through wet joints. Before pouring concrete into the wet joints, the existing bridge section was cleaned and leveled, then drilled for reinforcement, and then connected to the wider bridge. Traffic was not interrupted during construction. The disturbance caused by driving loads ran through the entire process of concrete solidification and hardening at the wet joint. Therefore, to make the test closer to the actual engineering conditions, the effect of traffic disturbance on the strength of high-strength (C60) concrete at different ages was investigated by applying the traveling disturbance to the new concrete at the initial, initial-final and final setting stages. Then, the splitting tensile strength tests were carried out to investigate the effect of traffic vibration on the splitting tensile strength of new-to-old concrete interfaces at bridge-widening wet joints.

2. Experimental Program

2.1. Measurement of Initial to Final Set Time

According to the test method in GB/T 50080-2016 [24], the setting time of concrete was measured. The tests were carried out using the HG-80 penetration resistance tester, as shown in Figure 1 to determine the initial and final setting times of the concrete. The concrete was sieved using a sieve with an aperture diameter of 5 mm to obtain the slurry. Then, it was poured into three test molds (see Figure 1). The upper diameter of the test molds was 16 cm, the lower diameter was 15 cm, and the depth was 15 cm. Three barrels were filled with concrete. The concrete in the three test molds was inserted with a vibrating rod at least 25 times and then smoothed and placed on the bottom plate of the penetration resistance meter. The concrete in the test mold did not exceed 5 mm from the opening of the test mold. The height of the penetration needle was adjusted to a distance of no more than 5 mm from the upper plane of the material cylinder according to the operating instructions for the concrete penetration resistance tester. After the completion of the adjustment, the pressure rod was used to conduct the test with a penetration depth of 2.5 cm. Each specimen underwent at least six penetration resistance tests. The last penetration resistance per unit area was not less than 28 MPa. The main material composition of C60 concrete provided by the concrete manufacturer is shown in Table 1. It was the same as the concrete used in the actual engineering bridge widening construction. The initial setting and final setting time of C60 concrete are shown in Table 2. It can be seen that the initial setting time of C 60 concrete was 3 h, and the final setting time was 7 h.

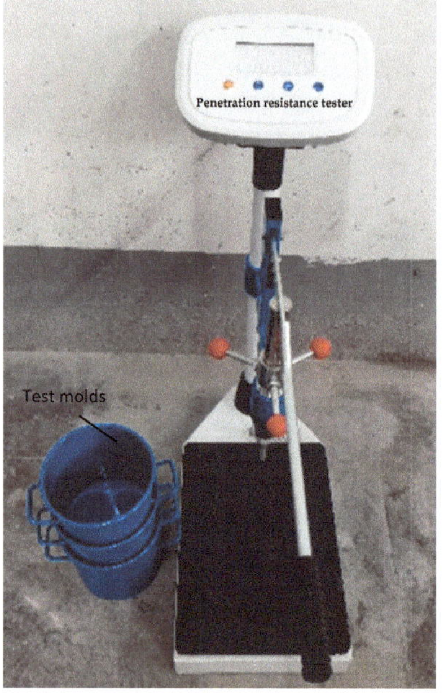

Figure 1. Test molds and a penetration resistance tester.

Table 1. Material composition of C60 concrete.

Cement	Sand	Stone	Water	Fly Ash	Slag	Water Reducer
340 kg/m^3	698 kg/m^3	1100 kg/m^3	152 kg/m^3	60 kg/m^3	90 kg/m^3	4.8 kg/m^3

Table 2. Initial and final set time. No indicators for C60 concrete were found in the specification. The initial and final set times were tested according to the specification GB/T 50080-2016 [19].

No.	Initial Sett	Final Set
1	2 h and 55 min	7 h and 5 min
2	2 h and 55 min	7 h and 5 min
3	3 h	7 h and 5 min

2.2. Determination of Traffic Disturbance Parameters

To truly and accurately study the effect of traffic disturbance on the concrete properties of bridge-widening wet joints, this article relies on the bridge-widening project of Lin'an to Jiande section of Linjin Expressway in China. The cross-sections of the widening bridge are shown in Figure 2. Due to the uninterrupted traffic, the traffic flow had a greater impact on the determination of the disturbance parameters. A dynamic test system was used to determine the bridge vibration parameters caused by traveling loads. The test system mainly consisted of a vibration pickup, a signal amplifier, a signal acquisition instrument, and a vibration testing and analysis system. The vibration signal of bridge vibration was converted into an analog signal by the vibration pickup, which was amplified by the signal amplifier and transmitted to the signal acquisition instrument. The signal acquisition instrument took the amplified analog signals for data acquisition and converted the analog signals into digital signals through the A/D conversion module. Sampling parameters were collected and controlled by a signal acquisition instrument INV3062W produced by Beijing Oriental Institute of Vibration and Noise Technology in Beijing, China and a field monitoring computer Dell P61F produced by Dell Technologies in Round Rock, TX, USA. Various bridge dynamic data collected in the field were analyzed by dynamic test analysis and modal analysis software Artemis Modal Pro v7.2.2.6 for bridge vibration test analysis and bridge vibration modal analysis. The vibration parameters were measured without interrupting the traffic. There were differences in the traffic flow in different periods, and the traffic load caused different degrees of forced vibration of the bridge. Therefore, the dynamic response of the bridge was measured to analyze, and the vertical vibration amplitude and frequency of the bridge caused by the traffic load were ultimately obtained. The instrument used for the test was the INV3062W acquisition instrument and the 941B vibration pickup as shown in Figure 3. The on-site testing images are shown in Figure 4. After calculation and analysis, it was found that the forced vibration frequency of the bridge was around 2 Hz, and the vibration amplitude was 1 mm when the traffic flow (1500 vehicles/h) was small. When the traffic flow (3500 vehicles/h) was large, the forced vibration frequency of the bridge was around 4 Hz, and the vibration amplitude was about 3 mm.

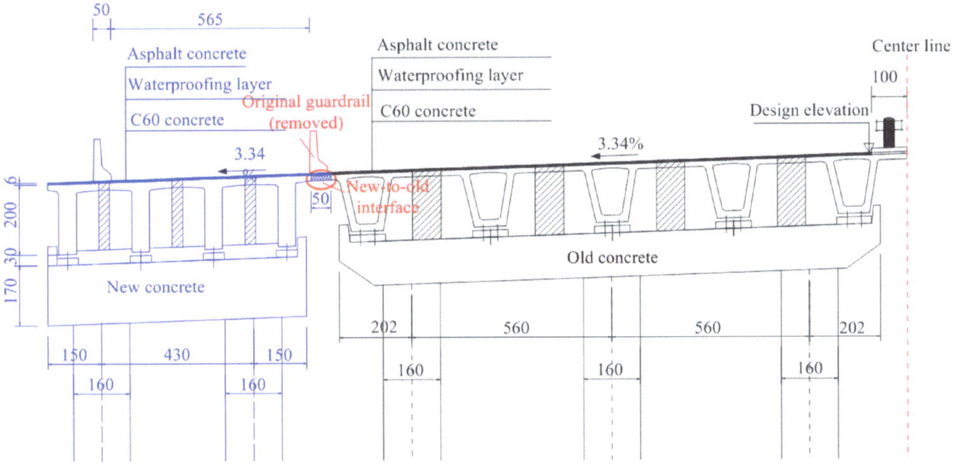

Figure 2. Cross-section of the widening bridge (Unit: mm) (The blue in Figure 2 for the new concrete bridge; the red in Figure 2 for the new-to-old interface).

2.3. *Concrete Compressive Strength Test Under Traveling Disturbance*

With reference to GB50152-2012 [25] and JGJ55-2011 [26], 12 groups of specimens were designed with a size of 150 mm × 150 mm × 150 mm. Each group has 3 specimens. The disturbance parameters of the most unfavorable working conditions measured at the project site were tested. An electromagnetic vertical vibration testbed (Figure 5) was used for traveling disturbance simulation with a vibration frequency of 4 Hz and amplitude of 3 mm. The specimen parameters for the concrete compressive strength test under traveling disturbance were designed as shown in Table 3. The specimen adopted the C60 concrete mix ratio used in the actual engineering bridge widening construction as shown in Table 1. In Table 3, C*i*-*j*d denotes the specimens were vibrated for *i* h and cured for *j* days, $i = 0, 3, 5, 7$ and $j = 3, 7, 28$.

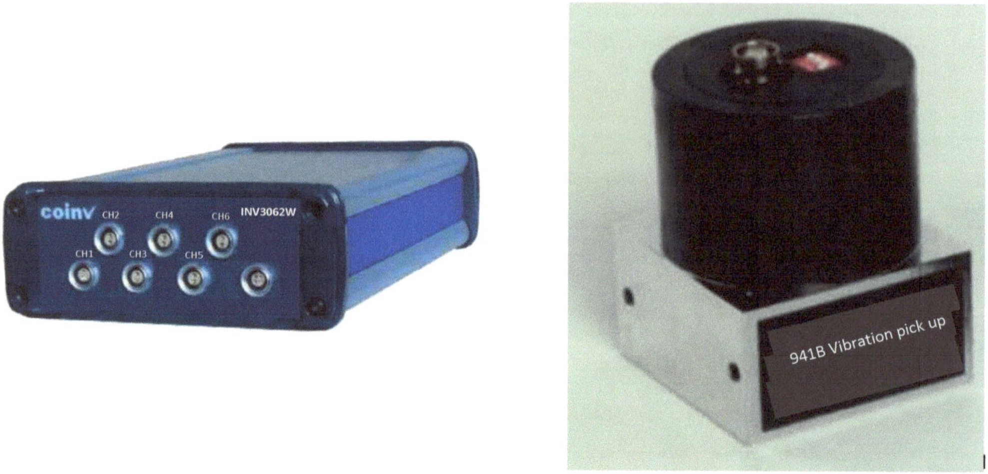

(**a**) INV3062W acquisition instrument (**b**) 941B vibration pickup

Figure 3. Data acquisition equipment.

Figure 4. On-site testing.

Table 3. Concrete specimen parameters for the concrete compressive strength test.

Specimen Group	Number	Vibration Time (h)	Curing Time (Day)	Specimen Group	Number	Vibration Time (h)	Curing Time (d)
C0-3d	3	0	3	C5-3d	3	5	3
C0-7d	3	0	7	C5-7d	3	5	7
C0-28d	3	0	28	C5-28d	3	5	28
C3-3d	3	3	3	C7-3d	3	7	3
C3-7d	3	3	7	C7-7d	3	7	7
C3-28d	3	3	28	C7-28d	3	7	28

With reference to GB/T 50107-2010 [27], the loading speed of the compressive strength test was taken as 0.8~1.0 MPa/s. The loading speed of the press was controlled to load continuously and uniformly. The testing instrument was a 30 t press, as shown in Figure 6. Each cubic compressive strength of concrete was calculated using Equation (1). The average of the compressive strengths of the three specimens was as the compressive strength of concrete.

$$f_{cu} = \frac{F}{A} \tag{1}$$

where f_{cu} is the cubic compressive strength (MPa); F is the ultimate load (N) of the specimen; A is the pressurized area of the specimen (mm^2).

Figure 5. Vertical vibration testbed.

Figure 6. Cube compressive strength testing device.

2.4. Splitting Tensile Strength Test of New-to-Old Concrete Interfaces Under Traveling Disturbance

With reference to GB50152-2012 [25] and JGJ55-2011 [26], 24 groups of new-to-old concrete surface splitting tensile strength specimens. Each group has 3 specimens. Each specimen consisted of two parts spliced together, simulating the inherent old concrete of the old bridge and the widened new concrete structure. The old concrete specimens were produced 1 year in advance. After production, they were placed in a standard rapid concrete curing box for 1 month before being taken out, and then they were placed in a natural state for curing. The old concrete was cut in half to obtain 150 mm × 150 mm × 75 mm specimens. New concrete specimens were made of C60 concrete with a size of 150 mm × 150 mm × 75 mm.

Old concrete surfaces need to be chiseled manually or mechanically, usually by hammer chiseling, wheel chiseling and high-pressure water jetting, before they are placed with new concrete. The varying degrees of roughness of the old concrete surface had a significant effect on the bond created between the two after the new concrete had been placed. Too shallow grooves would lead to the new concrete bond not being tight, and it was very easy to destroy by the external force. Too deep grooves would destroy the internal properties of the old concrete and high cost. In the actual project, a chiseling depth of 5–6 mm was usually adopted. Such surface roughness allowed for effective bonding between old and new concrete and more obviously simulated cracking loads and ultimate loads at the new-to-old concrete interface.

In this study, the old concrete specimens with good surface flatness were selected to carry out the roughness treatment of the bonding surface through the combination of manual chiseling and electric air hammer chiseling. The roughness of the bonding surface was controlled to be around 5–6 mm, and then the loose stones and dust on the surface were removed by using an angle grinder and a steel brush. The surfaces of the specimens were wet with water. After the specimen surface was free of moisture, brush the cement mortar with the same proportion as the new concrete. The well-mixed new concrete was poured into the test mold, and then placed on the standard vibration table for vibration. The surface of the new-to-old concrete specimen was smoothed, and it was placed on the vertical vibration test table to apply the traveling disturbances. The new-to-old concrete specimen was made of two pieces of new and old concrete spliced together as shown in Figure 7.

(**a**) Dimension of the specimen (**b**) Mold for specimen fabrication (**c**) Photo of the specimen

Figure 7. New-to-old concrete specimens.

The most unfavorable disturbance conditions measured at the project site were used for the experiments. The vibration frequency of the electromagnetic vibration test table was 4 Hz, and the vibration amplitude was 3 mm. The splitting tensile strength tests of the new-to-old concrete interface under traffic disturbance were carried out with the vibration time, curing time and loading mode as the main parameters. The curing time and vibration time were set the same as that of the compressive strength specimen. The loading mode was divided into downward (D) new-to-old concrete splicing joints and perpendicular (P) new-to-old concrete splicing joints. The test parameters of the specimens are listed in Table 4. The concrete of the specimens was made of C60 concrete, as shown in Table 1. In Table 3, P*i*-*j* d denotes the specimen was vibrated for *i* h and cured *j* days, $i = 0, 3, 5, 7$ and $j = 3, 7, 28$.

Table 4. Parameters for split tensile strength specimens.

Specimen Set	Vibration Time (h)	Curing Time (Day)	Specimen	Vibration Time (h)	Curing Time (Day)
P0-3d (D/P)	0	3	P5-3d (D/P)	5	3
P0-7d (D/P)	0	7	P5-7d (D/P)	5	7
P0-28d (D/P)	0	28	P5-28d (D/P)	5	28
P3-3d (D/P)	3	3	P7-3d (D/P)	7	3
P3-7d (D/P)	3	7	P7-7d (D/P)	7	7
P3-28d (D/P)	3	28	P7-28d (D/P)	7	28

With reference to GB/T 50107-2010 [27], the loading speed of the split tensile strength test was taken as 0.08 MPa/s~0.10 MPa/s. The loading speed of the press was controlled to load continuously and uniformly. The testing instrument was a 30 t press and the cub splitting tensile strength test setup is shown in Figure 8. The splitting tensile strength of the specimen can be calculated by Equation (2). The average of the splitting tensile strengths of the three specimens was the splitting tensile strength of concrete.

$$f_{ts} = 0.627\frac{P}{A} \quad (2)$$

where f_{ts} is the cub splitting tensile strength (MPa); P is the ultimate load when the specimen is damaged (N); A is the split area of the specimen (mm^2).

Figure 8. Cube splitting tensile strength test.

3. Results and Discussion

3.1. Compressive Strength Under Traveling Disturbance

At the beginning of loading, slight vertical cracks slowly began to appear on the surface of the pressurized part of the test specimen. With the gradual increase in the load, the cracks began to grow and spread around, mainly along the vertical development of the specimen, and gradually turned into larger vertical cracks, while the concrete specimen was also constantly deformed, with fine powder falling. When the concrete specimen reached the limit load, most of the specimens appeared in vertical penetration cracks and transverse cracks on the pressurized surface. After 28 d of curing and 3 h of vibration, the specimens suddenly exploded and made a loud noise, breaking into several small pieces and powder, as shown in Figure 9a. Except for the specimens that were cured for 28 d and vibrated for 3 h, the others were not broken, as shown in Figure 9b.

(**a**) Concrete was broken

(**b**) Concrete was not broken

Figure 9. Concrete failure modes.

The calculated results of the 12 sets of specimens measured in the compressive strength experiments are shown in Table 5. To intuitively analyze the effect of traveling load disturbance on the compressive strength of concrete at different stages, the results are plotted in Figure 10. Compared with the specimens without vibration, the compressive strength of the specimens cured for 3 days and vibrated for 3 h increased by 4.3%, the compressive strength of the specimens cured for 3 days and vibrated for 5 h increased by 5.2%, the compressive strength of the specimens cured for 3 days and vibrated for 7 h decreased by 13.7%. Compared with the specimens without vibration, the compressive strength of the specimens cured for 7 days and vibrated for 3 h increased by 5.7%, the compressive strength of the specimens cured for 7 days and vibrated for 5 h decreased by 3.9%, the compressive strength of the specimens cured for 7 days and vibrated for 7 h decreased by 20.4%. Compared with the specimens without vibration, the compressive strength of the specimens cured for 28 days and vibrated for 3 h increased by 11.9%, the compressive strength of the specimens cured for 28 days and vibrated for 5 h decreased by 4.4%, the compressive strength of the specimens cured for 28 days and vibrated for 7 h decreased by 19.9%. It can be concluded that the traffic disturbance had little effect on the compressive strength of the specimens at the initial and initial to final set stages, significantly decreased that at the final set stage; the impact of continuous disturbance on concrete can be mainly divided into compaction and destruction; proper vibration can promote a more uniform distribution of cementitious materials and aggregates, as well as accelerate the cement hydration reaction, making the concrete more compact; however, prolonged disturbance can cause the internal slurry of the concrete upward floating to sink, resulting in layered segregation and a significant decrease in concrete strength [28].

The compressive strengths of concrete specimens cured for 3, 7 and 28 days with 3 h disturbance were increased by 4.3%, 5.7% and 11.9%, respectively. During the initial set stage, concrete exhibited a Bingham body in a plastic state. When disturbed during this stage, the concrete skeleton can self-heal to a certain extent. The low-frequency disturbance in the initial setting stage made the concrete agglomerated cement particles dispersed, and the surface of the cement particles hydrated calcium silicate gel formed by the semi-permeable membrane was damaged, which in turn promoted the hydration of cement. The emergence of hydration products made the concrete strength continuously improve. The effects were greater than the destructive effect caused by the disturbance. During the initial set stage, the concrete internal aggregates became more compact. The distribution between aggregates and cementitious materials was relatively uniform, which can resist the destructive effect caused by disturbance and improve the strength of concrete.

Table 5. Compressive strength of concrete.

Specimen Group	Ultimate Load (kN)	f_{cu} (MPa)	Average of Compressive Strength (MPa)
C0-3d	1055.3 1098.0 994.5	46.9 48.8 44.2	46.6
C0-7d	1199.3 1248.8 1325.3	53.3 55.5 58.9	55.9
C0-28d	1451.3 1485.0 1541.3	64.5 66.0 68.5	66.3
C3-3d	1280.3 1309.5 1399.5	47.4 48.0 50.4	48.6
C3-7d	1055.3 1098.0 994.5	56.9 58.2 62.2	59.1
C3-28d	1604.3 1647.0 1759.5	71.3 73.2 78.2	74.2
C5-3d	1057.5 1098.0 1152.0	47.0 48.8 51.2	53.7
C5-7d	1188.0 1201.5 1237.0	52.8 53.4 55.0	53.7
C5-28d	1413.0 1422.0 1444.5	62.8 63.2 64.2	63.4
C7-3d	843.8 884.3 985.5	37.5 39.3 43.8	40.2
C7-7d	974.3 1003.5 1026.0	43.3 44.6 45.6	44.5
C7-28d	1176.8 1188.0 1219.5	52.3 52.8 54.2	53.1

The compressive strength of concrete specimens cured for 3 days with 5 h disturbance was increased by 5.2%, and the compressive strengths of concrete specimens cured for 7 and 28 days with 5 h disturbance were decreased by 3.9% and 4.4%. After the initial set and before the final set, concrete with disturbance shifted from the plastic state to the curing state. Vibration to promote the advantages of cement hydration reaction was no longer obvious. At this time the disturbance caused by internal cracks in the mortar and the skeleton displacement cannot be self-healing. Harmful cracks occurred inside the concrete, causing strength loss. The long-term disturbance caused delamination and segregation between concrete mortar and aggregate, thereby decreasing the compressive strength of concrete.

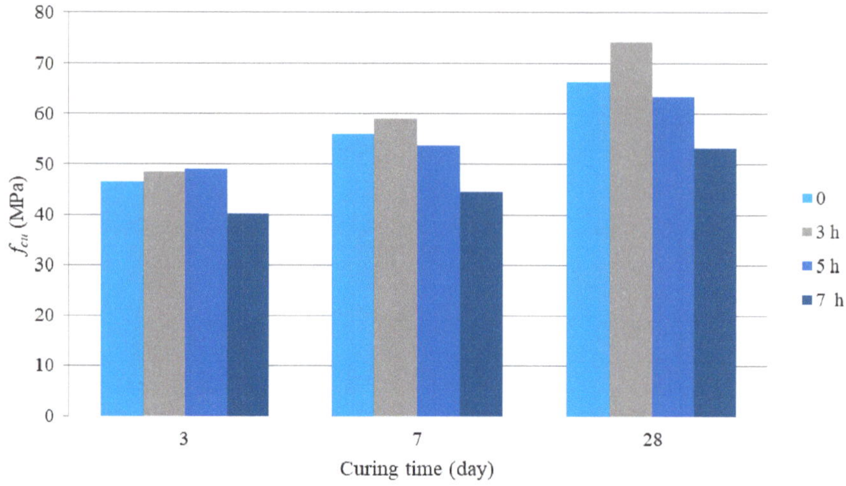

Figure 10. Compressive strength of concrete under travel disturbance.

The compressive strengths of concrete specimens cured for 3, 7 and 28 days with 7 h disturbance were decreased by 13.7%, 20.4% and 19.9%, respectively. Vibration to the final set of concrete, the concrete was hardened. The concrete developed a certain cohesion and resistance strength. Long-term disturbance had a greater destructive effect than the internal resistance of concrete, causing permanent damage to its performance. Moreover, prolonged disturbance led to the sinking of aggregates and the floating of cementitious materials, which cannot be uniformly combined; the phenomenon of delamination caused the formation of the bottom of the region of the coarse aggregate water-filled area; when the water evaporated with the curing time, it formed pores on the surface of the concrete, thereby reducing the strength of the concrete [29].

3.2. Splitting Tensile Strength Test Under Traveling Disturbance

Due to the low splitting tensile strength of concrete, with the increasing load, there was no obvious crack development on the surface of the concrete. When loaded to the ultimate load along the splice joints of the new-to-old concrete specimens, they cracked along the splice joints of the new-to-old concrete. The damage morphology is shown in Figure 11a. When loaded to the ultimate load along the direction perpendicular to the splice joints of the new-to-old concrete specimens, the specimens cured for 3 and 7 days still cracked along the new-to-old concrete splice joints, while the specimens cured for 28 days appeared new penetrating cracks along the direction perpendicular to the splice joints of the new-to-old concrete specimens. The damage morphology is shown in Figure 11b.

The results of the 24 sets of specimens measured in the split tensile strength experiments are shown in Tables 6 and 7. To intuitively analyze the effect of traffic disturbance on the split tensile strength of the new-to-old concrete interfaces, the measured data are plotted in Figures 12 and 13. The split tensile strengths of new-to-old concrete specimens without traffic disturbance gradually increased with the increase in curing time because of the internal slurry and aggregate coagulation and hardening of the specimens. Traffic disturbance on the hydration reaction of the cement had an impact on the hydration reaction of the cement. Different vibration times resulted in dislocations and cracks between the cementitious material and aggregates in the transition zone between new and old concrete interfaces, thereby reducing the splitting tensile strengths of the specimens.

(**a**) Damage along joints

(**b**) Damage along the direction perpendicular to joints

Figure 11. Damage morphology.

Table 6. Splitting tensile strength results in specimens loaded along the direction of the joints.

Specimen Group	Ultimate Load (kN)	f_{ts} (MPa)	Average of Compressive Strength (MPa)
P0-3d	81.0 81.0 83.3	2.26 2.26 2.32	2.3
P0-7d	90.0 92.3 94.5	2.51 2.57 2.63	2.6
P0-28d	108.0 119.3 123.8	3.00 3.32 3.45	3.3
P3-3d	74.3 74.3 78.8	2.07 2.07 2.19	2.1
P3-7d	77.5 79.3 83.3	2.16 2.21 2.32	2.2
P3-28d	94.4 99.0 103.7	2.63 2.76 2.89	2.8
P5-3d	75.0 76.5 76.8	2.09 2.13 2.14	2.1
P5-7d	80.0 83.6 86.1	2.23 2.33 2.40	2.3
P5-28d	101.6 104.8 104.8	2.83 2.92 2.92	2.9
P7-3d	66.0 67.8 79.3	1.84 1.89 2.21	2.0

Table 6. *Cont.*

Specimen Group	Ultimate Load (kN)	f_{ts} (MPa)	Average of Compressive Strength (MPa)
P7-7d	69.3 75.0 79.7	1.93 2.09 2.22	2.1
P7-28d	78.6 84.0 86.1	2.19 2.34 2.40	2.3

Table 7. Splitting tensile strength results in specimens loaded along the direction perpendicular to joints.

Specimen Group	Ultimate Load (kN)	f_{ts} (MPa)	Average of Compressive Strength (MPa)
P0-3d	76.5 76.5 81.0	2.13 2.13 2.26	2.2
P0-7d	87.8 87.8 90.0	2.45 2.45 2.51	2.5
P0-28d	148.5 150.8 150.8	4.14 4.20 4.20	4.2
P3-3d	76.5 76.5 78.8	2.13 2.13 2.19	2.2
P3-7d	69.6 75.7 82.9	1.94 2.11 2.31	2.1
P3-28d	131.3 136.0 137.4	3.66 3.79 3.83	3.8
P5-3d	78.6 79.7 81.8	2.19 2.22 2.28	2.2
P5-7d	75.7 80.0 80.0	2.11 2.23 2.23	2.2
P5-28d	135.0 139.5 144.0	3.76 3.89 4.01	3.9
P7-3d	76.5 76.5 78.8	2.13 2.13 2.19	2.2
P7-7d	72.0 72.0 74.3	2.01 2.01 2.07	2.0
P7-28d	108.0 108.0 112.5	3.01 3.01 3.14	3.1

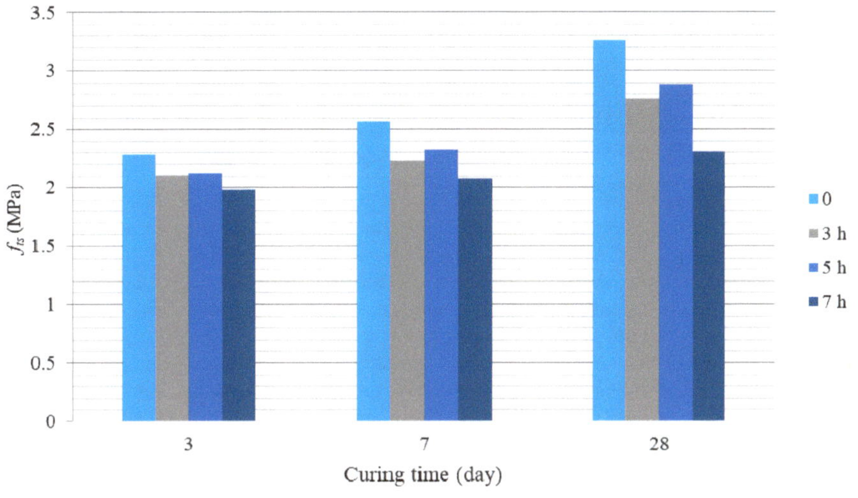

Figure 12. Comparison of splitting tensile strengths of specimens loaded along the direction of the joints.

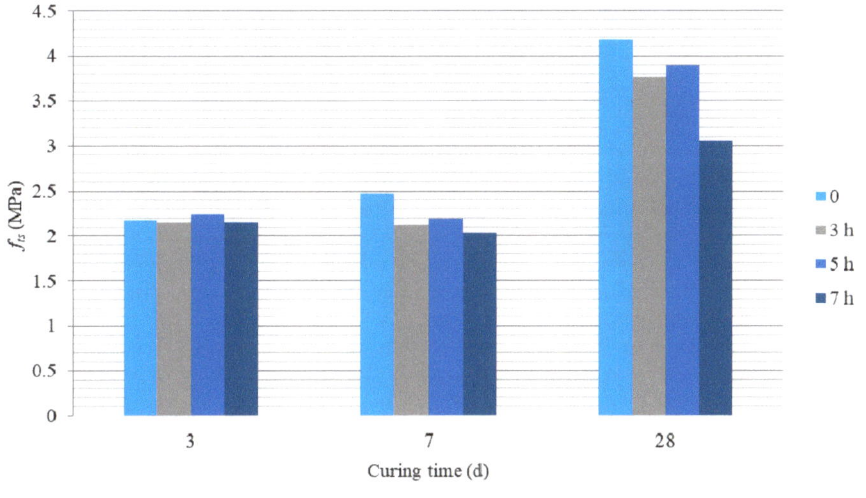

Figure 13. Comparison of splitting tensile strength of specimens loaded along the direction perpendicular to the joints.

Compared with the specimens loaded along the direction of the joints without vibration, the split tensile strengths of the specimens vibrated for 3 h and cured for 3, 7, and 28 days were decreased by about 8.7%, 12.1%, and 15.2%, respectively. The reason can be explained as follows. The specimens vibrated to the initial setting stage and failed to disperse the air bubbles and water secretion in the transition zone between the new and old concrete interfaces because of the vibration time. The hydrophilicity of the old concrete caused the water secretion and air bubbles in the new concrete to gather on the surface of the old concrete, which led to the high local water-cement ratio in the transition zone between the new and old concrete interfaces. The disturbance time was not long enough to dislodge the water secretion and air bubbles. The early vibration may cause the aggregate in the new concrete to form a kind of "point contact" with the stone bumps on the old concrete surface. The aggregate in the new concrete accumulated on the surface of the old concrete, which blocked the infiltration of the mortar with the bonding property, and thus affected the bond strength of the concrete. Compared with the specimens without vibration, the split tensile strengths of the specimens vibrated for 5 h and cured for 3, 7 and 28 days were decreased by about 8.7%, 11.5%, and 15.2%, respectively, because mortar aggregate segregation and internal skeleton displacement caused by vibration. Compared with the specimens vibrated for 3 h, the strengths of the specimens cured for 5 h increased because the vibration time (5 h) was sufficient to evenly vibrate the bubbles, bleeding, and other materials inside the transition zone between new and old concrete interfaces. Compared with the specimens without vibration, the split tensile strengths of the specimens vibrated for 7 h and cured for 3, 7, and 28 days were decreased by about 13.0%, 19.2%, and 30.3%, respectively, indicating that the disturbance-induced significantly decreased the splitting strength of the new-to-old specimen. Long-term vibration led to serious delamination of concrete mortar and aggregate segregation, resulting in the upper mortar layer of the specimen becoming thicker, and the bond strength of the specimen was seriously reduced. The vibration-induced delamination caused the moisture encapsulated on the surface of the aggregate to gradually sink, forming the phenomenon of internal osmosis; during the curing process, the water evaporated, and small pores appeared on the surface of the specimen and inside the specimen [29].

When loaded along the direction perpendicular to the joints, compared with the specimens loaded without vibration, the specimens cured for 3 and 7 days still cracked along the joint, the split tensile strengths of the specimens cured for 3 days changed little, while those of the specimens cured for 7 days and vibrated for 3 h, 5 h, and 7 h decreased by 16.0%, 12.0%, and 20.0%, respectively. The specimens cured for 28 days cracked in the direction perpendicular to the joint, and the split tensile strengths of the specimens vibrated for 3 h, 5 h, and 7 h decreased by 9.52%, 7.1%, and 26.2%, respectively. Vibrated for 3 h and 5 h and cured for 3 and 7 days, the vibration trends of the splitting tensile strengths of the specimens loaded along the direction perpendicular to the joints were almost the same as those of the specimens loaded along the direction of the joints. The shorter curing time resulted in less bond strength of the new-to-old concrete specimens. When loaded along the direction perpendicular to the joints, the specimen cured for 28 days was cracked perpendicular to the joint through the splice joint. The splitting tensile strength of the specimens cured for 28 days and loaded along the direction of the joints without vibration was 3.26 MPa. The smallest splitting tensile strength of the specimens cured for 28 d and loaded along the direction perpendicular to the joints was 3.05 MPa. It indicated that when loaded perpendicular to the direction of the joints, the bond of the old concrete also played a role, and the bond between the new and old concrete was strong enough to resist most of the disturbing forces.

4. Conclusions

To investigate the effect of traffic vibration on the compressive strength of high-strength concrete and the splitting strength of new-to-old concrete interfaces, the initial to final set time of high-strength concrete C60 was first investigated in this article. Then, the traffic disturbance parameters were determined. Later, the compressive strength of C60 concrete at different stages under traffic disturbance parameters was carried out. Finally, the splitting tensile strength of new-to-old concrete specimens at different stages with different loading modes was tested. This study can promote the widening and improvement of existing concrete highways and bridges, which can save resources and improve land use. The main conclusions can be drawn as follows:

(1) The initial setting time of C60 concrete was 3 h, and the final setting time was 7 h. The vibration frequency is 4 Hz, and the vibration amplitude is 3 mm.

(2) The compressive strengths of concrete specimens cured for 3, 7 and 28 days with 3 h disturbance were increased by 4.3%, 5.7%, and 11.9%, respectively. The compressive strength of concrete specimens cured for 3 days with 5 h disturbance was increased by 5.2%, and the compressive strengths of concrete specimens cured for 7 and 28 days with

5 h disturbance were decreased by 3.9% and 4.4%. The compressive strengths of concrete specimens cured for 3, 7, and 28 days with 7 h disturbance were decreased by 13.7%, 20.4%, and 19.9%, respectively.

(3) When loaded along the old and new concrete joint, the specimens cracked along the joint. Compared with the specimens cured for 3, 7, and 28 days without vibration, the split tensile strengths of the specimens vibrated for 3 h were decreased by about 8.7%, 12.1%, and 15.2%, respectively; those of the specimens vibrated for 5 h were decreased by about 8.7%, 11.5%, and 15.2%, respectively; those split tensile strengths of the specimens vibrated for 7 h were decreased by about 13.0%, 19.2%, and 30.3%, respectively.

(4) When loaded perpendicular to the joint, the specimens cured for 3 and 7 days still cracked along the joint, the split tensile strengths of the specimens cured for 3 days changed little, while those of the specimens cured and vibrated for 3 h, 5 h, and 7 h increased by 16.0%, 12.0%, and 20.0%, respectively. When loaded perpendicular to the joint, the specimens cured for 28 days cracked in the direction perpendicular to the joint, the split tensile strengths of the specimens vibrated for 3 h, 5 h, and 7 h increased by 9.52%, 7.1% and 26.2%, respectively.

The compressive strength of concrete decreases when it is disturbed by the whole process in the actual project. It is usually improved by adding admixtures and dopes to improve the maintenance conditions. The interfaces of the new-to-old concrete specimens were not reinforced with reinforcing steel in this study, so the bonding properties at the interface decreased considerably after disturbance. In the actual road bridge widening, reinforcement will be planted at the junction of old-to-new concrete, and the effect of traffic disturbance will be improved by controlling the roughness of the old concrete surface and the choice of interfacial agent during construction. In further study, the machine learning method will be developed to investigate the effect of parameters of concrete strength grade, different strength grades of old and new concrete, and the roughness of the old and new interfaces on the mechanical properties of new-to-old interfaces under traffic vibration.

Author Contributions: Conceptualization, P.G. and M.-L.Z.; methodology, P.G. and H.W.; software, H.W. and L.L.; validation, Z.L. and J.H.; formal analysis, H.W.; investigation, Z.L.; resources, M.-L.Z.; data curation, P.G.; writing—original draft preparation, P.G.; writing—review and editing, M.-L.Z.; visualization, H.W.; supervision, M.-L.Z.; project administration, P.G. and M.-L.Z. All authors have read and agreed to the published version of the manuscript.

Funding: This research has been supported by the Social Welfare Science Project of Nantong (MS2023064) and the Shandong Provincial Natural Science Foundation (ZR2024QE147).

Data Availability Statement: The original contributions presented in the study are included in the article, further inquiries can be directed to the corresponding author.

Conflicts of Interest: The authors declare no conflict of interest.

References

1. Ming, M.; Zheng, S.; Zhang, Y.; Zheng, Y.; Yang, S.; Song, M. Experimental study on the bond-slip behavior and stress transfer mechanism between shaped steel and high-performance fiber-reinforced concrete. *Structures* **2021**, *34*, 5013–5028. [CrossRef]
2. Zhou, T.; Liu, X.; Liu, H.; Li, Y.; Zhang, P.; Chen, H.; Chen, Z.; Li, L.; Shi, J.; Kou, J.; et al. Experimental study on fire resistance performance of concrete-filled steel plate composite walls. *Int. J. Steel Struct.* **2022**, *23*, 389–403. [CrossRef]
3. Tan, M.; Wu, Y.; Pan, W.; Liu, G.; Chen, W. Optimal design of a novel large-span cable-supported steel–concrete composite floor system. *Buildings* **2023**, *14*, 113. [CrossRef]
4. Deng, Y.; Li, L.; Lv, H.; Chen, X.; Ye, Z. 3D meso-scale numerical model and dynamic mechanical behavior of reinforced concrete. *Struct. Concr.* **2023**, *25*, 1819–1839. [CrossRef]
5. Huang, Y.; Zhao, P.; Lu, Y.; Liu, Z. Hysteresis performance of steel fiber-reinforced high-strength concrete-filled steel tube columns. *J. Constr. Steel Res.* **2024**, *219*, 108755. [CrossRef]
6. Shafighfard, T.; Kazemi, F.; Asgarkhani, N.; Yoo, D. Machine-learning methods for estimating compressive strength of high-performance alkali-activated concrete. *Eng. Appl. Artif. Intell.* **2024**, *136*, 109053. [CrossRef]
7. Yilima, Y.; Nayir, S. Machine learning based prediction of compressive and flexural strength of recycled plastic waste aggregate concrete. *Structures* **2024**, *69*, 107363.

8. Alyaseen, A.; Poddar, A.; Kumar, N.; Sihag, P.; Lee, D.; Kumar, R.; Singh, T. Assessing the compressive and splitting tensile strength of self-compacting recycled coarse aggregate concrete using machine learning and statistical techniques. *Mater. Today Commun.* **2024**, *38*, 107970. [CrossRef]
9. Zhou, J.; Zhou, Z.; Shi, Z.X. Traffic-induced vibrations at the wet joint during the widening of concrete bridges and non-interruption traffic control strategies. *Comput. Concr.* **2023**, *32*, 411–423.
10. Zhu, H.; Fu, Z.; Wang, Y.; Zhang, N. Study on splitting tensile strength of interface between the full lightweight ceramsite concrete and ordinary concrete. *Case Stud. Constr. Mater.* **2023**, *18*, e01829. [CrossRef]
11. Harsh, S.; Darwin, D. Effects of traffic induced vibrations on bridge deck repairs. In *NCHRP Synthesis of Highway Practice*; University of Kansas Center for Research, Inc.: Lawrence, KS, USA, 1984; pp. 1–68.
12. Ng, P.L.; Kwan, A. Effects of traffic vibration on curing concrete stitch: Part II—Cracking, debonding and strength reduction. *Eng. Struct.* **2007**, *29*, 2881–2892. [CrossRef]
13. Issa, M.A.; Yousif, A.A.; Issa, M.A. Effect of construction loads and vibrations on new concrete bridge decks. *J. Bridge Eng.* **2000**, *5*, 249–258. [CrossRef]
14. Dunham, M.R.; Rush, A.S.; Hanson, J.H. Effects of induced vibrations on early age concrete. *J. Perform. Constr. Facil.* **2007**, *21*, 179–184. [CrossRef]
15. Jiang, Z.; Ren, Q.; Yuan, Z. Effects of vehicle-bridge coupled vibration on early-age properties of concrete and its damage mechanism. *J. Build. Mater.* **2015**, *18*, 471–476.
16. Zhang, C.; Xu, B.; Zhao, Z.; Ni, T.; Liu, B. Influence of early age disturbance on mechanical properties of concrete based on acoustic emission technology. *Bull. Chin. Ceram. Soc.* **2020**, *40*, 3964–3975.
17. Li, L.; Xie, D.; Shi, L.; Zhou, F. Influence of three-dimensional disturbance on the mechanical properties of concrete in Nanjing Yangtze River Public-Railway Bridge. *China Concr. Cem. Prod.* **2019**, *3*, 85–88.
18. Dai, M.; Li, H. Influence of basalt fiber on the mechanical properties of concrete under disturbance condition. *Bull. Chin. Ceram. Soc.* **2017**, *36*, 642–648.
19. Pan, H.; Wang, F.; Zhao, Q. Influence of distribution in hardening stage on mechanical properties of concrete. *J. Build. Mater.* **2016**, *19*, 631–636.
20. Ahmed, M.D.; Abd, E.M.; Abd, E.; Mohamed, R.; Tag, E. Slant shear bond strength between self compacting concrete and old concrete. *Constr. Build. Mater.* **2017**, *130*, 73–82.
21. Wall, J.S.; Shrive, N.G. Factors affecting bond between new and old concrete. *ACI Mater. J.* **2011**, *85*, 449–456.
22. Wei, J.; Xing, J.; Fu, Z. Effect of traffic load induced bridge vibrations on concrete tensile properties. *J. Southeast Univ.* **2010**, *40*, 1057–1060.
23. Pan, H.; Zuo, J.; Song, R. Influence of maximum size of coarse aggregate on mechanical properties of concrete disturbed in early stage. *Bull. Chin. Ceram. Soc.* **2020**, *39*, 3131–3136.
24. *GB/T 50080-2016*; Standard for Test Method of Performance on Ordinary Fresh Concrete. National Standards of People's Republic of China; China Architecture and Building Press: Beijing, China, 2016.
25. *GB/T 50152-2012*; Standard for Test Method of Concrete Structures. National Standards of People's Republic of China; Ministry of Housing and Urban-Rural Development PRC: Beijing, China, 2012.
26. *JGJ 55-2011*; Specification for Mix Proportion Design of Ordinary Concrete. Ministry of Housing and Urban-Rural Development of the People's Republic of China; China Architecture & Building Press: Beijing, China, 2011.
27. *GB/T 50107-2010*; Standard for Evaluation of Concrete Compressive Strength. National Standards of People's Republic of China; China Standard Press: Beijing, China, 2010.
28. Li, L. Analysis on influence mechanism of disturbance on concrete compression strength. *North. Commun.* **2018**, *5*, 59–67.
29. Luo, Y.; Yu, C.-X.; Gu, A.-J. Study of the bond-slip behavior between rebar and concrete based on acoustic emission technology. *J. Exp. Mech.* **2011**, *26*, 343–349.

buildings

MDPI

Article

Nonlinear Static Analysis of Spherical Hinges in Horizontal Construction of Bridges

Lin Zhao [1], Xiaohu Sun [2,*], Zhe Wu [3], Ying Chen [3], Jian Liu [4] and Youzhi Wang [3]

1 Shandong Expressway Construction Management Group Co., Ltd., Jinan 250098, China; zhaolin_zl99@163.com
2 Shandong Tangzheng Testing Co., Ltd., Zibo 255000, China
3 School of Civil Engineering, Shandong University, Jinan 250061, China; 202235074@mail.sdu.edu.cn (Z.W.); chenying@sdu.edu.cn (Y.C.); wangyouzhi@sdu.edu.cn (Y.W.)
4 Wudi County Urban Construction Service Center, Binzhou 251900, China; lj1079247000@163.com
* Correspondence: xiaohusun88@163.com or xiaohu1980@163.com

Abstract: During the construction of parallel swivel bridges, the stress state of the spherical hinge under load is crucial. The stress results of the spherical hinge are of great significance to the subsequent structural design and even the safety and stability of the bridge structure. The refined finite element model of the spherical hinge was established using ABAQUS software. The vertical displacement and local stress state of the spherical hinge under vertical load were analyzed and discussed. The results indicate that the maximum principal compressive stress is less than the allowable stress and meets the requirements. When the spherical hinge is only subjected to the upper static load, the vertical stress of the upper and lower steel spherical hinges gradually increases along the center of the spherical hinge to the edge, in which the vertical stress at the edge of the spherical hinge is the largest and the stress of the lower spherical hinge is slightly smaller than that of the upper spherical hinge. The maximum compressive stresses of the upper spherical hinge and the lower spherical hinge are 63.89 MPa and 54.24 MPa, respectively. Under the upper static load, the displacements of the upper and lower spherical hinges are very small, with maximum displacements of 0.234 mm and 0.202 mm, respectively, thus meeting the design requirements.

Keywords: spherical hinge; vertical loading finite element simulation; static analysis

1. Introduction

Rotating construction can transform the complex aerial and water construction into level ground construction, which greatly reduces the construction difficulty and thus significantly shortens the construction period. During the construction process, it has less impact on the normal operation of the adjacent lines, thus achieving good economic and social benefits [1–4]. The essence of the rotating construction method is to build the bridge in a non-original position, and then rotate the bridge to the designed position through the rotation system [5–7]. According to the different construction methods, the rotating construction can be divided into three types: horizontal rotation [8], vertical rotation [9,10], and horizontal-vertical combination rotation, among which the horizontal rotation method is the most widely used in engineering practice, especially in the construction of bridges across rivers and existing lines in plain areas.

The spherical hinge is a key part of the rotating construction of a bridge, and the analysis of its stress state is crucial to ensure the safety and stability of the entire bridge superstructure during the rotation process. The nonlinear effects in spherical hinges refer to the complex phenomena such as plastic deformation, significant deformation influence, contact behavior, buckling characteristics, load path dependency, time-related changes, and environmental impacts, etc., which occur when the material is stressed. Currently, theoretical, numerical, and experimental validation methods have been used to analyze the stresses and displacements of spherical hinges. Parsons [11] analyzed the stresses and displacements of spherical hinges using the displacement method. On the basis of Parsons' study, Tuba and Chan [12] proposed the contact state assumption and solved the contact surface forces based on the displacement method. Shuting Li [3] applied the mathematical optimization analysis method to analyze the various types of errors appearing in the finite element analysis, analyzed the influence of these errors on the results of contact problem analysis, and summarized their regularities. Huang et al. [13] accurately predicted the impact of two main factors on the total deformation quality of the spherical hinge and discussed the results through optimization graphics. Hou [14] used the finite element method to conduct a study on the stress distribution of spherical hinges under static loading, with a V-shaped rigid bridge as the basic background. Liu et al. [15] proposed a new model for calculating the critical overturning moment of a swing bridge based on the stress distribution of the concrete spherical hinge joint, using non-Hertzian contact theory. Wang et al. [16] analyzed the impact of various factors on the total stress, friction stress, and slip distance of the rotating structure, and optimized the turntable parameters based on the analysis results. Zhao [17] considered the force conditions of the bracing feet around the spherical hinge in the rotating system under the eccentric state of the rotating bridge and validated the simulation analysis using the finite element method, proposing a new calculation formula that can accurately calculate the load borne by the bracing feet. Lan [18] studied the relationship between the weight and radius of the rotating spherical hinge under overloading, modeled and analyzed the stress distribution pattern of the spherical hinge using finite element software, and compared it with theoretical calculation values, which provided a solution to improve the load-bearing capacity of concrete. Fan et al. [19] established the corresponding finite element spatial model for the spherical hinge to analyze the central contact stress of the concrete-made spherical hinge and derived the functional relationship between the maximum contact stress and the design parameters of the spherical hinge.

In terms of experimental validation, Campbell [20] described the load transfer mechanism of a spherical hinge containing a polytetrafluoroethylene (PTFE) layer and featuring a bending-sliding compression surface under horizontal loading and analyzed its influence on the design of the spherical hinge. Chen et al. [21] proposed a new friction model to describe structural spherical hinges operating in quasi-static processes. The results indicated that the coefficient of stiffness deviation from the model can be explained microscopically by an asperity-based theory. Shi et al. [22] proposed a novel contact stress calculation method that integrated the cantilever structure of a spherical hinge with an elastic foundation model and verified it by 3D finite element models. Li et al. [23] proposed a reasonable range of values for rotational angular velocity and angular acceleration according to the length of the rotating cantilever by combining the model test data of the Xiangbei Shipyard Bridge. Yang [24] designed and simulated the spherical hinge for rotating construction and revealed its good overturning resistance for the wind resistance stability problem in the rotational process. Additionally, Yu et al. [25] and Xie et al. [26] have applied a new type of high-strength, high-toughness, and high-durability engineering material, reactive powder concrete (RPC), to the construction process of rotating bridge spherical hinge structures. Yuan et al. [27] found that the stiffness of concrete spherical hinges can be improved by filling them with RPC, which ensures the manufacturing precision and stable installation of the spherical hinges.

In summary, there have already been many studies on the stress and displacement analysis of spherical hinges. However, there are fewer static analyses of the spherical hinge under vertical loading. Additionally, there are fewer discussions comparing them with the European standard limits. The stresses and equilibrium state of spherical hinges are the key control elements for structural analysis, design optimization, and construction control of horizontal construction bridges. Therefore, a refined finite element model of the spherical hinge was established using ABAQUS 2018 software. The vertical displacement and local force state of the spherical hinge under vertical load were analyzed and discussed.

2. Project Overview and the Spherical Hinge

2.1. Project Overview

The main bridge of a high-speed interchange has a span of 2 × 70 m, giving a total length of 140 m, with diagonal crossings. It features a single-box, four-cell cross-section, V-shaped piers, and a rigid frame system, with a total bridge width of 33.7 m. The girders utilize a bi-directional prestressing system with both longitudinal and transverse reinforcement. The main bridge crosses the existing Jiaoji railway and adopts the horizontal construction method. The construction project involved the casting of (65 + 65) m long girders on both sides of the Jiaozhi railroad line using full-tangled brackets. After construction of the anti-collision guardrails on both sides of the bridge deck was completed, the entire V-shaped structure was rotated counterclockwise by 86° to its final bridge position. Subsequently, the alignment was adjusted, and the upper and lower plates of the rotation system were sealed and solidified. The weight of the rotating section was 18,800 tons.

2.2. Main Construction of the Rotating System

The rotation system of a horizontal construction mainly consists of three parts: the load-bearing system, the top push traction system, and the balance system. The structure of the rotating system is shown in Figure 1. The load-bearing system consists of a lower bearing platform, a spherical hinge, and an upper bearing platform. The lower bearing platform is the foundation that supports all the weight of the rotating structure, and the upper bearing platform forms the foundation of the bridge together with the completion of the rotating body. The upper and lower bearing platforms are reinforced concrete structures. The grade of concrete is C50, and the grade of reinforcement bars is HRB400. Vertical prestressing tendons are arranged on the upper bearing platform. The lower bearing platform is equipped with a lower spherical hinge of the rotating system, the lower spherical hinge supporting steel frame, the insurance foot, the ring slide, and the rotating jack reaction seat. Eight symmetrical supports are set around the upper turntable, each of which is double cylindrical and supported by a 24 mm thick steel walking plate. The processing accuracy of the walking plate is grade 3. The double cylinders are two steel pipes, each with a diameter of 800 mm. The supports are filled with C50 micro-expansive concrete. A turntable is provided in the portion where the spherical hinge and the gusset footing are connected to the upper disk. In the turntable, the ring steel beam is set up as the part of traction force exerted by the rotating body. At the same time, the bottom surface of the upper bearing platform is arranged with steel pipe concrete gusset footing and tray. Polytetrafluoroethylene (PTFE) plates are embedded between the upper and lower spherical hinges at the designed positions, with butter and PTFE powder applied between the plates. A positioning steel pin passes through the centers of the upper and lower spherical hinges for precise alignment. The mating surfaces of the upper and lower spherical hinges are tightly wrapped with tape on the outer periphery to ensure dust-proofing, waterproofing, and rust prevention. The foundation for the main piers of the swing bridge consists of 20 1.8 m diameter reinforced concrete bored piles with a length of 33 m.

The balance system consists of the spherical hinge structure itself, an upper turntable steel pipe concrete circular truss footing, gusset feet for double cylinders, and double cylinders for two steel pipes. Micro-expansion concrete is poured inside the truss footing. During installation, steel plates are supported under the leg walker as a gap between the turntable structure and the chute. After the drop frame is installed, it is ensured that there is a gap between the truss footing and the chute to allow the gusset feet to slide within the chute during rotation. Figure 2 gives a schematic diagram of the installation of the lower turntable, the truss footing, and the upper turntable of the rotating system.

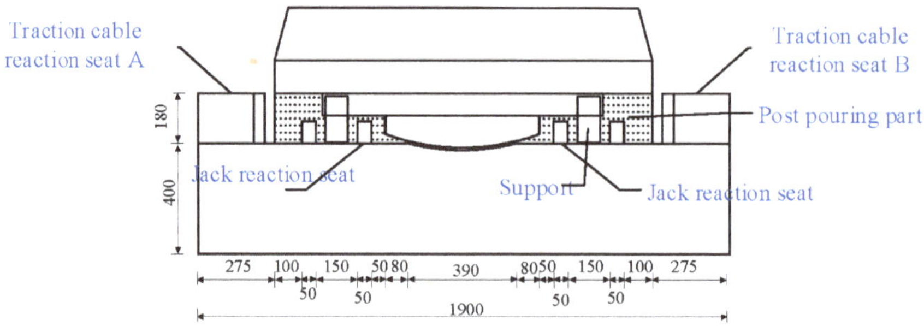

(**a**) Cross-bridge elevation view

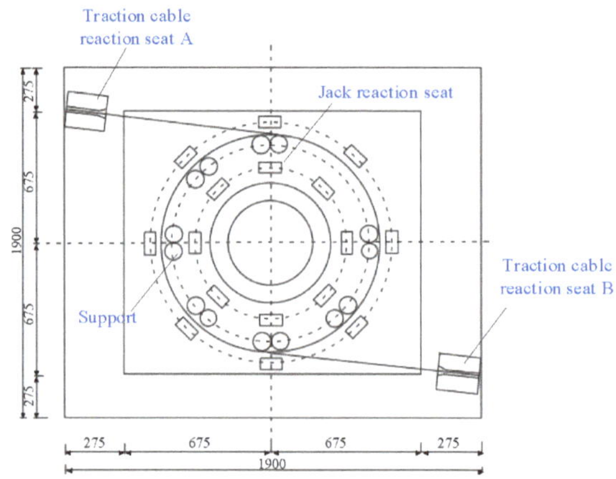

(**b**) Plan view

Figure 1. The structure of the rotating system (Unit: cm).

Figure 2. Schematic diagram of rotating system installation.

2.3. Parameters of the Spherical Hinge

The diameter of both the upper and lower spherical hinge planes is 3900 mm. The surface radius of the upper spherical hinge is 8000 mm, and that of the lower spherical hinge is 8040 mm. The thickness of both the upper and lower spherical hinges is 40 mm. To facilitate positioning and installation, a 325 mm diameter center pin is provided in the center of the spherical hinge, as shown in Figure 3. There is no interpenetration between the surfaces of the upper and lower turntable of the spherical hinge. Normal stresses can be transferred to each other between the upper and lower surfaces. There may be a gap between the upper and lower surfaces of the ball hinge, but there is no normal tension.

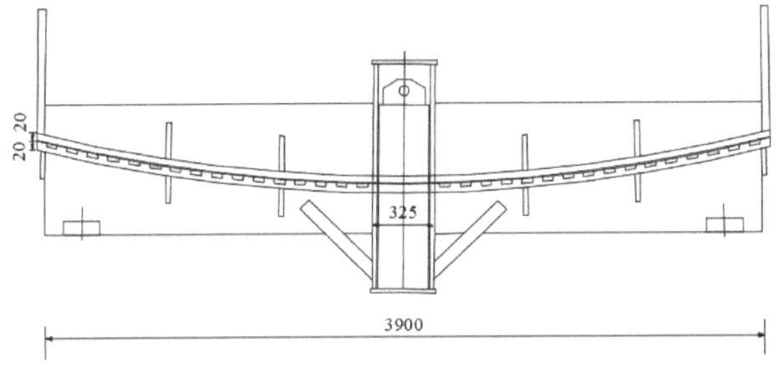

Figure 3. Upper and lower spherical hinge assembly (unit: mm).

The steel ball-hinge surface was made by factory prefabrication, while an appropriate number of concrete vibration holes were reserved on the lower spherical hinge surface to facilitate concrete construction below the spherical hinge surface. The pre-embedded sleeve of the spherical hinge center shaft was first precisely positioned and fixed, and then concrete was poured, thus facilitating the rotation of the center shaft. Later, the pouring of concrete for the lower spherical hinge was carried out. The steel rods for the rotating center shaft were placed into the lower turntable pre-embedded sleeve. Finally, the PTFE slide on the lower spherical hinge and the upper spherical hinge were installed. The PTFE slides were placed into the corresponding inlay holes in numerical order, ensuring that they were under high pressure.

2.4. Limitations of Spherical Bearings

BS EN 1337-7-2004 PART 7 [28] specifies the limitations for spherical hinges. The following provides a detailed introduction to its limitations.

Separation of the sliding surfaces may lead to wear due to contamination and increased deformation due to lack of confinement. To ensure that the sphere does not separate during rotation and that the eccentricity is within the projected area, Equation (1) must be satisfied. The limit of eccentricity for the spherical hinge in this study was calculated to be 487.5 mm based on Equation (1).

$$e_t \leq \frac{L}{8} \tag{1}$$

where e_t is the total eccentricity and L is the diameter of the projected area of the PTFE sheet.

Excessive pressure may lead to loss of the sliding function, which results in a state of structural failure or near structural failure. Therefore, this condition of Equation (2) is considered to be the ultimate limit state. According to Equation (2), the pressure limit for the spherical hinge is 7.7×10^8 kN.

$$N_{Sd} \leq \frac{f_k}{\gamma_m} \times A_r \tag{2}$$

where N_{Sd} is the design axial force at ultimate limit state; f_k is the characteristic value of compressive strength for PTFE sheets, referenced from tables in the code; γ_m is a coefficient, with a recommended value of 1.4; and A_r is the reduced contact area of the curved sliding surface, as given by Equation (3).

$$A_r = \lambda \times A \tag{3}$$

where A is the area of the projected curved sliding surface and λ is a coefficient given in tables in the code.

Spherical bearings must always ensure that the stainless steel plate in contact with the PTFE slider can completely cover the PTFE slider within the design rotation angle. In addition, there should be no direct contact between the upper plate of the bearing and the body of the bearing; a certain amount of space should be maintained to avoid restricting the rotation of the spherical bearing.

3. Static Analysis of the Spherical Hinge Under Vertical Load

3.1. Finite Element Models

The contact surface stress and contact surface deformation of the rotating spherical hinge under static load were established and analyzed using ABAQUS finite element software. To minimize the error between the numerical and actual results, the design dimensions of the finite element model are consistent with the actual dimensions. When defining contact surfaces, the surface of the upper spherical hinge was defined as the slave surface and the lower spherical hinge surface was defined as the master surface. For the contact mechanical behavior, penalty friction in ABAQUS software was applied to simulate the tangential behavior. The normal behavior was set to "hard" contact, the constraint method used "penalty", separation after contact was not allowed, and the contact stiffness was linear. The coefficient of friction was 0.6 based on experience. In accordance with the relevant requirements of Chinese standards [29,30], the mechanical parameters of C50 concrete and Q345 steel were tested. The equipment was manufactured by Xin Guang Testing Machine Manufacturing Co., Ltd in Jinan, China. Tables 1 and 2 list the parameters of the model materials. C50 concrete indicates that the standard value for the cube compressive strength of the concrete is 50 MPa. Q345 steel indicates that the yield strength of this steel is 345 MPa. According to Chinese standards [31], the standard value of the axial compressive strength of C50 concrete is 32.4 MPa, and the standard value of the axial tensile strength is 2.64 MPa.

Table 1. Parameter list of model concrete materials.

Material	Poisson's Ratio	Elastic Modulus (MPa)	Density (kg/m^3)	Cubic Compressive Strength (MPa)	Axial Tensile Strength (MPa)
C50 concrete	0.2	3.45×10^4	2.06×10^5	51.9	3.06

Table 2. Parameter list of model steel materials.

Material	Poisson's Ratio	Elastic Modulus (MPa)	Density (kg/m^3)	Yield Strength (MPa)	Ultimate Tensile Strength (MPa)
Q345 steel	0.3	2.60×10^3	7.85×10^3	345	610

The finite element model was established as shown in Figure 4. The dotted lines in Figure 4 are the central axes. The mesh of the spherical hinge bearing concrete was coarsely divided and the spherical hinge mesh was refined. Tetrahedral meshing was used for the upper and lower spherical hinge concrete and hexahedral meshing was applied to the upper and lower ball-hinges. The mesh size for spherical hinge bearing concrete was 100–150 mm. The mesh size for the spherical hinge was 10–15 mm. The finite element modeling using the above method ensures the accuracy of the calculation and facilitates the convergence of the solution.

Fixed restraints were applied at the bottom of the lower abutment to constrain its displacement and rotation degrees of freedom, and the upper abutment only carried the loads transferred from the bridge superstructure. A reference point was set at the top of the upper bearing platform. All degrees of freedom of the upper bearing platform were coupled to the established reference point. The Midas finite element model of the whole bridge established during the construction monitoring process was analyzed to obtain the vertical reaction force value at the top of the pier. This load value includes the dead load and live load transmitted from the upper part, amounting to 1.8×10^8 kN. This vertical reaction force value was loaded to the reference point in the same way as a spherical hinge lower center bearing. The load was applied without considering the unbalanced moment and only the vertical reaction force from the top of the pier was considered.

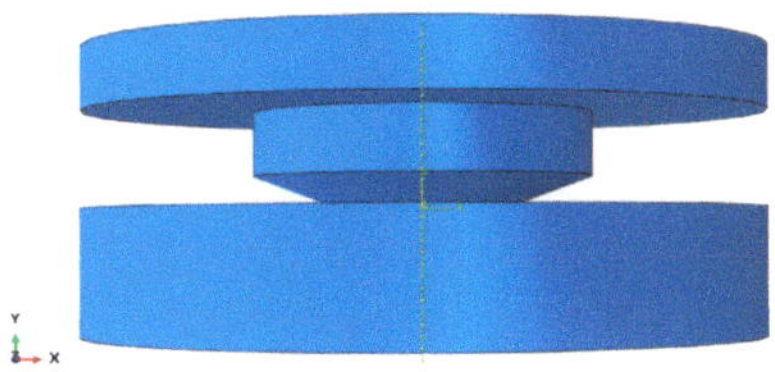

(**a**) Rotating system

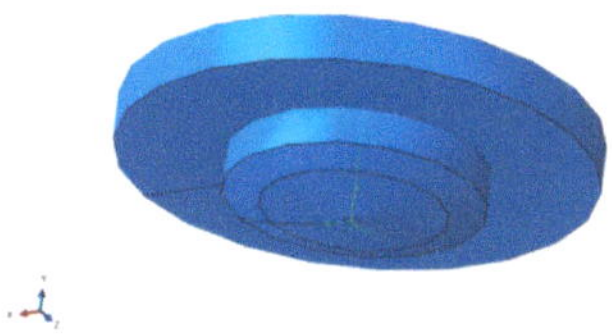

(**b**) Upper spherical hinge

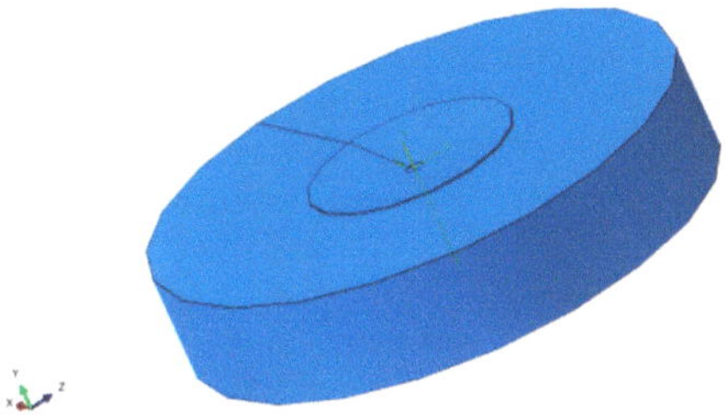

(**c**) Lower spherical hinge

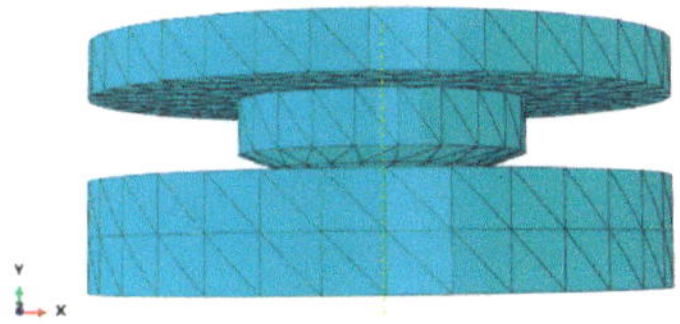

(**d**) Meshing

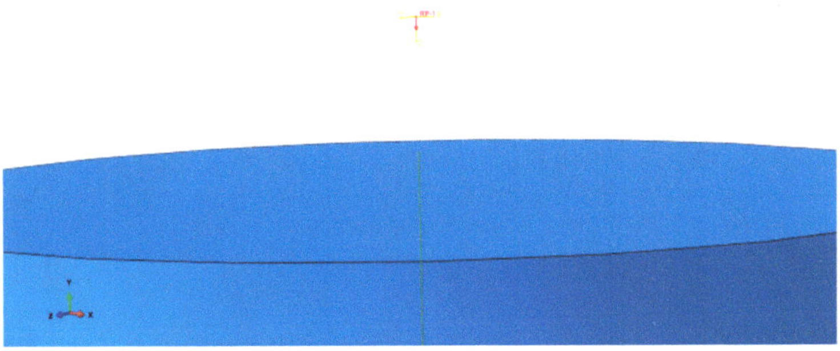

(**e**) Loading

Figure 4. Schematic diagram of the finite element model.

3.2. Finite Element Model Validation

Hertzian contact theory [32] assumes that the contact surface is an ideally smooth surface, the friction force can be neglected, and the deformation at the contact area is small, localized, and the boundary dimensions of the contact surface are much smaller than the geometric dimensions of the elastic body. According to Hertzian contact theory, the maximum contact pressure is given by Equation (4).

$$\sigma = \frac{3}{2}\frac{P}{\pi}\left[\frac{4(R_1 + R_2)}{3\pi P(k_1 + k_2)R_1R_2}\right]^{\frac{2}{3}} \tag{4}$$

where σ represents the maximum contact stress with the unit of MPa; P is the vertical load value with the unit of kN; R_1 and R_2 are the radii of the lower and upper spherical bodies, respectively, with the unit of mm; and $K_1 = (1 - {\mu_1}^2)/\pi E_1$, $K_2 = (1 - {\mu_2}^2)/\pi E_2$, where μ_1 and E_1 are the material constants of the lower spherical body, and μ_2 and E_2 are the material constants of the upper spherical body.

If we take $E_1 = E_2 = E$ and $\mu_1 = \mu_2 = 0.3$, then the maximum contact pressure is given by Equation (5).

$$\sigma = 0.388\left[\frac{PE^2(R_1 + R_2)^2}{R_1^2 R_2^2}\right]^{\frac{1}{3}} \quad (5)$$

Equation (5) is the commonly used calculation formula in engineering, where the maximum pressure stress on the contact surface occurs at the center of the contact area. For the contact mode of the rotating spherical hinge, simply taking R_1 as a negative value will suffice. The Hertz contact theory calculation result is 52.97 MPa, which fits the finite element results well with an error of 2.01%. This verifies the accuracy of the finite element model.

3.3. Finite Element Results

3.3.1. Local Force State of Spherical Hinge Under Vertical Loading

Figure 5 shows the mises stress in the spherical hinge. Figure 6 shows the vertical stress clouds of each component of the rotating system. It can be seen from Figure 6a that the vertical stress distribution between the upper bearing platform and the ball-hinge contact surface gradually increases from the middle to the radius direction, and the maximum vertical stress occurs on the circumferential boundary of the contact surface, which is 7.24 MPa and meets the specification requirements. At the same time, some tensile stresses are observed in the concrete surrounding the spherical hinge, indicating that care should be taken to ensure the quality of the concrete in this area during construction. It can be seen from Figure 6b that the vertical stress distribution of the upper spherical hinge is approximately the same as that of the upper bearing platform, with the minimum value appearing at the center and gradually increasing toward the edge, and the maximum stress at the edge is 57.86 MPa. From Figure 6c, it can be seen that the stress on the contact surface of the lower spherical hinge is approximately the same as that of the upper turntable, with a minimum stress of 10.14 MPa and a maximum stress of 54.06 MPa. It can be seen from Figures 5 and 6d that the magnitude of the maximum vertical stress on the contact surface in the lower bearing of the rotary system is relatively smaller than that in the upper bearing, with a magnitude of 3.88 MPa. Additionally, Figure 6d reveals that there is a small tensile stress in the concrete surrounding the spherical hinge of the lower bearing platform. By calculation using Equation (2), the pressure limit for the spherical hinge in this study is found to be 7.7×10^8 kN. The compressive stress limit for the spherical hinge calculated using Equation (5) is 90.4 MPa, and the stress values in Figures 5 and 6 are both within the limit.

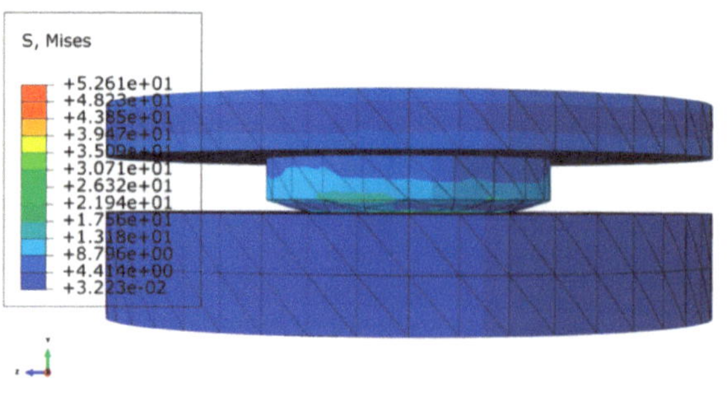

Figure 5. Stress cloud of the rotating system (unit: MPa).

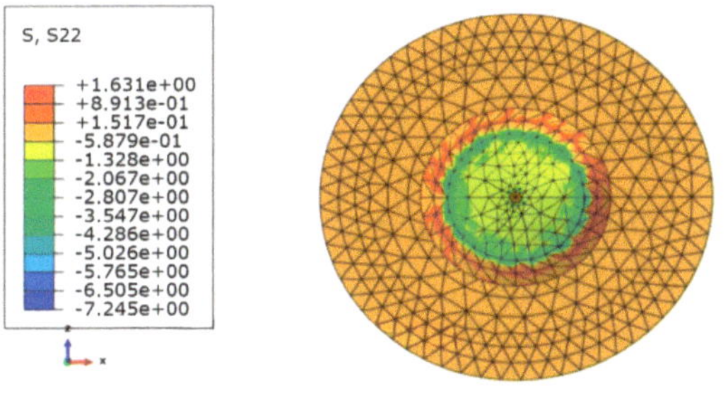

(**a**) The upper bearing platform

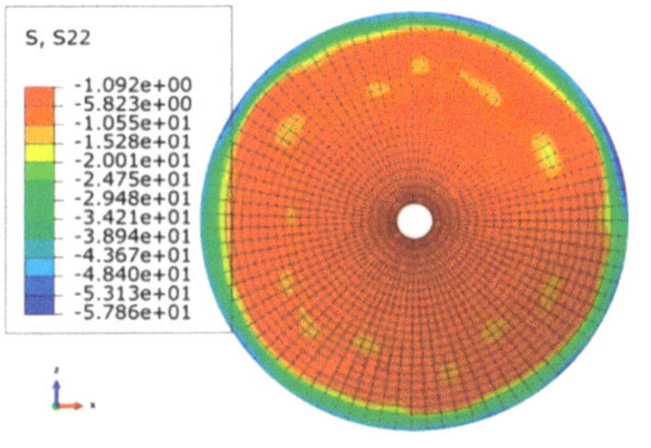

(**b**) The upper spherical hinge

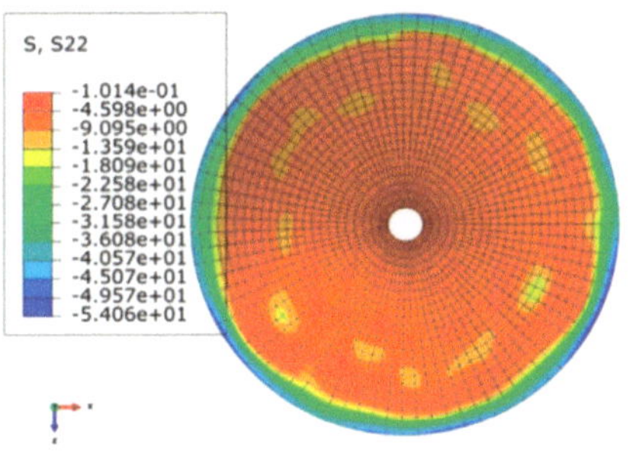

(**c**) The lower spherical hinge

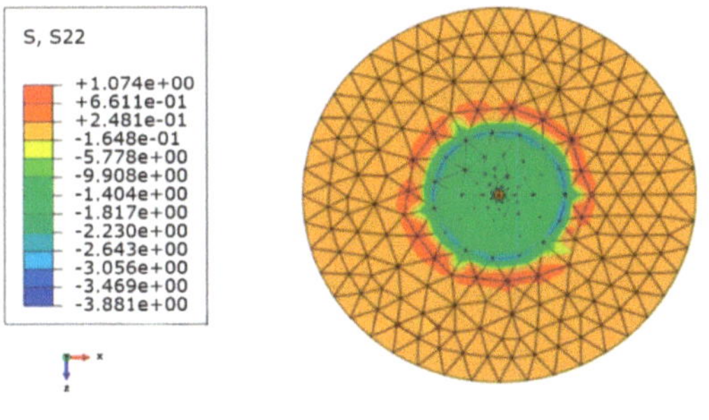

(**d**) The lower bearing platform

Figure 6. Vertical stress clouds of each component of the rotating system (unit: MPa).

3.3.2. Vertical Stress of the Spherical Hinge

To analyze the vertical stresses at different locations of the ball-hinge contact surface under vertical load, different points were taken at the same cross-section on the upper and lower ball-hinge contact surfaces to compare and analyze the change of vertical stress along the radius direction of the ball-hinge. Figure 7 shows the variation of the vertical stresses in the upper and lower spherical hinges with the distance from the center of the spherical hinge. For the upper and lower spherical hinges, the change trend of the vertical stress is from the center of the spherical hinge contact surface to the edge of the spherical hinge and reaches the maximum value at the edge of the spherical hinge. The maximum vertical stress in the upper spherical hinge is slightly higher than in the lower one, and the maximum values for both are less than the design strength values of the components.

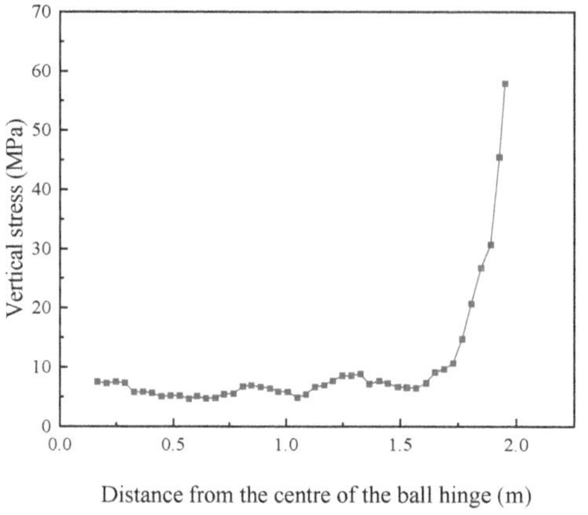

(**a**) The upper spherical hinge

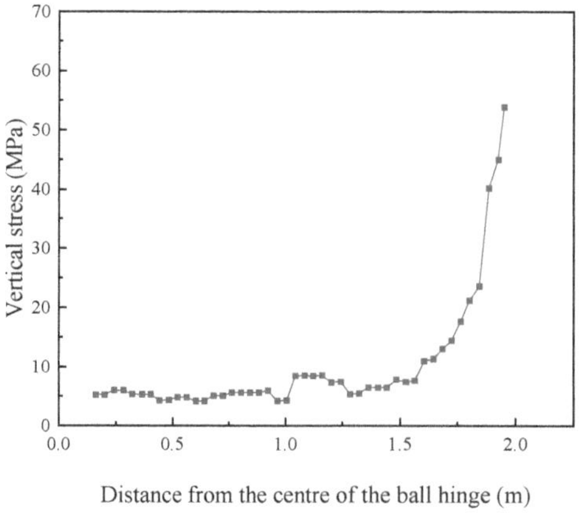

(**b**) The lower spherical hinge

Figure 7. Variation of vertical stress with distance from the center of the spherical hinge.

The maximum principal compressive stress distributions of the upper and lower bearing platforms as well as the spherical hinge are shown in Figure 8. As can be seen from Figure 8a, the maximum principal compressive stress values for the concrete material of the upper bearing are relatively low, ranging from 0.63 MPa to 8.75 MPa, with the highest principal stress occurring at the contact point between the upper spherical hinge and the bearing. As can be seen from Figure 8b, the maximum principal compressive stress in the upper spherical hinge ranged from 33 MPa to 63.89 MPa. The stress distribution followed a pattern of increasing, then decreasing, and finally increasing again as it moved from the center of the spherical hinge toward the periphery, with the maximum stress occurring at the edge. As can be seen from Figure 8c, the maximum principal compressive stresses in the lower spherical hinge range from 2.65 MPa to 54.24 MPa, demonstrating a gradual increase from the center toward the edge, with the maximum stress occurring at the edge. As can be seen from Figure 8d, the maximum principal compressive stresses for the concrete material of the lower bearing range from 0.4 MPa to 5.50 MPa, with the highest values occurring at the edge in contact with the spherical hinge. However, it can be observed that some tensile stress, amounting to 0.05 MPa, has developed in the concrete surrounding the lower bearing. The maximum principal stress in the entire rotating system is less than the allowable stress of the material, indicating that the structure is in a safe condition.

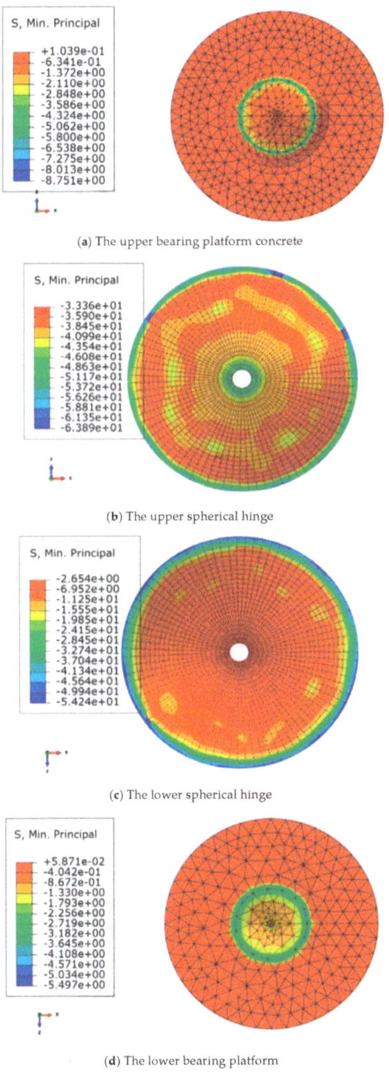

(**a**) The upper bearing platform concrete

(**b**) The upper spherical hinge

(**c**) The lower spherical hinge

(**d**) The lower bearing platform

Figure 8. Maximum principal compressive stress cloud of the rotating system (unit: MPa).

3.3.3. Vertical Displacement of the Spherical Hinge

The vertical displacement cloud of the lower spherical hinge under vertical load is shown in Figure 9. Under the action of vertical load, there is no obvious vertical displacement of the spherical hinge. The displacement distribution increases from the center of the spherical hinge toward the edge, with the maximum vertical displacement being 0.234 mm for the upper spherical hinge and 0.202 mm for the lower spherical hinge, both of which meet the design requirements.

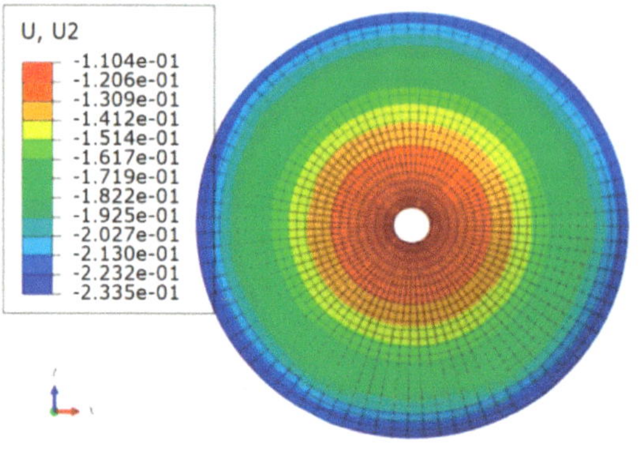

(**a**) The upper spherical hinge

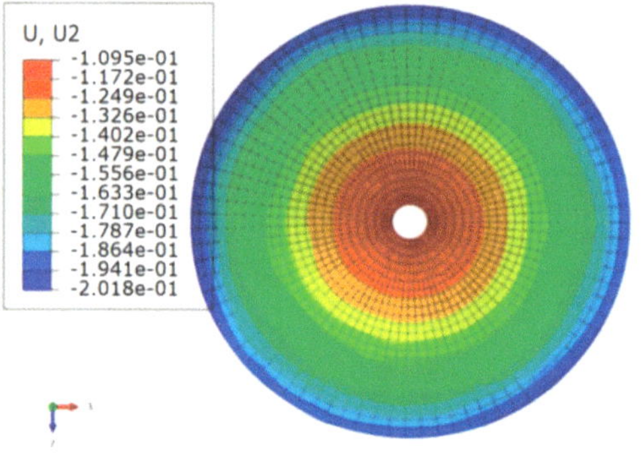

(**b**) The lower spherical hinge

Figure 9. Vertical displacement cloud of the spherical hinge (unit: mm).

4. Conclusions

(1) When the spherical hinge is only subjected to the upper static load, the vertical stress of the upper and lower steel spherical hinges increased gradually along the center of the spherical hinge to the edge, in which the vertical stress is the largest at the edge of the spherical hinge. Concurrently, the stress of the lower spherical hinge is slightly smaller than that of the upper spherical hinge, and the maximum principal compressive stresses of the upper spherical hinge and the lower spherical hinge are 63.89 MPa and 54.24MPa, respectively. Therefore, it can be concluded that the maximum principal compressive stresses of both are less than the allowable stress of the steel spherical hinge.

(2) The stress distribution of the spherical hinge is basically the same as that of the concrete of the upper and lower bearing platforms, and the contact edge of the concrete material with the spherical hinge stress reaches the maximum value. In addition, small tensile stresses are observed around the edges. Therefore, when

carrying out the design of the ball-hinge structure, it is necessary to strengthen the stress checking of the concrete contact part and ball-hinge to ensure the stability and safety of the structure.

(3) Under the upper static load, the displacement values of the upper and lower spherical hinges are very small, of which the maximum displacements of the upper and lower spherical hinges are 0.233 mm and 0.201 mm, respectively, and both of them meet the design requirements.

Author Contributions: Conceptualization, L.Z., X.S., Y.C., J.L. and Y.W.; Methodology, X.S., Z.W., Y.C. and Y.W.; Software, L.Z., X.S., Z.W., Y.C., J.L. and Y.W.; Validation, L.Z., X.S., Z.W., Y.C., J.L. and Y.W.; Formal analysis, L.Z., X.S., Z.W., Y.C., J.L. and Y.W.; Investigation, L.Z., X.S., Z.W., Y.C., J.L. and Y.W.; Resources, L.Z., X.S., Z.W., Y.C., J.L. and Y.W.; Data curation, L.Z., X.S., Z.W., Y.C., J.L. and Y.W.; Writing—original draft, L.Z., X.S., Z.W., Y.C., J.L. and Y.W.; Writing—review & editing, X.S., Z.W., Y.C. and Y.W.; Visualization, L.Z., X.S., Z.W., Y.C., J.L. and Y.W.; Supervision, X.S. and Y.W.; Project administration, L.Z., X.S. and Y.W.; Funding acquisition, X.S. and Y.W. All authors have read and agreed to the published version of the manuscript.

Funding: This research was funded by Shandong Provincial Natural Science Foundation: ZR2024QE147.

Data Availability Statement: The original contributions presented in the study are included in the article, further inquiries can be directed to the corresponding author.

Conflicts of Interest: Author Lin Zhao was employed by the company Shandong Expressway Construction Management Group Co., Ltd. Author Xiaohu Sun was employed by the Shandong Tangzheng Testing Co., Ltd. The remaining authors declare that the research was conducted in the absence of any commercial or financial relationships that could be construed as a potential conflict of interest.

References

1. Ma, Z.X. Analysis and study on spherical hing stress of horizontal rotation construction method for Continuous Beam. *J. Lanzhou Inst. Technol.* **2015**, *90*, 71–74.
2. Feng, Y.; Qi, J.; Wang, J.; Zhang, W.; Zhang, Q. Rotation construction of heavy swivel arch bridge for high-speed railway. *Structures* **2020**, *26*, 755–764. [CrossRef]
3. Jiang, L.C.; Gao, R. Deformation monitoring during removal of the supporting of T-type rigid frame bridge constructed by rotation method. *Procedia Eng.* **2010**, *4*, 355–360. [CrossRef]
4. Sun, Q.; Guo, X.; Zhang, D.; Guan, X.; Zheng, Y. Research on the application of horizontal rotation construction method with flat hinge in cable-stayed bridge construction. *Adv. Mater. Res.* **2011**, *1279*, 856–860. [CrossRef]
5. Siwowski, T.; Wysocki, A. Horizontal rotation via floatation as an accelerated bridge construction for long-span footbridge erection: Case study. *J. Bridge Eng.* **2015**, *20*, 05014014. [CrossRef]
6. Su, M.; Wang, J.; Peng, H.; Cai, C.; Dai, G. State-of-the-art review of the development and application of bridge rotation construction methods in China. *Sci. China-Technol. Sci.* **2021**, *64*, 1137–1152. [CrossRef]
7. Liu, T.; Fan, J.; Peng, Z. Central Load-Bearing Control in the Construction Process of the Concrete Spherical Joint Nandu River Swing Bridge: A Case Study. *Buildings* **2022**, *12*, 511. [CrossRef]
8. Li, W.; Zhang, Z.; Huang, C. Application of dynamic unstressed state method in vertical rotation construction of bridges. *Adv. Mater. Res.* **2011**, *1279*, 988–992. [CrossRef]
9. Li, W.; Huang, C.; Tang, C. Negative angle vertical rotating construction method of reinforced concrete arch bridge. *Struct. Eng. Int.* **2017**, *4*, 558–562. [CrossRef]
10. Allen, C.E. Bridge maintenance, safety, management, and life-cycle optimization. *Struct. Infrastruct. Eng.* **2011**, *7*, 807.
11. Parson, S.N. Investigation on determining thermal stress in massive concrete structure. *Mater. J.* **1996**, *93*, 96–101.
12. Aydın, Z.; Ayvaz, Y. Overall cost optimization of prestressed concrete bridge using genetic algorithm. *KSCE J. Civ. Eng.* **2013**, *17*, 769–776. [CrossRef]
13. Huang, Y.; Du, C.; Wang, L.; Zhang, Z.; Zhuang, X.; Liu, Q. Multi-objective preform optimization for spherical hinge mandrel based on response surface methodology. *IOP Conf. Ser. Mater. Sci. Eng.* **2019**, *504*, 012050. [CrossRef]
14. Hou, G.G. *Research on Key Technology of Overpass Swivel Construction in Shangyuan Road Baoji City*; Xi'an University of Technology: Xi'an, China, 2017.
15. Liu, T.; Yu, Q.; Fan, J.; Peng, Z.; Wang, E. Concrete spherical joint contact stress distribution and overturning moment of swing bridge. *Structures* **2020**, *28*, 1187–1195. [CrossRef]
16. Wang, J.; Sun, Q. Dynamic analysis of swivel construction method under multi-variable coupling effects. *Int. J. Struct. Integr.* **2019**, *10*, 580–598.
17. Zhao, H. Research on mechanical properties of steel spherical hinge. *Ind. Constr.* **2021**, *51*, 116–121.

18. Lan, Y. *Mechanics Analysis and Application Researchof Horizontal Rrotating Spherical Hinge;* Shijiazhuang Tiedao University China: Shijiazhuang, China, 2018.
19. Fan, J.; Liu, T.; Peng, Z.; Yin, Y.; Wang, J. Analysis on Contact Stress Parameters of Concrete Spherical Hinge Grinding Center of Swing Bridge. *J. Civ. Eng. Manag.* **2020**, *37*, 70–74.
20. Campbell, T.I. Design considerations for spherical bridges bearings under horizontal loads. *World Congr. Jt. Sealants Bear. Syst. Concr. Struct.* **1996**, *164*, 233–252.
21. Chen, L.; Ruan, X.; Guo, X.; Chen, W.; Fu, L. A new quasi-static friction model for the spherical hinge bearing in structural engineering. *Eng. Struct.* **2023**, *291*, 116414. [CrossRef]
22. Shi, X.; Liu, Z.; Guo, T.; Cai, C.; Jiang, C. Investigation on contact stress calculation method of spherical hinge structures for swivel construction. *Structures* **2024**, *69*, 107290. [CrossRef]
23. Li, Q.; Ma, X.; Zhang, P. Analysis of Reasonable Rotation Angular Speed and Angular Acceleration of Horizontal Rotation Construction Bridges. *Railw. Eng.* **2024**, *64*, 86–90.
24. Yang, Y. Research on Key Technique of Rotation Construction of Cast On-Site Prestress Concrete T-Shaped Rigid Frame Span Railway. Master's Thesis, Southwest Jiaotong University, Chengdu, China, 2021.
25. Ju, Y.; Yu, D.; Wang, D. A Study of the Partially Prestressed Tendons RPC Concrete Pole Connection Design. *Appl. Mech. Mater.* **2012**, *1799*, 534–537. [CrossRef]
26. Xie, J.; Zhao, X.; Yan, J. Experimental and numerical studies on bonded prestressed concrete beams at low temperatures. *Constr. Build. Mater.* **2018**, *188*, 101–118. [CrossRef]
27. Yuan, M.; Yan, D.; Zhong, H.; Liu, Y. Experimental investigation of high-cycle fatigue behavior for prestressed concrete box-girders. *Constr. Build. Mater.* **2017**, *157*, 424–437. [CrossRef]
28. *BS EN 1337-7:2004;* Structural Bearings—Part 7: Spherical and Cylindrical PTFE Bearings. British Standard Institution: London, UK, 2004.
29. *GB/T 50081-2002;* Standard for Test Method of Mechanical Properties on Ordinary Concrete. China Architecture & Building Press: Beijing, China, 2002.
30. *GT/B 228.1—2021;* Metallic Materials—Tensile Testing—Part 1: Method of Test at Room Temperature. Standardization Administration of the People's Republic of China: Beijing, China, 2021.
31. *GB 50010—2015;* Code for Design of Concrete Structures. China Architecture & Building Press: Beijing, China, 2015.
32. Tian, H.; Fang, Z.; Zhu, D. 133Years of Hertz Point Contact. *J. China Three Gorges Univ. (Nat. Sci.)* **2014**, *36*, 88–97.

buildings

MDPI

Article

Experimental Investigations on the On-Site Crack Control of Pier Concrete in High-Altitude Environments

Xiaochuan Hu [1], Lei Liu [1], Manping Liao [1], Ming Li [2,3], Cun Lu [2,3], Zaifeng Yao [4], Qiuming Huang [4] and Mei-Ling Zhuang [5,6,7,*]

1 The Civil Engineering Group Corporation of China Second Engineering Bureau Ltd., Beijing 101100, China; huxiaochuan@cscec.com (X.H.); liulei0905@cscec.com (L.L.); liaomanping@cscec.com (M.L.)
2 State Key Laboratory of High Performance Civil Engineering Materials, Jiangsu Research Institute of Building Science Co., Ltd., Nanjing 210008, China; liming@cnjsjk.cn (M.L.); lucun@cnjsjk.cn (C.L.)
3 Jiangsu Sobute New Materials Co., Ltd., Nanjing 211103, China
4 China Construction Second Engineering Bureau Ltd., Beijing 101101, China; yaozaifeng@cscec.com (Z.Y.); shiwei2j@cscec.com (Q.H.)
5 Water Resources Research Institute of Shandong Province, Jinan 250013, China
6 School of Civil Engineering, Shandong University, Jinan 250061, China
7 School of Transportation and Civil Engineering, Nantong University, Nantong 226019, China
* Correspondence: ml_zhuang99@163.com

Abstract: Concrete structures in high-altitude environments face many challenges. Establishing concrete crack control methods in high-altitude environments is crucial for enhancing the service capacity of concrete structures. In this study, a multi-field (hydration-temperature–humidity-constraint) coupling model was used to quantitatively assess the cracking risk of pier bodies at high altitude. On-site crack control tests were conducted on pier bodies using a micro-expansion anti-cracking agent to demonstrate the effectiveness of deformation shrinkage compensation in crack control at high altitudes. The results indicated that there was a risk of cracking in the pier body at high-altitude conditions, especially within 0.3 m from the pile cap and ±2.5 m from the center of the pier side surface. Compared with conventional piers, the micro-expansion anti-cracking agent approximately doubled the unit expansion deflection of piers at high temperatures while reducing the unit shrinkage deflection of piers by 11% to 12% at low temperatures. The concrete in conventional pier bodies was in a tension state after long-term hardening, while the concrete treated with the micro-expansive anti-cracking agent was in compression. Therefore, the deformation compensation effect of the micro-expansive anti-cracking agent was significant and reduced the risk of concrete cracking. In addition, early freezing had a significant impact on concrete strength, underscoring the importance of effective temperature control during the early stages of concrete placement in high-altitude environments.

Keywords: high altitude; concrete; on-site tests; strain compensation; cracking control

1. Introduction

The Sichuan–Tibet Railway commences from Chengdu, Sichuan Province, China, in the east and extends westward through Ya'an, Kangding, Changdu, Linzhi, and Shannan, culminating in Lhasa, Tibet [1,2]. The high altitude of this railroad, averaging over 3000 m (see Figure 1), and the extraordinary complexity of the environment present unique challenges to science and technology, making it one of the most challenging construction projects [3]. For example, at the Xindu Bridge site, positioned at an elevation of 3500 m above sea level, the highest temperature recorded was about 35 °C, and the lowest temperature can drop to −15 °C (see Figure 2). The average relative humidity in this area typically varies from 20% to 60%. In addition, the air pressure in the area is as low as 64 kPa, thunderstorms and hail are frequent, and the temperature difference between day and night can be as high as 35 °C. As per the established on-site monitoring system, data from

2020 indicated an average wind speed of 3 m/s, with a maximum wind speed of 16 m/s. Freeze–thaw is also an important factor affecting the performance of concrete [4–6], and there are about 88 freeze–thaw cycles (refers to the numbers in a year that the temperature drops from +3 °C to below −3 °C and then rises back to +3 °C) per year in this area. These environmental conditions pose a significant risk to the durability of bridge concrete, often leading to premature cracking of bridge concrete. Managing concrete failure in extreme environments is therefore an important technical challenge.

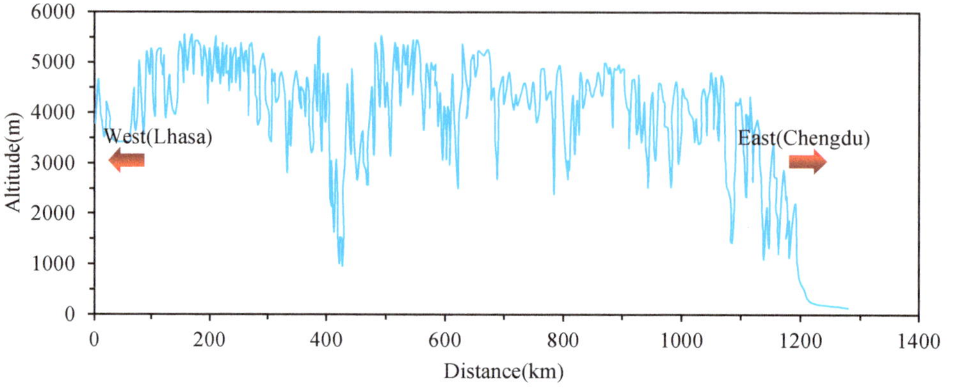

Figure 1. Altitude along Sichuan–Tibet Railway [1].

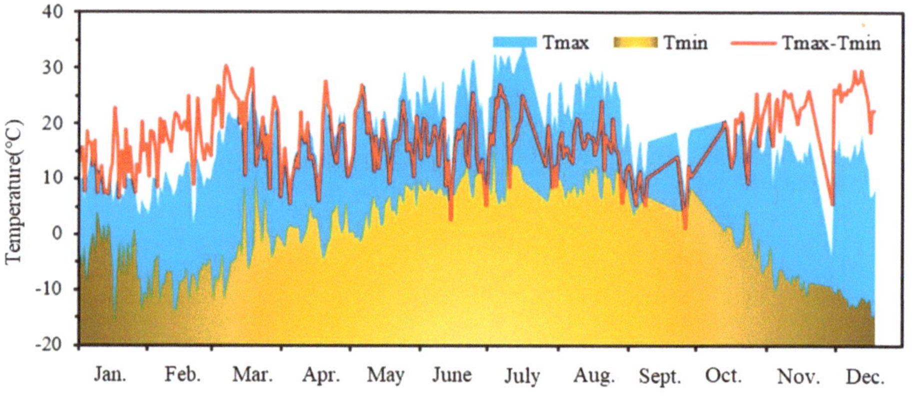

Figure 2. Temperature of Xindu Bridge in 2022.

Shrinkage cracking is a major technical challenge in civil engineering [7], and early shrinkage control of concrete has always been an important issue in preventing concrete from cracking [8,9]. Early shrinkage in concrete includes plastic shrinkage before hardening, self-shrinkage, drying shrinkage, and temperature drop shrinkage during the hardening stage. Among them, self-shrinkage and temperature drop shrinkage are considered the most influential factors. In high-altitude environments, the cumulative early shrinkage deformation can surpass 200 $\mu\varepsilon$ (i.e., 10^{-6}), making concrete more vulnerable to tensile failure, particularly under stringent constraint conditions [7,10]. To reduce the risk of concrete cracking due to these shrinkage factors, various measures have been used, such as shrinkage-reducing agents (SRAs), expansion agents, internal curing agents, high-strength fiber materials, external water-retaining coatings, or a combination of these methods [11–14]. However, challenges persist, such as the negative impact of shrinkage-reducing agents on early strength and initial setting time and the difficulty of controlling the mixing and production of high-strength fiber concrete. In addition, quality control of raw materials, optimization of concrete mix proportions, and construction techniques are also conducive to reducing concrete cracking. A lot of research and application work have been carried out, and great progress has been achieved in concrete shrinkage and crack control. Because of this, many projects in China, such as the Taihu Tunnel and Yangtze River Tunnel, have achieved good results in concrete shrinkage and crack control [15,16]. Unfortunately, previous studies have mainly focused on concrete crack control in conventional environments with low altitude, normal atmospheric pressure, and low wind speed [9,17]. Due to the limitations of experimental conditions, few studies have aimed to preliminarily reveal the effects of air pressure on strength, slump loss, and negative temperature on the mechanical properties and hydration characteristics of concrete [17–20]. Concrete crack control is influenced by various factors, including creep [21], materials, design, environment, and construction. However, the impact of extreme conditions at high altitudes, such as frequent temperature fluctuations and high wind speeds, is often overlooked. Crack control test results performed under standard conditions do not accurately reflect the actual service environment of the concrete structure. Currently, there is a notable lack of research on concrete crack control in real high-altitude environments. As a result, conducting on-site concrete tests to develop crack control strategies for extreme environments at high altitudes is crucial for enhancing the service capacity of concrete structures.

Based on the engineering environment near Xindu Bridge on the Sichuan–Tibet Railway, a multi-field (hydration-temperature–humidity-constraint) coupling model was used to quantitatively assess the risk of concrete cracking in a specific area of a pier under high-altitude conditions. Subsequently, on-site crack control tests were conducted on the pier to showcase the efficacy of managing concrete shrinkage deformation. Finally, the mechanisms of concrete crack control and early temperature regulation were discussed in depth, and recommendations for concrete crack management in extreme environments were provided. The findings can provide valuable insights and guidance for effectively addressing cracking problems of concrete structures in high-altitude environments.

2. Cracking Risk Analysis

2.1. Cracking Risk Analysis Method

A shrinkage model is used based on the multi-field (hydration-temperature–humidity-constraint) coupling mechanism [22,23]. In the evaluation of early cracking risk in concrete, various factors are considered in the model. It includes material, structure, construction, and environment. The concrete cracking risk coefficient η is calculated as

$$\eta = \sigma_t^T / f_t^T \tag{1}$$

where σ_t is the tensile stress of concrete at time T and f_t is the tensile strength of concrete at time T.

The early elastic modulus E and tensile strength f_t of concrete are dependent on the hydration degree α [24,25]:

$$E(\alpha) = E^{\infty}(\frac{\alpha - \alpha_0}{\alpha^{\infty} - \alpha_0})^p \tag{2}$$

$$f_t(\alpha) = f_t^{\infty}(\frac{\alpha - \alpha_0}{\alpha^{\infty} - \alpha_0})^q \tag{3}$$

where E^{∞} is the final elastic modulus of the concrete; f_t^{∞} is the final tensile strength of the concrete; p is 0.5 and q is 1; α_0 and α^{∞} represent the initial and final value of the degree of hydration α, respectively.

Research has shown that concrete will definitely crack when $\eta \geq 1.0$. There is a significant risk when $0.7 < \eta < 1.0$. Conversely, when $\eta \leq 0.7$, concrete is unlikely to crack, and the non-cracking guarantee exceeds 95%. The above cracking risk analysis methods have been widely used in various engineering projects, and the evaluation results align closely with real-world engineering outcomes.

2.2. Model Configuration

A one-second finite element model (FEM) of the bridge pier is established with dimensions of 4.5 m (x) × 3.8 m (y) × 7.5 m (z), as shown in Figure 3. The concrete at 0–1 m above the pier platform is C45, while that at 1–7.5 m above the pier platform is C35. In this study, the Solid65 element is chosen to simulate the concrete, and the linear elastic constitutive model is used. The bottom of the numerical model is fixed in all directions, and the displacement of the node at x = 0 in the x-direction is fixed. In order to simulate the maintenance measures (water energy film and insulation cover layer) after the concrete pouring of a pier, the surface heat dissipation coefficient of the concrete is 20 kJ/(m^2·h·K). The thermal conductivity coefficient of concrete is 1.5 W/(m·K). The ambient temperature and concrete pour temperature are 8 °C and 16 °C, respectively. The final elastic modulus E^{∞}, tensile strength f_t^{∞}, and other relevant parameters are given, as shown in Table 1.

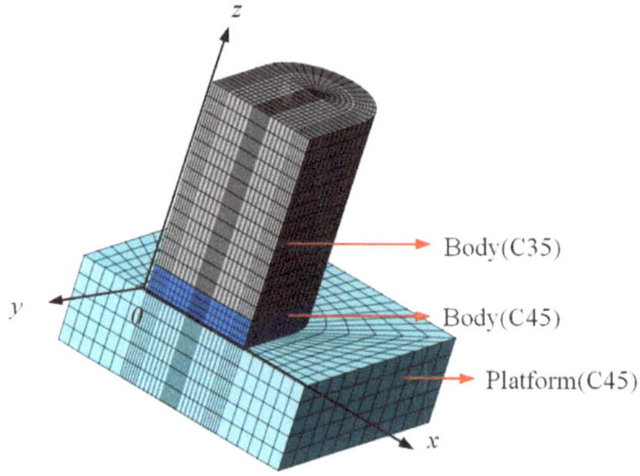

Figure 3. Numerical calculation model.

Table 1. Model parameters.

Parameter	C35	C45
Final tensile strength f_t^{∞}	2.4 MPa	2.6 MPa
Final elastic modulus E^{∞}	32 GPa	36 GPa
Poisson's ratio	0.23	0.23
Adiabatic temperature rise	43 °C	47 °C

2.3. Calculation of Cracking Risk

The focus of this study was to assess the risk of surface cracking of piers and to compare the risk of cracking in the vertical and horizontal directions of the pier bodies. It was found that in the vertical direction, the risk of surface cracking was lower at the center compared to 2 m away from the axis, meaning that cracking along line ① was less likely than that along line ② (see Figure 4). However, the cracking risk coefficient η of the pier body exceeded 0.7 at a distance of 0.3 m or more from the pier platform, indicating a high risk of cracking. Additionally, an analysis of transverse cracking risk on C35 and C45 concrete surfaces, with the transverse center as a reference point, indicated that the C35 surface had a higher risk of cracking than the C45 surface (see Figure 5). This meant that the degree of cracking along line ③ was higher than that along line ④. In the transverse direction (x direction) of the pier body, the cracking risk coefficient η exceeded 0.7 between −2.5 m and 2.5 m, and the local cracking risk coefficient η of C35 surfaces exceeded 1.0

(Figure 5). Therefore, there was a high potential for surface cracking over a horizontal span of more than 0.3 m from the pier body and approximately 2.5 m from the center of the side surface of the pier body.

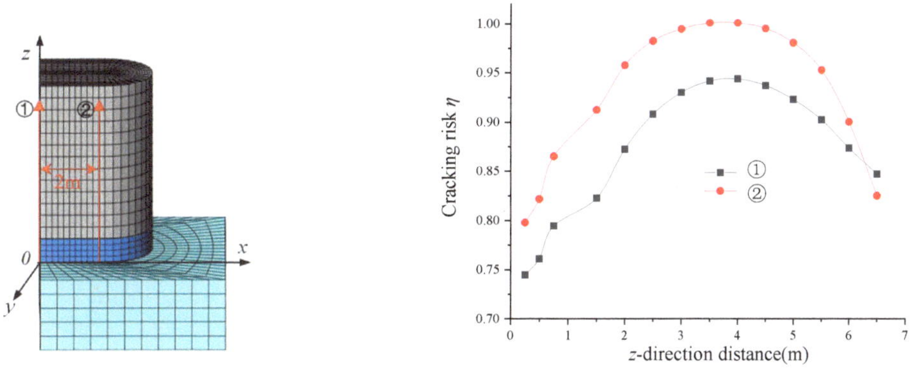

Figure 4. Cracking risk in the vertical direction of the pier body.

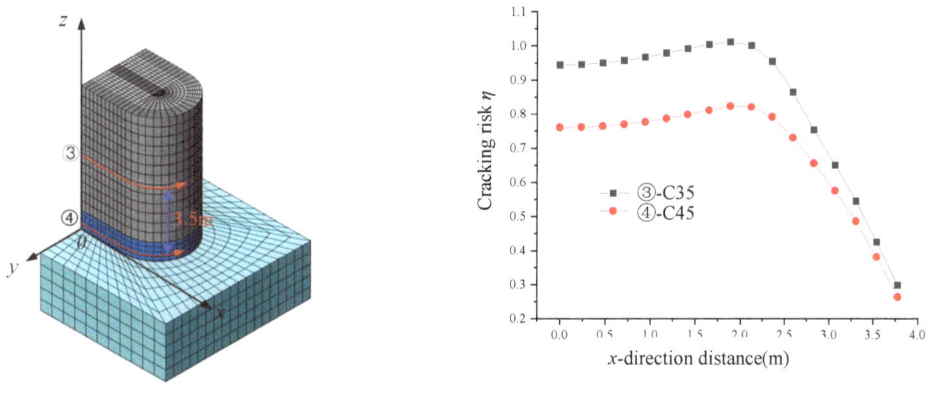

Figure 5. Cracking risk in the horizontal direction of the pier body.

3. On-Site Experiments

3.1. Materials and Methods

Concrete cracking and deformation shrinkage are closely related. The use of expanding components to generate volume expansion during the hydration process to compensate for the shrinkage of cement-based materials is an important method to enhance crack resistance and minimize the risk of cracking [7,8]. Therefore, an anti-cracking agent (HME®-V, produced by Jiangsu Sobute New Materials Co., Ltd. in Nanjing, China) with deformation compensation properties was utilized on site to develop an advanced high-altitude crack control strategy. The agent utilized specially designed CaO for early expansion, high-activity MgO for mid-term expansion, and low-activity MgO for later expansion, achieving phased and comprehensive shrinkage suppression and crack control in concrete [7,9,26].

In this study, two piers were poured for on-site testing to verify the effectiveness of the proposed strategy. One pier was poured on site as the reference pier (labeled as Ref) without adding any anti-cracking agent. In contrast, another pier (labeled as HME) was poured with the incorporation of HME®-V as the test pier. The bottom section of the pier body (0–1 m) was constructed with C45 concrete, while the portion above 1 m was made of C35 concrete. The concrete mix ratios are detailed in Table 2.

Table 2. Concrete mix ratio (kg/m^3).

No.	Cement	Fly Ash	Anti-Cracking Agent	Sand	Gravel	Water
C35-Ref	320	80	0	774	1070	150
C35-HME	300	68	32	774	1070	150
C45-Ref	329	141	0	686	1120	153
C45-HME	329	103	38	686	1120	150

In the on-site experiments, the cement utilized for pouring the bridge piers was P LH42.5 low heat cement, along with grade II fly ash. The aggregate consisted of three sizes of crushed stones: 5–10 mm, 10–20 mm, and 16–31.5 mm, achieving a continuous grading from 5 to 31.5 mm. Based on the designed mix ratios, cubic specimens measuring 150 × 150 × 150 mm were poured in the laboratory and cured in a standard curing room (with a relative humidity greater than 95% and a temperature of 20 ± 2 °C) on the construction site. After 28 days of curing, the uniaxial compressive and tensile strengths of these specimens met the requirements for both C35 and C45.

3.2. Monitoring Layout

In this study, the SBT-CDM I concrete temperature–strain wireless monitoring system was utilized to monitor the temperature and deformation (see Figure 6a). The monitoring system was equipped with a data transmission unit (DTU) module with 4G wireless transmission capabilities, and the data can be transmitted to a cloud server through this DTU. The client interface provided access to the data. Vibrating wire strain gauges with temperature and strain sensing capabilities were used as the sensor, as depicted in Figure 6b. The sensors at the core of the pier body and pier cap were arranged as illustrated in Figure 6b. The strain sensors on the outer side of the pier body and pier cap were located 50 mm away from the surface and were specifically placed in the west and north directions only. The sensor layout is detailed in Figure 6c.

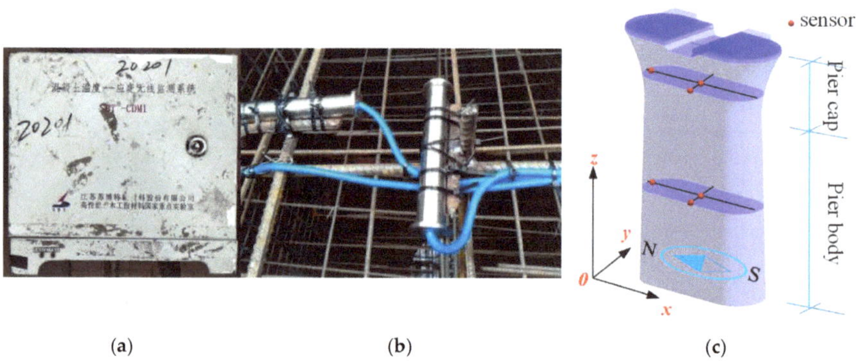

Figure 6. Monitoring system: (**a**) wireless monitoring system; (**b**) sensors; (**c**) sensor layout.

3.3. On-Site Testing

Pier 1 and Pier 2 were constructed as reference and test piers, respectively, to investigate the impact of adding an anti-cracking agent on temperature and deformation compensation and to verify the effectiveness of crack control. Pier 1 had a height of 7.5 m and was made with regular concrete (no HME®-V), while Pier 2 was 7 m high and constructed with anti-cracking (HME®-V) concrete. The concrete had a slump measurement of about 195 mm in the laboratory and showed good workability when placed on site.

Post-demolding, the piers were covered, insulated, and moisturized using the thermal insulation and moisturizing covering materials (geotextile and geomembrane). Intermittent watering and curing procedures were carried out during this period. The on-site testing is illustrated in Figure 7.

Figure 7. On-site test.

4. Testing Results

4.1. Temperature Results

The temperature monitoring results of Pier 1 are illustrated in Figure 8. The concrete temperature at the time of entry into the mold was 16.5 °C. The results indicated that the C35 pier body experienced a maximum temperature increase of 36.4 °C, peaking at 52.9 °C. Similarly, the C35 pier cap reached a maximum temperature increase of 37.7 °C with a peak at 54.2 °C. Both C35 and C45 concrete took approximately 2.4–2.5 days to reach their highest temperature. In this testing, the temperature difference between the concrete core and the outside (5 mm away from the surface) was approximately 25 °C. During the initial 7-day temperature drop phase, the average cooling rate at the center point of the pier body and pier cap varied between 1.4 °C/d and 2.3 °C/d, indicating a relatively gradual cooling process.

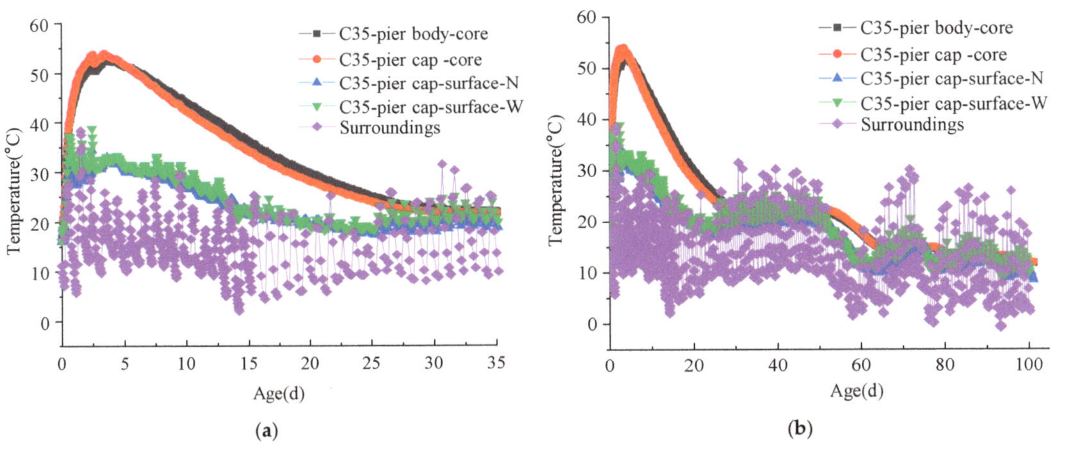

Figure 8. Monitored temperature of reference pier: (**a**) early temperature history; (**b**) long-term temperature history. Note that N and W denote the north and west directions, respectively, as defined in Figure 6c.

The temperature monitoring results of Pier 2 are presented in Figure 9. The temperature of concrete at the time of entry into the mold was 22.8 to 23.5 °C. The C35 pier body experienced a maximum temperature rise of 36.5 °C, peaking at 59.9 °C. Similarly, the C35 pier cap reached a maximum temperature increase of 35.6 °C, reaching 58.4 °C. The highest concrete temperature was observed between 2.5 and 2.7 days, with a maximum temperature difference of 35 °C between the concrete core and the exterior. After 7 days, the average cooling rate of the core during the temperature decrease phase was 2.0–2.3 °C/d. The initial temperature of the test pier upon mold entry was 6–7 °C higher than that of the reference pier. While the maximum temperature rise in the C35 pier body was similar to the reference pier, the peak temperature of the experimental pier was 6–7 °C higher. Moreover, the temperature differential between the concrete core and the outside, as well as the cooling rate, were greater in the test pier compared to the reference pier. After about 30 days of casting, the concrete temperature of two piers reached a dynamic equilibrium with the ambient temperature and gradually decreased with the ambient temperature.

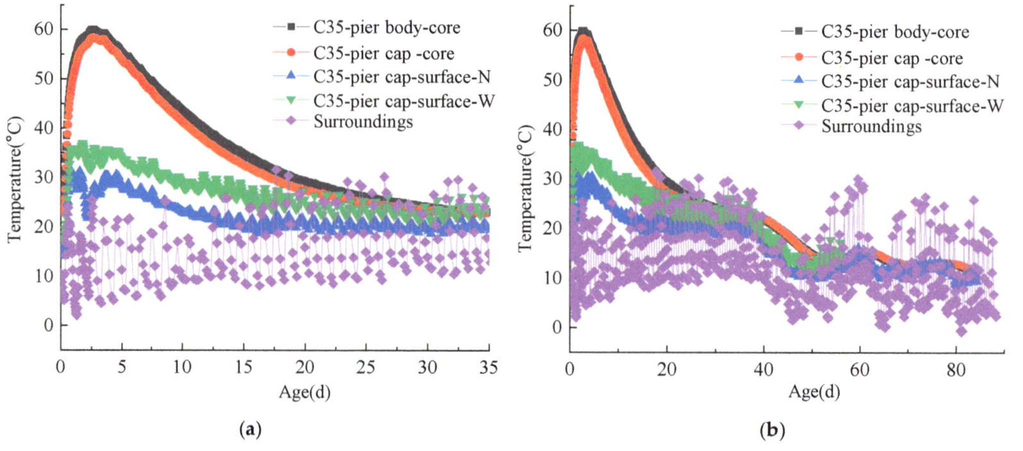

Figure 9. Monitoring temperature of test pier: (**a**) early temperature; (**b**) long-term temperature.

4.2. Deformation Results

The monitored strains of Pier 1 are shown in Figure 10. The results revealed that during the temperature rise stage, the unit expansion strain along the length direction was 10.6 με/°C for the C35 pier body and 10.3 με/°C for the pier cap (see Figure 10c). Conversely, in the temperature drop stage, it was measured at 9.5 με/°C for the pier body and 8.4 με/°C for the pier cap (see Figure 10d). During the temperature rise stage, the unit expansion strain along the thickness direction or y direction (defined in Figure 10c) at the center of the C35 pier body and pier cap was 12.2 με/°C and 12.7 με/°C, respectively (see Figure 10c). In the temperature drop stage, the unit shrinkage strain was 12.6 με/°C for the pier body and 12.2 με/°C for the pier cap (see Figure 10d). At around 40 days of age, a sudden change in strain was observed at the center of the C35 pier body along the length direction or x direction (see Figure 10b), possibly due to concrete shrinkage cracking. This cracking may not necessarily extend to the concrete surface without external forces, as indicated by similar engineering monitoring results. As the concrete temperature gradually decreased and stabilized (see Figure 8b), the residual strain of the concrete became negative, indicating a shrinkage state. Negative strain denotes a tensile state, while positive strain indicates a compressive state, suggesting that the concrete was bearing tensile stress.

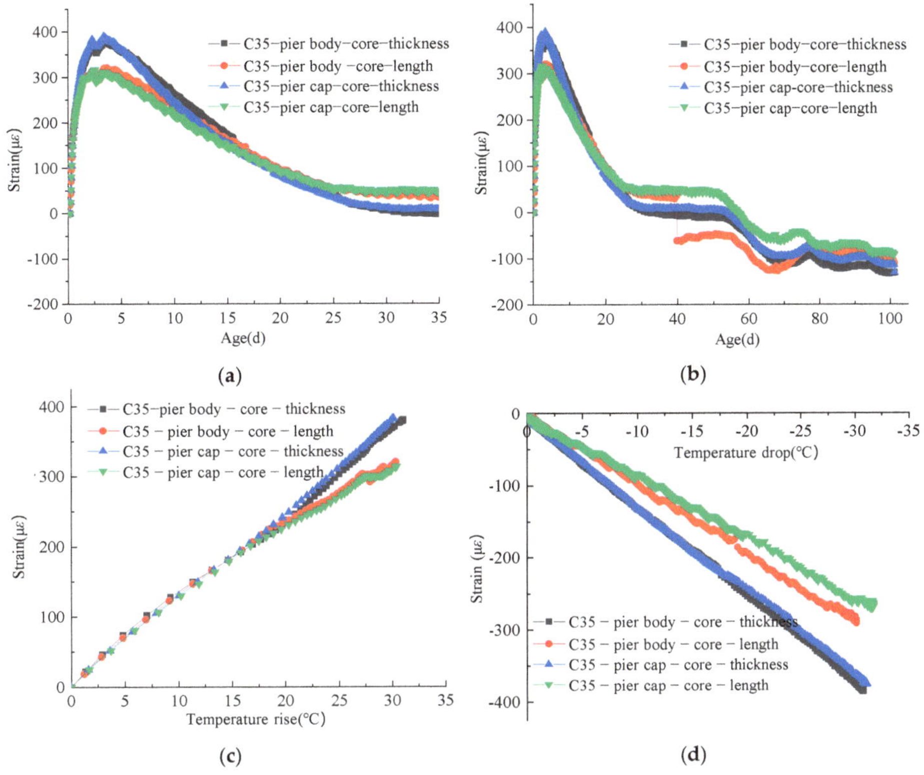

Figure 10. Monitoring strain of the reference pier: (**a**) early strain; (**b**) long-term strain; (**c**) expansion strain during temperature rise stage; (**d**) shrinkage strain during temperature drop stage. Note that the thickness direction and length direction are consistent with the y and x directions in Figure 3.

The deformation of Pier 2 is illustrated in Figure 11. The unit expansion deformation of the C35 concrete in the center of the pier body and pier cap along the length direction was 17.5 με/°C and 17.0 με/°C, respectively, during the temperature rise stage (see Figure 11c), with corresponding unit shrinkage deformations of 8.4 με/°C and 7.2 με/°C during the temperature drop stage (see Figure 11d). The unit expansion deformation of C35 at the center of the pier body and pier cap along the thickness direction was 25.3 με/°C and 25.1 με/°C, respectively, during the temperature rise stage, with unit shrinkage deformations of 11.1 με/°C and 10.8 με/°C, respectively, during the temperature drop stage. Compared with the reference pier, the unit temperature expansion deformation occurs during the temperature rise stage of the test pier, especially along the thickness direction. Conversely, during the temperature drop stage, the unit temperature shrinkage deformation of the test pier decreased by about 11% to 12% compared with that of the reference pier. The deformation curves of the test pier exhibited a smooth progression, without any sudden changes or significant signs of cracking. As observed in Figure 11d, the residual deformation of the concrete was positive (exceeding 200 με), indicating a state of compression. Considering that concrete had a poor ability to withstand tensile stresses but a strong ability to withstand compressive stresses, this compensating effect was very significant and is conducive to improving the crack resistance of concrete.

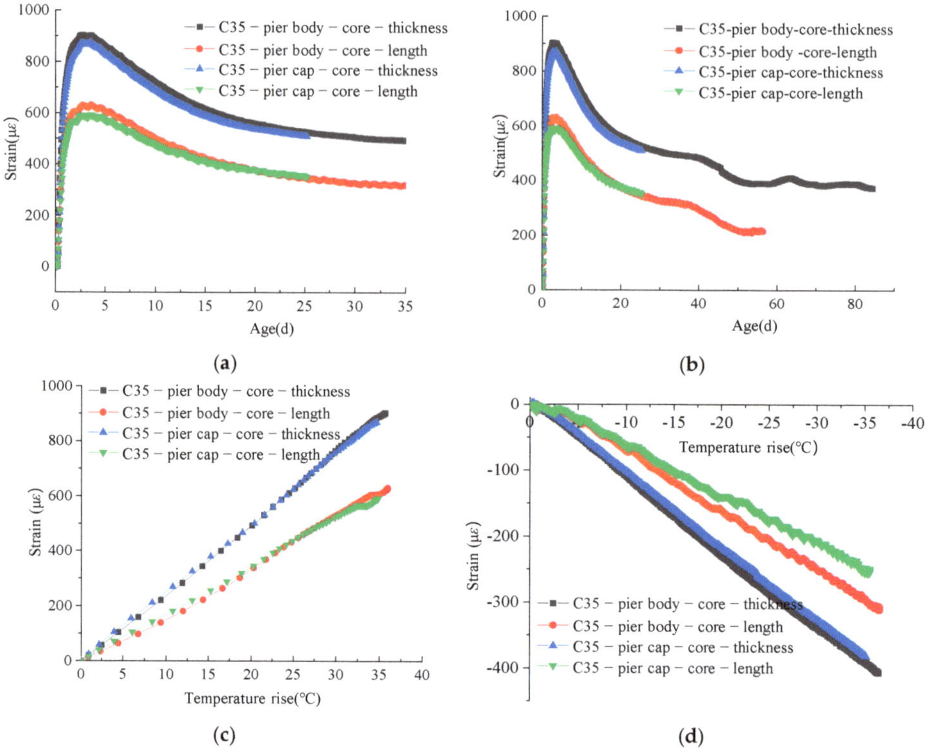

Figure 11. Monitoring strain of the test pier: (**a**) early deformation; (**b**) long-term deformation; (**c**) expansion strain during temperature rise stage; (**d**) shrinkage strain during temperature drop stage.

4.3. Comparison of On-Site Cracking

Cracking cases at high altitude have revealed that many cracks occur in the thickness direction of pier bodies. The cracking of pier bodies was investigated to analyze the distribution characteristics of concrete cracking. Vertical cracks were observed on both the front and back sides near the center of the pier body. On the west side, a crack located approximately 1.8 m from the top of the pier cap had a length of about 2.2 m, a width ranging from 0.1 mm to 0.28 mm, and a depth between 20 mm and 30 mm. Conversely, the crack on the east side was situated around 2.5 m away, with a width varying from 0.1 mm to 0.26 mm and a depth between 32 mm and 35 mm. The position and size of the cracks on both sides are basically the same, and the path of crack development is tortuous. The distribution of these cracks is illustrated in Figure 12. Notably, no cracks were detected in the test pier so far, and there was no abrupt strain within the concrete, remaining in a compressed state, which indicated a significant crack control effect.

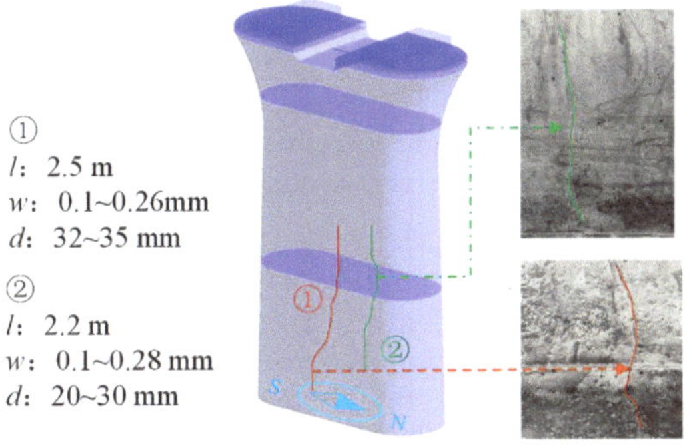

Figure 12. On-site cracking.

5. Discussion and Suggestions

5.1. Strain Compensation and Crack Control Mechanism

GaO and MgO in anti-cracking agents react with water to produce $Ga(OH)_2$ and $Mg(OH)_2$, respectively, which leads to volume expansion. The expansion could compensate for the self-shrinkage, drying shrinkage, and temperature-drop shrinkage of concrete. The early and middle stages of concrete exhibit significant self-shrinkage and temperature drop shrinkage, necessitating substantial expansion for compensation. In contrast, a minor expansion was required in the later stages to stabilize the expanding pre-compression stresses developed in the early stages. The specially designed GaO reacts rapidly and exhibits a higher expansion rate, thereby playing a crucial role in early expansion [9]. The rapid hydration of CaO results in the peak expansion of concrete typically occurring within 1–3 days after pouring, leaving limited compensation for temperature drop shrinkage in the middle and later stages [7,27]. The MgO expansion agent has the advantage of a designable expansion process, and different MgO activities produce different expansion amounts and rates. High-activity MgO expansion agents demonstrate relatively rapid expansion development and early stability in expansion. Consequently, high-activity MgO in anti-cracking agents can effectively compensate for temperature drop shrinkage and self-shrinkage during the temperature drop phase [28,29]. The expansion development of low-activity MgO following the water reaction is slower, with a delayed expansion stability time. Thus, this delayed expansion can be utilized to counteract the shrinkage of concrete in the later stages [9]. Therefore, HME®-V achieves a staged full-process compensation for concrete shrinkage [9,26]. By compensating for deformation throughout the full process, the tensile stress caused by shrinkage inside the concrete is reduced, thereby lowering the risk of cracking, as illustrated in Figure 13.

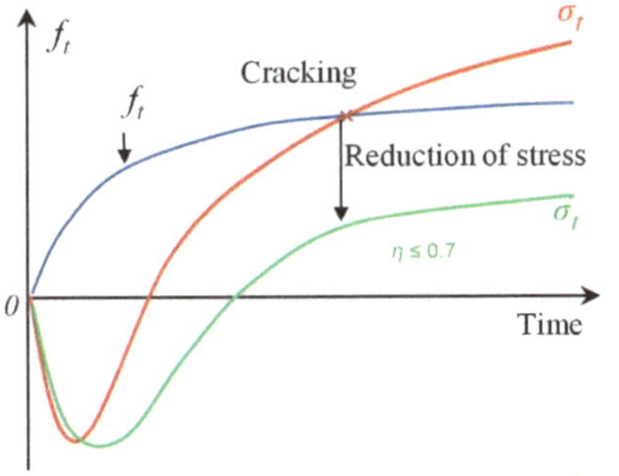

Figure 13. Strain compensation control technology for concrete cracking [9].

5.2. Early Temperature Control of Concrete

Freeze damage to early-age concrete has a significant impact on its subsequent strength development and is not conducive to crack prevention and control. Figure 14 illustrates the uniaxial compressive strength (UCS) development of a specific concrete mix exposed to frost damage during various early curing periods. Comparing the concrete samples with the same freezing time, the compressive strength of standard cured concrete aged 3 days was 46.4 MPa, while the strength of the concrete subjected to 1-day freezing after 1 day of setting (1d-1d) and 2-day freezing after 2 days of setting were 41.2 MPa and 42.5 MPa, respectively, representing reductions of 11.2% and 8.4%. By 28 days of age, their strengths were 68.0 MPa, 59.1 MPa, and 61.2 MPa, respectively, with a decrease of below 15%. These results indicated that frozen concrete exhibited slower strength development, with earlier freezing ages having a more pronounced negative impact due to low temperatures. Concrete samples with a 3-day setting period and 1-day frost damage (3d–1d) showed more complete internal hydration development after 3 days of standard curing compared with 1-day and 2-day frozen specimens. Consequently, only 1-day exposure to low temperatures had a minor impact on their strength development, aligning their curve with that of concrete under standard curing conditions. On the other hand, for the same freezing age, longer durations of freezing led to greater strength reductions and slower development. Notably, a 3-day freezing duration significantly affected concrete strength development. Therefore, insulating early-age concrete was crucial for crack control in harsh conditions and enhancing structural service performance.

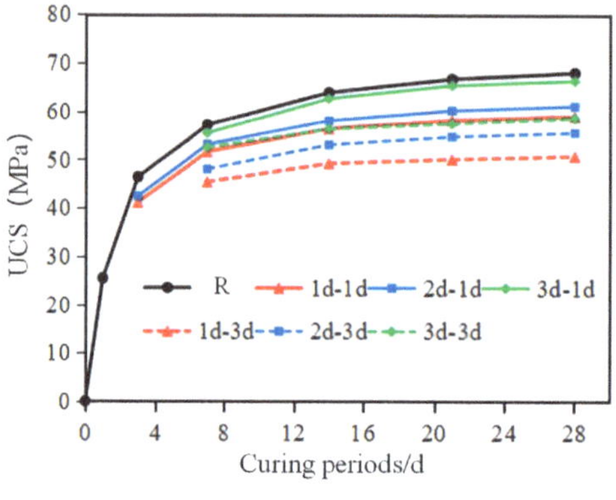

Figure 14. Strength of concrete at different curing periods under a single freeze–thaw cycle. Note that R represents the reference specimen.

Cotton quilts and warm air sheds are commonly used in engineering to control the temperature of concrete in low-temperature environments. However, operations at high altitudes often face drawbacks such as low turnover utilization, limited insulation, and high labor intensity. It is critical to find ways to overcome these challenges in extreme high-altitude environments. Nanoaerogels, known for their low heat conduction in both gas and solid states, are considered the most effective solid materials for insulation. A SiO_2 aerogel coating with a thermal conductivity of 0.028 W/(m·K) was used to conduct thermal insulation tests on formwork in this study. A 7 mm paint spray on the formwork resulted in effective temperature control in the initial 72 h following concrete pouring, as illustrated in Figure 15. Even after 10 turnover cycles with the formwork, the coating remained undamaged. Therefore, utilizing early-setting concrete formwork coating insulation has been proven to be a viable method for temperature control in high-altitude concrete projects.

Figure 15. Field test of thermal insulation of SiO_2 aerogel coating formwork.

6. Conclusions

In this study, the effects of extreme environmental conditions at high altitude on the cracking of bridge piers were investigated by quantitative analysis using a multi-field coupled model. Through on-site crack control tests in Sichuan, the effectiveness of deformation shrinkage compensation crack control technology was demonstrated. Additionally, the mechanisms of deformation compensation crack control and early temperature control of concrete were revealed. The key findings were outlined as follows:

(1) The quantitative risk assessment of the coupled hydration–temperature–moisture–constraint multi-field model indicated that the C35 concrete on the surface of the pier body had a higher risk factor for cracking compared with C45 concrete. The cracking risk coefficient exceeded 0.7 in the transverse direction of the pier body within the −2.5 m to 2.5 m range, with local C35 surface exceeding 1.0. The risk of cracking was also high in the height range of 0.3 m or more from the piers and horizontally from the center of the lateral surfaces of the piers.

(2) The on-site tests indicated that bridge piers treated with an anti-cracking agent exhibited approximately twice the unit expansion deformation during temperature rise compared with conventional piers. These treated piers experienced a decrease in shrinkage deformation by approximately 11% to 12% during temperature drop stages. Untreated concrete reached a tensile state after long-term hardening, while concrete treated with an anti-cracking agent maintained a positive expansion state during residual deformation. The anti-cracking agent demonstrated a deformation compensation effect, effectively reducing the likelihood of concrete cracking at high altitudes.

(3) The early freezing period of concrete had a significant impact on its strength. The earlier the freezing age, the more obvious the strength decline at low temperatures. Additionally, when the initial freezing age was the same, the longer the freezing time of concrete, the more significant the strength decrease, and the slower the rate of strength development. Therefore, in extreme environments at high altitudes, early temperature control of concrete was crucial.

The on-site crack resistance tests were influenced by various environmental factors. The local wind field in the plateau is strong, resulting in significant drying shrinkage of the surface concrete. The anti-cracking agent utilized in this study has limitations in effectively reducing water evaporation. Further research is necessary to enhance the control of surface moisture evaporation. Additionally, the effect of the anti-cracking agent on reducing temperature peaks was not evident in this study. It is necessary to further explore the role of anti-cracking agents in reducing the heat of hydration in future studies to enhance its practical application.

Author Contributions: Conceptualization, X.H.; Methodology, M.L. (Ming Li) and M.-L.Z.; Software, L.L.; Validation, L.L., M.L. (Manping Liao) and M.-L.Z.; Formal analysis, X.H., C.L. and Q.H.; Investigation, X.H., M.L. (Manping Liao), M.L. (Ming Li) and Z.Y.; Resources, M.L. (Manping Liao), Z.Y. and Q.H.; Data curation, L.L., C.L. and M.-L.Z.; Writing—original draft, X.H.; Writing—review & editing, M.-L.Z.; Visualization, M.L. (Ming Li) and C.L.; Supervision, Z.Y. and Q.H. All authors have read and agreed to the published version of the manuscript.

Funding: The authors greatly acknowledge the financial support from the CSCEC Technology R&D Program Funding Projects (CSCEC-2021-S-1); China Construction Science and Technology Innovation Platform Grant (CSCEC-PT-017); Outstanding Young Engineer Training Program of China Beijing Association for Science and Technology; and Shandong Provincial Natural Science Foundation (ZR2024QE147).

Data Availability Statement: The original contributions presented in the study are included in the article, further inquiries can be directed to the corresponding author.

Conflicts of Interest: Authors Xiaochuan Hu, Lei Liu and Manping Liao were employed by the company The Civil Engineering Group Corporation of China Second Engineering Bureau Ltd. Authors Ming Li and Cun Lu were employed by the company State Key Laboratory of High Performance Civil Engineering Materials, Jiangsu Research Institute of Building Science Co., Ltd., and Jiangsu Sobute New Materials Co., Ltd. Authors Zaifeng Yao and Qiuming Huang were employed by the company China Construction Second Engineering Bureau Ltd. The remaining author declares that the research was conducted in the absence of any commercial or financial relationships that could be construed as a potential conflict of interest.

References

1. Guo, C.B.; Zhang, Y.S.; Jiang, L.W.; Si, J.S.; Meng, W.; Du, Y.B.; Ma, C.T. Discussion on the Environmental and Engineering Geological Problems along the Sichuan-Tibet Railway and Its Adjacent Area. *Geoscience* **2017**, *31*, 877–889.
2. Peng, J.; Cui, P.; Zhuang, J. Challenges to Engineering Geology of Sichuan—Tibet railway. *Chin. J. Rock Mech. Eng.* **2020**, *39*, 2377–2389. [CrossRef]
3. Lu, C.F.; Cai, C.X. Challenges and Countermeasures for Construction Safety during the Sichuan-Tibet railway Project. *Engineering* **2019**, *5*, 833–838. [CrossRef]
4. He, R.; Lu, N. Air void system and freezing-thawing resistance of concrete composite with the incorporation of thermo-expansive polymeric microspheres. *Constr. Build. Mater.* **2024**, *419*, 135535. [CrossRef]
5. He, R.; Lu, N. Hydration, fresh, mechanical, and freeze-thaw properties of cement mortar incorporated with polymeric microspheres. *Adv. Compos. Hybrid Mater.* **2024**, *7*, 92. [CrossRef]
6. He, R.; Nantung, T.E.; Lu, N. Unraveling Microstructural Evolution in Air-Entrained Mortar and Paste: Insights from MIP and Micro-CT Tomography Amid Cyclic Freezing-Thawing Damage. *J. Build. Eng.* **2024**, *94*, 109922. [CrossRef]
7. Liu, J.P.; Tian, Q. *Early Deformation and Crack Control of Modern Concrete*; China Communications Press: Beijing, China, 2020.
8. Liu, J.; Tang, J.; Han, F. Toughening and Crack Prevention of Modern Concrete: Mechanisms and Applications. *China Civ. Eng. J.* **2021**, *54*, 47–54, 63.
9. Liu, J.P.; Tian, Q.; Wang, Y.J.; Li, H.; Xu, W. Evaluation Method and Mitigation Strategies for Shrinkage Cracking of Modern Concrete. *Engineering* **2021**, *7*, 348–357. [CrossRef]
10. Zhang, J.; Hou, D.; Chen, H. Experimental and Theoretical Studies on Autogenous Shrinkage of Concrete at Early ages. *J. Mater. Civ. Eng.* **2011**, *23*, 312–320.
11. Mo, L.; Deng, M.; Wang, A. Effects of MgO-based Expansive Additive on Compensating the Shrinkage of Cement Paste under Non-wet Curing Conditions. *Cem. Concr. Compos.* **2011**, *34*, 377–383. [CrossRef]
12. Saliba, J.; Rozière, E.; Grondin, F.; Loukili, A. Influence of Shrinkage-reducing Admixtures on Plastic and Long-term Shrinkage. *Cem. Concr. Compos.* **2011**, *33*, 209–217. [CrossRef]
13. Sun, W.; Chen, H.; Luo, X. The Effect of Hybrid Fiber Sand Expansive Agent on the Shrinkage and Permeability of High-performance Concrete. *Cem. Concr. Compos.* **2001**, *31*, 595–601.
14. Zhang, J.; Li, V.C. Influence of Fibers on Drying Shrinkage of Fiber Reinforced Cementitious Composite. *J. Eng. Mech.-ASCE* **2001**, *127*, 37–44.
15. Li, M.; Xie, B.; Xu, W.; Wang, Y.J.; Tian, Q. Comparison on the Crack Control Technique Schemes of Cast in Situ Tunnel Concrete. *Concrete* **2021**, *3*, 137–144.
16. Wang, J.; Li, M.; Wang, Y.J.; Li, Q.S.; Li, M. Research and Application of Anticracking and Seepage Prevention Approaches for A Working Shaft Sidewall Concrete in Yangtze River Tunnel in Jiangyin, Jingjiang. *Tunn. Constr.* **2022**, *42*, 1514–1520.
17. Li, X.; Fu, Z.; Luo, Z.; Ke, G.J. Effect of Low-pressure in Plateau Environment on Air Content of Fresh Concrete. *J. Southeast Univ. (Nat. Sci. Ed.)* **2014**, *44*, 1046–1051.

18. Ge, X.; Ge, Y.; Li, Q.F.; Cai, X.P.; Yang, W.C.; Du, Y.B. Effect of Low Air Pressure on the Durability of Concrete. *Constr. Build. Mater.* **2018**, *187*, 830–838.
19. Huo, J.Y.; Wang, Z.J.; Chen, H.X.; He, R. Impacts of Low Atmospheric Pressure on Properties of Cement Concrete in Plateau Areas: A literature Review. *Materials* **2019**, *12*, 1384. [CrossRef] [PubMed]
20. Nassif, A.Y.; Petrou, M.F. Influence of Cold Weather during Casting and Curing on the Stiffness and Strength of Concrete. *Constr. Build. Mater.* **2013**, *44*, 161–167.
21. Zdenek, P. Bazant. Prediction of concrete creep and shrinkage: Past, present and future. *Nucl. Eng. Des.* **2001**, *203*, 27–38.
22. Li, H.; Liu, J.P.; Wang, Y.J.; Yao, T.; Tian, Q.; Li, S.C. Deformation and Cracking Modeling for Early-age Sidewall Concrete Based on the Multi-field Coupling Mechanism. *Constr. Build. Mater.* **2015**, *88*, 83–94. [CrossRef]
23. Tian, Q.; Li, M.; Yao, T.; Li, H.; Wang, Y.J.; Xu, W. Multi-field Coupling Shrinkage Cracking Model of Concrete and Its Application in Civil Engineering. In Proceedings of the 3rd International RILEM Conference on Microstructure Related Durability of Cementitious Composites, Nanjing, China, 24–26 October 2016.
24. Cervera, M.; Oliver, J.; Prato, T. Thermo-chemo-mechanical Model for Concrete I: Hydration and Aging. *J. Eng. Mech.* **1999**, *125*, 1018–1027. [CrossRef]
25. Ulm, F.J.; Coussy, O. Modeling of Thermo-Chemo-Mechanical Couplings of Concrete at Early Ages. *J. Eng. Mech.* **1995**, *121*, 785–794. [CrossRef]
26. Lu, A.; Li, H.; Wang, Y. Effects of Temperature History on Expansion Properties of CaO- and MgO-bearing Expansive Agent for Concrete. In Proceedings of the 1st International Innovation in Low-Carbon Cement & Concrete Technology, London, UK, 24–26 June 2019.
27. Li, M.; Xu, W.; Wang, Y.J.; Tian, Q.; Liu, J.P. Shrinkage crack inhibiting of cast in situ tunnel concrete by double regulation on temperature and deformation of concrete at early age. *Constr. Build. Mater.* **2020**, *240*, 117834. [CrossRef]
28. Revilla-Cuesta, V.; Evangelista, L.; de Brito, J.; Skaf, M.; Ortega-Lopez, V. Mechanical performance and autogenous and drying shrinkage of MgO-based recycled aggregate high-performance concrete. *Constr. Build. Mater.* **2022**, *314*, 125726. [CrossRef]
29. Wang, L.; Li, G.H.; Li, X.; Guo, F.X.; Lu, X.; Hanif, A. Influence of reactivity and dosage of mgo expansive agent on shrinkage and crack resistance of face slab concrete. *Cem. Concr. Compos.* **2022**, *126*, 104333. [CrossRef]

buildings

MDPI

Article

Inversion Analysis for Thermal Parameters of Mass Concrete Based on the Sparrow Search Algorithm Improved by Mixed Strategies

Yang Wang [1], Yang Gao [2,*], Kaixing Zhang [1], Mei-Ling Zhuang [3], Runze Xu [1], Xiumin Yan [4] and Youzhi Wang [2,*]

[1] Shandong Expressway Jinan Round City West Highway Co., Ltd., Jinan 250300, China; wangyang92730@163.com (Y.W.); dxhxmb@163.com (K.Z.); renze_xu123@163.com (R.X.)
[2] Shandong University School of Civil Engineering, Shandong University, Jinan 250061, China
[3] Water Resources Research Institute of Shandong Province, Jinan 250013, China; ml_zhuang99@163.com
[4] School of Transportation and Civil Engineering, Nantong University, Nantong 226019, China; 2233110327@stmail.ntu.edu.cn
* Correspondence: 202215015@mail.sdu.edu.cn (Y.G.); wyz96996@163.com (Y.W.)

Abstract: In the traditional mass concrete temperature field calculation, the accuracy of the thermal parameters is extremely important. However, the actual thermal parameters of mass concrete may have some errors with the laboratory-measured values or specification values due to the site ambient temperature, concrete surface insulation measures, cooling water flow, etc. Therefore, it can be combined with the measured temperature of the field temperature sensors using the sparrow search algorithm (SSA) for the inverse analysis of thermal parameters. Firstly, to address the problem that SSA has low convergence accuracy and easily falls into local optimum, a mixed strategy was adopted to improve the algorithm, including Logistic Chaos mapping initialization of the population, the introduction of adaptive weighting factors, and the use of the Cauchy mutation strategy. Then, the performance test was carried out to compare the performance of the algorithm with three different intelligent algorithms and reflect the superiority of the SSA that was improved by mixed strategies (SSAIMSs). Finally, the proposed method was applied to the thermal parameter inversion of a mass concrete pile cap. The inversion results demonstrated that SSAIMSs can improve the accuracy and speed of thermal parameter inversion, and the calculated results of the thermal parameters and temperatures obtained using the SSAIMSs matched well with the measured results in the field, which can meet the accuracy requirements of the actual engineering.

Keywords: thermal parameter; inversion analysis; mass concrete; sparrow search algorithm; mixed strategies

1. Introduction

After the casting construction of concrete dams, bridge piers, pile caps, and other large-volume concrete, due to its internal heat of hydration and external environmental temperature and other factors, it is easy to produce a certain degree of temperature stress. If the tensile strength of the concrete is unable to resist the temperature stress, cracks will occur, which may seriously threaten the safety of concrete structures [1–5]. Although much research and effort have been made before the pouring construction stage, the cracking problem still exists. In addition to improving materials and construction methods, it is vital to predict and monitor the internal temperature of the concrete structures to reduce the generation of cracks after the pouring construction. Compared with the previous manual on-site thermometer measurements, there is now an increasing tendency to bury temperature sensors to read the temperature in real-time [6–9]. New fiber-optic sensors are emerging, but to comprehensively and accurately obtain the internal temperature changes of mass concrete, it is not sufficient to attain the current temperature data from temperature sensors. It is also necessary to use finite element software to simulate the mass concrete, predict the temperature changes throughout the pouring process, and combine it with the measured temperature data to prevent cracking due to excessive temperature stress. Finite element temperature field analysis of mass concrete requires concrete thermal parameters, which are generally measured in the laboratory or provided by the specification. However, due to the ambient temperature, concrete surface insulation measures, cooling water flow rate, and other conditions of the site, the measured data of these parameters are often in large error with the laboratory measurements or specification values [10]. Therefore, by combining the temperature measurements from temperature sensors and the measured values of thermal parameters in the field, a new intelligent algorithm can be used to invert the required thermal parameters.

As the inversion of thermal parameters requires repeated use of the temperature field calculation procedure, the traditional calculation method requires more manpower, energy, and time, and there is a possibility of large calculation errors, so there are many scholars who choose to use intelligent algorithms to carry out the inversion of thermal parameters of concrete [11–14]. Zhang et al. [15] used BP neural network algorithms to carry out the inversion of the thermal parameters of the pile cap mass concrete, and the calculated values were in good agreement with the field-measured values. Mao et al. [16] used a cross-global artificial bee colony algorithm to invert the thermal parameters of arch dams considering the multi-stage water passage of the cooling water pipe, and the results showed that the algorithm had good adaptability in arch dam species. Wang et al. [17] used hybrid particle swarm optimization (HPSO) and fiber optic temperature monitoring data to perform parameter inversion for concrete with different materials, and the results showed that the algorithm could improve the accuracy of parameter inversion. Hu et al. [18] proposed an intelligent inversion model using field-distributed monitoring data and numerical simulation and used an improved whale swarm algorithm to find the optimal solution. The effectiveness of the intelligent inversion model is verified in terms of noise reduction effect, calculation convergence speed, and inversion accuracy. Su et al. [19] proposed an inversion analysis method for thermal parameters of lock head based on the BP neural network due to engineering limitations or other reasons, and the results of the case analysis showed that using uniform design theory to generate the parameter set to be inverted could improve the convergence speed of neural network inverse analysis. Sun and He [20] introduced the Metropolis acceptance criterion from a simulated annealing algorithm to improve the genetic algorithm and perform parameter inversion. The results showed that it basically conformed to the temperature change process of the cushion concrete of the sixth dam section of the Hohhot Pumped Storage Power Station.

Many novel intelligent algorithms simulate a process in nature, generally with the goal of solving optimization problems, and are sometimes very useful in practical engineering applications. The sparrow search algorithm (SSA) is a novel intelligent algorithm proposed by Xue and Shen [21] in 2020, which simulates the predatory and anti-predatory behaviors of sparrows. The SSA has been widely acclaimed for its high efficiency and speed, but like most other intelligent algorithms, it is easy to fall into local optimal solutions and suffers from insufficient accuracy. Therefore, some scholars have improved the SSA and applied it to examples. Zheng et al. [22] proposed a Sine-SSA-BP mode to improve the SSA and optimize inland vessel trajectory prediction. Zhu et al. [23] proposed a new optimization algorithm called adaptive sparrow search algorithm (ASSA) to conduct three practical case studies, and the final results showed that the proposed ASSA was the most efficient. Yao et al. [24] introduced two improvement points to obtain the improved sparrow search algorithm (ISSA) to predict the river runoff, and the results showed that the proposed model was significantly better than other baseline models. Li et al. [25] proposed a novel framework using bidirectional gated recurrent unit (Bi-GRU) and SSA, and the observations showed that the proposed method performed better than other methods in terms of accuracy and robustness through three case studies. Behera and Saikia [26] proposed a system with an anti-windup mixed-order generalized integrator (AWMOGI) and an ISSA, which was validated by the simulation using MATLAB software and OPAL-RT real-time simulation testbed. Khedr et al. [27] proposed a Modified Sparrow Search Algorithm-based Mobile Sink Path Planning for WSNs (MSSPP) to generate shorter routes of travel for MS and minimize data collection delays, and the results indicated that MSSPP improved performance and was more effective than other related methods. Ma et al. [28] introduced a two-dimensional logistic chaotic system, a Levy flight strategy, and nonlinear adaptive weighting for the sparrow search algorithm, which was able to optimize the working action trajectory of industrial robots and increase efficiency.

In this paper, the SSA was used as the inverse for thermal parameters of mass concrete, and a mixed improvement strategy was adopted to improve the SSA for the shortcomings of the SSA algorithm. Firstly, the Logistic chaos mapping was used to initialize the population at the initialization stage. Then, an adaptive weighting factor was introduced to improve the SSA. Finally, the Cauchy mutation strategy was adopted for the location of the optimal individuals. To verify the effectiveness of the sparrow search algorithm improved by mixed strategies (SSAIMS), it was tested by 12 standard test functions with the other three algorithms. At the same time, it was applied to the actual engineering of mass concrete of a pile cap in Shandong, China, which can provide reference and inspiration for other mass concrete construction projects.

2. Principles and Steps of SSAIMSs

2.1. SSA

The SSA is inspired by the various behaviors of sparrows in the process of foraging and anti-predation. It is necessary to rank the sparrow population using fitness advantages and disadvantages after initializing the population. There are different divisions of labor within the sparrow population, which can be roughly classified into three types when conducting a spatial search: producers, scroungers, and alerters. The producers are the better-adapted individuals who search for food for the sparrow population; the scroungers are the less well-adapted individuals who may compete with the producers for food after they have found the location of the food; and the alerters are the randomly-adapted individuals that are on the lookout for the food and signal to flee if they find any dangers such as natural enemies.

The producers provide approximate foraging directions for the entire sparrow population, and the location update criterion for the producers is

$$X_{i,j}^{k+1} = \begin{cases} X_{i,j}^{k} \cdot \exp\left(-\frac{i}{q t_{\max}}\right), & R_2 < \delta_{st} \\ X_{i,j}^{k} + Q \cdot L, & R_2 \geq \delta_{st} \end{cases} \quad (1)$$

During the search for food, the scroungers snatch the food found by the producers. If the snatch fails, the location update criterion for the scroungers is

$$X_{i,j}^{k+1} = \begin{cases} Q \cdot \exp\left(\frac{X_{worst} - X_{i,j}^{k}}{i^2}\right), & i > \frac{n}{2} \\ X_P^{k+1} + \left|X_{i,j}^{t} - X_P^{k+1}\right| \cdot A^{+} \cdot L, & i \leq \frac{n}{2} \end{cases} \quad (2)$$

The alerters monitor the foraging area space of the entire population and immediately signal danger when there is danger, followed by rapid movement towards the safe area space. The location update criterion for the alerters is

$$X_{i,j}^{k+1} = \begin{cases} X_{best}^{k} + \eta \cdot \left|X_{i,j}^{k} - X_{best}^{k}\right|, & f_i > f_g \\ X_{i,j}^{k} + K \cdot \left(\frac{\left|X_{i,j}^{k} - X_{worst}^{k}\right|}{(f_i - f_w) + \varepsilon}\right), & f_i = f_g \end{cases} \quad (3)$$

2.2. Sparrow Search Algorithm Improved by Mixed Strategies (SSAIMS)

2.2.1. Initialization of Population by Using Logistic Chaos Mapping

The Sparrow Search Algorithm adopts a random strategy for the generation of sparrow populations. Although the random number method can randomly assign values to the populations in the search space, it usually cannot guarantee that the initial populations fully and evenly cover the search space. Otherwise, it is difficult to find out the various possibilities of the solution, which often leads to the generation of problems such as low diversity and uneven distribution of sparrow populations and ultimately reduces the quality of the solution due to premature convergence. Chaotic mapping can produce chaotic sequences that are more uniformly distributed than random generation and are therefore used in sparrow search algorithms to generate sparrow populations. Currently, the frequently used mappings are tent mapping, Chebyshev mapping, Singer mapping, logistic mapping, sine mapping, circle mapping, and so on. In this paper, Logistic chaos mapping is introduced to improve the Sparrow Search Algorithm, as shown in Equation (4).

$$X_{k+1} = \mu X_k(1 - X_k) \tag{4}$$

2.2.2. Adaptive Weight Factor

The weighting factor plays a large role in objective function optimization, and appropriate weights can speed up the convergence speed of the algorithm and improve its accuracy. To solve the problem of slow convergence speed and low precision in the optimization process, the adaptive weight factor ω is introduced, as shown in Equation (5), according to the inspiration of Liu and He [29]. In the very first iteration stage of the SSA, the sparrow population can traverse the entire search space with a larger weight, which is conducive to the algorithm to improve global development capabilities and accelerate the convergence speed. In the middle and late stages of the SSA iteration, the algorithm converges gradually, and a smaller weight can be used to explore the small region finely to improve the convergence accuracy; finally, in the final stage of the sparrow algorithm iteration, it can be assigned a relatively large perturbation to solve the problem that SSA is easy to fall into local optimum.

$$\omega = \begin{cases} \delta_1 \cdot \left(\cos\left(k \cdot \frac{\pi}{\delta_2}\right) + \delta_3\right), \ k \le k_s \\ \rho_1 \cdot \sin(\rho_2 \cdot k \cdot \pi) + \rho_3, \ k > k_s \end{cases} \tag{5}$$

When $k_{\max} = 500$, the adaptive weight factor curve is shown in Figure 1.

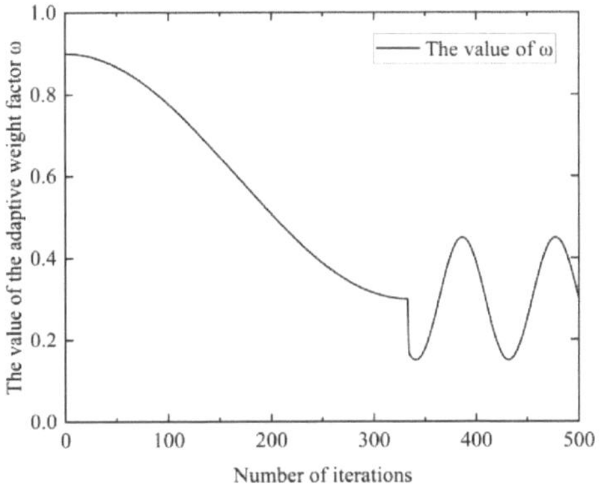

Figure 1. Plot of the values of the adaptive weighting factor ω.

In the standard SSA, the value of the factor $\exp\left(-\frac{i}{qt_{\max}}\right)$ in Equation (1) is in the range of (0, 1), which is positive for the convergence of the algorithm. However, as the number of iterations continues to increase, the value of the factor approaches 0 more and more quickly, which makes the producers converge to the origin, and the algorithm is prone to fall into the local optimum with the low speed of the convergence procedure [30]. In this paper, the adaptive weight factor ω is used to replace this factor, which makes the SSA converge faster and does not easily fall into the local optimum. The improvement is shown in Equation (6).

$$X_{i,j}^{k+1} = \begin{cases} X_{i,j}^{k} \cdot \omega, \ R_2 < \delta_{st} \\ X_{i,j}^{k} + Q \cdot L, \ R_2 \geq \delta_{st} \end{cases} \tag{6}$$

2.2.3. Cauchy Mutation Strategy

The Cauchy mutation is derived from the Cauchy distribution of a continuous type probability distribution with a one-dimensional Cauchy density function concentrated near the origin, which is functionally defined as Equation (7).

$$f(x) = \frac{1}{\pi} \cdot \frac{a}{a + x^2}, \ x \in (-\infty, +\infty) \tag{7}$$

When $a = 1$, it is called the standard Cauchy distribution.

The one-dimensional Cauchy density distribution function is shown on the image to be concentrated near the origin. According to this feature, the current optimal individual is perturbed by Cauchy mutation to promote its development in a better direction, which is conducive to solving the problem of falling into local optimum. The equation of the Cauchy mutation strategy is

$$X_{new} = X_{best} \cdot [1 + Cauchy(0,\ 1)] \tag{8}$$

After the Cauchy mutation, it has a positive effect on the SSA to jump out of the local optimum, but there is no comparison of whether the updated position is better than the original position. Therefore, in order to determine whether the position information should be updated or not, a greedy selection strategy is introduced to compare the fitness values of the old and the new. At the same time, it can enhance the speed and accuracy of the algorithm convergence and improve the algorithm's performance in optimization seeking. The greedy selection strategy is shown in Equation (9).

$$X_{i,j}^{k+1} = \begin{cases} X_{new}, \ f(X_{new}) < f(X_{best}) \\ X_{best}, \ f(X_{new}) \geq f(X_{best}) \end{cases} \tag{9}$$

2.3. *Performance Testing*

To test the performance of the SSAIMSs, this paper uses 12 commonly used test functions to perform simulation experiments with the standard particle swarm optimization algorithm (PSO) [31,32], simulated annealing algorithm (SA) [33,34], grey wolf optimization (GWO) [35,36]. The experiment introduces 12 test functions with different optimization characteristics, as shown in Table 1, of which the first four are high-dimensional single-peak functions, which have only one optimal solution and are used to test the convergence speed and accuracy of the algorithms; the middle four are high-dimensional multi-peak functions, and the last four are low-dimensional multi-peak functions, which have multiple local optimal solutions and can be used to test whether the algorithms are prone to fall into local optimum. At the same time, the tests are conducted on the same experimental environment of Windows 10 and the experimental simulation platform of MATLAB R2020a to fairly compare the performance of various algorithms. The number of generated populations n is 30, and the maximum number of specified iterations $k_{\max}$ is 500. To reduce the impact of randomness on various algorithms and make the results scientific and credible, 20 independent experiments are carried out for each algorithm, and then the mean value and variance of the results of the 20 experiments are compared.

Table 1. Test Functions.

Function	Dimension	Search Scope	Theoretical Value
$F_1(x) = \sum_{i=1}^{n} x_i^2$	30	[−100, 100]	0
$F_2(x) = \sum_{i=1}^{n} \lvert x_i \rvert + \prod_{i=1}^{n} \lvert x_i \rvert$	30	[−10, 10]	0
$F_3(x) = \max_i\{\lvert x_i \rvert, 1 \le i \le n\}$	30	[−100, 100]	0
$F_4(x) = \sum_{i=1}^{n} i x_i^4 + random[0, 1)$	30	[−1.28, 1.28]	0
$F_5(x) = \sum_{i=1}^{n} \left[x_i^2 - 10\cos(2\pi x_i + 10)\right]$	30	[−5.12, 5.12]	0
$F_6(x) = -20\exp\left(-0.2\sqrt{\frac{1}{n}\sum_{i=1}^{n} x_i^2}\right) - \exp\left(\frac{1}{n}\sum_{i=1}^{n}\cos(2\pi x_i)\right) + 20 + e$	30	[−32, 32]	0
$F_7(x) = \frac{1}{4000}\sum_{i=1}^{n} x_i^2 - \prod_{i=1}^{n}\cos\left(\frac{x_i}{\sqrt{i}}\right) + 1$	30	[−600, 600]	0
$F_8(x) = \frac{\pi}{n}\left\{10\sin(\pi y_1) + \sum_{i=1}^{n-1}(y_i - 1)^2\left[1 + 10\sin^2(\pi y_{i+1})\right] + (y_n - 1)^2\right\} + \sum_{i=1}^{n} u(x_i, 10, 100, 4)$ $y_i = 1 + \frac{x_i + 1}{4} \quad u(x_i, a, k, m) = \begin{cases} k(x_i - a)^m & x_i > a \\ 0 & -a < x_i < a \\ k(-x_i - a)^m & x_i < -a \end{cases}$	30	[−50, 50]	0
$F_9(x) = \left(\frac{1}{500} + \sum_{j=1}^{25}\frac{1}{j + \sum_{i=1}^{2}(x_i - a_{ij})^6}\right)^{-1}$	2	[−65, 65]	1
$F_{10}(x) = \left[1 + (x_1 + x_2 + 1)^2(19 - 14x_1 + 3x_1^2 - 14x_2 + 6x_1x_2 + 3x_2^2)\right] \times \left[30 + (2x_1 - 3x_2)^2 \times (18 - 32x_1 + 12x_1^2 + 48x_2 - 36x_1x_2 + 27x_2^2)\right]$	2	[−2, 2]	3
$F_{11}(x) = -\sum_{i=1}^{5}\left[(X - a_i)(X - a_i)^T + c_i\right]^{-1}$	4	[0, 10]	−10.1532
$F_{12}(x) = \sum_{i=1}^{10}\left[(X - a_i)(X - a_i)^T + c_i\right]^{-1}$	4	[0, 10]	−10.5363

The test results of the four algorithms are shown in Table 2. It can be seen that for most of the test functions, SSAIMSs are superior to the other three algorithms, and the variance index is slightly worse than the SA only in F9. For the high-dimensional single-peak function of F1–F4, although SSAIMS does not find the theoretical optimal solution of 0, it has higher accuracy and better stability, and it is closer to 0 than the other algorithms. For the high-dimensional multi-peaked functions of F5–F8, SSAIMS can find the theoretical optimal solution of 0 on the F5 and F7 functions. While the calculated fitness value for F6 is not 0, the variance is 0, which is very stable. For the low-dimensional multi-peaked functions of F9–F12, SSAIMS is also very superior and almost never falls into the local optimum and finds the actual optimal solution. It can be concluded that the SSAIMS is more effective than the other three algorithms and can be applied to the inversion of thermal parameters of mass concrete.

Table 2. Performance comparison of four algorithms.

Function	PSO		SA		GWO		SSAIMS	
	Mean Value	Variance	Mean Value	Variance	Mean Value	Variance	Mean Value	Variance
F1	6.60×10^{-5}	4.78×10^{-9}	1.89×10^{-17}	2.65×10^{-33}	3.75×10^{-57}	9.23×10^{-60}	1.31×10^{-71}	3.44×10^{-141}
F2	4.16	1.40	5.60×10^{-5}	1.16×10^{-9}	6.39×10^{-17}	2.01×10^{-33}	2.84×10^{-28}	1.08×10^{-54}
F3	5.70×10^{-2}	7.27×10^{-4}	8.96×10^{-3}	6.25×10^{-5}	4.55×10^{-18}	7.55×10^{-35}	6.63×10^{-32}	8.32×10^{-62}
F4	13.8	76.5	0.146	2.13×10^{-3}	2.15×10^{-3}	1.02×10^{-6}	1.15×10^{-3}	3.60×10^{-7}
F5	161	1.17×10^{3}	3.38	2.31	2.62	12.4	0	0
F6	2.66	6.93×10^{-2}	7.22×10^{-3}	3.19×10^{-4}	1.06×10^{-13}	2.13×10^{-28}	8.88×10^{-16}	0
F7	0.117	1.82×10^{-3}	4.97×10^{-2}	2.46×10^{-3}	4.51×10^{-3}	7.37×10^{-5}	0	0
F8	1.33×10^{-4}	2.16×10^{-7}	5.47×10^{-19}	2.64×10^{-36}	1.49×10^{-2}	1.32×10^{-2}	1.47×10^{-32}	7.36×10^{-66}
F9	3.17	5.77	0.998	6.20×10^{-26}	3.06	9.88	0.998	2.51×10^{-22}
F10	3	3.19×10^{-29}	8.40	122	3	2.16×10^{-9}	3	1.61×10^{-30}
F11	−7.64	10.6	−4.01	5.54	−9.14	4.32	−9.52	3.90
F12	−9.86	4.50	−5.52	12.1	−10.1	3.29	−10.3	1.36

3. Fundamentals of Inversion of Thermal Parameters of Mass Concrete

3.1. Calculation Criterion for Temperature Field of Concrete

In the days just after the concrete has been poured, the concrete undergoes a heat of hydration reaction when there is an internal heat source. The poor thermal conductivity of concrete and a large amount of internal heat of hydration generated cannot be propagated out quickly for the time being, resulting in rapid heating of the concrete at an early stage, which may result in cracking due to temperature stresses, which requires special attention in the case of mass concrete. The equation for the thermal conductivity of concrete is

$$\frac{\partial T}{\partial t} = \frac{\lambda}{c\rho}(\frac{\partial^2 T}{\partial x^2} + \frac{\partial^2 T}{\partial y^2} + \frac{\partial^2 T}{\partial z^2}) + \frac{\partial \theta_0}{\partial t} \tag{10}$$

The equation for the adiabatic temperature rise of concrete is expressed by Equation (11), which is a more intuitive representation of the heat generated by the heat of hydration of concrete.

$$\frac{\partial T}{\partial t} = \frac{\lambda}{c\rho}(\frac{\partial^2 T}{\partial x^2} + \frac{\partial^2 T}{\partial y^2} + \frac{\partial^2 T}{\partial z^2}) + \frac{\partial \theta_0}{\partial t} \tag{11}$$

The equation for calculating the equivalent exothermic coefficient of a concrete structure covered with an insulating material is expressed by Equation (12), which can be measured and then calculated in the field. Concrete surfaces are susceptible to the site environment as they are exposed to frequent direct sunlight.

$$\frac{\partial T}{\partial t} = \frac{\lambda}{c\rho}(\frac{\partial^2 T}{\partial x^2} + \frac{\partial^2 T}{\partial y^2} + \frac{\partial^2 T}{\partial z^2}) + \frac{\partial \theta_0}{\partial t} \tag{12}$$

The concrete boundary condition is

$$-\lambda \frac{\partial T_c}{\partial m} = \beta(T_c - T_a) \tag{13}$$

Mass concrete in the pouring is often used to cool water pipes for water cooling to prevent its internal heating from being too fast and too high. According to academician Zhu [37] and Zhou et al. [38], Equations (14)–(17) can be used to calculate the temperature of the concrete under the simultaneous consideration of the effects of the cooling water pipe and the external ambient temperature.

$$T(t) = T_{ij} + (T_i - T_{ij})\phi_i(t) + \theta_0 \psi_i(t) + \eta_i(t) \tag{14}$$

$$\phi_i(t) = \exp(-p_i t) \tag{15}$$

$$\psi_i(t) = \frac{r}{r - p_i}(\exp(-p_i t) - \exp(-rt)) \tag{16}$$

$$\eta_i(t) = (T_{ia} - T_{ij})\sum_k \{\exp[-p_i(t - t_k)] - 1\} \cdot \Delta\mathrm{erf}\left(\frac{h}{2\sqrt{at_k}}\right) \tag{17}$$

3.2. Selection of Thermal Parameters of Concrete

The heat of hydration of concrete is an important factor in the generation of temperature stresses; therefore, it is crucial to obtain the thermal parameters of concrete. Calculation of thermal parameters of concrete is generally related to parameters such as thermal conductivity α, thermal conductivity λ, adiabatic temperature rise θ_0, equivalent surface heat dissipation coefficient β_s, rate of reaction of heat of hydration r, density ρ and specific heat capacity c. For the density ρ and specific heat capacity c of concrete, the changes are very small during the pouring process, which can be obtained by laboratory and field measurements without inversion, while the thermal conductivity coefficient α can be calculated based on $\alpha = \frac{\lambda}{c\rho}$ to derive α. In general, the four parameters of λ, θ_0, β_s, and r after concrete pouring are greatly affected by the environment, which is not easy to obtain in the laboratory and can be used for the inversion of thermal parameters.

3.3. Selection of the Objective Function

The objective function is established as shown in Equations (18) and (19).

$$F(X) = \frac{\sqrt{\sum_{m=1}^{M}\sum_{n=1}^{N}\left(T'_{mn} - T_{mn}\right)^2}}{MN} \tag{18}$$

$$X = [x_1, x_2, x_3, x_4] \tag{19}$$

The objective function is established so that the parametric inversion problem becomes an optimization problem, and the objective function $F(X)$ is converged to the minimum by constant computation. Then, the inversion parameters are continuously updated in MATLAB software. When the objective function reaches the maximum number of iterations t_{max} of the algorithm, the calculation is stopped, and the parameter inversion result is the final output. The flowchart of the SSAIMSs combined with the temperature field calculation procedure is shown in Figure 2.

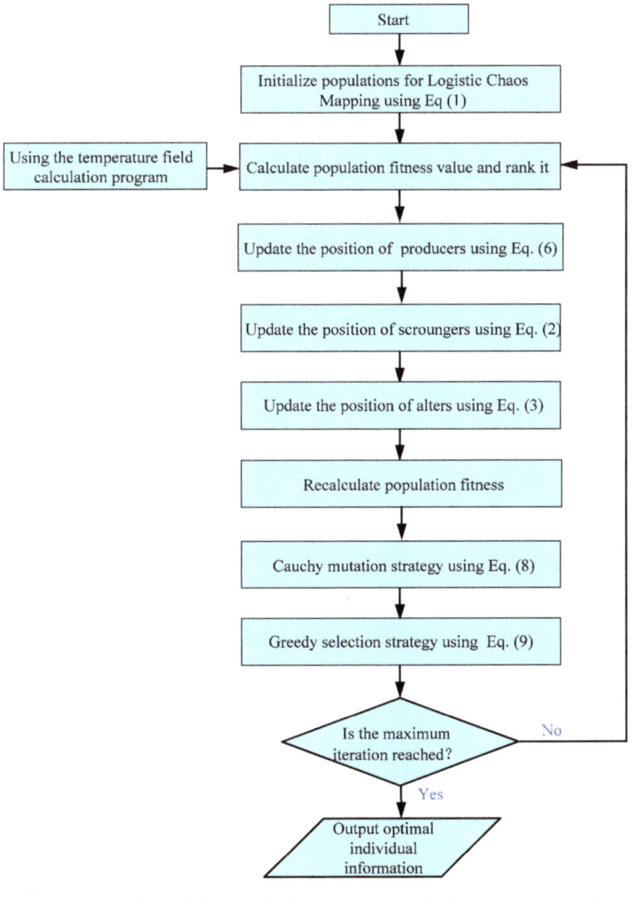

Figure 2. Flowchart of the SSAIMS combined with the temperature field calculation procedure.

4. Example Application

A twin-tower, single cable-stayed, pre-stressed concrete, large-span suspension cable-stayed bridge was constructed at a site in Shandong, China. The total length of the main bridge was 394.6 m, with a two-way six-lane carriageway, a roadbed width of 34.5 m, and a roadway width of 30 m. The main road was arranged as a 0.75 m wide earth shoulder, 3 m wide hard shoulder, 3.75 m wide carriageway, 0.75 m wide curb strip, and 3.0 m wide central divider, with symmetrical left and right widths. In addition, the main bridge was located on the straight section with +1.930% and −1.70% corresponding longitudinal slopes, respectively, and 2% cross slope of the bridge deck. The structural system was a tower–beam consolidation and tower–pier separation system, The structural schematic of the main bridge is shown in Figure 3.

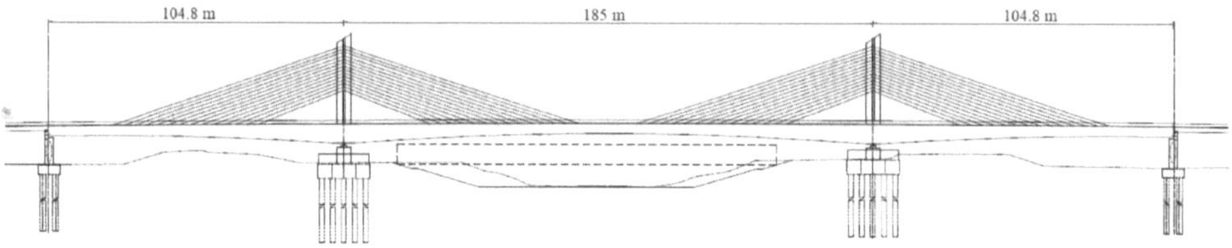

Figure 3. The structural schematic of the main bridge.

The main materials used in this bridge include concrete, rebars, pre-stressed steel strands, stay cables, asphalt, and so on. Among them, all kinds of concrete were tested to meet the Chinese standard specifications, and the concrete grades used in each part of the bridge are shown in Table 3. The rebars used in the design were HPB300 and HRB400 grades. The performance of the pre-stressed steel strands showed high strength and low relaxation, with a nominal diameter of 15.2 mm, a nominal area of 139 mm^2, a standard value of tensile strength of 1860 MPa, a modulus of elasticity of 1.95×10^5 MPa, and a relaxation rate of 2.5 percent. The stay cables adopted epoxy steel strands. Besides, there were 28 pairs of stay cables on each tower, with a total of 56 pairs on the whole bridge, with a longitudinal standard spacing of 4 m on the main girder and a vertical spacing of 1.2 m on the tower. After the bridge was completed, the bridge deck was paved with 4 cm SMA-13 fine-grain asphalt.

Table 3. Concrete grades are at different positions on the bridge.

Main Girder	Main Tower	Main Pier and Side Pier	Pile Cap	Pile Foundation	Bearing Cushion Layer	Bottom Sealing of Pile Cap
C55	C50	C40	C35	C30	C50	C25

The main construction process of the main pier foundation is as follows: (1) insertion and drilling of steel shields; (2) construction of bored piles; (3) assembling of steel sheet pile cofferdams; (4) positioning of cofferdams; (5) pouring of bottom sealing concrete; (6) pumping and bearing platform construction; (7) construction of the pier body. The main pier pile cap adopts an octagonal pile cap, with a plan dimension of 25.5 m × 18.588 m, 5 m thick, using C35 concrete, which was a kind of mass concrete. The main pier foundation adopted a pile foundation, which was designed as a friction pile. There were 27 bored piles with a diameter of 1.8 m and a length of 83 m underneath.

When the ambient temperature is too high, the hydration reaction of the concrete accelerates, resulting in the internal temperature of the concrete. Because the internal heat of hydration heat can not be dissipated in time, the internal temperature stress will be generated. When the temperature stress exceeds the tensile strength of the concrete, a large number of cracks will be generated to reduce the performance of the concrete, so it is necessary to use cooling water and re-measure the thermal parameters of the concrete. When the ambient temperature is too low, the rate of hydration of the concrete decreases, which affects the strength development of the concrete and may lead to expansion of the concrete, irreversible cracks, and loss of strength. It is, therefore, necessary to insulate the concrete and re-evaluate the thermal parameters in a timely manner. The project site has moderate temperatures and abundant thermal energy resources, with average annual temperatures ranging from 11.0 °C to 14.3 °C. In addition, the lowest temperatures are found in January and February, and the highest temperatures are found in July. In addition, the temperatures are the lowest in January and February and highest in July. The mass concrete pile caps were placed in early August. Due to the high temperature in summer, the thermal parameters of the concrete to be inverted, such as thermal conductivity λ, adiabatic temperature rise θ_0, equivalent surface heat dissipation coefficient β_s, and heat of hydration reaction rate r, are easily affected by the ambient temperature, which plays an important role in the temperature change inside the concrete. The pile cap information is shown in Figure 4.

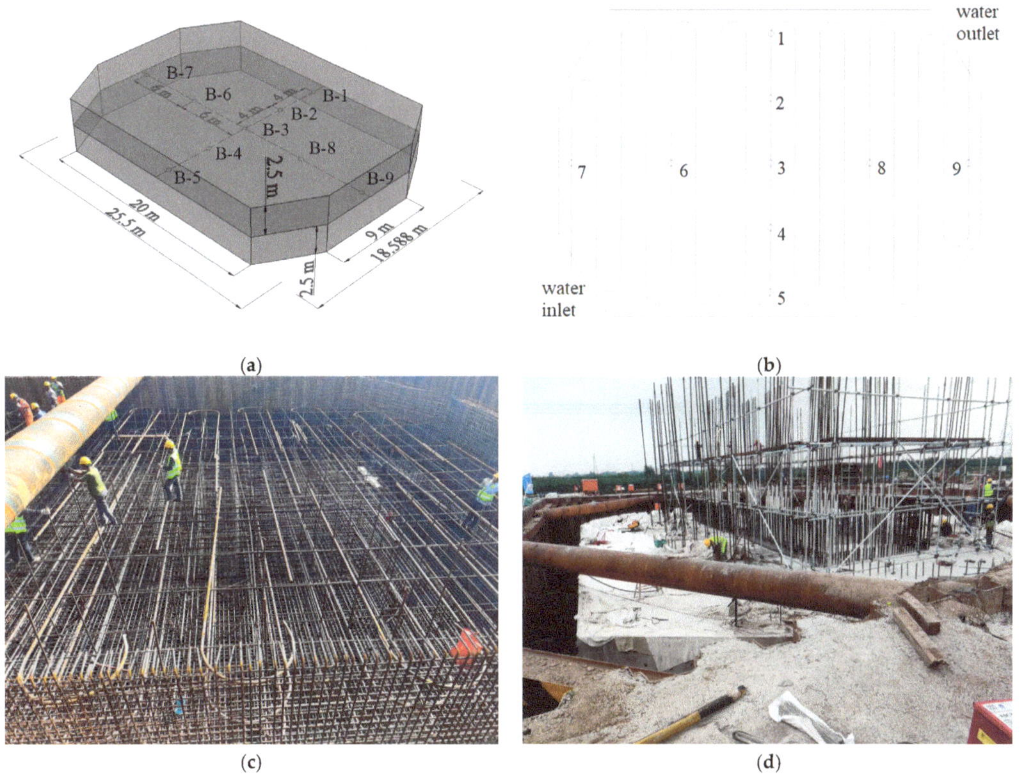

Figure 4. Information diagram of the pile cap. (**a**) Dimensions of pile cap and layout of measurement points at level B; (**b**) Layout of measurement points and cooling water pipes, the numbers can represent the positions of temperature sensors on the three layers A, B, and C; (**c**) Rebar binding of pile cap; (**d**) On-site after pouring the pile cap.

To measure the temperature, temperature sensors were tied on the surface of the reinforcement bars on site before pouring. A total of three layers were strapped, with the bottom layer of layer A at 1 m above the ground level, the middle layer of layer B at 2.5 m above the ground level, and layer C at 4 m above the ground level. Nine sensors were tied in each layer, with points 7, 6, 3, 8, and 9 spaced 3 m apart and points 1, 2, 3, 4, and

5 spaced 4 m apart, all on axes. In addition, to ensure the accuracy and reliability of the temperature data, the inlet and outlet temperatures, the concrete top surface temperature, and the atmospheric temperature were measured simultaneously. The pile cap was poured with three rows of cooling water pipes with the same height position as the temperature sensors, 60 mm diameter, 3 mm thickness, zigzagging back and forth horizontally, and 1.5 m spacing. After the pouring of the pile cap, the temperature was continuously measured by using the temperature sensor. It was feasible to measure the temperature on site, but it was easier and more accurate to take automatic readings and record the temperature data on a computer by using the temperature sensor software Vircom 10.9. Then, it was necessary to manually check the data and draw graphs for comparative analysis in the following steps. The temperature sensor information is shown in Figure 5.

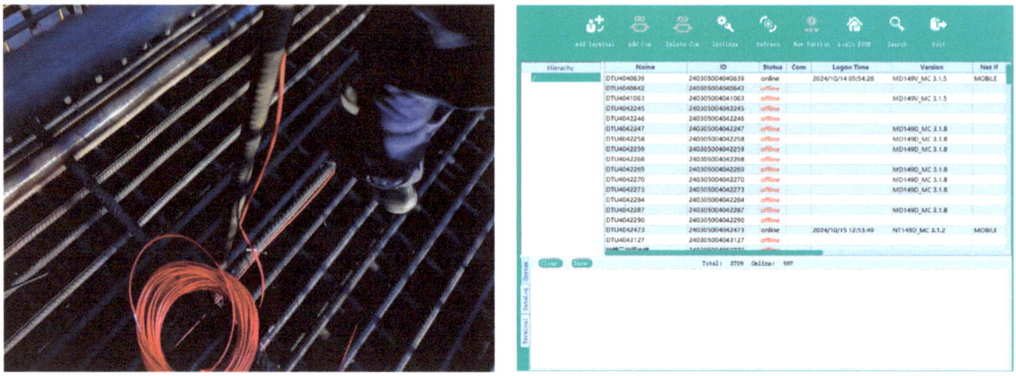

(a) (b)

Figure 5. Information diagram of the temperature sensor. (**a**) Binding of the temperature sensor; (**b**) One of the temperature sensor software.

4.1. Finite Element Modelling

According to the construction plan of the site, this paper establishes a finite element model by Midas FEA NX 2022 v1.1 as in Figure 6. It used an eight-node hexahedral cell, the upper half of which was the pile cap and the lower half of which was the foundation, and the temperature sensor measurement points were distributed in the model cell nodes. As the mesh near the cooling water pipes might have a large temperature difference, the cell nodes were modified to second-order cells after modeling to facilitate the analysis of the temperature magnitude and refine the cells. Finally, the total number of cells was 3100, and the total number of nodes was 41,305.

4.2. Inversion of Thermal Parameters and Comparison of Results

To verify the actual performance of the various algorithms in the engineering example, the thermal parameter ranges, as in Table 4, were selected for inversion in MATLAB software according to the preliminary parameter trial calculation and the actual requirements of the site. As can be seen from Figure 7, the fitness values of all algorithms decreased as the number of iterations increased, but different algorithms had different speeds of decreasing fitness values and different final values. The PSO had the highest first-generation fitness value and the slowest convergence, falling into a local optimum around the 10th generation, with a final fitness value of 0.4977. The SA converged around the 16th generation, with a final fitness value of 0.4966; the GWO was also susceptible to falling into a local optimum but with a lower final fitness value of 0.4965; in contrast, the SSAIMSs performed the best, converging around generation 12, with the lowest fitness value of 0.4955. Overall, it can be seen that the SSAIMSs converge significantly faster, with the highest convergence accuracy, which can reduce the number of iterations required to meet the accuracy requirement, save the arithmetic time, power, and time, and perform thermal parameter inversion analysis more effectively.

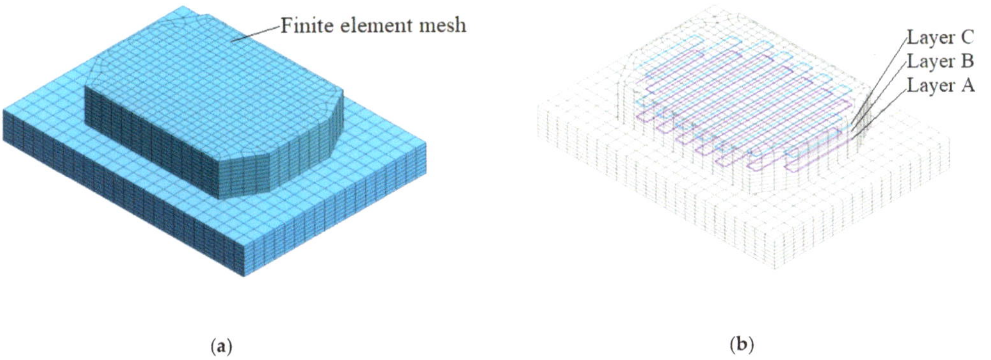

Figure 6. Finite element modeling diagram of the pile cap and foundation. (**a**) Finite element meshing diagram; (**b**) Cooling water pipe layout for finite element modeling.

Table 4. Range of thermal parameters of C35 concrete.

Thermal Parameters of C35 Concrete	Unit	Range
Thermal conductivity λ	KJ/(m·h·°C)	[8, 16]
Adiabatic temperature rise θ_0	°C	[50, 80]
Equivalent surface heat dissipation coefficient β_s	KJ/(m^2·h·°C)	[10, 100]
Hydration heat reaction rate r	h^{-1}	[0.01, 0.04]

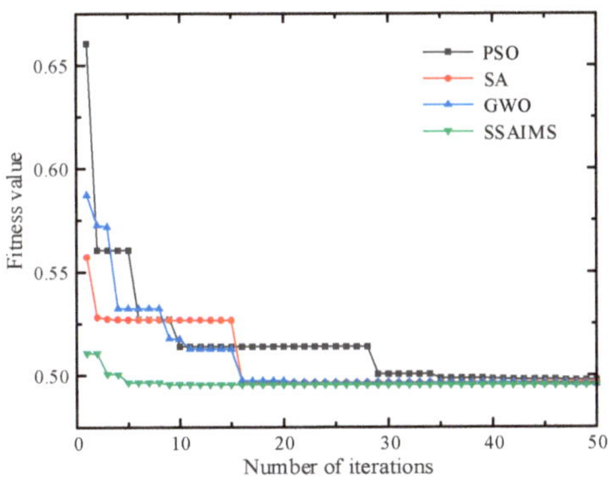

Figure 7. Plot of the change process of the fitness values of the four algorithms.

After iterative operation, in addition to the final fitness value of the four algorithms, the thermal parameter values corresponding to the fitness value could also be obtained. To compare the validity and reliability of the thermal parameter values of the four algorithms, the actual measured and calculated thermal parameter values in the field were compared with them, as shown in Table 5. It lists the absolute values of the relative errors between the thermal parameter values calculated by the four algorithms and the actual measured values in the field. As can be seen in Table 5, the relative error between the thermal parameters calculated by the SSAIMSs and the actual measured values in the field was smaller in absolute value than the other three algorithms. Compared with the other three algorithms, the thermal parameters derived from the SSAIMSs were closer to reality, and the absolute values of the relative errors were all within 6% so that the accuracy could meet the practical requirements.

Table 5. Comparison of inversion results of thermal parameters of four algorithms and field measurements.

Comparison of Algorithms	Thermal Conductivity λ/[KJ·(m·h·°C)$^{-1}$]	Adiabatic Temperature Rise θ_0/°C	Equivalent Surface Heat Dissipation Coefficient β_s/ [KJ·(m^2·h·°C)$^{-1}$]	Hydration Heat Reaction Rate r/(h^{-1})
On-site measurement	10.6	74.5	49.8	0.0162
PSO	12.2 (15.1%)	85.7 (15.0%)	86.2 (73.9%)	0.0153 (5.56%)
SA	9.7 (8.49%)	72.8 (2.28%)	48.3 (3.01%)	0.0194 (19.8%)
GWO	10.3 (2.83%)	76.8 (3.09%)	53.8 (8.03%)	0.0179 (10.5%)
SSAIMS	10.4 (1.89%)	76.6 (2.82%)	50.9 (2.21%)	0.0171 (5.56%)

The four thermal parameters obtained using the SSAIMS were brought into the finite element modeling software for analysis, where the cooling water temperature was left unchanged at the same round of 10 °C. When observing and analyzing the temperature results, the 1/4 model was used to visually obtain the temperature magnitude. The temperature distribution inside and on the surface of the model at the age of 5 days is shown in Figure 8.

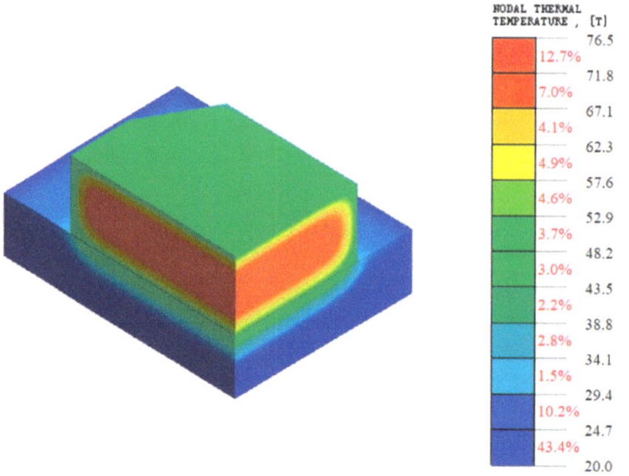

Figure 8. Temperature distribution of finite element 1/4 model at the age of 5 days. (unit: °C).

Temperature sensor readings at locations B-1, B-2, B-3, B-6, and B-7 in layer B (the middle layer) were compared with the temperatures derived from finite element simulations. Due to the rapid temperature change due to the internal heat of hydration after casting, the two were compared daily for the first 9 days and then every 2 days thereafter. The comparison results are shown in Figure 9. As can be seen from Figure 9, on day 3, the maximum absolute value of the difference between the two temperatures at position B-3 was 4.6 °C, and the corresponding relative error was 6.18%, which was relatively small. In the early stage, with cooling water, the temperature errors at all locations were larger than in the later stage, when there was no heat or water supply. In addition to the slight errors between the thermal parameters calculated by the algorithm and the actual situation, as well as the accuracy issues of finite element software calculations, the main reasons may be the unstable water supply in the early stage and the large temperature difference between day and night on site. When the ambient temperature is too high in the middle of the day, it is important to combine the finite element software to calculate the need to temporarily increase the amount of water supply. When the atmospheric temperature is too low at night, the concrete should be covered. Overall, the temperature sensor readings measured on site were in good agreement with the temperature calculated by the finite element model, and the trend was basically the same, which can better reflect the temperature changes on the construction site. Therefore, the thermal parameters of the inversion performance obtained using the SSAIMSs were more effective and can be used in the actual engineering sites.

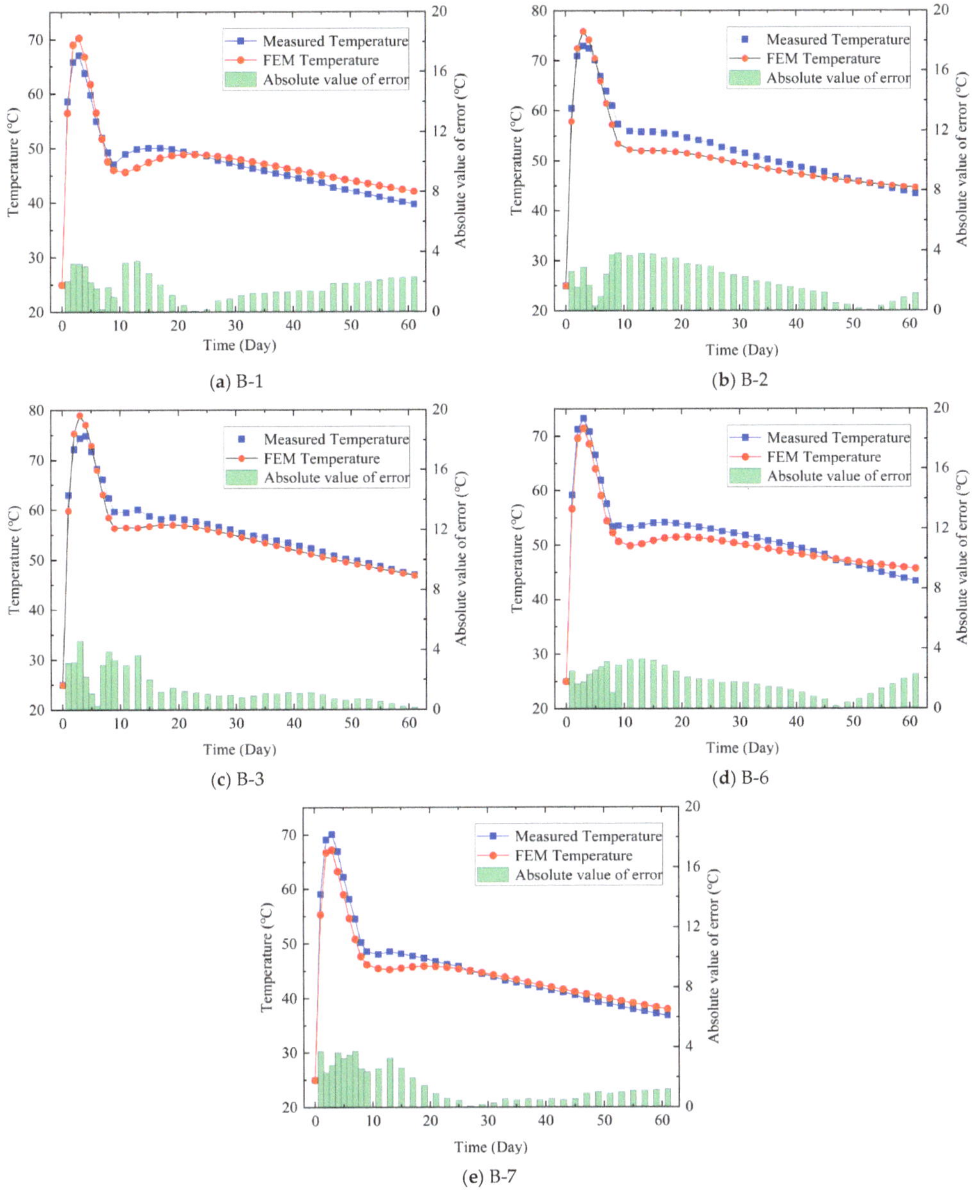

Figure 9. Comparison of the measured and finite element temperatures at different measurement points.

5. Conclusions

(1) To speed up the convergence of the SSA algorithm, reduce the possibility of falling into a local optimum as well and improve the computational accuracy, three mixed improvement strategies, Logistic chaos mapping initialization of the population, the

adaptive weighting factor, and Cauchy mutation, were used to improve the SSA, which was called;
(2) The performance test was carried out to compare the performance of the algorithm with three different intelligent algorithms (PSO, SA, and GWO) and reflected the superiority of the SSA improved by mixed strategies (SSAIMS). The results show that the SSAIMS was better than the other three algorithms in general. Therefore, it can be used to invert the thermal parameters of mass concrete;
(3) The SSAIMS was used to invert the thermal parameters of an octagonal mass concrete pile cap. The error between the results of the SSAIMS and the field-measured values was no more than 6%, which was better than the other three algorithms and verified the accuracy of the inversion of thermal parameters inversion of the SSAIMS;
(4) The inverse thermal parameters were brought into the finite element software for analysis. The points of the B layer were selected to compare with the measured temperature on site. The results indicated that the maximum error was 4.6 °C at the initial stage of using cooling water, which may be due to the unstable water throughput and the large temperature difference between day and night at the site. Therefore, we should combine finite element software to increase the water supply at noon and take insulation measures at night. Meanwhile, the error was smaller at the later stage without cooling water, and the overall trend was close. Therefore, the SSAIMSs can be used for practical engineering applications in the field.

Author Contributions: Methodology, Y.W. (Yang Wang) and M.-L.Z.; software, Y.G., M.-L.Z. and Y.W. (Youzhi Wang); formal analysis, Y.W. (Yang Wang), K.Z., M.-L.Z., R.X. and X.Y.; investigation, Y.W. (Yang Wang) and Y.G.; data curation, Y.G. and X.Y.; writing—original draft preparation, Y.W. (Yang Wang) and Y.G.; writing—review and editing, Y.G., M.-L.Z. and Y.W. (Youzhi Wang); visualization, M.-L.Z. and Y.W. (Youzhi Wang); supervision, Y.G. and Y.W. (Youzhi Wang); project administration, Y.G., K.Z., R.X. and Y.W. (Youzhi Wang); funding acquisition, Y.W. (Youzhi Wang). All authors have read and agreed to the published version of the manuscript.

Funding: The research is supported by the National Natural Science Foundation of China (51578325) and Shandong Provincial Natural Science Foundation (ZR2024QE147).

Data Availability Statement: The data presented in this study are available on request from the corresponding author due to privacy.

Conflicts of Interest: Author Yang Wang, Kaixing Zhang and Runze Xu were employed by the company Shandong Expressway Jinan Round City West Highway Co., Ltd. The remaining authors declare that the research was conducted in the absence of any commercial or financial relationships that could be construed as a potential conflict of interest.

List of Notations

SSA	sparrow search algorithm
SSAIMS	SSA improved by mixed strategies
$X_{i,j}^{k+1}$	The position information matrix of the j-th dimensional (k + 1)-th generation of the producer i
$X_{i,j}^{k}$	The position information matrix of the j-th dimensional k-th generation of the producer i
k	The current number of iterations
$k_{\max}$	The maximum number of iterations
q	A random number between (0, 1)
Q	A random number obeying a normal distribution
L	A $1 \times d$ vector with all elements being 1
R_2	The warning values between [0, 1]
st	The safety values between [0.5, 1]
X_p^{k+1}	The optimal position matrix searched by the producer in generation k + 1
X_{worst}	The current global worst position matrix
A^+	A 1 × d matrix with elements 1 or −1 and $A^+ = A^T(AA^T)^{-1}$

X_{best}^{k}	The matrix of the location of the individual with the best fitness in the current iteration number
η	A standard normally distributed random number controlling the step size
K	A random number between $[-1, 1]$
ε	A very small number such that the denominator is not zero
f_i	The individual fitness value of the current alerter
f_g	The current global best fitness value
f_w	The current global worst fitness value
μ	Logistic parameter and $\mu = 4$
X_k	A random number between [0, 1]
δ_1	One of the constant factors and $\delta_1 = 0.3$
δ_2	One of the constant factors and $\delta_2 = 2t_{\max}/3$
δ_3	One of the constant factors and $\delta_3 = 2$
ρ_1	One of the constant factors and $\rho_1 = 0.3$
ρ_2	One of the constant factors and $\rho_2 = 11/t_{\max}$
ρ_3	One of the constant factors and $\rho_3 = 2t_{\max}/3$
k_s	The specified iteration number
X_{best}	The optimal individual position matrix before the Cauchy mutation
X_{new}	The optimal individual position matrix after the Cauchy mutation
Cauchy (0, 1)	The standard Cauchy distribution
$f(x)$	The fitness value of the x position
T	The temperature, °C
c	The specific heat capacity of concrete, kJ/(kg·°C)
t	The time, h
ρ	The density of concrete, kg/m^3
λ	The thermal conductivity of concrete, KJ/(m·h·°C)
θ_0	The adiabatic temperature rise of concrete, °C
$Q(t)$	The temperature rises due to the heat of hydration, °C
r	The rate of reaction of the heat of hydration after the completion of pouring
t_0	The time of the start of the heat of hydration, h
β_s	The equivalent exothermic coefficient of the concrete structure covered with thermal insulation material, $W/(m^2 \cdot °C)$
R_s	The equivalent thermal resistance, $m^2 \cdot h \cdot °C/KJ$
β	The original surface exothermic coefficient of concrete, $KJ/(m^2 \cdot h \cdot °C)$
h_i	The thickness of the thermal insulation layer, m
λ_i	The thermal conductivity of the thermal insulation layer, KJ/(m·h·°C)
T_a	The value of ambient temperature, °C
T_c	The temperature at the concrete measurement point
m	The normal direction of the outer surface of the structure
$T(t)$	The average temperature of the concrete at time t
T_{ij}	The water temperature of the i-th round of cooling water
T_i	The temperature of the concrete at the beginning of the i-th round of cooling water
$\phi_i(t)$	The water cooling function of the i-th round of cooling water
$\psi_i(t)$	the water cooling temperature rise function of the i-th round of cooling water
p_i	the water cooling parameter
$\eta_i(t)$	the ambient temperature effect function of the i-th round of cooling water
T_{ia}	The ambient temperature at the i-th round of cooling water
t_k	The specified smaller time
Δerf	The error function
a	The concrete thermal conductivity coefficient
h	The equivalent distance from the cooling water pipe to the concrete surface
$F(X)$	The objective function that needs to be optimized to solve the problem
T'_{mn}	The calculated value of the temperature field calculation procedure for the concrete at location n at time m
T_{mn}	The actual measured value of the concrete in the field at location n at time m
M	The total number of temperature measurement points
N	The total time that the temperature measurement was carried out

References

1. Zhu, B.F. Current situation and prospect of temperature control and cracking prevention technology for concrete dam. *J. Hydraul. Eng.* **2006**, *37*, 1424–1432.
2. Liu, Y.; Xin, J.; Zhang, G.; Zhang, L.; Li, S.; Zhu, Z.; Zhai, H. Intelligent Monitoring Technology of Anti-thermal Cracking for Mass Concrete. *J. Chin. Ceram. Soc.* **2023**, *51*, 1228–1233.
3. Dissanayaka, M.; Yapa, H. Numerical prediction of early age concrete temperature via 3D finite difference simulation. *J. Natl. Sci. Found. Sri Lanka* **2021**, *49*, 539–550. [CrossRef]
4. El-Mir, A.; Assaad, J.J.; Nehme, S.G.; El-Hassan, H. Correlating strength and durability to time-temperature profiles of high-performance mass concrete. *Case Stud. Constr. Mater.* **2022**, *16*, e01055. [CrossRef]
5. Do, T.A.; Tia, M.; Nguyen, T.H.; Hoang, T.T.; Tran, T.D. Assessment of Temperature Evolution and Early-Age Thermal Cracking Risk in Segmental High-Strength Concrete Box Girder Diaphragms. *KSCE J. Civ. Eng.* **2022**, *26*, 166–182. [CrossRef]
6. Zhou, S.; Deng, F.; Wei, Y.; Yu, L.; Wu, X.; Fu, Z. Research on concrete temperature-monitoring technique based on passive RFID. *China Civ. Eng. J.* **2017**, *50*, 65–73, 82.
7. Farhan, K.Z.; Shihata, A.S.; Anwar, M.I.; Demirboğa, R. Temperature and humidity sensor technology for concrete health assessment: A review. *Innov. Infrastruct. Solut.* **2023**, *8*, 276. [CrossRef]
8. Kampli, G.; Chickerur, S.; Chitawadagi, M. Real-time in-situ strength monitoring of concrete using maturity method of strength prediction via IoT. *Mater. Today Proc.* **2023**, *88*, 110–118. [CrossRef]
9. Kim, S.; Jung, D.; Kim, J.Y.; Mun, J.H. Study on Early Age Concrete's Compressive Strengths in Unmanaged Curing Condition Using IoT-Based Maturity Monitoring. *Buildings* **2024**, *14*, 798. [CrossRef]
10. Wang, Z.; Zhu, Y.; Wu, Q.; Zhang, Y. Thermal parameters of concrete by test and back analysis. *Rock Soil Mech.* **2009**, *30*, 1821–1825, 1830.
11. Wang, F.; Zhao, C.; Zhou, Y.; Zhou, H.; Liang, Z.; Wang, F.; Seman, E.A.; Zheng, A. Multiple Thermal Parameter Inversion for Concrete Dams Using an Integrated Surrogate Model. *Appl. Sci.* **2023**, *13*, 5407. [CrossRef]
12. Zhao, C.; Wang, F.; Jiang, X.; Song, R.; Zhang, A.; Liu, X. Thermal parameter identification of concrete dams based on hybrid particle swarm optimization using distributed optical fiber monitoring data. *Swarm Evol. Comput.* **2024**, *87*, 101582. [CrossRef]
13. Huang, Y.; Ding, Q.; Wang, Y.; Liao, Z.; Yin, X. Evolution of thermal parameters of wet-screened dam concrete after different freeze–thaw deterioration. *Struct. Concr.* **2022**, *23*, 2185–2197. [CrossRef]
14. Zhao, H.; Fan, G.; Wei, Z.; Wang, J.; Huang, D. Investigation of thermal conductivity and related parameters of early-age cement paste. *Int. J. Heat Mass Transf.* **2020**, *155*, 119798. [CrossRef]
15. Zhang, Y.; Ma, C.; Li, C.; Gao, S. Inverse analysis on thermal parameters of mass concrete based on uniform design and BP neural network. *J. Civ. Environ. Eng.* **2021**, *43*, 148–157.
16. Mao, D.; Zhang, A.; Wang, F.; Zhou, Y.; Tan, T. Inverse Analysis of Arch Dam Thermal Parameters Based on Cross-Global Artificial Bee Colony Algorithm. *J. Yangtze River Sci. Res. Inst.* **2023**, *40*, 162–169.
17. Wang, F.; Zhou, Y.; Zhao, C.; Zhou, H.; Chen, W.; Tan, Y.; Liang, Z.; Pan, Z.; Wang, F. Thermal parameter inversion for various materials of super high arch dams based on the hybrid particle swarm optimization method. *J. Tsinghua Univ. Sci. Technol.* **2021**, *61*, 747–755.
18. Hu, Y.; Bao, T.; Ge, P.; Tang, F.; Zhu, Z.; Gong, J. Intelligent inversion analysis of thermal parameters for distributed monitoring data. *J. Build. Eng.* **2023**, *68*, 106200. [CrossRef]
19. Su, C.; Wang, L.; Dong, Y.; Dao, V. Back analysis on thermal parameters of ship lock head based on BP neural network. *South--North Water Transf. Water Sci. Technol.* **2015**, *13*, 354–357.
20. Sun, W.; He, Y. Inversion analysis of thermal parameters of cushion concrete and reasons of early stage cracks based on improved genetic algorithms. *Eng. J. Wuhan Univ.* **2014**, *47*, 193–200.
21. Xue, J.; Shen, B. A novel swarm intelligence optimization approach: Sparrow search algorithm. *Syst. Sci. Control Eng.* **2020**, *8*, 22–34. [CrossRef]
22. Zheng, Y.; Li, L.; Qian, L.; Cheng, B.; Hou, W.; Zhuang, Y. Sine-SSA-BP Ship Trajectory Prediction Based on Chaotic Mapping Improved Sparrow Search Algorithm. *Sensors* **2023**, *23*, 704. [CrossRef] [PubMed]
23. Zhu, Y.; Yousefi, N. Optimal parameter identification of PEMFC stacks using Adaptive Sparrow Search Algorithm. *Int. J. Hydrogen Energy* **2021**, *46*, 9541–9552. [CrossRef]
24. Yao, Z.; Wang, Z.; Wang, D.; Wu, J.; Chen, L. An ensemble CNN-LSTM and GRU adaptive weighting model based improved sparrow search algorithm for predicting runoff using historical meteorological and runoff data as input. *J. Hydrol.* **2023**, *625*, 129977. [CrossRef]
25. Li, X.; Ma, X.; Xiao, F.; Xiao, C.; Wang, F.; Zhang, S. Time-series production forecasting method based on the integration of Bidirectional Gated Recurrent Unit (Bi-GRU) network and Sparrow Search Algorithm (SSA). *J. Pet. Sci. Eng.* **2022**, *208*, 109309. [CrossRef]
26. Behera, M.K.; Saikia, L.C. A Novel Resilient Control of Grid-Integrated Solar PV-Hybrid Energy Storage Microgrid for Power Smoothing and Pulse Power Load Accommodation. *IEEE Trans. Power Electron.* **2023**, *38*, 3965–3980. [CrossRef]
27. Khedr, A.M.; Al Aghbari, Z.; Raj, P.P.V. MSSPP: Modified sparrow search algorithm based mobile sink path planning for WSNs. *Neural Comput. Appl.* **2023**, *35*, 1363–1378. [CrossRef]

28. Ma, F.; Sun, W.; Jiang, Z.; Suo, S.; Wang, X.; Liu, Y. Industrial Robot Trajectory Optimization Based on Improved Sparrow Search Algorithm. *Machines* **2024**, *12*, 490. [CrossRef]
29. Liu, C.; He, Q. Golden Sine Chimp Optimization Algorithm Integrating Multiple Strategies. *Acta Autom. Sin.* **2023**, *49*, 2360–2373.
30. Wang, M.; Tang, M. Hybrid GreyWolf Optimization Algorithm with Opposition-Based Learning. *J. Front. Comput. Sci. Technol.* **2017**, *11*, 673–680.
31. Gad, A.G. Particle Swarm Optimization Algorithm and Its Applications: A Systematic Review. *Arch. Comput. Methods Eng.* **2022**, *29*, 2531–2561. [CrossRef]
32. Jain, M.; Saihjpal, V.; Singh, N.; Singh, S.B. An Overview of Variants and Advancements of PSO Algorithm. *Appl. Sci.* **2022**, *12*, 8392. [CrossRef]
33. Ghannadi, P.; Kourehli, S.S.; Mirjalili, S. A review of the application of the simulated annealing algorithm in structural health monitoring (1995–2021). *Frat. Ed Integrita Strutt. Struct. Integr.* **2023**, *17*, 51–76. [CrossRef]
34. Amine, K. Multiobjective Simulated Annealing: Principles and Algorithm Variants. *Adv. Oper. Res.* **2019**, *2019*, 8134674. [CrossRef]
35. Hatta, N.M.; Zain, A.M.; Sallehuddin, R.; Shayfull, Z.; Yusoff, Y. Recent studies on optimisation method of Grey Wolf Optimiser (GWO): A review (2014–2017). *Artif. Intell. Rev.* **2019**, *52*, 2651–2683. [CrossRef]
36. Makhadmeh, S.N.; Alomari, O.A.; Mirjalili, S.; Al-Betar, M.A.; Elnagar, A. Recent advances in multi-objective grey wolf optimizer, its versions and applications. *Neural Comput. Appl.* **2022**, *34*, 19723–19749. [CrossRef]
37. Zhu, B. The equivalent heat conduction Eq. of pipe cooling in mass concrete considering influence of external temperature. *J. Hydraul. Eng.* **2003**, *34*, 49–54.
38. Zhou, J.; Huang, Y.; He, X.; Tian, K. Dynamic Prediction of Concrete Temperature for the Plugging of Bottom Diversion Outlet of Xiangjiaba Hydropower Project. *J. Yangtze River Sci. Res. Inst.* **2015**, *32*, 119–122.

buildings

MDPI

Article

Experimental and Numerical Simulation Investigations on the Bearing Capacity of Stepped Variable-Section DX Piles under Vertical Loading

Jinsheng Cheng [1], Lei Tong [1], Chuanzhi Sun [2,*], Hanbo Zhu [2] and Jibing Deng [3]

[1] Suqian City Urban Construction Investment (Group) Co., Ltd., Suqian 223800, China; cheng_js@126.com (J.C.); tonglei8899@126.com (L.T.)
[2] School of Civil Engineering and Architecture, Suqian College, Suqian 223800, China; 19156@squ.edu.cn
[3] China Construction Fifth Engineering Bureau Co., Ltd., Changsha 410004, China; cscec5bhdgs@cscec.com
* Correspondence: schzh_xzh@163.com

Abstract: As a new type of pile, the bearing characteristics of stepped variable-section DX piles (multi-joint extruded and expanded piles) are quite complicated; thus, their design concepts and pile-forming processes are still in the exploration stage, and their application in actual engineering is not particularly mature. The settlement law and load transfer law of the variable section DX pile have not been studied deeply, and the values of the parameters of engineering design are not clear, which are the problems to be solved for the variable section DX pile. To solve the above problems, the present study on the bearing characteristics of stepped variable-section DX piles under vertical loading is of great scientific significance and engineering practical value. In this study, the bearing capacity of a DX pile with two variable steps was first analyzed experimentally. Then, the bearing capacity of variable cross-section DX piles and equal cross-section piles were simulated under the same soil conditions. Later, the numerical simulation results were compared with the experimental results to verify the validity and accuracy of the numerical models established in ABAQUS software. Finally, the bearing capacity of stepped variable-section DX piles in different soil layers was analyzed numerically to compare the effect of different soils on the compressive bearing capacity of piles. The results indicated that the load-bearing plates had a greater influence on the bearing capacity of the stepped variable-section DX piles. At the optimum variable section ratio, which was close to 0.9, DX piles had a good bearing capacity. The relative errors of the numerical simulation ultimate loads were below 10%, which verified the accuracy of the developed numerical model. The simulated ultimate load of the equal-section pile was the smallest. The vertical compressive bearing capacity and the effect of controlling settlement under the same level of load of the variable section DX pile in sandy soil were both better than those in silt soil. There was little difference between the bearing capacities of the piles with a load-bearing plate. The bearing capacity of the pile with two load-bearing plates was the best, which can be used in practical engineering.

Keywords: DX piles; stepped variable-section; load–settlement; soils; bearing capacity

1. Introduction

In structural engineering, pile foundation as a traditional form of foundation has been used until now, and plays an increasingly important role. It is mainly used in the case of poor geological conditions or high building requirements, with small settlement and high bearing capacity [1–5]. Among them, grouted piles are widely used because of their advantages, such as the high bearing capacity of a single pile, easy control of construction depth, low cost, and low noise [6–10]. The reinforced concrete equal cross-section grouted piles, which are more frequently used in China [11–15], mainly rely on the end bearing force at the bottom of the pile and the friction force at the side of the pile to carry the load. The way to improve the bearing capacity of piles is to increase the pile end area and increase the pile side friction resistance. The usual practice is to increase the pile diameter and increase the pile side area, but this practice is usually more difficult to construct and increases the project cost.

Figure 1a shows a stepped variable-section pile. Variable cross-section piles have an obvious squeezing effect on the soil layer during settlement, increasing the pile's lateral friction resistance [16,17]. When the pile is under pressure, the step form can play a certain bearing role in each soil layer, and then, combined with the pile end grouting process, it can improve the bearing performance of the pile. Variable cross-section form fully applies the axial force of the pile body gradually decreases from top to bottom under vertical load, and the distribution law of bending moment shear force under transverse load is also big at the top and small at the bottom, so as to save the material used for the pile body and reduce the cost of the project. Figure 1b shows a DX pile. DX piles can greatly increase the single-pile bearing capacity because the load-bearing plate improves the contact area between the pile body and the soil, which enhances the bearing performance, and the rotary excavation and squeezing and expanding equipment has the effect of squeezing and compacting the soil around the pile.

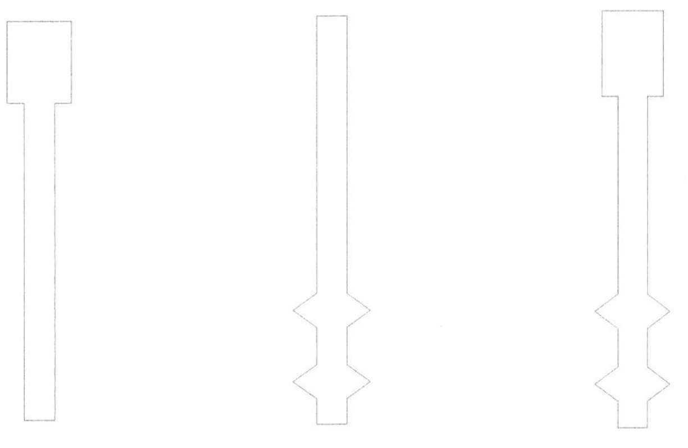

(**a**) A stepped variable-section pile (**b**) A DX pile (**c**) A stepped variable-section DX piles

Figure 1. Schematic diagram of piles.

On the basis of ensuring the safety and stability of pile foundations, it is a new challenge for the engineering community to meet the strict requirements of high quality, fast schedule and low cost on the project through theoretical and technological innovation. With this aim, many new forms of piling have been developed [18,19]. A stepped variable section DX pile [20] (named DX Pile after the inventor, Mr. Dexin He) is a new type of variable-section pile based on the combination of the advantages of both stepped variable section piles and DX piles, as shown in Figure 1c. It combines the advantages of both stepped variable-section piles and DX piles. It can fully mobilize the potential of pile–soil interaction and greatly improve the pile bearing capacity while reducing the cost. However, the design of stepped variable cross-section DX piles still adopts the recommended formula of the specification [21]. To meet the specification requirements, it is necessary to use the method of increasing the diameter of the pile body; this is not conducive to the construction progress and cost control. Therefore, how to ensure the stepped variable cross-section DX piles can ensure safety under vertical and horizontal loads as well as control the construction cost is a worthy research topic.

The stepped variable cross-section DX pile is subjected to pressure when it is affected by many factors, such as the pile type, material, and the influence of soil quality [22]. The bearing mechanism of the stepped variable cross-section DX pile is very complex under loading [23]. The study of pile–soil interaction on complex pile types is quite complicated. It is very difficult to obtain accurate calculation results by the theoretical method of derivation of equations. The finite element numerical simulation calculation method has been widely used in the study of pile bearing mechanism, group pile effect, and pile–soil action. Finite element simulation can consider the aspects of nonlinearity, non-uniformity, and pile–soil contact properties of the soil so it can better reflect the actual situation. Ogura H. [24–26] studied a pile type similar to bamboo piles and carried out experiments in sandy soil to analyze the relationship between load and settlement. The results indicated that its bearing capacity was much higher than that of a straight-bore pile of the same diameter. Zhang et al. [27] proposed a numerical simulation method to consider the damaged area and parameters surrounding the pile based on field tests, indoor simulations, and damage theory. Based on the field test, Dong et al. [28] used ABAQUS finite element software to numerically simulate the experimental process of vertical load capacity test of a single pile of equal-section pile and variable-section pile. The results indicated that the average lateral friction resistance of variable section piles was larger than that of equal section piles, and the effect of lateral friction resistance of variable section piles is more advantageous than that of equal section piles.

As verified by several engineering examples, the step-type variable cross-section DX pile technology has the features of unique construction technology, advanced machinery, and simple construction process; reliable quality of bearing plate cavity and high bearing performance of single pile; and reduction of pile body materials to reduce the project cost [29]. However, the design of the step-type variable cross-section DX pile still adopts the recommended formula of the specification. To meet the specification requirements, the method of increasing the diameter of the pile body is adopted, which not only increases the difficulty of construction but also improves the project cost. Therefore, how to ensure the safety of stepped variable section DX piles under vertical or horizontal loads while controlling the construction cost is a topic worth studying. Therefore, based on the above purpose, the present study on the bearing characteristics of stepped variable-section DX piles under vertical compression has important scientific significance and practical value in engineering.

As a new type of pile, the bearing characteristics of stepped variable-section DX piles are quite complicated; thus, the design concept and pile-forming process are still in the exploration stage, and their application in actual engineering is not particularly mature. The settlement law and load transfer law of the variable section DX pile have not been studied deeply, and the values of the parameters of engineering design are not clear, which are the problems to be solved for the variable section DX pile. In this study, the bearing capacity of a DX pile with two variable steps was first analyzed experimentally. Then, combined with the experimental results in reference [30], numerical simulation analysis were carried out to analyze the bearing process of variable cross-section DX piles and equal cross-section piles under the same soil conditions by using ABAQUS finite element software. Later, the numerical simulation results were compared with the experimental results to verify the validity and accuracy of the numerical models. Finally, the bearing capacity of stepped variable-section DX piles in different soils was analyzed numerically to compare the effect of different soils on the compressive bearing capacity of piles. This study can promote the application and dissemination of stepped variable-section DX piles in practical engineering.

2. Design of Specimens and Experimental Program

2.1. Physical and Mechanical Properties of the Soil Samples

The soil used for the test was powdered soil from a site in Suqian as the model material for the foundation soil. After each layer of soil was compacted, soil samples (see Figure 2) were taken out around the model piles using the ring knife method to conduct geotechnical tests. Then, the physical and mechanical property indicators of the soil samples were obtained in Table 1.

Figure 2. Geotechnical test.

Table 1. Physical and mechanical properties of the soil samples.

Density	Specific Gravity of Soil Particles	Moisture Content	Cohesive Force	Angle of Internal Friction	Modulus of Compression	Poisson's Ratio
1.59 g/cm^3	2.72 g/cm^3	16.23%	15.12 kPa	25.4°	10.06 MPa	0.4

2.2. Design of Test Piles

The model design first observed the condition of geometrical similarity according to the actual engineering needs and then integrated the difficulty and cost of the test production, the actual conditions of the production equipment, the accuracy requirements of the test measurements, and other aspects of the considerations, which can ensure the accuracy of the test piles. Piles S1–S5 were from reference [30], and one pile, S6, was the experimental model in this study, which was made in the same batch as piles S1–S5. The diameter of the piles body was 50 mm, and the length of the piles was 1100 mm. The diameter of the load-bearing plate was 100 mm. The variable ratio b is the ratio of the lower pile diameter to the upper pile diameter. The location of the variable section was 500 mm from the top of the pile. The variable section pile adopted a stepped pile type. The number of load-bearing plates was 2 for pile S6, 1 for piles S2–S5, and 0 for pile S1. The position of the load-bearing plate of the pile S6 was 180 mm and 360 mm from the bottom of the pile. The parameters of the test piles are shown in Table 2. The material parameters of each pile body are shown in Table 3. Figure 3 shows the design drawing of piles S1–S6.

Table 2. Design parameters of test piles (Unit: mm).

No.	L	D_1	D_2	b	l_1	l_2	D_3
S1	1100	50	50	1	—	—	—
S2	1100	50	50	1	—	180	100
S3	1100	50	45	0.9	500	180	100
S4	1100	50	40	0.8	500	180	100
S5	1100	50	35	0.7	500	180	100
S6	1100	50	40	0.8	500	360/180	100

Note: L was the length of the pile; l_1 was the distance from the variable section to the pile top; l_2 was the distance from the load-bearing plate to the pile bottom; D_1 was the diameter of the pile top; D_2 was the diameter of the pile bottom; D_3 was the diameter of the load-bearing plate.

Table 3. Material parameters of pile body.

Pile Material	Density	Elastic Modulus	Poisson's Ratio
Pine wood	0.58 g/cm^3	5.4 GPa	0.3

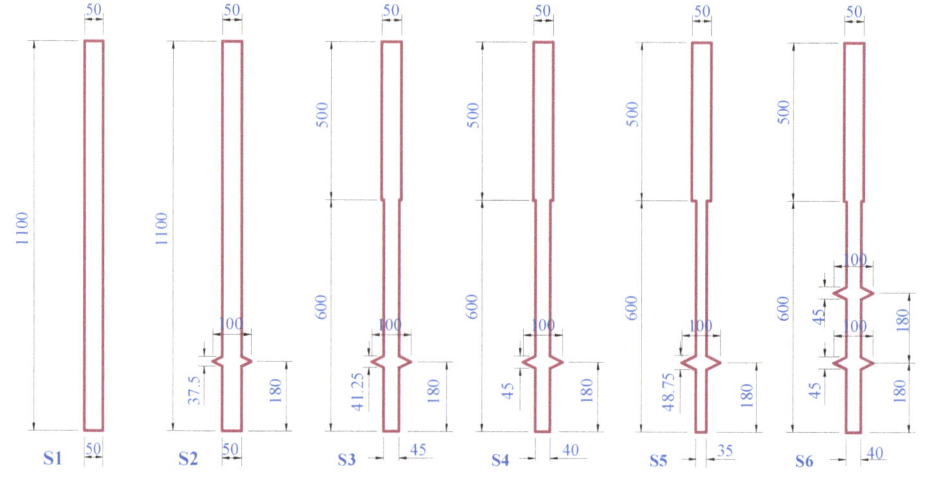

Figure 3. Design drawing of piles S1–S6.

2.3. Test Setup and Loading

The soil was first filled to a predetermined pile end height location. Then, the model pile was placed and secured at the designated location, and then the soil fill was continued to a predetermined pile top height. The model pile and soil pressure box were buried, as shown in Figure 4.

(**a**) A model pile

(**b**) A soil pressure box

Figure 4. Buried photos.

To better simulate the actual environment for the test, the model box and the loading device were specially designed for this model test. The bottom edge of the model box measured 1.8 m × 1.8 m, and the height was 2 m. The bottom plate of the model box was cross-welded with channels, and the four directions of the middle plate of the model box each extended out to connect the channels of the loading device. The peripheral skeleton of the model box was made of equilateral angle steel welded together, and the wall of the box was made of steel plate with a thickness of 5 mm, which was designed as a flat type for filling and unloading the soil. The loading device was designed and produced by the lever principle. The column was connected to the chassis, and then the lever was connected to the column with an articulated connection. The other end of the lever hung an iron frame for placing the loading weights. To prevent the lever from tilting sideways during the loading process so that the centralized force was eccentric, a limit on both sides of the lever was designed. The centralized force was applied by means of a removable screw mounted underneath with a spherical head. Under the condition of meeting the distance requirement, up to five model piles can be placed in two directions in the plane. When the test of one pile was finished, the mounting screw could be removed to carry out the loading test on another model pile. The model loading device diagram is shown in Figure 5. The vertical load model test loading mode adopted the slow sustaining load method in reference [5]. The vertical load was applied step by step incrementally during the test, and then the next level of load was added after each level of load reached relative stability until the model pile reached the ultimate load.

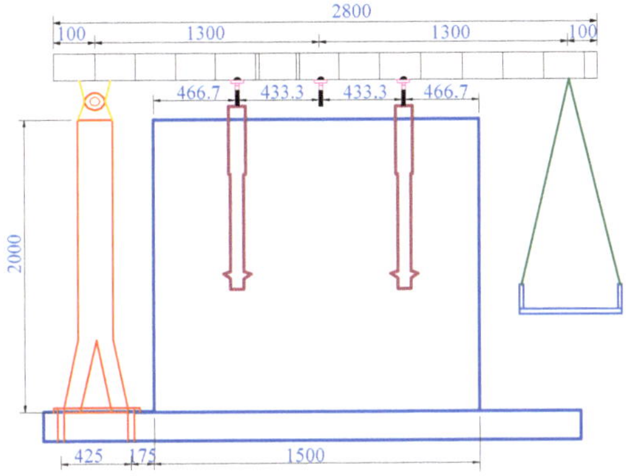

Figure 5. Loading device.

An electronic displacement meter with an accuracy of 0.01 mm was used to measure the settlement of the pile tops. Magnetic bearings held the displacement gauge to the channel steel, and the measurement point of the displacement gauge was set at the steel plate at the top of the pile. During the test loading, the corresponding displacement readings were recorded after each stage of load stabilization. Resistance strain gauges were pasted on each model pile to test the strain values. Ten pairs of strain gauges were arranged symmetrically on both sides of each pile body, totaling 20 gauges, and finally, the average value of the two sets of data was taken as the final value. The indicators of the strain gauges are shown in Table 4. The earth pressure measurement points were in the soil layer under the bearing plate, variable section, and pile end. When the soil was laid in layers, it was necessary to bury the earth pressure box at the measurement points, such as the bottom of the pile, the bottom of the bearing plate, and the variable section of each model pile, as shown in Figure 6. The main technical specifications of the earth pressure box used are shown in Table 5. A schematic diagram of the distribution of strain gauges in the pile, and the location of the earth pressure box is given in Figure 6.

Table 4. Indicators of strain gauges.

Model Number	Dimension	Resistance	Sensitivity Factor	Accuracy Class	Spring Tab
BK120-50AA	50 mm × 3 mm	119.5 ± 0.1 Ω	2.05 ± 1%	A	Half bridge

Table 5. Indicators of earth pressure boxes.

Scales	Bridge Pressure	Dimension	Sensitivity Factor
2 MPa	2 V	Φ20 mm × 10 mm	0.55

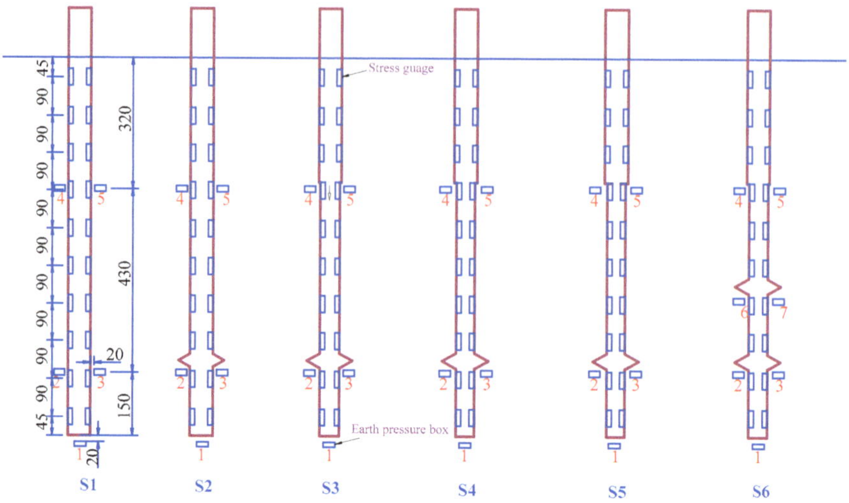

Figure 6. Distribution of strain gauges and locations of earth pressure boxes.

2.4. Experimental Results

2.4.1. Load–Settlement Curves

Load–settlement curves of piles S1–S6 are shown in Figure 7. Compared with the equal-section pile, the DX piles were obviously stronger in terms of settlement control. The presence of the load-bearing plate made the contact area between the pile and the soil body increase, and the lateral friction resistance of the pile body was greatly increased, thus increasing the bearing capacity of the pile. In the early stage of loading, when the load on the top of the pile was small, the settlement of stepped variable-section DX piles was similar to that of the equal section pile, and the advantages of stepped variable-section DX piles were gradually revealed with the increase of the load.

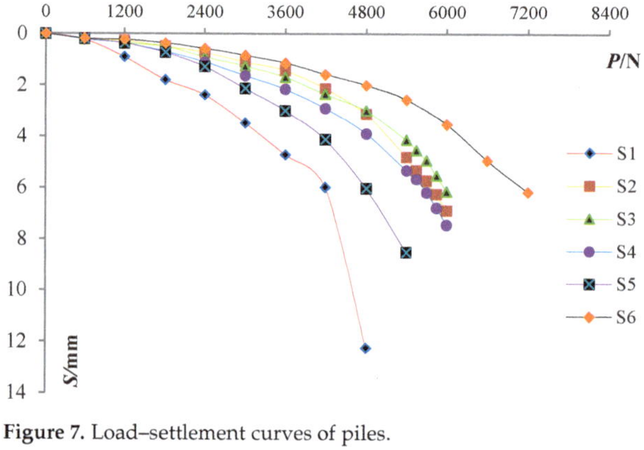

Figure 7. Load–settlement curves of piles.

2.4.2. Ultimate Load Capacity and Material Load Capacity per Unit Volume

To analyze and compare the efficiency of material use per unit volume of each pile, Q/V was introduced. In addition, the ratio of vertical bearing capacity Q_{si} of stepped variable-section DX piles to standard piles Q_{s1} and the ratio of volume V_{si} of stepped variable-section DX piles to standard piles V_{s1} (i is the number of each pile) were introduced to compare the changes in bearing capacity and material volume between DX piles, as shown in Table 6.

Table 6. Ultimate load capacity and material load capacity per unit volume.

Index	S1	S2	S3	S4	S5	S6
Q (N)	4200	5850	6000	5700	4800	7200
V (cm^3)	1766	1864	1655	1470	1307	1568
Q/V (N/cm^3)	2.4	3.1	3.6	3.9	3.7	4.5
Q_{si}/Q_{s1}	1.00	1.38	1.42	1.35	1.14	1.71
V_{si}/V_{s1}	1.00	1.06	0.94	0.83	0.74	0.88

From Table 6, the ultimate load of DX piles was higher than that of the equal section pile S1. The total volume of S2 piles with a DX load-bearing plate was increased by 6%, but its bearing capacity was increased by 38%. The bearing capacity of pile S3 was the most prominent among the DX piles with a DX load-bearing plate, and the ultimate load of the pile decreased gradually as the pile diameter ratio became smaller, among which the ultimate load of pile S5 decreased more obviously. It indicated that the factor of variable cross-section pile diameter ratio had a large influence on the bearing capacity of the DX pile. There was an optimal variable cross-section ratio, whose value was near 0.9, the DX piles exert a large bearing capacity. Pile S6 was a DX pile with two DX load-bearing plates. Its bearing capacity was higher than that of pile S3 with a DX load-bearing plate, indicating that the load-bearing plates played a greater role in improving the ultimate load of the pile.

The Q/V had an obvious correlation with the variable cross-section pile diameter ratio. As the variable cross-section ratio decreased, the ultimate capacity per unit volume first increased and then decreased. From the index of ultimate load per unit volume of material, the variable section pile diameter ratio cannot be too large to maximize the efficiency of material use. There was an optimal variable section ratio b, whose value was higher when b was near 0.8. The value of Q/V Pile S6 with two load-bearing plates was greater than the other DX piles with a load-bearing plate, and that of the equal-section straight pile S1 was the smallest, indicating that the load-bearing plates had a greater influence on the ultimate capacity per unit volume of the stepped variable-section DX piles.

2.4.3. Distribution Curves of Lateral Friction Resistance

Figure 8 shows the distribution of lateral friction resistance of pile S6. Compared with pile S4 in reference [30], variable section S6 pile had two load-bearing plates, and its lateral friction resistance was fully exerted. The sudden change of lateral friction resistance at the bearing discs can illustrate the great role of the bearing plate in the bearing of the pile body. With the gradual increase of loading, the relative displacement between the pile and the soil was also increased, the lateral friction resistance became larger, and the bearing plate was more fully utilized. When the pile settlement reached a certain level, there was a gap between the pile and the soil, and the soil was loosened, resulting in less lateral friction resistance at the top of the load-bearing plate.

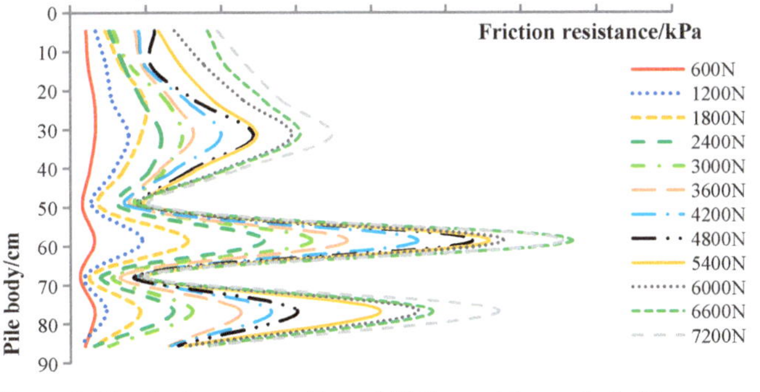

Figure 8. Distribution curves of lateral friction resistance of pile S6.

2.4.4. Earth Pressure

Figure 9 shows the variation curves of soil pressure at the pile end under vertical loading. Under the same vertical load, the soil pressure at the pile end of equal-section pile S1 was larger than that of DX piles S2–S6. The friction resistance of the pile body and load-bearing plate resistance of the DX pile bear a large portion of the pile body, which played a great role. The soil pressure at the pile end was very small, which can effectively control the settlement at the pile end. Among the various DX piles, the S6 pile had two load-bearing plates. Its soil pressure at the end of the pile was the smallest, and the advantage of settlement control was obvious. The soil pressures at the end of piles S2, S3, and S4 were small, indicating that a reasonable variable section was favorable to the bearing capacity of the piles. Pile S5, with a smaller *b* value, had larger soil pressure at the pile end and an obvious loss of bearing capacity.

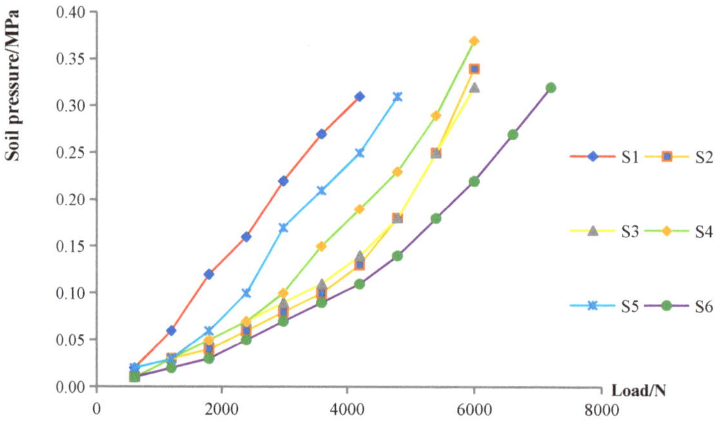

Figure 9. Soil pressure curves at the end of piles.

3. Finite Element Modeling of Pile Bodies

3.1. Basic Assumptions

(1) The pile body was rigid so that the pile body was not compressed by vertical pressure. The pile body was assumed to be modeled by linear elastic.
(2) The soil was assumed to be an isotropic elastic–plastic body, and the Mohr–Coulomb yield criterion was used.
(3) The friction coefficients between the units did not change during the modeling process.
(4) The effect of disturbance of the soil layer by factors such as the construction process was not considered.

3.2. The Constitutive Model of the Soil Body

The primary issue in the analysis of pile–soil problems was the selection of a constitutive model for the soil properties. The soil was assumed to be an isotropic elastic–plastic body, and the Mohr–Coulomb yield criterion was used.

The three widely used classical elastoplastic models are the Mohr–Coulomb model, the Druker–Prager model, and the Cambridge model. The Mohr–Coulomb model was used to simulate the pile–soil interaction. It can describe the material properties of geotechnical materials, such as the isotropic strain hardening and softening properties, and has been widely used in the simulation of geotechnical engineering. In the simple stress state, the Mohr–Coulomb yield criterion is

$$f(\sigma_1, \sigma_2, \sigma_3) = \frac{1}{2}(\sigma_1 - \sigma_3) + \frac{1}{2}(\sigma_1 + \sigma_3)\sin\psi - c\cos\psi = 0 \quad (1)$$

where σ_1, σ_2, σ_3 refer to the first principal stress, the second principal stress, and the third principal stress; c and Ψ must refer to the cohesion and the angle of internal friction, respectively.

As shown in Figure 10, the friction angle φ also indicates the shape of the yield surface of the material on the π-plane, and when $\varphi = 0°$, the Mohr–Coulomb model is transformed into the Tresca model, which is independent of the peripheral pressure, and at this time, the yield surface on the π-plane exhibits an ortho-hexagonal shape. At $\varphi = 90°$, the Mohr–Coulomb model is transformed into the Rankine model when the yield surface of the π-plane presents a positive triangle and $R_{mc} \to \infty$. If $0° \leq \varphi \leq 90°$, then the yield surface presents a hexagonal shape with equal sides but not equal angles, which is called the Mohr–Coulomb hexagon.

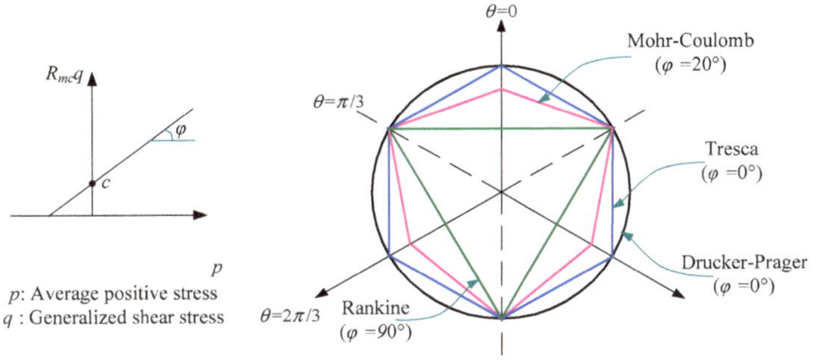

Figure 10. Meridian plane and deviatoric stress plane yield surface shape.

3.3. *The Constitutive Mode of the Pile Body*

The pile body was rigid so that the pile body was not compressed by vertical pressure. The pile body was assumed to be modeled by linear elastic. The constitutive relation is

$$\sigma_{ij} = \left[\frac{2G\mu}{1-2\mu}\delta_{ij}\delta_{kl} + G\left(\delta_{ik}\delta_{jl} + \delta_{il}\delta_{jk}\right) \right] \varepsilon_{kl} = D_{ijkl}\varepsilon_{kl} \tag{2}$$

where E is the elastic modulus of the material, μ is the Poisson's ratio of the material, $G = E/2(1+\mu)$, D_{ijkl} is the component of the elasticity tensor, and the matrix form of D_{ijkl} is

$$D = \frac{E}{(1+\mu)(1-2\mu)} \begin{bmatrix} 1-\mu & \mu & \mu & 0 & 0 & 0 \\ & 1-\mu & \mu & 0 & 0 & 0 \\ & & 1-\mu & 0 & 0 & 0 \\ & & & \frac{1}{2}-\mu & 0 & 0 \\ & & & & \frac{1}{2}-\mu & 0 \\ & & & & & 0 \end{bmatrix} \tag{3}$$

3.4. *Finite Element Modeling*

According to the actual dimensions of the pile and the soil body of the model test, the soil body around the pile was a rectangular body of 1.8 m × 1.8 m × 2 m, the outer diameter of the pile body was 0.05 m, the length of the pile was 1.1 m, and the depth of the pile below the soil layer was 0.9 m. The model with 1/4 symmetry was established. The pile body was modeled as shown in Figure 11a. The model space was selected as three-dimensional (3D), and its type was deformable. The shape in the basic features was selected as solid and its type was rotated. The model space of the soil body was also selected as 3D, and its type was deformable. The shape in the basic features was selected as solid and its type was stretched. The soil body of the rectangular body was modeled, and then the geometric elements were split to divide the soil part where the pile was located, and the soil body was established, as shown in Figure 11b. A linear elastic model was used for the pile, and an isotropic elastoplastic body was selected for the soil. The material modeling parameters of the pile and soil are shown in Tables 1 and 3.

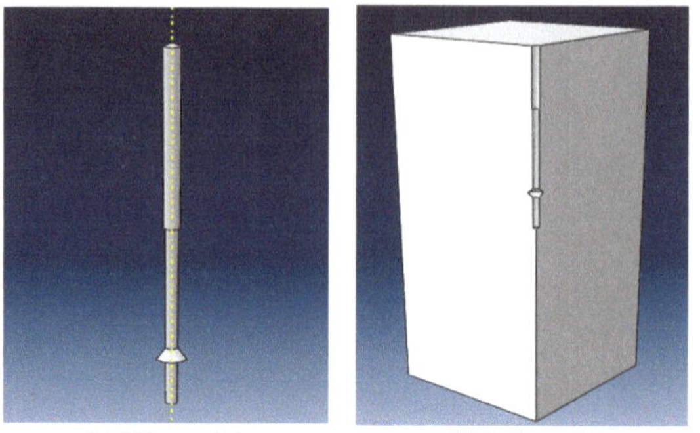

(**a**) Pile model (**b**) Soil model

Figure 11. Establishment of the pile–soil model.

After defining the material properties of the parts, they were assembled in the assembly module, where the parts were assembled into a whole by means of different positioning relationships. The contact module was utilized to define the contact units. Face-to-face contact was selected to create contact surfaces for each component, and then a master–slave algorithm was used to create contact pairs. A softer soil material was selected for the follower surface. The soil surface and the pile surface were set up as bound lifts that would not be displaced relative to each other. ABAQUS software has three main meshing techniques: free meshing, structured meshing, and swept meshing. Both the pile model and the soil model were meshed by the swept mesh technique using linear C3D8R units. The meshing was completed, as shown in Figure 12. Applying loads and boundary constraints was performed in the loading module. The boundary conditions for the numerical model were vertical displacement and bottom radial selection constraints for the soil model, and radial displacement selection constraints for the side parts of the soil model. To study the stress–strain relationship of the model, the magnitude curve should be added after selecting the load type. The type of the curve was a tabular magnitude curve, which determined the increase of the load at each time point, and the size and sequence of loading were based on the experimental pile.

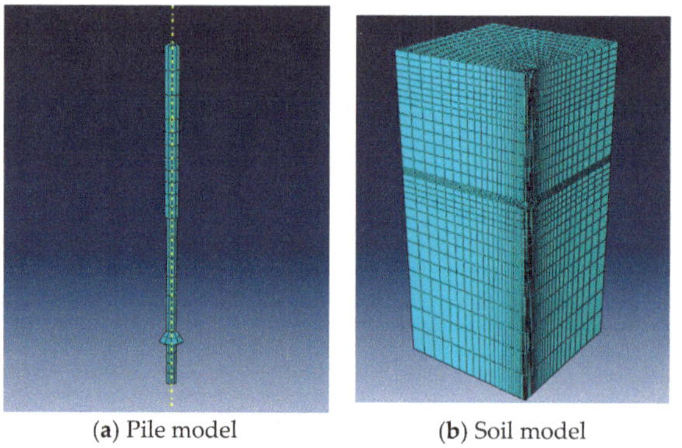

(**a**) Pile model (**b**) Soil model

Figure 12. Schematic diagram of meshing.

3.5. Validation of Numerical Results

Numerical simulation analysis was conducted on the equal-section pile S1, the DX pile S2, and stepped variable-section DX piles S4 and S6. Figure 13 shows the comparison of experimental and numerical simulation load–settlement curves. The experimental values and simulation values of the ultimate load existed within a certain allowable error; this was because the numerical simulation of the parameter value was generally idealized, while in the actual experiment, there were many uncontrollable factors, such as uneven soil layer densities, etc., which caused errors, but the experimental and numerical simulation load–settlement curves overlapped. The experimental and numerical simulation ultimate loads were compared in Table 7. From Table 7, it can be seen clearly that the relative errors of the numerical simulation results were below 10%, which validated the accuracy of the finite element model.

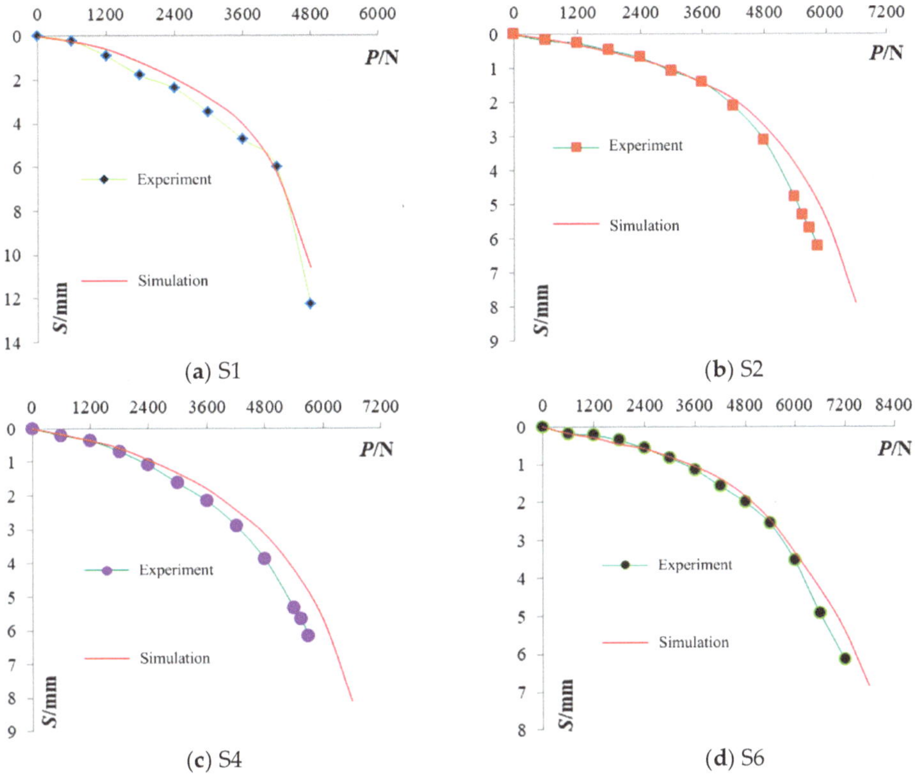

Figure 13. Comparison of experimental and numerical simulation load–settlement curves.

Table 7. Comparison of experimental and numerical simulation ultimate loads.

No.	Experiment E_e/N	Simulation S/N	Relative Error $E_r = (S - E_e)/E_e$
S1	4200	4100	−2.4%
S2	5850	6300	7.7%
S4	5700	6200	8.8%
S6	7200	7500	4.2%

3.6. *Finite Element Results*

3.6.1. Stress Analysis of the Pile Body

When the vertical load at the pile top was 4.2 kN, the stress cloud of each pile body was simulated, as shown in Figure 14. Under vertical pressure, the stress of the equal-section pile S1 gradually decreased from top to bottom, the stress of DX pile S2 also decreased from top to bottom, and the decrease at the bearing plate was obviously accelerated, indicating that a large portion of the load was transmitted to the soil around the pile through the bearing plate. Variable-section piles S4 and S6 had reduced cross-sectional area at the variable section and higher stresses, and the stress decreasing above and below their load-bearing plates was obvious. Compared with the equal-section pile S1, the axial force in the area below the bearing plate of the variable cross-section DX pile was extremely small, indicating that most of the load had been transferred to the soil around the piles, and the soil pressure at the bottom of the piles was small, which can effectively control the settlement of the piles, and reflected the advantages of variable cross-section DX piles in bearing.

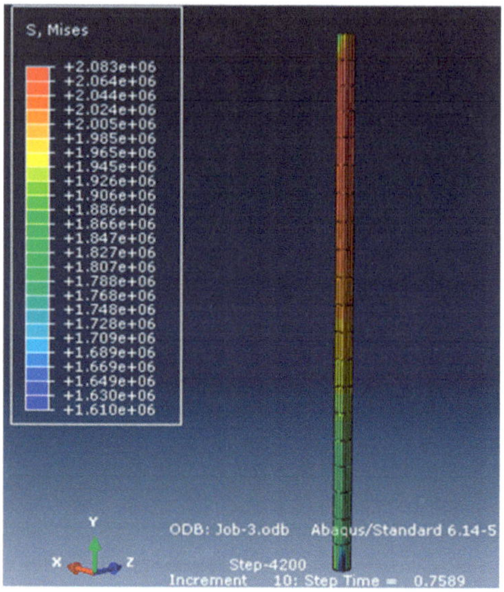

(**a**) S1

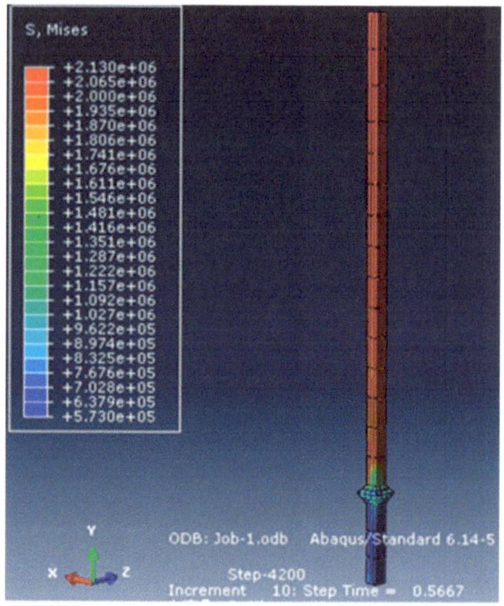

(**b**) S2

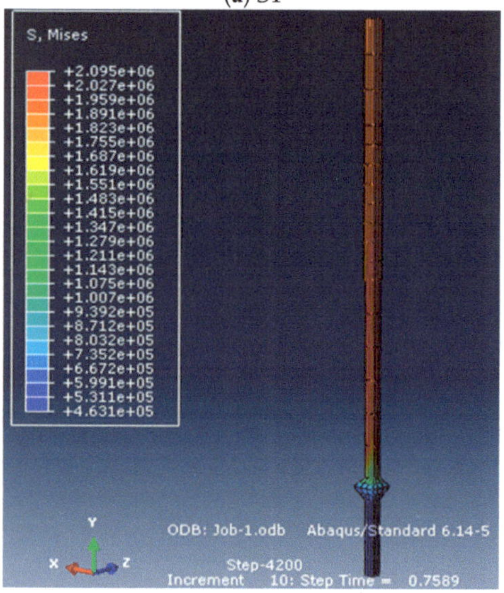

(**c**) S4

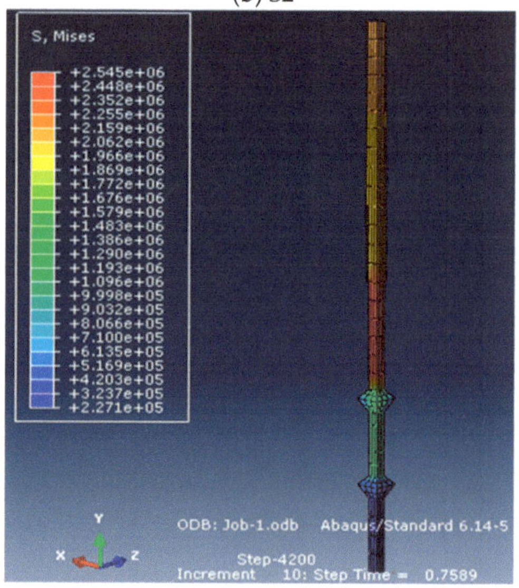

(**d**) S6

Figure 14. Stress cloud maps of piles.

3.6.2. Soil Pressure around Piles

When the vertical load at the pile top was 4.2 kN, the stress cloud of soil around piles was simulated, as shown in Figure 15. The soil around the pile took more load transfer and played a great role in bearing. The soil stress at the load-bearing plate and variable section of the variable section DX pile was larger, while the soil stress on the pile side of the equal section pile was very small, indicating that in the process of pile bearing, due to the existence of the variable section and the load-bearing plate, it made a large part of the load to be transferred to the soil around the pile through the end-bearing action, which reduced the axial force of the pile end, and thus played a great role in the pile bearing. The soil pressure at the pile end of the equal-section straight piles was the largest, indicating that the equal-section straight pile was easy to settle and subside, while the soil pressure at the pile end of the DX piles was significantly increased, indicating that the stepped variable-section DX piles were advantageous for the control of pile settlement, and with the increase in the number of load-bearing plates, the bearing capacity of the stepped variable-section DX pile became better.

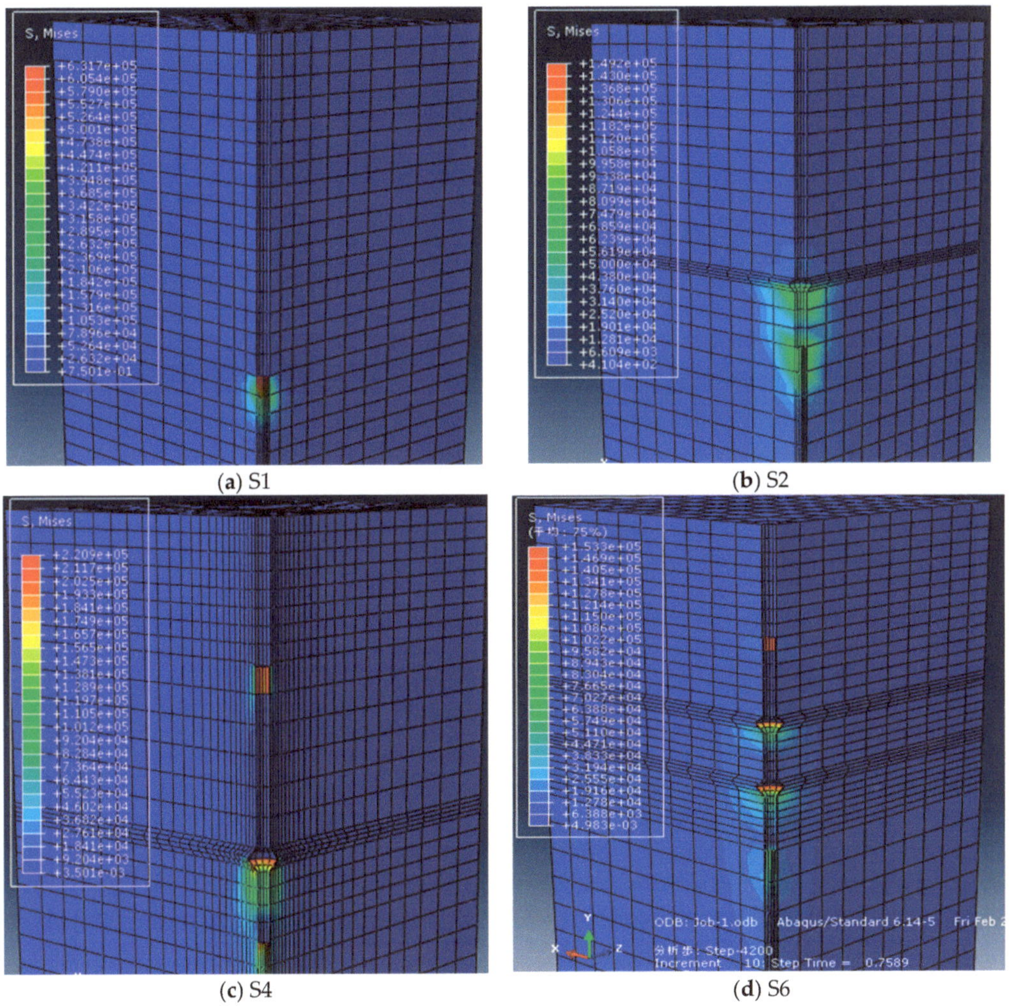

(**a**) S1 (**b**) S2 (**c**) S4 (**d**) S6

Figure 15. Stress cloud maps of soil around piles.

3.6.3. Compressive Bearing Capacity of a Pile in Different Soil Layers

Various parameters of the soil were varied to simulate and analyze the pile-bearing process in different soil layers. The parameters of silt soil and sandy soil are shown in Table 8. The load–settlement curves of piles in pulverized soil and sandy soil are shown in Figure 16. The settlement of pile S4 was the largest in the preloading period, and that of pile S2 was accelerated in the later period. There was not much difference between the two ultimate loads. The bearing capacity of stepped variable-section DX piles in sandy soil was stronger than that in silt soil. The settlements of the three-stepped variable-section DX piles in sandy soil were smaller than those in silt soil, indicating that the piles were more effective in resisting compressive bearing in sandy soil than in pulverized soil.

Table 8. Parametric soil test parameters for silt and sandy soils.

Name	Density /(g/cm^3)	Specific Gravity of Particles /(g/cm^3)	Water Content	c /kPa	Ψ /°	E /MPa	μ
Silt soil	1.59	2.72	16.23%	15.12	25.4	10.06	0.4
Sandy soil	1.79	2.65	1.65%	4.27	16.2°	34.3	0.4

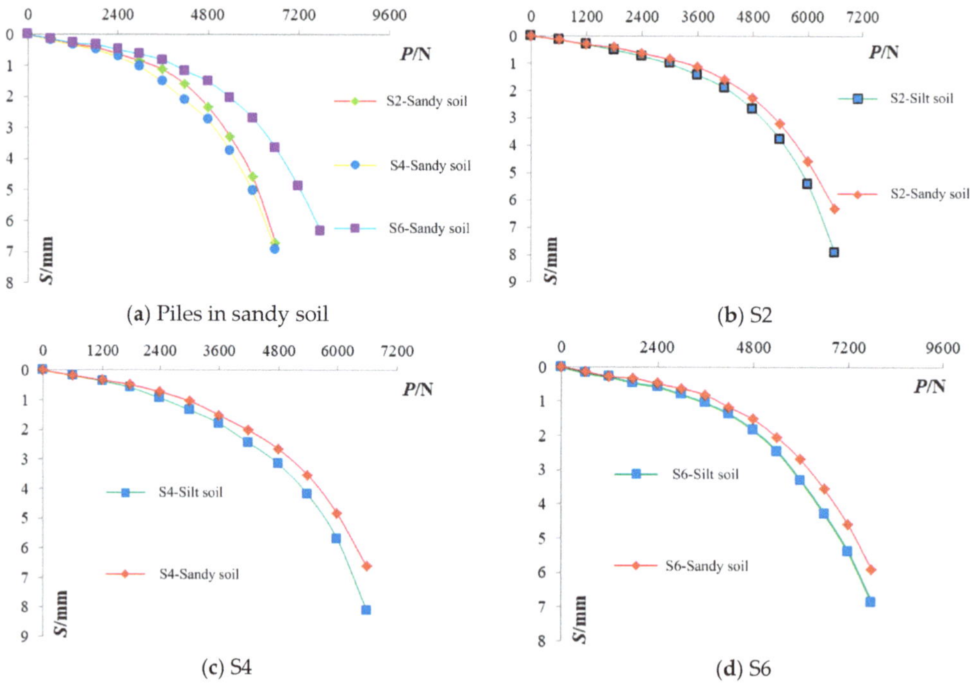

(a) Piles in sandy soil (b) S2 (c) S4 (d) S6

Figure 16. Load–settlement curves of the piles in different soil layers.

4. Conclusions

The bearing capacity of a DX pile with two variable steps was first analyzed experimentally. Then, the bearing capacity of variable cross-section DX piles and equal cross-section piles were simulated under the same soil conditions. Later, the numerical simulation results were compared with the experimental results to verify the validity and accuracy of the numerical models. Finally, the bearing capacity of stepped variable-section DX piles in different soils was analyzed numerically to compare the effect of different soils on the compressive bearing capacity of piles. This study can promote the application and dissemination of stepped variable-section DX piles in practical engineering.

The main conclusions were as follows.

1. Piles S1–S5 were from reference [30], and one pile, S6, was the experimental model in this study, which was made in the same batch with piles S1–S5. The diameter of the piles body was 50 mm, and the length of the piles was 1100 mm. The diameter of the load-bearing plate was 100 mm. The variable ratio b was the ratio of the lower pile diameter to the upper pile diameter, which took the values of 0.7, 0.8, 0.9, and 1. The location of the variable section was 500 mm from the top of the pile. The variable section pile adopted a stepped pile type. The number of load-bearing plates was 2 for pile S6, 1 for piles S2–S5, and 0 for pile S1;
2. The load-bearing plates had a greater influence on the bearing capacity of the stepped variable-section DX piles. At the optimum variable section ratio, which is close to 0.9, DX piles had a high bearing capacity. The bearing capacity of stepped variable-section DX piles with two variable steps was improved and its bearing capacity was best. The sudden change of lateral friction resistance at the bearing discs can illustrate the great role of the bearing plate in the bearing of the pile body. The soil pressure at the end of the pile with two variable steps was the smallest, and its settlement control was obvious;
3. The experimental values and simulation values of the ultimate load existed within a certain allowable error; this was because the numerical simulation of the parameter value was generally idealized, while in the actual experiment, there were many uncontrollable factors such as uneven soil layer densities, etc., which caused errors, but the experimental and numerical simulation load–settlement curves overlapped. The relative errors of the numerical simulation ultimate loads were below 10%, which verified the accuracy of the developed numerical model;
4. The simulated ultimate load of the equal-section pile was the smallest. The effect of settlement control of equal-section piles was poor. Due to the existence of the load-bearing plate, the ultimate load of the piles with a bearing plate was greatly improved. The piles with two load-bearing plates had the highest bearing capacity;
5. The vertical compressive bearing capacity and the effect of controlling settlement under the same level of load of the variable section DX pile in sandy soil were both better than those in silt soil. There was little difference between the bearing capacities of the piles with a load-bearing plate. The bearing capacity of the pile with two load-bearing plates was the best.

Author Contributions: Conceptualization, C.S.; methodology, J.D; software, J.C.; validation, J.D.; formal analysis, L.T.; investigation, L.T.; resources, J.C.; data curation, L.T.; writing—original draft preparation, J.C.; writing—review and editing, C.S.; visualization, H.Z.; supervision, C.S.; project administration, C.S.; funding acquisition, C.S. All authors have read and agreed to the published version of the manuscript.

Funding: This research has been supported by the Natural Science Research Project of Jiangsu Province Colleges and Universities (21KJD560002 and 23KJA560007), China; Research and Innovation Team Project of Suqian College (2021TD04), China; Suqian Sci & Tech Program (H202313), China; Jiangsu Civil Architecture Society project ((2023) No. 4 Item 9), China; the Youth Fund Project of Suqian College (2023XQNA03), China; and the Fifth Provincial Research Funding Project of "333 High-level Talent Training" in 2020 (BRA2020241), China.

Data Availability Statement: The data presented in this study are available in the article here.

Conflicts of Interest: Author Jinsheng Cheng and Lei Tong were employed by the company Suqian City Ur-ban Construction Investment (Group) Co. Author Jibing Deng was employed by the company China Construction Fifth Engineering Bureau Co. The remaining authors declare that the research was conducted in the absence of any commercial or financial relationships that could be construed as a potential conflict of interest.

References

1. Zhao, X.Q.; Gong, X.N.; Duan, Y.; Guo, P. Load-bearing performance of caisson-bored pile composite anchorage foundation for long-span suspension bridge: 1-g model tests. *Acta Geotech.* **2023**, *18*, 3743–3763. [CrossRef]
2. Hassan, I.; Mohamedelhassan, E. Enhancing the load-bearing capacity of H-pile foundation by using electrokinetic treatment. *Int. J. Civ. Eng.* **2023**, *21*, 991–1005. [CrossRef]
3. Kumar, M.; Kumar, D.R.; Khatti, J.; Samui, P.; Grover, K.S. Prediction of bearing capacity of pile foundation using deep learning approaches. *Front. Struct. Civ. Eng.* **2024**, *18*, 870–886. [CrossRef]
4. Gao, F.; Wang, W.; Lv, Q.; Cheng, X. Experimental study on the bearing characteristics of rigid-flexible long-short pile composite foundations in thick collapsible loess areas. *KSCE J. Civ. Eng.* **2024**, *28*, 1690–1701. [CrossRef]
5. Kumar, A.; Kumar, S. Settlement based load-bearing in a combined pile–raft foundation. *Geotech. Geol. Eng.* **2024**, *42*, 1405–1421. [CrossRef]
6. Igoe, D.; Spagnoli, G.; Doherty, P.; Weixler, L. Design of a novel drilled-and-grouted pile in sand for offshore oil & gas structures. *Mar. Struct.* **2014**, *39*, 39–49.
7. Zhou, J.; Gong, X.; Naggar MH, E.; Zhang, R. Field study on the behavior of pre-bored grouted planted pile with enlarged grout base. *Acta Geotech.* **2021**, *16*, 3327–3338. [CrossRef]
8. Wan, Z.; Dai, G.; Gong, W. Study on the response of postside-grouted piles subjected to lateral loading in calcareous sand. *Acta Geotech.* **2022**, *17*, 3099–3115. [CrossRef]
9. Cao, S.; Wang, Q.; Ma, J.; Xiao, Z.; Li, C.; Yang, Y.; Wang, J. Experimental Investigation of Vertical Bearing Characteristics of Composite Post-grouting Piles in Sandy Soil. *Int. J. Civ. Eng.* **2024**, *22*, 303–315. [CrossRef]
10. Salem, T.N.; El-Basset, O.H.A.; Hassan, R. 3D Numerical Analysis of Post-Grouted Piles. *Indian Geotech. J.* **2024**, *54*, 1562–1583. [CrossRef]
11. Wan, Z.; Dai, G.; Gong, W. Field study on post-grouting effects of cast-in-place bored piles in extra-thick fine sand layers. *Acta Geotech* **2019**, *14*, 1357–1377. [CrossRef]
12. Wen, L.; Kong, G.; Li, Q.; Zhang, Z. Field tests on axial behavior of grouted steel pipe micropiles in marine soft clay. *Int. J. Geomech.* **2020**, *20*, 06020006. [CrossRef]
13. Huang, Y.; Zhuang, X.; Wang, P.; Zong, Z.L. Axial behavior of pressure grouted helical piles installed in marine soft clay based on full-scale field tests. *Geotech. Geol. Eng.* **2022**, *40*, 5799–5812. [CrossRef]
14. Zhou, Z.; Xu, F.; Lei, J.; Bai, Y.; Chen, C.; Xu, T.; Zhang, Z.; Zhu, L.; Liu, T. Experimental study of the influence of different hole-forming methods on the bearing characteristics of post-grouting pile in Loess Areas. *Transp. Geotech.* **2021**, *27*, 100423. [CrossRef]
15. Zhang, Y.; Bi, J.; Bai, X.; Yan, N.; Sang, S.; Lin, Z.; Wang, B.; Chen, J.; He, L. Test on bearing capacity of mudstone foundation and rock-socketed pile. *Sci. Technol. Eng.* **2023**, *23*, 722–730.
16. Zhuang, Y.; Song, K.; Easa, S.; Song, Y. Soil Interaction of H-shaped steel-RC stepped pile of integral abutment bridge: Experimental evaluation. *KSCE J. Civ. Eng.* **2023**, *27*, 1174–1190. [CrossRef]
17. Chen, L.; Zhuang, Y.; Song, K.; Easa, S.M.; Zhu, H. Novel HS-RC stepped pile foundation: Experimental and numerical evaluation. *Ocean Eng.* **2024**, *299*, 117196. [CrossRef]
18. Zyka, K.; Mohajerani, A. Composite piles: A review. *Constr. Build. Mater.* **2016**, *107*, 394–410. [CrossRef]
19. Wu, J.; Sun, F.; El, N.M.H.; Zhao, S.; Wang, K. Analytical solution for consolidation of sand-filled nodular pile (SFNP) foundation and its engineering application. *Int. J. Numer. Anal. Methods Geomech.* **2022**, *46*, 3236–3255. [CrossRef]
20. Wang, M.; He, D.; Tang, S. New pile foundation technology of 21 century: DX pile. *Eng. Sci.* **2021**, *14*, 4–12.
21. *JGJ 94-2008*; Industry Standard of the People's Republic of China. Technical Code for Building Pile Foundation. China Construction Industry Press: Beijing, China, 2008.
22. Qian, C.; Chen, L.; Tans, S.; Chai, X. Model test method of DX pile. *J. Beijing Jiaotong Univ.* **2012**, *36*, 85–89.
23. Wu, P.; Zhang, D.; Liu, Y.; Li, H.; Wang, Y. Model test on bearing characteristics of DX rock-socketed pile with bearing plate in strata. *J. Beijing Jiaotong Univ.* **2012**, *47*, 26–34.
24. Ogura, H. Study on bearing capacity of nodular cylinder pile by scaled model test. *J. Struct. Eng.* **1987**, *374*, 87–97.
25. Ogura, H. Study on bearing capacity of nodular cylinder pile by full-scale test of jacked piles. *J. Struct. Eng.* **1988**, *386*, 66–77.
26. Ogura, H. A theoretical analysis on load-settlement behavior of nodular piles. *J. Struct. Eng.* **1988**, *303*, 152–164.
27. Zhang, Y.; Wang, F.; Bai, X.; Yan, N.; Sang, S.; Kong, L.; Zhang, M.; Wei, Y. Numerical Simulation of Bearing capacity of Bored Piles in Mudstone Based on Zoning Assignment of Soil around Piles. *Buildings* **2022**, *12*, 1877. [CrossRef]
28. Dong, Q.; Xiao, Z.; Tang, D. Experimental Analysis of Vertical Bearing capacity of Piles with Different Variable Sections. *J. Wuhan Inst. Technol.* **2021**, *43*, 436–441.

29. Guo, P. Experimental Study on Bearing Characteristic and Failure Mechanism of Rock-Socketed DX Pile. Master's Thesis, Beijing Jiaotong University, Beijing, China, 2017.
30. Wang, Z.; Liu, J.C.; Gu, C.; Sun, C.Z. Model Test Study on vertical static bearing behaviors of DX pile with stepped variable cross-section. *Sci. Technol. Eng.* **2018**, *18*, 67–72.

buildings

MDPI

Article

Bridge Surface Defect Localization Based on Panoramic Image Generation and Deep Learning-Assisted Detection Method

Tao Yin [1,2], Guodong Shen [2], Liang Yin [2] and Guigang Shi [1,*]

[1] School of Civil Engineering, Anhui Jianzhu University, Hefei 230009, China; flitetest@163.com
[2] Anhui Transport Consulting and Design Institute Co., Ltd., Hefei 231299, China; shenguodong_jgy@163.com (G.S.); yliang_jgy@163.com (L.Y.)
* Correspondence: shiguigang@ahjzu.edu.cn

Abstract: Applying unmanned aerial vehicles (UAVs) and vision-based analysis methods to detect bridge surface damage significantly improves inspection efficiency, but the existing techniques have difficulty in accurately locating damage, making it difficult to use the results to assess a bridge's degree of deterioration. Therefore, this study proposes a method to generate panoramic bridge surface images using multi-view images captured by UAVs, in order to automatically identify and locate damage. The main contributions are as follows: (1) We propose a UAV-based image-capturing method for various bridge sections to collect close-range, multi-angle, and overlapping images of the surface; (2) we propose a 3D reconstruction method based on multi-view images to reconstruct a textured bridge model, through which an ultra-high resolution panoramic unfolded image of the bridge surface can be obtained by projecting from multiple angles; (3) we applied the Swin Transformer to optimize the YOLOv8 network and improve the detection accuracy of small-scale damages based on the established bridge damage dataset and employed sliding window segmentation to detect damage in the ultra-high resolution panoramic image. The proposed method was applied to detect surface damage on a three-span concrete bridge. The results indicate that this method automatically generates panoramic images of the bridge bottom, deck, and sides with hundreds of millions of pixels and recognizes damage in the panoramas. In addition, the damage detection accuracy reached 98.7%, which is improved by 13.6% when compared with the original network.

Keywords: bridge inspection; UAV; damage detection; 3D reconstruction; deep learning

1. Introduction

Bridges are inevitably prone to defects and cracking under long-term loads, long-age phenomena of concrete, and prestressing losses with time [1,2]; therefore, damage detection and assessment for bridges constitute essential tasks in bridge safety maintenance, the routine activities to which bridge management authorities must allocate significant human and financial resources. Although conventional manual inspection methods play a crucial role in bridge inspection, their drawbacks, including low efficiency, high costs, and high risk, remain inherent limitations. In recent years, rapid advancements and the widespread application of various UAVs and robots as well as the proliferation and cost reduction of technologies such as light detection and ranging (LiDAR), stereo cameras, and inertial navigation systems have facilitated the widespread adoption of UAVs and robotic technologies across various industries. In the field of bridge defect detection, the utilization of UAVs to access complex, narrow, high-altitude, and hazardous environments and the automated acquisition of images have enabled the inspection of locations such as high piers, tall towers, cables, and bridge undersides, where traditional manual inspections are difficult. All of these methods are regarded as having promising application prospects. Concurrently, rapid advancements and applications of visual analysis methods such as image processing and deep learning have been witnessed in recent years. In the realm of structural detection, research on employing these methods for automated defect identification and analysis has garnered considerable attention. Consequently, this study proposes a method utilizing UAVs to capture images of bridge surfaces and apply deep learning-based visual recognition techniques to identify typical bridge defects, thereby achieving rapid defect detection as an alternative to manual inspection methods.

The rapid acquisition of complete and high-precision images of bridge surfaces for the detection of damage is key in fieldwork, serving as the basis for deciding whether the damage can be accurately recognized. However, because bridges generally cross mountains and rivers and are large and locally complex, it is difficult to collect complete data. Research has applied robots for photography to collect images depicting bridge damage. For instance, Hou et al. employed a climbing robot to inspect the cables of a large-span cable-stayed bridge, using four cameras mounted on the robot to detect surface damage on the cables [3]. Wang et al. developed a magnetic adhesion-based tracked inspection robot equipped with magnetic adhesion and crawling traction modules, enabling the robot to crawl adaptively on the surfaces of steel structure bridges with varying curvatures [4]. Qin et al. designed a novel hexapod climbing robot and established a kinematic model for the exterior inspection of large bridges [5]. While these climbing robots can be used for relatively simple surfaces or steel structures, challenges persist in crawling on complex and non-magnetic surfaces such as concrete. The use of UAVs for damage detection has also gained significant attention in recent years. Compared to various types of robots, UAVs offer lower costs and greater flexibility, with almost no limitations on the geometry of the detection area [6,7]. For example, Jang et al. proposed a hybrid image-scanning method based on UAVs to capture and detect cracks in concrete structures [6]; the method combines a hybrid of both vision and infrared thermography images, which turned out to be able to improve crack detectability while minimizing false alarms. Tang et al. proposed a 3D modeling method that combines oblique photography with inclined photography so that clearer textures, more complete lines, and higher accuracy can be observed for damage detection [7]. The aforementioned studies demonstrate that UAVs can serve as convenient tools for acquiring bridge surface images. However, most of these studies have focused on localized positions on structures such as bridges and did not comprehensively inspect the entire surface. For bridge surface defect detection, a complete inspection of the entire bridge is necessary. Therefore, the first aspect of our research is studying UAV-based photography methods for each part of a bridge such that the sequential images captured within the limited field of view of the UAV's camera can provide complete coverage of the surface of each part of the bridge.

Recognizing damage in massive, high-resolution bridge surface images through visual analysis is another focus of bridge inspection. With the rapid advancement of computer vision technology in recent years, these methods have increasingly been applied in structural damage assessment [8] and detection [9,10]. For example, Cataldo et al. [8] proposed a low-cost vision-based measurement system for assessing the structural damage caused by seismic loads on buildings, and the results demonstrated that the method has a high potential for extracting and analyzing modal parameters. Concrete cracking is the most common structural defect, and because cracks can be used to improve the finite element model for more accurate analysis of bridge performance, crack detection is considered crucial. Some of the earliest research involved the application of such methods for crack detection. Li et al. [9] proposed a Two-RTSs system for UAV hovering accuracy and applied Faster R-CNN network for crack detection; the achieved accuracy, recall, and F1 score were 92.03%, 96.26%, and 94.10%, respectively. Li et al. [10] proposed a robust and direct crack detection method based on geometric correction and calibration algorithms to overcome the problem of inaccurate crack identification due to the deviation of the viewing angle, and the results showed that the four-point laser-based method has a higher accuracy in crack-width identification compared with the lens imaging conceptual method, with a measurement accuracy of more than 95%. Zhang et al. [11] achieved higher accuracy with their proposed CrackNet network, which lacks any pooling layers and predicts each pixel individually, enabling pixel-level recognition accuracy. Ni et al. proposed a CNN-based dual-scale detection method to detect wide and thin cracks and a sub-pixel crack-width measurement method based on Zernike moments to detect the width of fine cracks, enabling the accurate detection of cracks with pixel sizes less than 5 [12]. In addition to crack detection, research on the detection of other structural defects, such as concrete spalling, corrosion, and loose bolts, has also been conducted. Various deep learning models such as Faster R-CNN, you look only once (YOLO) v3, and SSD networks [13] have gradually been applied to achieve the high-precision automatic detection of bolt loosening [14], corrosion [15], and various types of concrete damage [16]. For example, Xu et al. proposed a nested attribute-based meta-learning method for the identification of 10 representative defect types, overcoming the low robustness of traditional networks on limited datasets [17]. The above studies have demonstrated the high accuracy, efficiency, and application potential of vision-based damage detection methods in structural inspection; however, these learning-based methods are also characterized by high dependence on their training sets and the use of precision-designed superior networks. A substantial number of results in existing studies have been obtained based on small-batch datasets built by researchers, some of which use images collected in the laboratory, which does not necessarily represent the real surface conditions of in-service bridges. Meanwhile, with the rapid development of deep learning networks, many outstanding novel networks have been designed and proposed. Thus, it is also important to train and test the performance of state-of-the-art networks on datasets containing a large number of real, in-service bridge surface images in order to evaluate whether they can be applied for the detection of multiple damage types.

Based on the above review, this study proposes a bridge inspection method that applies UAVs to capture multi-angle photographs to obtain a complete surface image of a bridge, from which an overall 3D model of the bridge can be reconstructed. It then applies a deep learning-based damage recognition algorithm for damage detection. The contributions of this study are as follows: (1) A UAV-based image-capturing method is proposed for various bridge sections to collect close-range, multi-angle, and overlapping bridge surface images. (2) A 3D reconstruction technique is introduced, utilizing multi-view images to create a textured model of the bridge. This allows for the generation of ultra-high resolution, panoramic, unfolded images of the bridge surface through multi-angle projections. (3) A bridge damage dataset was established based on images collected from in-service bridges, using a Swin Transformer to optimize the YOLOv8 network to improve the network's detection accuracy for small-scale damage. Sliding window segmentation was then applied to automatically identify the damage in the ultra-high resolution panoramic image.

The structure of this article is outlined as follows: Section 1 introduces the research background. Section 2 presents the framework of the proposed method and provides detailed explanations of its core methodologies and technologies. Section 3 discusses the application of the proposed method to detect defects in a concrete bridge. Section 4 concludes the article.

2. Framework of the Proposed Method

This section introduces the framework of the proposed method, which can be divided into three parts (Figure 1): the overall bridge image acquisition method based on UAV close-range photography; the refinement model of the 3D reconstruction method based on stereo geometry; and the identification of typical bridge damage based on improved YOLOv8.

Section 2.1 discusses the overall image acquisition method based on UAV close-range photography, focusing on ways to select appropriate photography angles and flight paths for each part of the bridge. Section 2.2 introduces the proposed 3D bridge reconstruction method based on the structure from motion (SFM) and multi-view stereo (MVS) algorithms, focusing on accurately setting the camera position for image sequences with a finite overlap rate. Section 2.3 describes the establishment of a dataset based on in-service bridge defect images as well as improving and training state-of-the-art object detection networks to achieve the automatic recognition of typical bridge defects.

When applying the proposed method for the detection of bridge surface damage, the flow is also shown in Figure 1. Firstly, the method in Section 2.1 is used to control the UAV to take photos of each part of the bridge with a certain speed, distance, and photo overlap rate in a complete coverage style to obtain close-range and high-resolution images. Then, the method proposed in Section 2.2 is used to reconstruct the 3D model of the bridge from the original image, and the ultra-high resolution panoramic image of the bridge surface is projected, which can fuse a large number of fragmented images into a complete panoramic image. Finally, the network proposed in Section 2.3 is used to recognize the damage in the panoramic image and label the location of the damage to complete the detection of the bridge surface damage.

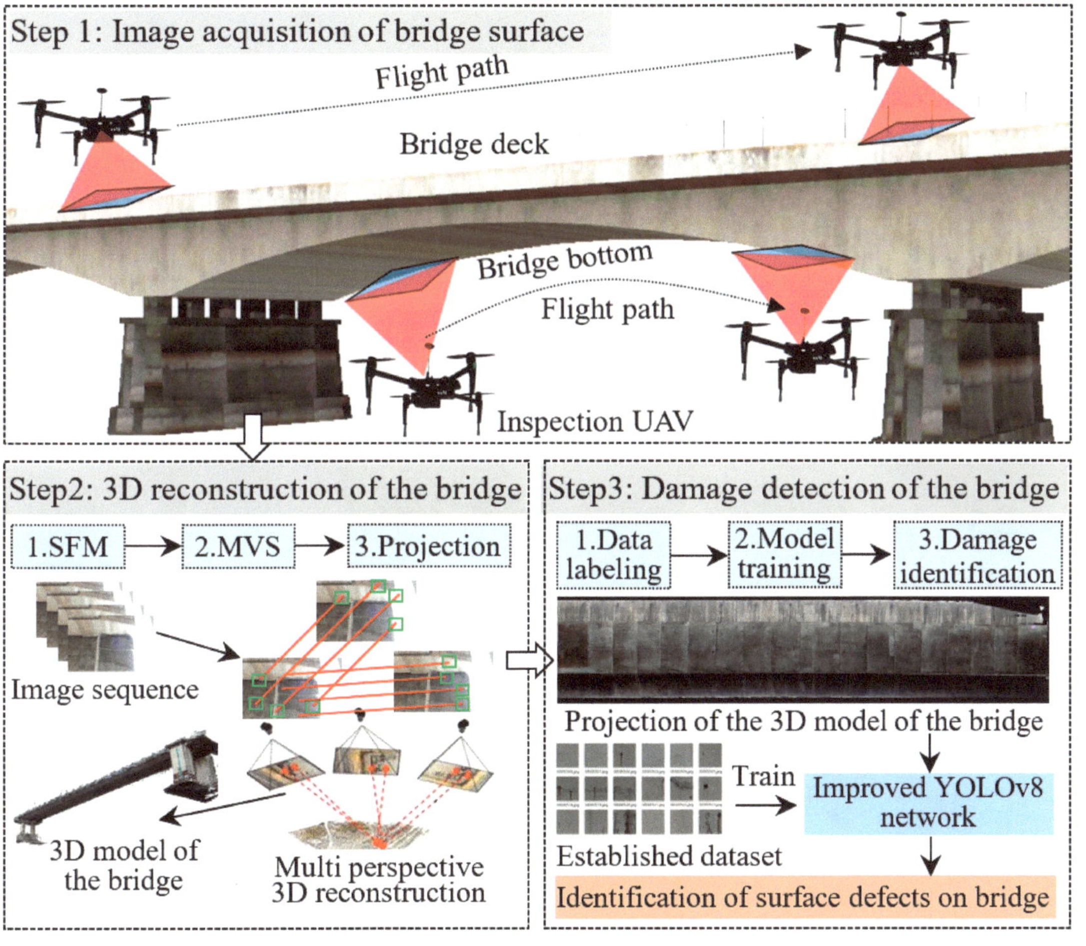

Figure 1. Framework of the proposed method.

2.1. Close-Range Photography

When applying UAVs to capture images around bridges, it is recommended that the UAVs have the following characteristics: (1) As there are obstacles and limited headroom around bridges, especially at the bottoms of small bridges, the UAVs should have obstacle-sensing and obstacle-avoidance capabilities in multiple directions. (2) When flying around bridges, especially underneath, it is difficult for UAVs to receive GPS positioning signals.

Therefore, they should have other positioning capabilities such as visual positioning. (3) The camera carried by the UAV should have high resolution and small distortion to enhance the efficiency of data acquisition. Considering the above conditions, this study adopted the M300RTK equipped with a P1 camera from DJI Company, Shenzhen, China, to acquire bridge surface images. This UAV has six-direction vision and laser-ranging functions, enabling it to perceive obstacles even in complex environments. The P1 camera was used for aerial mapping, which not only has low distortion but also has 45 million effective pixels and can acquire images with a resolution of 8192 × 5460 pixels.

After determining the UAV to be used, it was necessary to determine the appropriate photography method for different bridge sections. Bridges are generally divided into two major sections: the upper structure, including bridge towers, cables, and bridge decks, and the lower structure, mainly including the bottom and piers of the bridge. As the upper structure is generally in an open environment with fewer obstacles, and the UAV can receive sufficient GPS signals for accurate positioning, we adopted a capture method similar to aerial photogrammetry, which involves setting up parallel routes covering the whole bridge using a route planning tool. To capture multiple perspectives, the UAV takes photos from multiple angles, including directly below, in front, at the back, on the left side, and on the right side. In photographing the lower structure, it is difficult to plan a route for automatic photography, as there is typically no GPS signal underneath the bridge. Thus, we mounted the camera on top of the UAV, such that it could take images directly upward or tilted upward while flying under the bridge. Specifically, three route types were used to cover the range of the bridge bottom: several parallel routes distributed directly below the bridge to capture images at a 90° angle vertically from the bottom upwards and two tilted routes on the left and right sides to capture images at a 45° angle tilted toward the bridge. The piers around the bridge were photographed with reference to the method used for UAVs when photographing buildings. The photography methods for the bridge deck, bottom, and piers are shown in Figure 2.

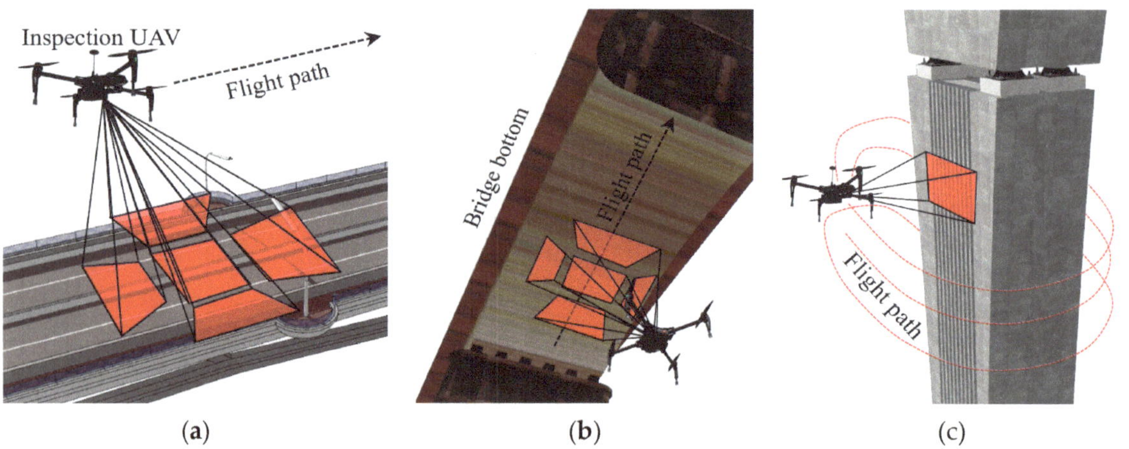

Figure 2. Photography methods for various bridge sections. (**a**) Photographing bridge deck. (**b**) Photographing bridge bottom. (**c**) Photographing bridge pier.

The images acquired by the UAV should not only cover the entire surface of the bridge but also have a degree of overlap with respect to each other. This makes it possible to reconstruct a 3D model of the bridge. Thus, the flight speed of the UAV should be maintained in a suitable range during photography such that the rate of image overlap is no less than 60%. The distance between the UAV and its target must also be calculated in advance, based on the required image resolution. According to the pinhole camera imaging model, this distance should satisfy the following formula:

$$\frac{f}{H} = \frac{size_{sensor}}{L} \tag{1}$$

where L is the length of the bridge covered by the image, H is the object distance, f is the focal length of the UAV's camera, and $size_{sensor}$ is the size of the camera sensor.

2.2. Fine 3D Reconstruction

This section describes a method for constructing a 3D model of the bridge from the acquired images and projecting a panoramic unfolded view of its surface using this model. The 3D reconstruction method based on multiple views is an essential computer vision technique that has been widely researched and applied in recent years, which has been applied in various fields such as reverse engineering, photogrammetry, and aerial mapping [18–20]. In this section, the structure from motion (SFM) algorithm and the multi-view stereo (MVS) algorithm [21] are applied to reconstruct a 3D point cloud and a model of the bridge from the original images acquired with the UAV. The detailed textures in the original image are mapped onto the model to obtain a fine 3D model of the bridge. The 3D reconstruction algorithm applied in this step can be divided into three steps: sparse reconstruction, depth map estimation, and dense reconstruction. The sparse reconstruction stage involves image feature extraction, feature matching, matching optimization, triangulation, attitude estimation, and beam leveling. The image feature extraction stage adopts the A-Sift method, which simulates various affine distortions through establishing a hemispherical space and performs SIFT feature point matching on this basis to achieve affine invariance. The feature matching stage uses the Euclidean distance to compute the detected image features, and as the acquired images are sequential, the matching computation is performed only on the neighboring images. The matching results are verified for possible overlapping image feature associations via geometric calibration, using the RANSAC algorithm to estimate the transformation matrix and verifying the validity of the matching through mapping the projective geometry in the image pairs [22]. After feature extraction and matching, sparse point cloud reconstruction is performed using an incremental reconstruction method. The method first selects an appropriate initial image initialization model and then calculates the bitmap in an image-by-image manner using the PnP method, matching the 2D feature points of the image into the 3D coordinates. At the same time, the triangulation and beam-leveling methods are used; the former is used to extend the point set to increase the scene range, and the latter is used to increase pose redundancy through the use of additional triangulation points to further optimize the overall model's pose and avoid obvious drift phenomena in the reconstruction process. At this point, the setting of the image pose and the reconstruction of the sparse point cloud can be completed. The depth information of all points in the image is further calculated using the MVS algorithm to obtain a dense point cloud, applying the method proposed by Schönberger et al. [21], which is a depth map fusion-based method that generates depth and normal maps for all the registered images, fuses the depth and normal maps into a dense point cloud with normal information, and, finally, fuses the dense surface of the point cloud using Poisson estimation. The next step is to project the pixel details onto the surface of the 3D model through texture mapping of the image, which finally yields a 3D model of the bridge with realistic texture [23]. As the UAV is close to the bridge surface when taking photos, usually in the range of 5 to 10 m, the 3D model is a fine model with highly detailed textures.

After obtaining the 3D model, it is projected from the top, bottom, left, and right sides of the bridge to obtain panoramic unfolded views of the entire surface. The steps of the aforementioned method are illustrated in Figure 3.

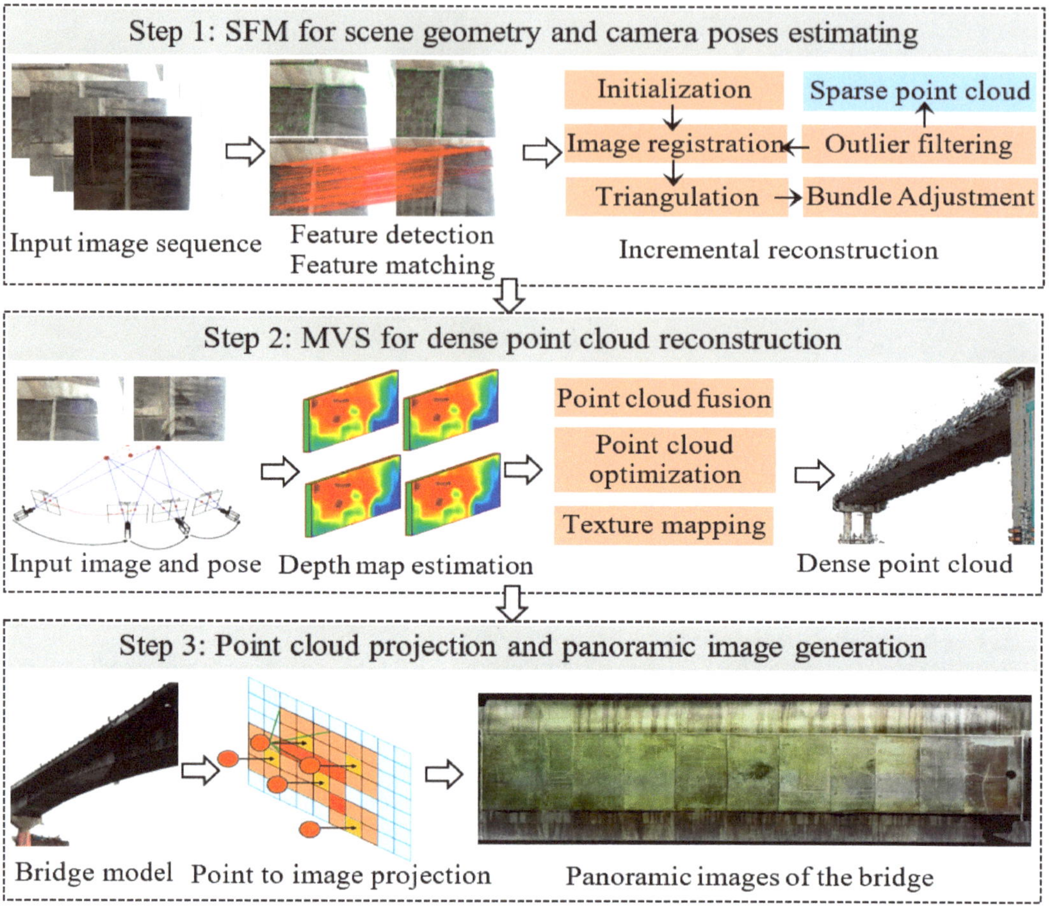

Figure 3. Steps of the 3D reconstruction method for bridges.

2.3. *Identifying Bridge Damage*

As the most rapidly developing and widely used type of machine vision method, deep learning-based image recognition has become the mainstream technique for structural damage identification. The key aspects of this type of method are establishing datasets and setting up networks, and so, this section introduces these two aspects. Mainstream deep learning networks for image processing are currently divided into three main types: classification networks, object detection networks, and segmentation networks. Object detection and segmentation networks can frame and label objects in an image, where the difference is that segmentation networks segment the pixels of objects from the background. Due to the need for rapid and qualitative inspection in daily inspections, segmentation networks generally have slower inference speeds than object detection networks because of their higher complexity. In this study, an object detection network is used as the identification network for bridge defects [24].

With the rapid development of deep learning networks for detection tasks, the performance of object detectors has greatly improved. Mainstream object detection networks mainly include one-stage and two-stage detectors, with the former framing detection as a "one-step" process and the latter defining it as a "coarse-to-fine" process. Representative two-stage detector networks include Faster R-CNN and Feature Pyramid Networks

(FPNs) [25], which split the detection task into two steps: localization followed by recognition. Localization retains as much useful foreground information as possible, filters out the background information that is not useful for the subsequent task, and then distinguishes the foreground information. As the foreground information is distinguished, this classification network is finer. A one-stage detector relies on a network output containing both coordinates and a tensor with category confidence, greatly simplifying the two-stage framework and leaving the localization and classification to the RPN, thus improving the inference speed and simplifying the training steps. The YOLO series networks are typical two-stage detectors, and for early networks of this type, the inference speed is fast, but the accuracy is insufficient [26–28]. With the rapid development and maturity of these algorithms, the YOLO series networks have gone through eight generations, and the latest network, YOLOv8, has been significantly improved in terms of accuracy and speed. Therefore, we selected the YOLOv8 network for bridge damage identification.

2.3.1. Overview of YOLOv8

The YOLOv8 network is an updated network based on the optimization of the YOLO series networks. Its improvements include the introduction of a new backbone network, the inclusion of anchor-free detection headers, and the use of a new loss function, enhancing the performance and flexibility of the network. The backbone network and neck part of YOLOv8 replace the C3 structure of the earlier YOLOv5 with a C2f structure with richer gradient flow, adjusting the number of channels differently for different scale models. The specific changes are as follows: the kernel of the first convolutional layer was changed from 6×6 to 3×3; all C3 were replaced with C2f modules; more skipping layer connections and additional Split operations were added, thus fine-tuning the model structure and expanding the size of the network from the original four dimensions to five; and a p6 network with a resolution of up to 1280 pixels was added, making it easier to apply the network to different tasks. In the detection head part, the YOLOv8 network changed from the original coupling head to a decoupling head and from the anchor-based method of YOLOv5 to an anchor-free method; this change removes the need to preset the anchor. Only the target center point and the width and height of the feature maps at different scales need to be regressed, reducing time consumption and computing capacity. This can also avoid missed or repeated detection problems caused by unreasonable anchor settings. In the loss function, mainstream object detection networks generally adopt the dynamic allocation strategy, such as YOLOX's simOTA [29] or TOOD network's Task-Aligned Assigner [30]. The dynamic allocation strategy can dynamically adjust the weights according to the progress of training and the characteristics of the samples. In the early stage of training, the model may have difficulty differentiating between positive and negative samples, so it should pay more attention to samples that are easy to misclassify. As training progresses, the model gradually becomes better able to distinguish between samples, slowly reducing the weights of difficult samples while increasing the weights of easy-to-distinguish samples. The dynamic allocation strategy can be adjusted according to the training loss or other metrics, which can be better adapted to different datasets. On the other hand, YOLOv5 still adopts a static allocation strategy, which is usually based on experience and cannot fully utilize sample information, possibly leading to poor training results. YOLOv8 employs the Task-Aligned Assigner, which selects positive samples based on the weighted classification and regression scores, expressed as $t = s^{\alpha} \times u^{\beta}$, where s represents the predicted score corresponding to the annotated category, and u is the intersection over union (IOU) between the predicted box and the ground truth box.

2.3.2. Improvement of YOLOv8

Defects such as fine cracks, small pieces of spalled concrete, and exposed reinforcement challenge the original YOLOv8 network when detecting damage on bridge surfaces. In order to overcome this problem, this study added a module for small-target feature extraction to the original YOLOv8 network, enhancing the network's recognition accuracy for small lesions. Network strategies and modules for tiny target detection have been proposed in recent years, of which Featurized Image Pyramid (FPN), sliding window segmentation, and attention mechanisms are representative [31]. Notably, a transformer module incorporating self-attention has been widely applied in visual recognition in recent years [32]. In this section, the Swin Transformer module [33], which further optimizes the computational complexity of the original transformer structure, is added to the YOLOv8 network to improve its performance for small-target detection.

The Swin Transformer has a hierarchical design containing four stages, each of which reduces the resolution of the input feature maps and expands the sensory field in a layer-by-layer manner, similar to a CNN. While the original transformer module computes self-attention for N tokens with a computational complexity of $O(N^2)$, the Swin Transformer divides N tokens into N/n groups, each consisting of n tokens, reducing the complexity to $O\left(N \times n^2\right) = O(N)$. However, after group-wise computation, there will be a lack of interaction between groups, posing challenges in extracting global information. Therefore, after each stage, the Swin Transformer integrates and compresses the feature vectors of 2×2 groups, ensuring that the network's receptive field increases gradually with each stage, similar to CNN-based structures. Additionally, through the use of shifted windows, adjacent patches interact with each other. In this section, we attempt to replace the backbone network of YOLO v8 with the Swin Transformer and evaluate the extent of the network's performance improvement on a bridge defect recognition dataset. The modified network architecture is illustrated in Figure 4.

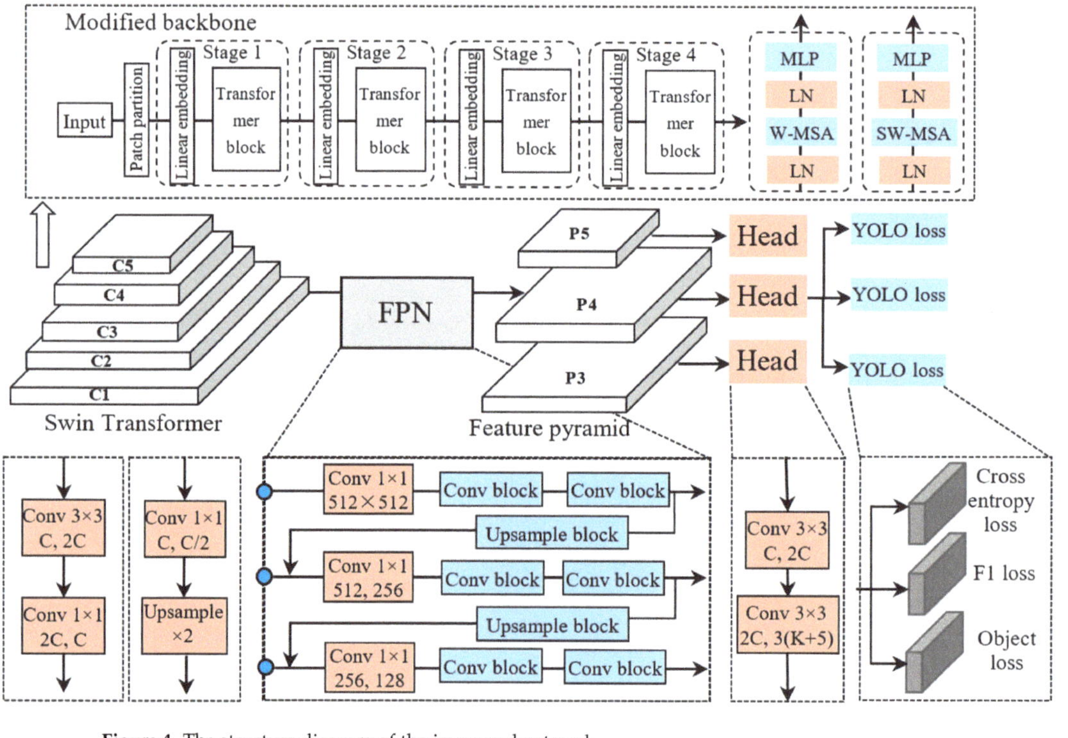

Figure 4. The structure diagram of the improved network.

2.3.3. Establishment of the Bridge Damage Dataset

To make the dataset used in the network training process as close as possible to the actual situation of the bridges targeted by the network deployment, the images in the dataset were taken from images of typical damages on several bridges captured by UAVs or digital cameras. The types of damage include three categories: concrete cracks, spalling, and exposed internal reinforcement due to spalling. The bridges from which the images were collected and typical damage images are shown in Figure 5. A 800D camera, from Canon Company, Tokyo, Japan, a M200 UAV and a M300 UAV from DJI Company, Shenzhen, China, were used for data acquisition, and 500, 1000, and 2000 images captured with the three devices, respectively, were selected for manual labeling. Considering the different resolutions of the three devices, the acquired images were first cropped and resized to the same 1280 pixels as the input size of the network. Finally, 3000 images containing the damage were acquired. Several weather conditions, including sunny, cloudy, and foggy days, were also selected to acquire bridge images with different exposures, considering that different weather conditions result in different levels of image exposure and shadow masking.

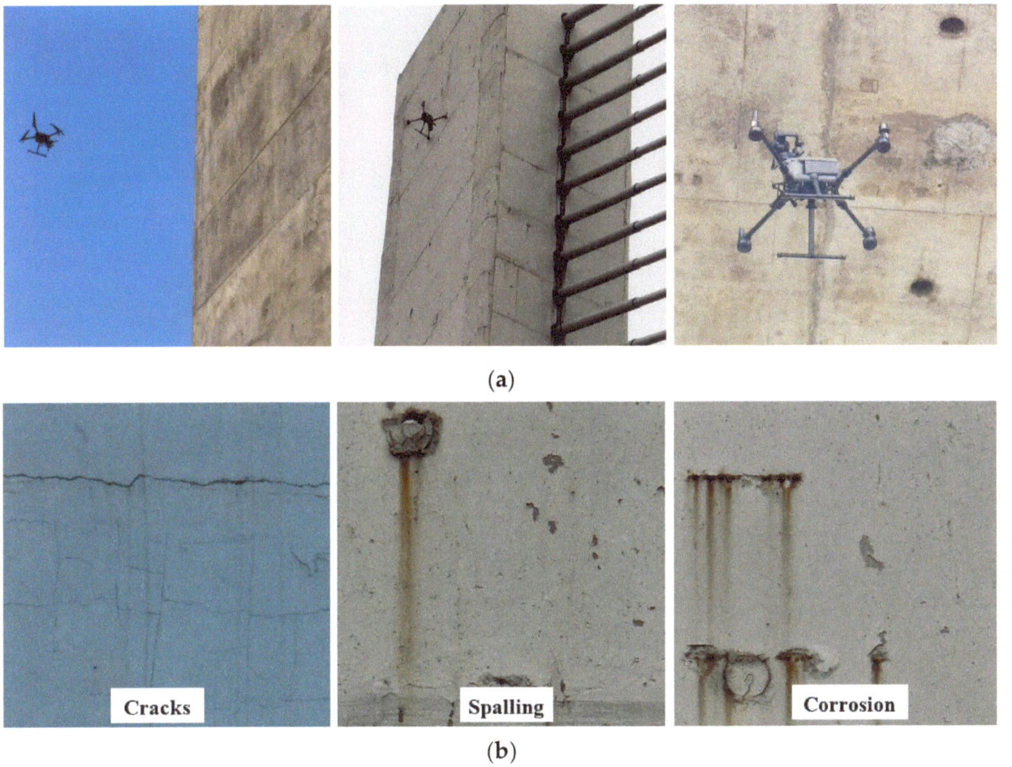

Figure 5. Establishment of a dataset for typical bridge damage. (**a**) Image acquisition of surface defects on three in-service bridges using UAVs. (**b**) Images of three typical types of bridge damage.

In addition to the 3000 original images containing defects, we expanded the dataset in order to enhance the generalization ability of the trained network through applying data augmentation techniques to introduce random disturbances such as random noise and occlusion. The augmented dataset thus comprised 9000 images with defects. Finally, the dataset was partitioned into training, testing, and validation sets in a ratio of 7:2:1, respectively.

3. In-Service Bridge Damage Detection

To test the practicality and accuracy of the proposed method, it was tested on a three-span concrete bridge using UAV-based data acquisition and the 3D reconstruction of multi-view images, followed by training and testing of the improved network.

3.1. Overview of the Tested Bridge

We tested a concrete bridge on an urban circular highway with a length of 170 m and a main span of 70 m. The bridge consists of four closely spaced bridges, each of which is 13 m wide, for a total width of 60 m. The bridge crosses a river that is approximately 100 m wide, with an underbelly clearance of 15 m. The bridge has been in service for over 20 years and presents a series of surface defects. Previous manual-based inspection methods showed that the bridge was subjected to long-term loading, deterioration of concrete properties, and loss of prestressing, resulting in extensive surface cracking, concrete spalling, and corrosion of reinforcing steel, some of which were repaired and filled in early on, but some new defects have continued to appear ever since. As the four bridges are closely spaced, and their width exceeds the length of the bridge inspection vehicle's cantilever, the inspection vehicle could not inspect the bottom of the main span. An overview of the bridge is shown in Figure 6.

Figure 6. Overview of the tested bridge.

3.2. Data Collection and 3D Reconstruction

The photography method proposed in Section 2.1 was first used to capture the bridge surface; specifically, three photography methods were used, according to the three main bridge sections: the deck, bottom, and piers. As there were no obstacles around the bridge, the same multi-angle photography method as the aerial photogrammetry was adopted for the deck; that is, five angles of vertical down, 45° tilt-forward, tilt-backward, tilt-left, and tilt-right were photographed at each position. Using a planning tool set, the UAV's route was directly planned to cover the entire bridge with several routes. The overlap of the photos was set to 70%. For the bottom of the bridge and the piers, given the limited clearance height of the bottom, the obstacles, and the lack of GPS signals, the photographs were taken in a completely manual control mode. The photographs of the bottom were similar to those of the deck and were also taken from multiple angles according to multiple parallel routes. The UAV maintained a safe distance of about 5 m from the bottom of the bridge. For the piers, encircling capture was carried out; that is, the UAV flew and took photos around the piers at multiple heights and angles. The photography process of the UAV is shown in Figure 7. The UAV and camera used in the tests are consistent with those introduced in Section 2.1.

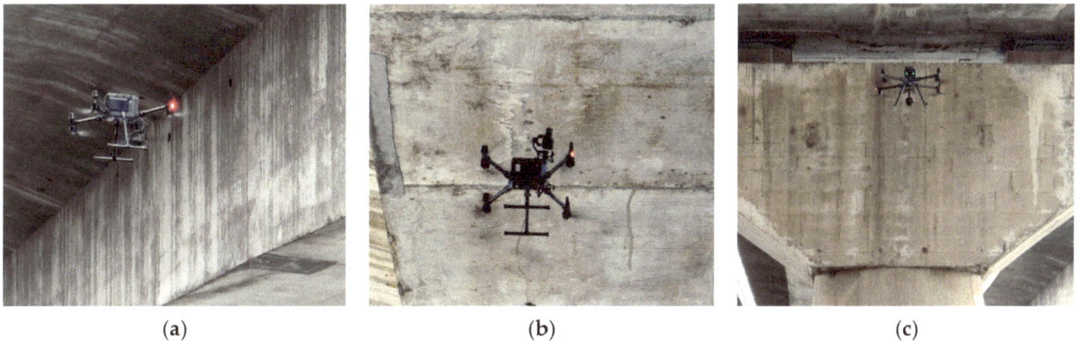

Figure 7. Photography schematic diagram of different sections of the bridge using a UAV. (**a**) Bridge side. (**b**) Bridge bottom. (**c**) Bridge pier.

The data collection process took approximately 1.5 h, resulting in a total of 1608 surface images of the target span. Utilizing the 3D reconstruction method proposed in Section 2.2, the original images were reconstructed, yielding a sparse 3D point cloud, a dense point cloud, and a 3D bridge model, as depicted in Figure 8.

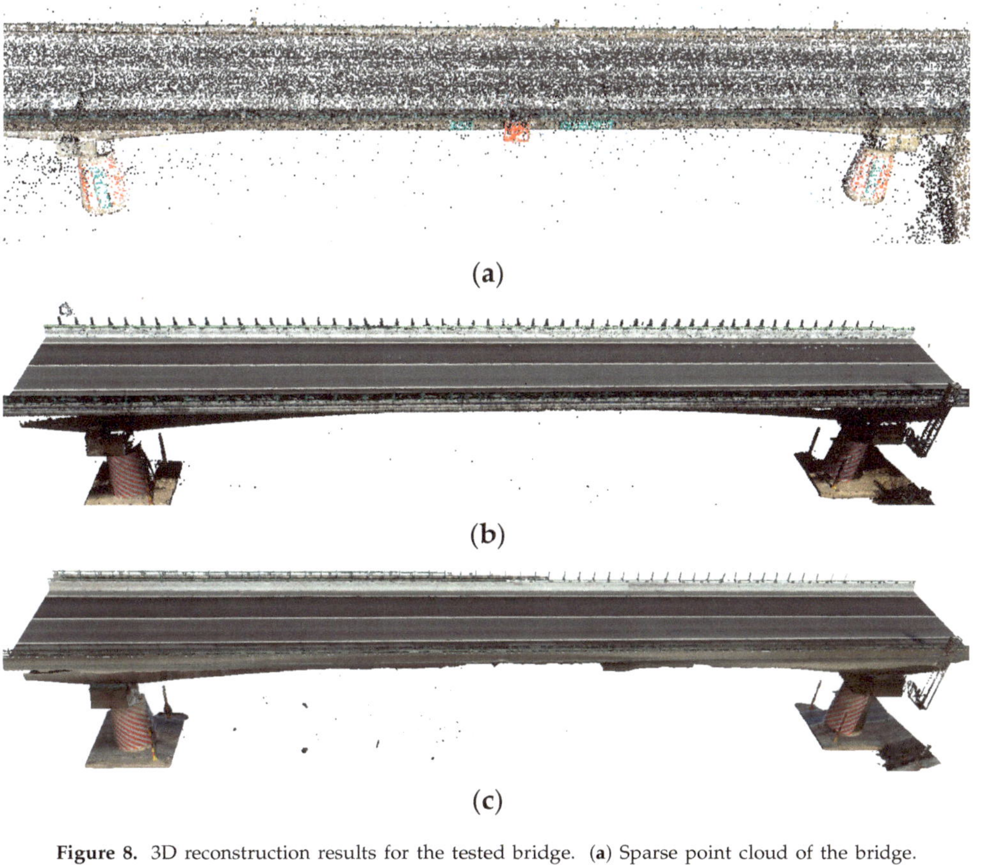

Figure 8. 3D reconstruction results for the tested bridge. (**a**) Sparse point cloud of the bridge. (**b**) Dense point cloud of the bridge. (**c**) 3D model of the bridge.

The model was further projected from multiple directions to panoramically unfold the bridge deck, sides, and bottom, as shown in Figure 9. The resolution of the bridge deck panorama was 26,715 × 5397 pixels, with a true resolution of 3.25 mm/pixel; the resolution of the bottom panorama was 76,241 × 13,011 pixels, with a true resolution of 1.26 mm/pixel; and the resolution of the side panorama was 77,079 × 11,605 pixels, with a true resolution of 1.19 mm/pixel. These results demonstrate that the proposed UAV-based photography method can be used to obtain a complete 3D bridge model in a global range and a panoramic unfolding map at the millimeter scale.

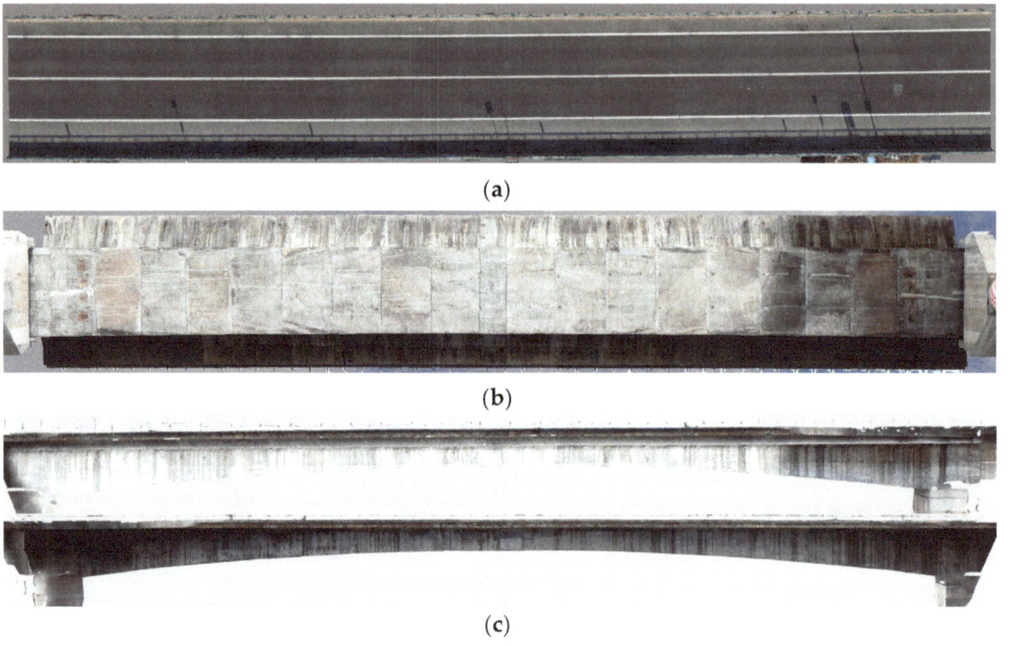

Figure 9. Panoramic unfolded view of the tested bridge. (**a**) Bridge deck. (**b**) Bridge bottom. (**c**) Bridge side.

3.3. Bridge Damage Identification Results

After obtaining the panoramic unfolding maps of each face of the bridge, the next step was to automatically identify damage in the panoramic unfolding maps using the method described in Section 2.3. For the improved network to recognize bridge damages, it was first necessary to train the network on the typical damage dataset. To verify that the designed network had a higher detection accuracy than the original network, both the designed network and the original YOLOv8 network were trained on the created dataset, and the final accuracy achieved by both networks was compared. Because the YOLOv8 network has four scales, from small to large, and a larger scale means more layers in the network and higher accuracy, the largest scale of YOLOv8-p6-x was chosen as the base network.

Network training was performed on a computer equipped with an Nvidia RTX3090 GPU, an i7 13,700 K CPU, and 64 GB of RAM, and the framework used for training was Ultralytics. The parameters set during network training were as follows: The optimizer was set to istochastic gradient descent, the base learning rate was set to 0.01, the base weight decay was 0.0005, the optimizer momentum was 0.937, the batch size was set to 16, the learning rate schedule was set to linear, the number of training epochs was 500, and the input size was 1280 × 1280 pixels. The parameter variation curves of the two networks during training are shown in Figure 10.

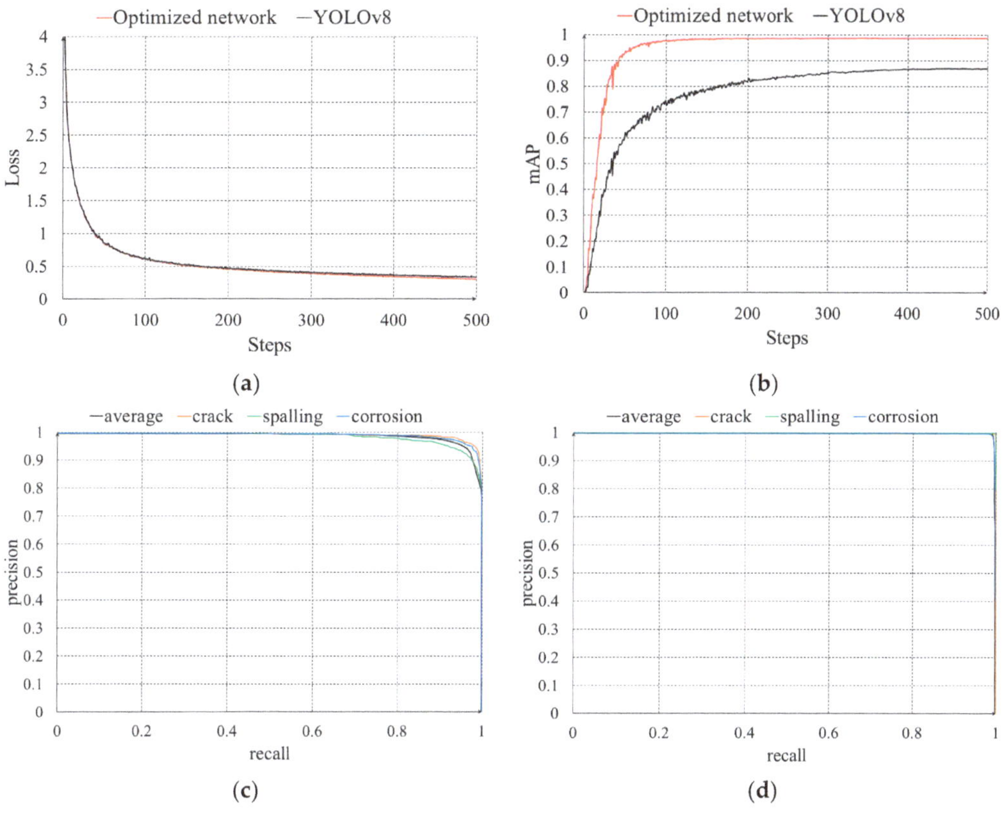

Figure 10. Parameter variation curves of the two networks during training. (**a**) Loss curves of the two networks. (**b**) mAP curves of the two networks. (**c**) P-R curves of YOLOv8. (**d**) P-R curves of the optimized network.

The loss curve and accuracy curve trends of the training process show that both networks were correctly trained and eventually stabilized. After stabilization, the original YOLOv8 network achieved a mAP of 0.869, with accuracies of 0.897 for cracks, 0.808 for spalling, and 0.903 for rebar exposure. The improved network achieved a mAP of 0.987, with accuracies of 0.990 for cracks, 0.981 for spalling, and 0.991 for rebar exposure. These results show that the improved network improved the average accuracy of damage detection relative to the original network by 13.6%, proving that the improved network performs better. Additionally, the precision–recall curves of the two networks indicate that the improved network exhibited higher precision.

After the network was correctly trained, it was applied for damage detection in the panoramic bridge images. Because the panoramic images had high resolution, and it was thus not possible to input them directly into the network, a sliding window with the same size as the network input was used to sequentially slide and segment the panoramic images to obtain a sequence of sub-images. These sub-images were input into the network for damage detection. The recognition results on the sub-images are shown in Figure 11a, and after all the sub-images were recognized correctly, they were stitched together in sequence to obtain a panoramic image of the bridge with damage markers, as shown in Figure 11b. Using this panoramic image, it is possible to quickly count the number of damage points on the bridge and to analyze the distribution density of the damages in each section, providing a basis for analysis of the health status of the bridge.

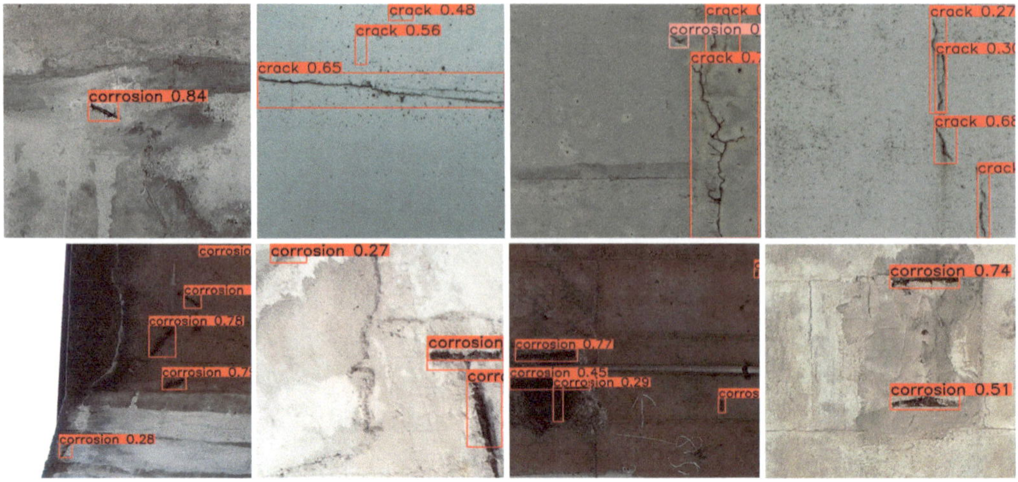

(a)

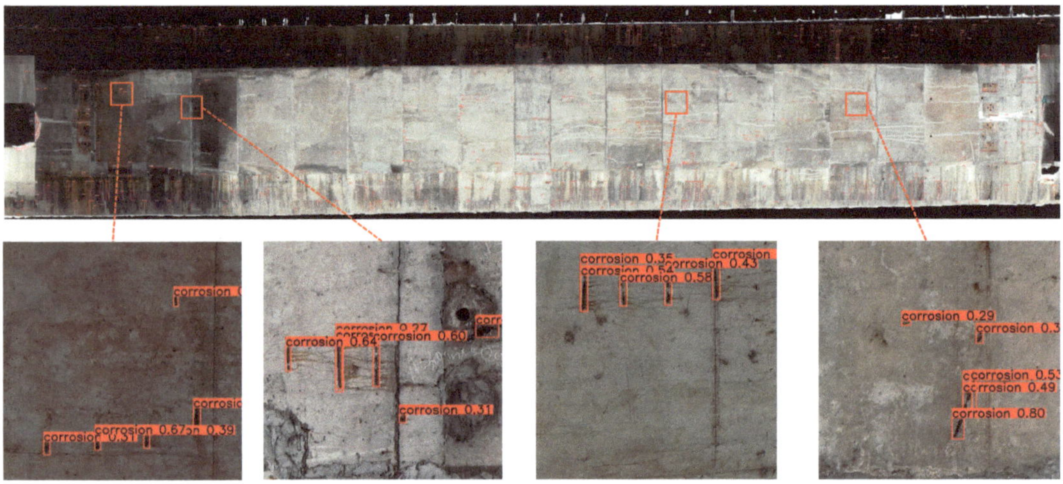

Enlarged image of area 1 Enlarged image of area 2 Enlarged image of area 3 Enlarged image of area 4

(b)

Figure 11. Damage detection results in sub-images and panoramic images. (**a**) Damage detection results for sub-images cut by sliding windows in panoramic images. (**b**) Damage detection in panoramic image of bridge bottom.

4. Conclusions and Future Work

This study proposed a bridge damage detection method based on UAV photography, multi-view 3D reconstruction, and deep learning-based object detection. Specifically, we used close-range bridge surface images captured by a UAV from multiple angles in a covering manner as the data source, which allowed for the establishment of a refined 3D model of the bridge that could be projected to obtain a panoramic, unfolding image of the bridge surface. We then adopted a deep learning-based object detection method to rapidly and automatically identify typical damage on the global scope of the bridge using the obtained panoramic map. The proposed method contains the whole process from data acquisition to preprocessing and damage identification, and the method can be used as a guide in the inspection of similar bridges to enhance the inspection efficiency and reduce the cost. In addition, unlike existing methods, the proposed method does not directly use the acquired original images for damage detection but reconstructs a large number of fragmented images into a complete panoramic view of the bridge surface, which avoids omissions and enables the location of the damage to be known from the panoramic view. Specific conclusions are as follows:

(1) To quickly acquire detailed, high-resolution surface images, we propose using a UAV to photograph bridges from multiple angles in close range, providing overlapping images with full coverage of the bridge. The results of photographing in-service bridges proved that the proposed method allows millimeter-level images of complete bridge surfaces to be acquired, which can be used for damage analysis;

(2) To generate panoramic images of bridge surfaces, we used multi-view stereo algorithms to reconstruct a 3D model of the bridge based on a large number of original images. The bridge was then panoramically unfolded through projection of the model. The test results proved that the method can panoramically unfold the bridge surface at the millimeter scale, and the resolution of the resulting panoramic image can be up to hundreds of millions of pixels. In the tested bridge, the resolution of the panorama at the bottom of the bridge was 1.26 mm/pixel, and the size of the image amounted to 76,241 $\times$ 13,011 pixels;

(3) To automatically detect damage, the state-of-the-art YOLOv8 network was improved using a Swin Transformer. The results of training on a bridge damage dataset and testing on data obtained from an in-service bridge demonstrated that the improved network achieved a 13.6% increase in bridge damage recognition accuracy when compared with the original network, reaching 98.7% in the bridge test. After damage identification, the panorama presents the location of each instance of damage on the bridge surface.

The proposed damage identification method is expected to supplement or even replace existing manual-based damage detection methods for many similar concrete bridges, but drawbacks and limitations of the present method have also been identified in its application, including non-automatic data acquisition, limitations on the types of bridges to be detected, and long inspection times. First, due to the influence of weak GPS signals and multiple obstacles at the bottom of bridges, the UAV still needs to rely on the control of skilled manipulators when performing inspections at the bottom of bridges, and such manipulation skills are not possessed by most workers, which makes it difficult to popularize the use of UAV-based bridge damage inspections in a large number of applications. With the extensive research on automated UAV-based inspection techniques in GPS-constrained environments (e.g., indoors), it has become possible to make UAVs fly automatically in such environments, so future work will investigate the use of sensors such as LIDAR, stereo vision, and simultaneous localization and mapping (SLAM)-based algorithms to make UAVs fly automatically at the bottom of bridges and get rid of the dependence on manual control. Secondly, the current proposed method was only trained and tested on concrete bridges, but the applicability of the proposed method could be expanded through the consideration of more bridge types (e.g., steel bridges) and damage types (e.g., corrosion and cable damage). In addition, since 3D reconstruction-based methods require a large amount of computational time, the currently proposed method cannot achieve fast solving and output of the detection results. Investigating deep learning-assisted 3D reconstruction methods, which utilize learning-based methods to directly match images and generate depth maps, will expectedly lead to a reduction in computational time, and this work will also be a future focus of research.

Author Contributions: T.Y., conceptualization, methodology, validation, writing—original draft, and writing—review and editing; G.S. (Guodong Shen), conceptualization and validation; L.Y., validation and writing—review and editing; G.S. (Guigang Shi), methodology, supervision, resources, and funding acquisition. All authors have read and agreed to the published version of the manuscript.

Funding: The research presented was financially supported by the Key technology projects in the transportation industry (No.: 2021-MS1-055) and Anhui provincial natural science research project—major project (No.: KJ2019ZD53).

Data Availability Statement: The data supporting this study's findings are available from the corresponding author upon reasonable request.

Conflicts of Interest: Authors Tao Yin, Guodong Shen and Liang Yin were employed by the company Anhui Transport Consulting and Design Institute Co., Ltd. The remaining author declares that the research was conducted in the absence of any commercial or financial relationships that could be construed as a potential conflict of interest.

References

1. Yang, J.; Guo, T.; Li, A. Experimental investigation on long-term behavior of prestressed concrete beams under coupled effect of sustained load and corrosion. *Adv. Struct. Eng.* **2020**, *23*, 2587–2596. [CrossRef]
2. Bonopera, M.; Chang, K.C. Novel method for identifying residual prestress force in simply supported concrete girder-bridges. *Adv. Struct. Eng.* **2021**, *24*, 3238–3251. [CrossRef]
3. Hou, S.; Dong, B.; Wang, H.; Wu, G. Inspection of surface defects on stay cables using a robot and transfer learning. *Autom. Constr.* **2020**, *119*, 103382. [CrossRef]
4. Wang, Y.; Zhang, X.; Zhang, M.; Sun, L.; Li, M. Self-compliant track-type wall-climbing robot for variable curvature facade. *IEEE Access* **2021**, *10*, 51951–51963. [CrossRef]
5. Qin, Y.; Dong, S.; Pang, R.; Xia, Z.; Zhou, Q.; Yang, J. Design and kinematic analysis of a wall-climbing robot for bridge appearance inspection. In *IOP Conference Series: Earth and Environmental Science;* IOP Publishing: Bristol, UK, 2021; Volume 638, p. 012062. [CrossRef]
6. Jang, K.; Kim, N.; An, Y.K. Deep learning–based autonomous concrete crack evaluation through hybrid image scanning. *Struct. Health Monit.* **2019**, *18*, 1722–1737. [CrossRef]
7. Tang, Z.; Peng, Y.; Li, J.; Li, Z. UAV 3D Modeling and Application Based on Railroad Bridge Inspection. *Buildings* **2023**, *14*, 26. [CrossRef]
8. Cataldo, A.; Roselli, I.; Fioriti, V.; Saitta, F.; Colucci, A.; Tatì, A.; Ponzo, F.C.; Ditommaso, R.; Mennuti, C.; Marzani, A. Advanced video-based processing for low-cost damage assessment of buildings under seismic loading in shaking table tests. *Sensors* **2023**, *23*, 5303. [CrossRef]
9. Li, R.; Yu, J.; Li, F.; Yang, R.; Wang, Y.; Peng, Z. Automatic bridge crack detection using Unmanned aerial vehicle Faster, R.-C.N.N. *Constr. Build. Mater.* **2023**, *362*, 129659. [CrossRef]
10. Li, J.; Li, X.; Liu, K.; Yao, Z. Crack identification for bridge structures using an unmanned aerial vehicle (UAV) incorporating image geometric correction. *Buildings* **2022**, *12*, 1869. [CrossRef]
11. Zhang, A.; Wang, K.C.; Fei, Y.; Liu, Y.; Tao, S.; Chen, C.; Li, J.Q.; Li, B. Deep learning–based fully automated pavement crack detection on 3D asphalt surfaces with an improved CrackNet. *J. Comput. Civ. Eng.* **2018**, *32*, 04018041. [CrossRef]
12. Ni, F.; Zhang, J.; Chen, Z. Zernike-moment measurement of thin-crack width in images enabled by dual-scale deep learning. *Comput.-Aided Civ. Infrastruct. Eng.* **2019**, *34*, 367–384. [CrossRef]
13. Han, X.; Chang, J.; Wang, K. You only look once: Unified, real-time object detection. *Procedia Comput. Sci.* **2021**, *183*, 61–72. [CrossRef]
14. Jiang, S.; Zhang, J.; Wang, W.; Wang, Y. Automatic inspection of bridge bolts using unmanned aerial vision and adaptive scale unification-based deep learning. *Remote Sens.* **2023**, *15*, 328. [CrossRef]
15. Jiang, S.; Wu, Y.; Zhang, J. Bridge coating inspection based on two-stage automatic method and collision-tolerant unmanned aerial system. *Autom. Constr.* **2023**, *146*, 104685. [CrossRef]
16. Bao, Q.; Xie, T.; Hu, W.; Tao, K.; Wang, Q. Multi-type damage localization using the scattering coefficient-based RAPID algorithm with damage indexes separation and imaging fusion. *Struct. Health Monit.* **2023**, *23*, 1592–1605. [CrossRef]
17. Xu, Y.; Bao, Y.; Chen, J.; Zuo, W.; Li, H. Surface fatigue crack identification in steel box girder of bridges by a deep fusion convolutional neural network based on consumer-grade camera images. *Struct. Health Monit.* **2019**, *18*, 653–674. [CrossRef]
18. Hu, F.; Zhao, J.; Huang, Y.; Li, H. Structure-aware 3D reconstruction for cable-stayed bridges: A learning-based method. *Comput.-Aided Civ. Infrastruct. Eng.* **2021**, *36*, 89–108. [CrossRef]
19. Ma, Z.; Liu, S. A review of 3D reconstruction techniques in civil engineering and their applications. *Adv. Eng. Inform.* **2018**, *37*, 163–174. [CrossRef]
20. Pintore, G.; Mura, C.; Ganovelli, F.; Fuentes-Perez, L.; Pajarola, R.; Gobbetti, E. State-of-the-art in automatic 3D reconstruction of structured indoor environments. *Comput. Graph. Forum* **2020**, *39*, 667–699. [CrossRef]
21. Schönberger, J.L.; Frahm, J.M. Structure-from-motion revisited. In Proceedings of the IEEE Conference on Computer Vision and Pattern Recognition, Las Vegas, NV, USA, 27–30 June 2016; pp. 4104–4113.
22. Li, Z.; Shan, J. RANSAC-based multi primitive building reconstruction from 3D point clouds. *ISPRS J. Photogramm. Remote Sens.* **2022**, *185*, 247–260. [CrossRef]
23. Dundar, A.; Gao, J.; Tao, A.; Catanzaro, B. Fine detailed texture learning for 3d meshes with generative models. *IEEE Trans. Pattern Anal. Mach. Intell.* **2023**, *45*, 14563–14574. [CrossRef] [PubMed]

24. Padilla, R.; Netto, S.L.; Da Silva, E.A. A survey on performance metrics for object-detection algorithms. In Proceedings of the 2020 IEEE International Conference on Systems, Signals and Image Processing (IWSSIP), Niteroi, Brazil, 1–3 July 2020; pp. 237–242. [CrossRef]
25. Lin, T.Y.; Dollár, P.; Girshick, R.; He, K.; Hariharan, B.; Belongie, S. Feature pyramid networks for object detection. In Proceedings of the IEEE Conference on Computer Vision and Pattern Recognition, Honolulu, HI, USA, 21–26 July 2017; pp. 2117–2125.
26. Yu, Z.; Shen, Y.; Shen, C. A real-time detection approach for bridge cracks based on YOLOv4-FPM. *Autom. Constr.* **2021**, *122*, 103514. [CrossRef]
27. Jiang, P.; Ergu, D.; Liu, F.; Cai, Y.; Ma, B. A Review of Yolo algorithm developments. *Procedia Comput. Sci.* **2022**, *199*, 1066–1073. [CrossRef]
28. Zhang, Y.; Liu, C. Network for robust and high-accuracy pavement crack segmentation. *Autom. Constr.* **2024**, *162*, 105375. [CrossRef]
29. Ge, Z.; Liu, S.; Wang, F.; Li, Z.; Sun, J. Yolox: Exceeding yolo series in 2021. *arXiv* **2021**, arXiv:2107.08430. [CrossRef]
30. Feng, C.; Zhong, Y.; Gao, Y.; Scott, M.R.; Huang, W. Tood: Task-aligned one-stage object detection. In Proceedings of the 2021 IEEE/CVF International Conference on Computer Vision (ICCV), Montreal, BC, Canada, 11–17 October 2021; IEEE Computer Society. pp. 3490–3499. [CrossRef]
31. Patil, S.M.; Raut, C.M.; Pande, A.P.; Yeruva, A.R.; Morwani, H. An efficient approach for object detection using deep learning. *J. Pharm. Negat. Results* **2022**, *13*, 563–572. [CrossRef]
32. Han, K.; Wang, Y.; Chen, H.; Chen, X.; Guo, J.; Liu, Z.; Tang, Y.; Xiao, A.; Xu, C.; Xu, Y.; et al. A survey on vision transformer. *IEEE Trans. Pattern Anal. Mach. Intell.* **2022**, *45*, 87–110. [CrossRef]
33. Liu, Z.; Lin, Y.; Cao, Y.; Hu, H.; Wei, Y.; Zhang, Z.; Lin, S.; Guo, B. Swin transformer: Hierarchical vision transformer using shifted windows. In Proceedings of the IEEE/CVF International Conference on Computer Vision, Montreal, BC, Canada, 11–17 October 2021; pp. 10012–10022.

buildings

MDPI

Article

Research on Construction Risk Assessment of Long-Span Cantilever Casting Concrete Arch Bridges Based on Triangular Fuzzy Theory and Bayesian Networks

Zhengyi He [1], Yi Xiang [1,*], Linshu Li [1], Mei Wei [2], Bonan Liu [1] and Shuyao Wu [1]

1 School of Civil Engineering, Hunan City University, Yiyang 413000, China; hezhengyi@hncu.edu.cn (Z.H.)
2 School of Digital Arts, Hunan Art and Crafts Vocational College, Yiyang 413000, China
* Correspondence: xiangyi@hncu.edu.cn

Abstract: Considering the complex construction processes involved, there are significant risks during the construction of long-span cantilever casting arch bridges. In this study, a risk assessment method for the construction process of cantilever casting concrete arch bridges was developed. The compositional elements and characteristics of safety risks in the construction of cantilever casting concrete arch bridges were clarified, and a safety risk source list that includes seven major risk sources and thirty-three minor risk sources was formed. Then, a Bayesian model for the risk analysis of cantilever casting concrete arch bridge construction was established, and a method was proposed to determine the prior and posterior probabilities of the Bayesian network using triangular fuzzy numbers. This method fully utilizes the experience of experts while avoiding the subjectivity of expert opinions. A cantilever casting concrete open spandrel arch bridge (Bridge A) with a total span length of 287 m was taken as an example, and a safety risk assessment was conducted during its construction process. The calculation results show that the construction safety risk level of Bridge A was level III. This engineering application verified the feasibility of determining key node parameters of the Bayesian network using triangular fuzzy numbers.

Keywords: construction risk assessment; cantilever casting concrete arch bridges; triangular fuzzy number; Bayesian network

1. Introduction

The cantilever casting construction method [1–3], which is widely used in concrete construction engineering both domestically and internationally, is increasingly being used to build large-span reinforced concrete arch bridges [4]. With the innovation of bridge structures and the continuous increase in their spans, a series of uncertain factors will arise. Due to the complex processes, high difficulty, and high uncertainty involved, huge losses may occur during the construction of reinforced concrete arch bridges using the cantilever casting method. At present, due to the immature technology in construction, there are still risks such as hanging baskets falling off, tower misalignment, and cable breakage, as shown in Figure 1, which then threaten the construction safety of the cantilever casting concrete arch bridges.

Moreover, there is currently relatively little systematic analysis and research on the construction risks of large-span cantilever casting arch bridges. Therefore, for large-span reinforced concrete arch bridges constructed using the cantilever casting method, construction risk analyses are necessary and very meaningful.

To date, domestic and foreign scholars have achieved a certain research foundation in the evaluation of bridge risks [6–10]. Khan et al. [11] proposed the seismic hazard analysis of fan-shaped cable-stayed bridges using the concept of a damage probability matrix. Novak et al. [12] comprehensively considered the structural factors of bridges after a series of disaster events and constructed a detailed probability model. Ahn et al. [13] analyzed the records of third-party damage and financial losses during bridge construction and constructed a quantitative relationship between damage and risk indicators. Taejun et al. [14] conducted a probabilistic risk assessment on prestressed concrete (PSC) box girder railway bridges and constructed an implicit limit state function for a target PSC railway bridge. Bao et al. [15] integrated the Analytic Hierarchy Process and Grey Relational Analysis to establish a multi-level comprehensive evaluation model, which was then used to assess risks during the construction process of large-span bridges. Wang et al. [16] proposed a fuzzy TOPSIS method based on the alpha-level set for bridge risk assessment. In an analysis of three numerical examples, the proposed fuzzy TOPSIS method showed distinct performance advantages compared to other fuzzy TOPSIS methods. Sun et al. [17] proposed a fuzzy framework based on a transformation system and aggregation process to evaluate the impact of explosion events caused by hazardous materials on bridges. Based on seismic risk assessment results for hundreds of bridges with the complete structural information, Andres et al. [18] analyzed the uncertainty in quantifying expectations at different levels of knowledge using classification methods and machine learning models. Jelena et al. [19] developed a fuzzy logic controller for estimating the degree of damage to bridges.

Figure 1. Photos of some risk locations during the construction process of cantilever casting concrete arch bridges [5].

Overall, research on bridge risk analysis and assessment in foreign countries has received widespread attention. In terms of analysis methods, the bridge risk assessment methods mainly rely on probability analysis and hierarchical analysis. Currently, a single analysis method is often used, and a combination of methods is rarely used. In terms of the analysis objective, there is much research content related to the risk analysis of modern bridge construction, but most of it has focused on large-span cable-stayed bridges [20,21] and continuous rigid-frame bridges [22,23]. In contrast, research results related to risk analyses in cantilever casting concrete arch bridge construction are few and far between. If the existing analysis methods are used to comprehensively analyze the entire construction process of a cantilever casting concrete arch bridge, it will inevitably lead to a significant decrease in the accuracy of the evaluation results; even the process of formulating construction risk control measures may be affected.

This study focuses on the construction risk assessment of cantilever casting concrete arch bridges. Firstly, the risk sources during the construction process of cantilever casting concrete arch bridges are analyzed, and a method for determining the weight of risk sources is proposed. Then, the principle of using Bayesian networks for risk assessment and the process in its entirety are introduced. Finally, this method is applied to the construction of a bridge. The proposed method can be used to predict the probability and risk level of a potential risk source occurring during the construction process for cantilever casting concrete arch bridges. Through taking scientific and reasonable control measures based on the risk level, the probability of risk occurrence and the losses caused by accidents can be reduced.

2. Risk Source Identification Process

The construction process of cantilever casting concrete arch bridges is complex, with a large time span and many problems that may occur throughout the entire construction process. The risk sources faced in different construction time periods are completely different. Hence, identifying risk sources is the first step in risk analysis for bridge construction. Cantilever casting concrete arch bridges are different from other types of arch bridges, and the construction-process-related risk sources should be fully considered.

Construction teams in China have relatively limited experience with cantilever casting concrete arch bridges, and the equipment systems, such as cable-stayed suspension systems and hanging basket systems, are complex. In addition, construction is generally carried out at high altitudes, with high-quality cantilever casting concrete and high construction technology requirements. These characteristics greatly increase the safety risks during construction. Therefore, the safety risks in the construction of cantilever casting concrete arch bridges are characterized by their uncertainty, objectivity, diversity, complexity, and dynamism.

Based on this, risk source identification should include four steps, as follows:

① Collecting engineering data;

② Decomposing construction processes;

③ Identifying risk sources during the construction process;

④ Determining major risk sources.

Firstly, the construction process is broken down based on the hierarchy of unit engineering, sub-unit works, and item projects, along with the main processes. A breakdown of the construction technology for cantilever casting concrete arch bridges is shown in Table A1. Then, the construction methods, operational procedures, mechanical equipment, and building materials of the evaluation unit are clearly defined. The typical types of accidents that occur in the target units are analyzed, and they are summarized and processed to establish a relevant risk source survey list, which is divided into seven primary risk sources and thirty-three secondary risk sources. The primary risk sources can mainly be divided into safety risks in bridge-approach pile foundation construction, arch seat construction, cable-stayed suspension systems, formwork or temporary supports, construction hanging baskets, main arch closure sections, and prefabricated T-beam construction. The 33 secondary risk sources are detailed risks from the primary risk sources. The full list of risk sources for the construction of cantilever casting concrete arch bridges is shown in Table A2.

3. The Principle of Risk Assessment during the Construction Process

3.1. Principle of Bayesian Network Risk Assessment

The principle of Bayesian network risk assessment [24,25] for probabilistic inference is to build a Bayesian network, for which Bayesian probability is the foundation. Bayesian probability is divided into the prior probability, conditional probability, total probability, and posterior probability, which are mainly converted through the Bayesian equation. The Bayesian equation is shown in Equation (1):

$$P(N_i/M) = \frac{P(M/N_i)P(N_i)}{\sum\limits_{i=1}^{n} P(M/N_i)P(N_i)}, \quad (1)$$

where $P(N_i)$ is the prior probability, $P(N_i/M)$ is the posterior probability, and $P(M/N_i)$ is the conditional probability.

The nodes of a Bayesian network can represent a series of random variables, and directed edges represent causal relationships between these variables. In this paper, a node represents a hazard source as a random variable, and a directed edge represents a relationship between hazard sources [26].

Bayesian networks are needed to construct corresponding models based on the characteristics of each evaluation object when they are applied. The composition conditions of the Bayesian network during the bridge construction process are the related risk sources, and the nodes in the network are regarded as the basic units for evaluating the specific risk events. The correlation between a child node and its parent nodes is based on logical settings between job programs. When setting up this model, it is necessary to determine the node dependency relationship and obtain the prior risk probability in order to provide support for subsequent analysis.

The entire Bayesian network modeling process is as follows:

(1) Establish dependency relationships between nodes in the Bayesian network;

(2) Determine the prior probability of risk events;

(3) Determine the conditional probability between nodes.

There are various styles of risk events, each with its own unique characteristics, so the probability distribution form also has various types. Therefore, the prerequisite for accurate risk probability assessment is to select the appropriate prior probability of risk events. In most cases, relevant statistical data on accidents are lacking, so the prior probability of accidents is generally obtained based on expert experience combined with relevant engineering data analysis.

Similarly, as most scenarios during construction cannot be simulated or replicated, the lack of accident data and the unique nature of the projects determine that expert experience is highly reliable. Based on the expert experience method, the conditional probability between nodes can be obtained; the large amount of knowledge and long-term practical experience provided by the experts can then be used to accurately determine the potential risks of the project.

3.2. Method for Determining the Weight of Risk Sources

After identifying the risk sources, we must determine their importance. There may be multiple major risk sources in a construction project; therefore, they should be prioritized to meet the project's safety requirements. The determination of major risk sources mainly relies on expert judgment, but this method is affected to a certain degree by subjective influence.

Therefore, in this study, the triangular fuzzy number method [27] was used to identify major risk sources in the construction of cantilever casting concrete arch bridges; the respective proportions for different types of risk sources were determined during the evaluation and analysis process. The proportions (weights) for different types of risk sources refer to the node parameters in Bayesian network risk assessment. Fuzzy theory can be used to solve the above problems.

In fuzzy theory, the level at which an element belongs to a fuzzy set can be described by its membership degree, which is often represented by a membership function. There are also certain differences in the results obtained during the analysis process with different membership functions; thus, an appropriate membership function can allow for better solutions to specific problems. The fuzzy set of hazard sources determined for the construction process contains many continuous values, which are generally applied to piecewise linear functions in the parameterization process. Finally, the triangular membership function was chosen to identify major risk sources in the construction of cantilever casting concrete arch bridges. During the evaluation and analysis process, different types of risk sources were evaluated, and their respective weights were determined.

The trigonometric membership function h_x is represented by Equation (2).

$$h_x = \begin{cases} \frac{x-a}{b-a} & x \in [a,b] \\ \frac{x-c}{b-c} & x \in [b,c] \\ 0 & other \end{cases}, \quad (2)$$

where a, b and c are used to represent the left, middle, and right intervals of h_x.

In this study, the safety risk level of cantilever casting concrete arch bridges was divided into seven levels; namely, extremely low (VL), low (L), slightly low (FL), medium (M), slightly high (FH), high (H), and extremely high (VH). The probability range and triangular fuzzy number corresponding to each risk level are shown in Table 1.

Table 1. Risk levels and their corresponding membership functions.

Risk Level	Probability	Triangular Fuzzy Number
		F = (*a*, *b*, *c*)
VL	<1%	F = (0.0, 0.0, 0.1)
L	1~15%	F = (0.0, 0.1, 0.3)
FL	15~35%	F = (0.1, 0.3, 0.5)
M	35~65%	F = (0.3, 0.5, 0.7)
FH	65~85%	F = (0.5, 0.7, 0.9)
H	85~99%	F = (0.7, 0.9, 1.0)
VH	>99%	F = (0.9, 1.0, 1.0)

Firstly, the score S_i was determined as shown in Equation (3).

$$S_i = [\sum_{j=1}^{n} a_{ij}, \sum_{j=1}^{n} b_{ij}, \sum_{j=1}^{n} c_{ij}] \tag{3}$$

The total score of all evaluation objects was then calculated as shown in Equation (4).

$$\sum_{i=1}^{n} S_i = [\sum_{i=1}^{n}\sum_{j=1}^{n} a_{ij}, \sum_{i=1}^{n}\sum_{j=1}^{n} b_{ij}, \sum_{i=1}^{n}\sum_{j=1}^{n} c_{ij}] \tag{4}$$

The fuzzy relative weights based on the obtained scoring results were calculated as shown in Equation (5).

$$w_i = \frac{S_i}{\sum\limits_{i=1}^{n} S_i}[\frac{\sum\limits_{j=1}^{n} a_{ij}}{\sum\limits_{i=1}^{n}\sum\limits_{j=1}^{n} a_{ij}}, \frac{2\sum\limits_{j=1}^{n} b_{ij}}{n(n-1)}, \frac{\sum\limits_{j=1}^{n} c_{ij}}{\sum\limits_{i=1}^{n}\sum\limits_{j=1}^{n} c_{ij}}] \tag{5}$$

Each triangular fuzzy number M was defuzzified as shown in Equation (6).

$$M = M(a,b,c) \rightarrow M = (a + 2b + c)/4 \tag{6}$$

After processing with Equation (6), w_i was normalized to obtain the weight of the evaluation object.

3.3. *Evaluation of Safety Risk Levels*

At present, the methods most commonly used in risk assessment are numerical analysis, the risk matrix method, the risk graph method, and so on. Based on practical experience, the evaluation process integrates the probability of a risk event and the associated losses; this has a higher degree of conformity with the actual risk and can effectively meet the evaluation requirements in this regard. Due to the numerous risk factors involved in the construction of cantilever casting concrete arch bridges and the strong hierarchical nature of the risk source identification results based on the decomposition of the operational procedures, the risk matrix method is applicable. The expression of the risk matrix method is shown in Equation (7):

$$R = P \times L, \tag{7}$$

where R is the degree of risk, P is the probability of risk events occurring, and L is the consequences or losses due to a risk event.

The risk matrix method for the evaluation of cantilever casting concrete arch bridges is divided into the following three steps.

First of all, based on fuzzy theory and Bayesian networks, a risk assessment model for the construction process is established, and the risk probability is calculated based on this model. Second, based on a detailed investigation of this information, the degree of damage caused by different accidents involving bridge hanging baskets is determined, and then the relevant risk and loss levels are determined. Third, the probability P of risk event occurrence is combined with the loss level L of the risk event, and the risk R is determined based on Equation (7). Next, the risk acceptability criteria are set, and their acceptability level is evaluated. Finally, relevant rectification measures are determined.

The specific meaning of the probability of a safety risk in the construction of cantilever casting concrete arch bridges is the probability of the occurrence of a hazard, which can be calculated by combining expert ratings, triangular fuzzy numbers, and Bayesian networks. The results are shown in Table 2.

Table 2. Probability of occurrence of safety risks in the construction of cantilever casting concrete arch bridges.

Frequency Description	Accident Probability
Rare	$0.01 \geq P > 0$
Occasional	$0.3 \geq P > 0.01$
Possible	$0.9 \geq P > 0.3$
Frequent	$1 \geq P > 0.9$

The degree of risk loss in the construction of cantilever casting concrete arch bridges refers to the degree of construction accident losses, mainly considering casualties and economic losses. The results are shown in Table 3.

Table 3. The degree of loss associated with safety risks in the construction of cantilever casting concrete arch bridges.

Description of the Severity of the Accident	Description of Accident Losses	Accident Severity Score
Catastrophic	Significant economic losses, project delays	$L = 100$
Serious	Casualties, certain economic losses, project delays	$99 \geq L \geq 90$
Medium	No casualties, no impact on other indicators	$89 \geq L \geq 30$
Slight	Almost no impact	$29 \geq L > 0$

The acceptability levels and response measures for safety risks are also divided into four levels, as shown in Table 4.

Table 4. Security risk levels and acceptability levels.

Risk Level	Acceptability Level of Risk	Countermeasures
IV	Unacceptable	Immediate shutdown and rectification required
III	Rectification required	Focus on and address risks
II	Acceptable	Pay attention to risk prevention
I	Negligible	Routine management

Finally, the safety risk assessment matrix for the construction of a cantilever casting concrete arch bridge is shown in Table 5. For this table, the level of risk was determined based on the likelihood of risk occurrence and the severity of the event.

Table 5. Risk assessment matrix for construction safety of cantilever casting concrete arch bridges.

Probability	Extent of the Loss			
	Catastrophic	Serious	Medium	Slight
Rare	IV	IV	IV	III
Occasional	IV	IV	III	II
Possible	III	III	II	I
Frequent	II	II	I	I

4. Example of a Bayesian Risk Assessment Project for Cantilever Casting Concrete Arch Construction

4.1. Project Overview

The proposed super-large bridge, Bridge A (see Figure 2), has a total span length of 287 m and a total length of 303 m, for which the aperture arrangement is 3 × 20 m + 187 m + 2 × 20 m. The main bridge is a reinforced concrete box arch bridge with a net span of 180 m, with a net rise-to-span ratio of 1/6 and an arch axis coefficient of 1.99. The construction scheme is cable suspension and cantilever pouring, with the main arch seat constructed in an open-cut form. The upper structure of the bridge approach comprises prestressed concrete T-beams, while the lower structure comprises column piers and pile foundations. The bridge abutment is a gravity U-shaped abutment, with an open-cut expanded foundation.

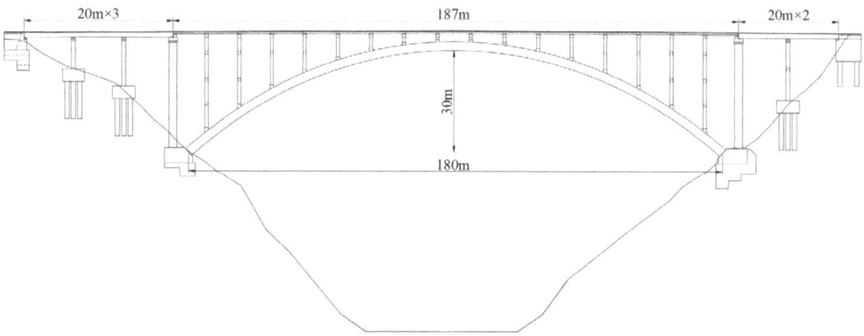

Figure 2. Vertical view of Bridge A.

4.2. The Process of Establishing a Bayesian Risk Source Analysis Model

The risk assessment process for the entire cantilever casting arch bridge construction process is shown in Figure 3.

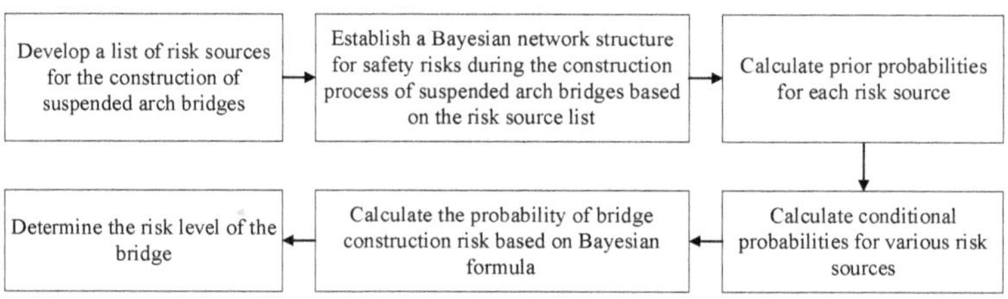

Figure 3. Risk assessment process diagram for cantilever casting concrete arch bridge construction.

The Bayesian network structure diagram for Bridge A relies on the project's structural system and the corresponding characteristics of the cantilever casting concrete arch bridge. Using the risk source identification method proposed in this paper, the corresponding Bayesian network was established, as shown in Figure 4.

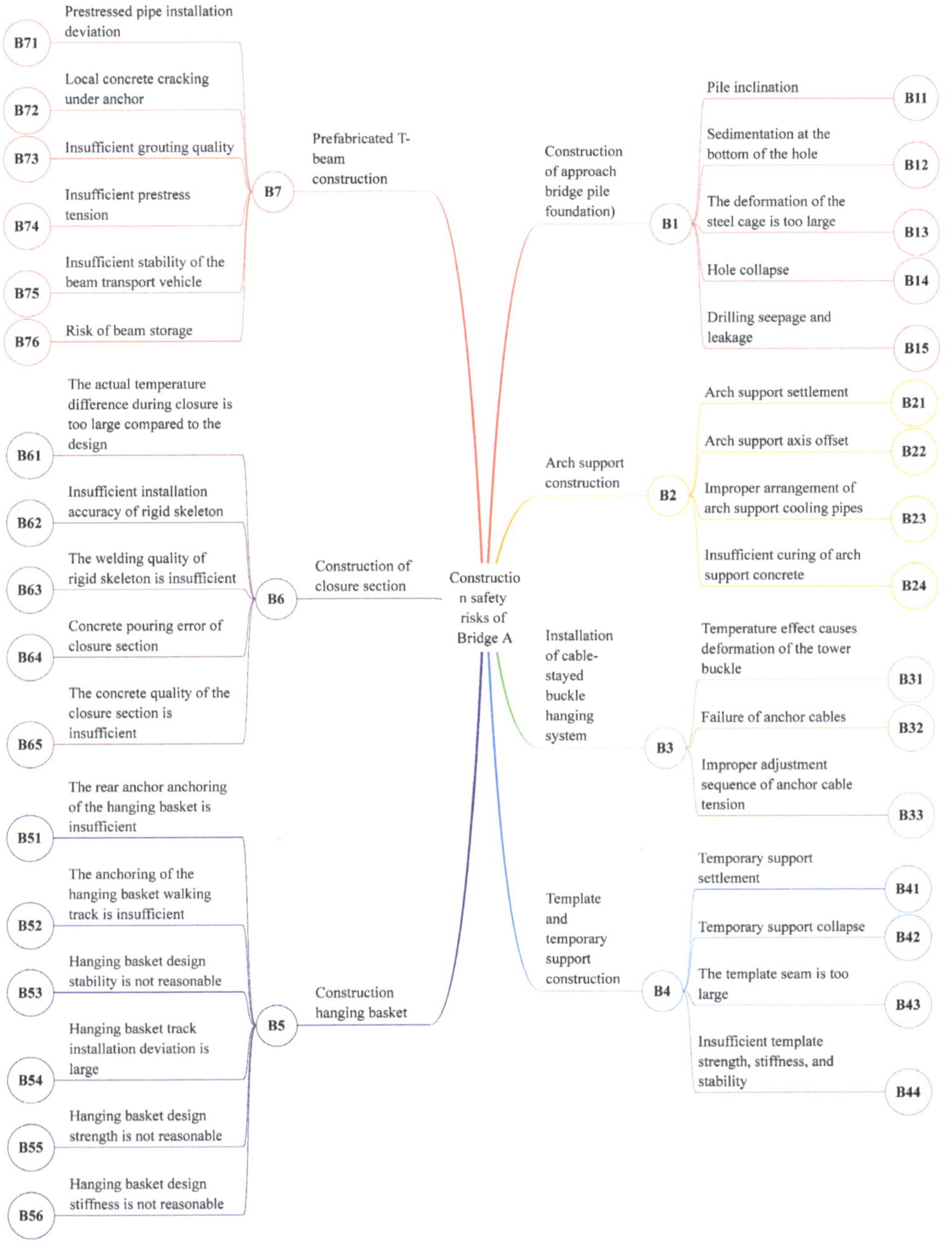

Figure 4. Bayesian network structure diagram for Bridge A.

After a Bayesian network was established, it was necessary to solve for the prior probability and conditional probability values of the risks.

Firstly, the prior probability was calculated.

The hanging basket construction risk is taken as an example from the list of risk sources for cantilever casting concrete arch bridge construction to illustrate the method for determining the prior probability of risk events.

The concept of prior probability is of great significance in the process of building Bayesian networks, and it also directly affects the accuracy of probabilistic risk results. There are various styles of risk events, each with its own unique characteristics; therefore, the probability distribution form also has various types. Therefore, the prerequisite for an accurate risk probability assessment is to select the appropriate prior probability of risk events. In most cases, accident data are lacking, but data can generally be obtained based on expert experience combined with relevant engineering data analysis.

According to Table 1, the values given by experts are triangular fuzzy numbers (*a*, *b*, and *c*), which are difficult to directly use as prior probability values for calculating the risk probability of hanging basket construction in the application process. Therefore, it is necessary to adopt appropriate methods to convert these opinions into exact values, corresponding specifically to solving fuzzy numbers. In this study, the triangular membership function was selected to defuzzify the expert opinions; specifically, the mean area method was adopted. The defuzzification formula in this method can be specifically expressed as $(a + 2b + c)/4$.

During the research process, the evaluation opinions of five experts were collected, taking the prior probability calculation of "inadequate anchoring after hanging the basket" as an example. Experts A–E rated the risk sources of inadequate anchoring after the construction of the hanging basket as "slightly high", "slightly high", "high", "high", and "high". The mean area method was then used to defuzzify these ratings and obtain clear scores. According to the relationship between risk assessments and triangular fuzzy numbers in Table 1, the corresponding scores S_k for experts A–E were 0.7, 0.7, 0.875, 0.875, and 0.875, respectively.

Similarly, the arithmetic mean values of the fuzzy numbers (a_v, b_v, and c_v) provided by the experts were calculated according to Table 1; these were 0.62, 0.82, and 0.96, respectively. The similarity *SL* of the average fuzzy values was applied to calculate the weights for the experts' responses, using the calculation formula shown in Equation (8). The resulting similarity *SL* values were 0.75, 0.75, 0.833, 0.833, and 0.833, respectively. The expert weight coefficients were then calculated using these *SL* values, as shown in Equations (9) and (10).

$$SL(S_k, S_v) = 1 - \frac{(|a_k - a_v| + |b_k - b_v| + |c_k - c_v|)}{\sum_{k=1}^{n} (|a_k - a_v| + |b_k - b_v| + |c_k - c_v|)}, \quad (8)$$

where a_v, b_v, and c_v are the mean values of a_k, b_k, and c_k, respectively. S_k is the expert evaluation value, and S_v is the arithmetic mean of fuzzy numbers evaluated by experts.

$$\omega_k = \frac{SL(S_k, S_v)}{\sum_{k=1}^{n} SL(S_k, S_v)} \quad (9)$$

$$(\omega_A, \omega_B, \omega_C, \omega_D, \omega_E) = (0.187, 0.187, 0.208, 0.208, 0.208) \quad (10)$$

where ω_k is the expert weight coefficient, and k = A, B, C, D, and E.

The prior probability for the risk source in which the rear anchoring of the hanging basket $P_{B51-pior}$ is insufficient is shown in Equation (11).

$$P_{B51-pior} = \sum_{k=1}^{n} \omega_k \times S_k = 0.809 \quad (11)$$

The risk sources of the construction hanging basket were then taken as the research object to determine the conditional probabilities of risk events for each sub-node risk source.

Firstly, a relative attribute judgment matrix for the sub-node risk sources was constructed. Based on the relative magnitudes of the seven levels in Table 1, the sub-node risk sources B51~B56 were evaluated. Based on the obtained results, the corresponding attribute judgment matrix was constructed, as shown in Figure 5.

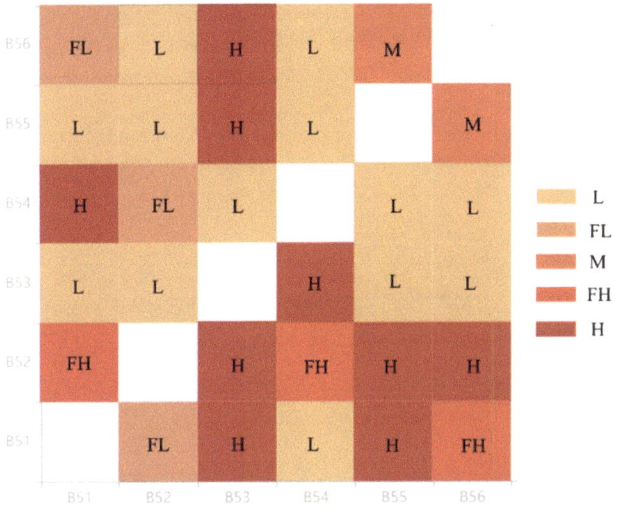

Figure 5. Relative attribute judgment matrix for sub-nodes of risk source B5.

According to Table 1, the seven evaluation levels were converted into the corresponding triangular fuzzy numbers to obtain the corresponding fuzzy measure matrix. Then, the scores for each evaluation object were calculated according to Equations (2)–(4), as shown in Equation (12). The relative fuzzy weight calculation result is shown in Equation (13).

$$\begin{bmatrix} S_{51} \\ S_{52} \\ S_{53} \\ S_{54} \\ S_{55} \\ S_{56} \end{bmatrix} = \begin{bmatrix} \sum_{j=1}^{6} a_{1j}, \sum_{j=1}^{6} b_{1j}, \sum_{j=1}^{6} c_{1j} \\ \sum_{j=1}^{6} a_{2j}, \sum_{j=1}^{6} b_{2j}, \sum_{j=1}^{6} c_{2j} \\ \sum_{j=1}^{6} a_{3j}, \sum_{j=1}^{6} b_{3j}, \sum_{j=1}^{6} c_{3j} \\ \sum_{j=1}^{6} a_{4j}, \sum_{j=1}^{6} b_{4j}, \sum_{j=1}^{6} c_{4j} \\ \sum_{j=1}^{6} a_{5j}, \sum_{j=1}^{6} b_{5j}, \sum_{j=1}^{6} c_{5j} \\ \sum_{j=1}^{6} a_{6j}, \sum_{j=1}^{6} b_{6j}, \sum_{j=1}^{6} c_{6j} \end{bmatrix} = \begin{bmatrix} 1.1, 1.9, 2.9 \\ 0.2, 0.9, 1.9 \\ 2.4, 3.3, 4.1 \\ 1.0, 1.7, 2.7 \\ 1.7, 2.5, 3.3 \\ 1.5, 2.3, 3.2 \end{bmatrix} \tag{12}$$

$$\begin{bmatrix} \omega_{51} \\ \omega_{52} \\ \omega_{53} \\ \omega_{54} \\ \omega_{55} \\ \omega_{56} \end{bmatrix} = \begin{bmatrix} 0.14, 0.15, 0.16 \\ 0.025, 0.07, 0.10 \\ 0.30, 0.26, 0.22 \\ 0.12, 0.13, 0.14 \\ 0.21, 0.19, 0.18 \\ 0.19, 0.18, 0.17 \end{bmatrix} \tag{13}$$

The final risk weights (the conditional probabilities of the Bayesian network nodes) $P_{B5-conditional}$ were then obtained via defuzzification using the mean area method, as shown in Equation (14).

$$P_{B5-conditional} = (\overline{\omega_{B51}}, \overline{\omega_{B52}}, \overline{\omega_{B53}}, \overline{\omega_{B54}}, \overline{\omega_{B55}}, \overline{\omega_{B56}}) = (0.15, 0.068, 0.26, 0.136, 0.198, 0.18) \quad (14)$$

The risk probability for the construction hanging basket risk source P_{B5-all} was obtained by combining the prior probability $P_{B5-pior}$ and the conditional probability $P_{B5-conditional}$ with the Bayesian total probability formula.

$$P_{B5-all} = P_{B5-pior} \cdot P_{B5-conditional} = 0.313 \quad (15)$$

For the other six primary risk sources, the same method was used to analyze their impact on the overall construction risk of Bridge A. The relative attribute judgment matrix for the primary risk sources was determined through expert scoring, as shown in Table 6.

Table 6. Expert evaluation matrix for primary risk sources.

	B1	**B2**	**B3**	**B4**	**B5**	**B6**	**B7**
B1	—	L	L	FL	FL	FL	H
B2	H	—	M	FH	FL	FH	H
B3	H	M	—	H	M	FH	FH
B4	FL	FL	FL	—	FL	FL	M
B5	FH	FH	M	H	—	FH	FH
B6	FH	FL	FL	FH	FL	—	FH
B7	L	L	FL	M	FL	FL	—

After the score calculation and defuzzification, the weights of the primary risk sources were obtained, as shown in Equation (16). According to Equation (16), the importance of the seven primary risk sources during the construction process of Bridge A is in the order B5 > B3 > B6 > B2 > B4 > B1 > B7. The weight coefficients are the values of conditional probability $P_{A-conditional}$ in the Bayesian network.

$$P_{A-conditional} = (\overline{\omega_{B1}}, \overline{\omega_{B2}}, \overline{\omega_{B3}}, \overline{\omega_{B4}}, \overline{\omega_{B5}}, \overline{\omega_{B6}}, \overline{\omega_{B7}}) = (0.09, 0.15, 0.21, 0.10, 0.23, 0.17, 0.05) \quad (16)$$

The calculation process for the conditional probabilities of the other secondary risk sources is similar to that for the seven primary risk source conditional probabilities mentioned above. The conditional probabilities of secondary risk sources during the construction process of this bridge were thereby obtained, as shown in Figure 6.

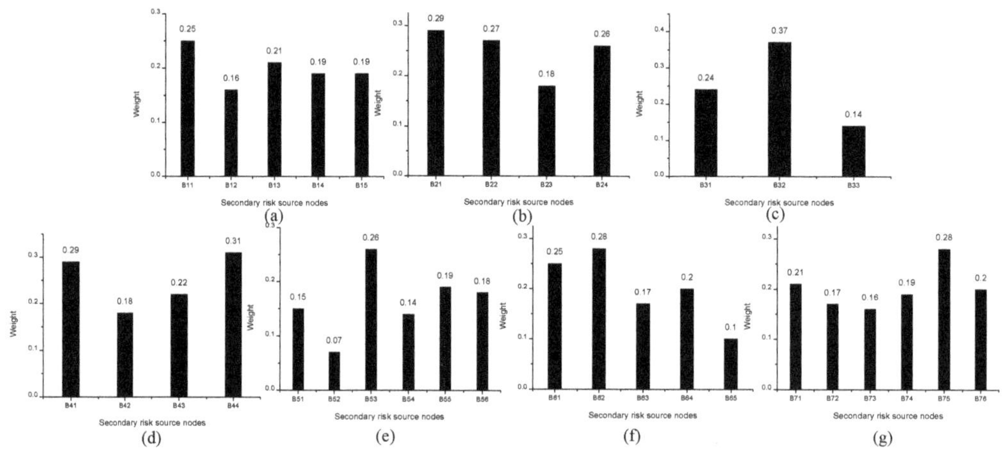

Figure 6. The conditional probabilities for the secondary risk sources corresponding to each primary risk source. (**a**–**g**) represent the primary risk source numbers B1–B7, respectively.

Combining the conditional probabilities and prior probabilities for the primary risk sources, the construction risk for Bridge A P_A was ultimately obtained.

$$P_A = P_{A-pior} \cdot P_{A-conditional} = \sum_{1}^{7} P_{Bi-pior} \cdot P_{Bi-conditional} = 0.191 \quad (17)$$

A construction risk assessment of Bridge A was carried out using the risk loss assessment model established in Section 3.3. The risk level assessment results for the primary risk sources for Bridge A are shown in Table 7.

Table 7. Assessment results of primary risk sources for Bridge A.

Primary Risk Sources	Probability	Frequency	Loss Level	Risk Level
B1	0.0943	Occasional	Serious	III
B2	0.1452	Occasional	Serious	III
B3	0.2280	Occasional	Serious	III
B4	0.0548	Occasional	Serious	III
B5	0.313	Possible	Serious	IV
B6	0.1971	Occasional	Serious	III
B7	0.0539	Occasional	Serious	III

From Table 7, it can be seen that the risk frequency level for Bridge A is largely "occasional", the level of loss is "serious", and the corresponding risk level is largely level III. The evaluation results indicate that there are significant risks during the construction process. Strict control of risks in the project should be implemented, and reasonable and effective risk reduction measures should be formulated.

5. Conclusions

In this study, a construction risk assessment method for cantilever casting concrete arch bridges based on triangular fuzzy theory and Bayesian networks was proposed. The method was validated using an actual engineering construction, Bridge A, as an example. The main conclusions are as follows:

(1) Through a study of the construction process of cantilever casting concrete arch bridges, the main process of risk assessment in the construction process was summarized through collecting engineering data, decomposing the construction processes, identifying the risk sources during the construction processes, and determining the major risk sources. The construction procedure of cantilever casting concrete arch bridges was broken down, and a list of risk sources was established. The list comprises seven primary risk sources and thirty-three secondary risk sources.

(2) A risk assessment model for cantilever casting concrete arch bridge construction based on a Bayesian network was constructed, and the prior probability and conditional probability in the Bayesian network were calculated using triangular fuzzy numbers, laying the foundation for other calculations of construction risk probabilities through Bayesian networks in the future. The method of combining Bayesian networks and triangular fuzzy numbers not only can fully utilize the experience of experts, but also avoids the subjectivity of expert opinions. The established evaluation system has a clear hierarchy and clear results.

(3) Taking Bridge A as an example, the risk analysis process, calculation model, and reliability evaluation of cantilever casting concrete arch bridges were carried out, and the practicality of the proposed method of determining key node parameters of Bayesian networks using triangular fuzzy numbers was verified. The risk analysis showed that the risk frequency level and the level of loss of Bridge A can be described as "occasional" and "severe", respectively, and the corresponding risk level is level III, indicating the existence of significant risks during the construction process. Therefore, strict risk control for Bridge

A should be implemented during the construction process, and reasonable and effective risk reduction measures should be formulated.

Author Contributions: Writing—original draft, Z.H.; Writing—review and editing, Y.X., L.L. and S.W.; Methodology, M.W.; Data curation, B.L.; Funding acquisition, Z.H. and M.W. All authors have read and agreed to the published version of the manuscript.

Funding: This work was supported by the Natural Science Foundation of Hunan Province, China (grant numbers 2024JJ7170 and 2024JJ7077) and the Research Foundation of Education Bureau of Hunan Province, China (grant numbers 23B0732, 22A0561 and 23A0559).

Data Availability Statement: All data generated or analyzed during this study are included in this published article.

Conflicts of Interest: The authors declare no conflicts of interest.

Appendix A

Table A1. The breakdown table of construction operations for cantilever casting concrete arch bridges.

Project	Unit Construction Project	Part Project	Item Project
Main bridge	Substructure	Arch support	Excavation of foundation pit Reinforcement binding Cooling pipe layout Concreting Concrete curing
	Superstructure	Main arch	Installation of diagonal pull buckle system Construction of block 1 at the arch foot position Installation and walking of construction hanging baskets Reinforcement binding Concreting Installation of rigid skeleton Main arch closure
		Spandrel column	Reinforcement binding Concreting
		Main beam	Prefabrication of main beam Installation of main beam
		Bridge floor system and appurtenance	Installation of expansion joint Installation of supports Installation of concrete anti-collision guardrails
Approach bridge	Substructure	Foundation	Bored holes for cast-in-place piles Installation of steel reinforcement cage for cast-in-place piles Concrete pouring for cast-in-place piles Excavation of bridge abutment foundation Reinforcement binding of the bridge abutment Concreting of the bridge abutment
		Pier	Reinforcement binding Concreting
	Superstructure	Main beam of approach bridge	Prefabrication of main beam of approach bridges Installation of main beam of approach bridges
		Bridge floor system and appurtenance	Installation of expansion joint Installation of supports Installation of concrete anti-collision guardrails

Table A2. List of risk sources for cantilever casting concrete arch bridge construction.

Primary Risk Sources	Secondary Risk Sources
Construction of approach bridge pile foundation	Pile inclination Sedimentation at the bottom of the hole The deformation of the steel cage is too large Hole collapse Drilling seepage and leakage
Arch support construction	Arch support settlement Arch support axis offset Improper arrangement of arch support cooling pipes Insufficient curing of arch support concrete
Installation of cable-stayed buckle hanging system	Temperature effect causes deformation of the tower buckle Failure of anchor cables Improper adjustment sequence of anchor cable tension
Template and temporary support construction	Temporary support settlement Temporary support collapse The template seam is too large Insufficient template strength, stiffness, and stability
Construction of hanging basket	The rear anchor anchoring of the hanging basket is insufficient The anchoring of the hanging basket walking track is insufficient Hanging basket design stability is not reasonable) B54 (Hanging basket track installation deviation is large) Hanging basket design strength is not reasonable Hanging basket design stiffness is not reasonable
Construction of closure section	The actual temperature difference during closure is too large compared to the design Insufficient installation accuracy of rigid skeleton The welding quality of rigid skeleton is insufficient Concrete pouring error of closure section The concrete quality of the closure section is insufficient
Prefabricated T-beam construction	Prestressed pipe installation deviation Local concrete cracking under anchor Insufficient grouting quality Insufficient prestress tension Insufficient stability of the beam transport vehicle Risk of beam storage

References

1. Hu, F.; Huang, P.; Dong, F.; Blanchet, A. Stability safety assessment of long-span continuous girder bridges in cantilever construction. *J. Intell. Fuzzy Syst.* **2018**, *35*, 4027–4035. [CrossRef]
2. Liu, Z.; Zhou, J.; Wu, Y.; You, X.; Qu, Y. Linear control method for arch ring of oblique-stayed buckle cantilever pouring reinforced concrete arch bridge. *Adv. Civ. Eng.* **2021**, *2021*, 6633717.1–6633717.14. [CrossRef]
3. Fang, R.G. Discussion on the measurements in cantilever construction with guyed hanging basket. *Appl. Mech. Mater.* **2014**, *3307*, 2797–2800.
4. Wu, Y.; Wang, X.; Fan, Y.; Shi, J.; Luo, C.; Wang, X. A study on the ultimate span of a concrete-filled steel tube arch bridge. *Buildings* **2024**, *14*, 896. [CrossRef]
5. Wu, Y.; Zhang, B.; Zhou, J.; Yan, R. Analysis of Reasonable Construction State of the Long-span Concrete Cable-stayed Bridge Based on Stress-free Forward Iteration Method. *Highw. Eng.* **2020**, *45*, 15–20, 72.
6. Bashar, T.; Wassel, A.B.; Teruhisa, M. Inspection and risk assessment of mechanically stabilized earth walls supporting bridge abutments. *J. Perform. Constr. Fac.* **2018**, *32*, 04017131.
7. Payam, T.; Denis, M. Seismic risk assessment of four-span bridges in montreal designed using the Canadian bridge design code. *J. Bridge Eng.* **2014**, *19*, A4014002.1–A4014002.9.
8. Mehdi, D.N.; Reza, R.; Ahmad, D. A Network-Based Importance Measurement Index for Bridge Security Risk Assessment and Prioritisation. *Balt. J. Road Bridge Eng.* **2022**, *17*, 1–30.
9. Messore, M.M.; Capacci, L.; Biondini, F. Life-cycle cost-based risk assessment of aging bridge networks. *Struct. Infrastruct. Eng.* **2021**, *17*, 515–533. [CrossRef]

10. Pouraminian, M.; Ghaemian, M. Shape optimisation of concrete open spandrel arch bridges. *Gradevinar* **2015**, *67*, 1177–1185.
11. Khan, R.A.; Datta, T.K.; Ahmad, S. Seismic risk analysis of modified fan type cable stayed bridges. *Eng. Struct.* **2006**, *28*, 1275–1285. [CrossRef]
12. Vazul, B.; Balthasar, N. Risk Analysis of Bridges for Hazard Scenarios. *Neurochem. Neuropharmacol.* **2012**, *1*, 73–101.
13. Ahn, S.; Kim, T.; Kim, J.M. Sustainable risk assessment through the analysis of financial losses from third-party damage in bridge construction. *Sustainability* **2020**, *12*, 3435. [CrossRef]
14. Taejun, C.; Jeong-Bae, L.; Seong-Soo, K. Probabilistic risk assessment for the construction phases of a PSC box girder railway bridge system with six sigma methodology. *KSCE J. Civ. Eng.* **2011**, *15*, 119–130.
15. Bao, L.; Wang, W.; Liu, K. Multi-objective risk analysis and documents assessment of bridge construction based on AHP-GRAY. *Adv. Sci. Lett.* **2011**, *4*, 2543–2546. [CrossRef]
16. Wang, Y.; Elhag, T.M. Fuzzy TOPSIS method based on alpha level sets with an application to bridge risk assessment. *Expert Syst. Appl.* **2006**, *31*, 309–319. [CrossRef]
17. Sun, P.; Hou, X.; Zheng, W. Risk assessment for bridge structures against blast hazard via a fuzzy-based framework. *Eng. Struct.* **2021**, *232*, 111874.1–111874.14. [CrossRef]
18. Andres, A.; Ricardo, M.; Gerard, J. Exposure knowledge impact on regional seismic risk assessment of bridge portfolios. *Bull. Earthq. Eng.* **2022**, *20*, 7137–7159.
19. Jelena, A.; Dagang, L. Risk assessment of bridges using fuzzy logic controller. In Proceedings of the IX International Conference on Structural Dynamics (EURODYN 2014), Porto, Portugal, 30 June–2 July 2014; pp. 3073–3078.
20. Zhong, J.; Jeon, J.; Ren, W. Risk assessment for a long-span cable-stayed bridge subjected to multiple support excitations. *Eng. Struct.* **2018**, *176*, 220–230. [CrossRef]
21. Choi, H.; Mahadevan, S. Construction project risk assessment using existing database and project-specific information. *J. Constr. Eng. Manag.* **2008**, *134*, 894–903. [CrossRef]
22. Bao, L.; Wang, J.; Yu, L.; Shi, Y. Long-span continuous rigid frame bridge beam crack closure construction risk analysis and preventive measures. *J. Shenyang Jianzhu Univ. Nat. Sci.* **2012**, *28*, 94–99.
23. Huang, Z.; Shi, Z.; Zhang, Z.; Yuan, C. Probability analysis of risk during the construction process of high pier and long span continuous rigid frame bridges. *J. China Foreign Highw.* **2013**, *33*, 198–201.
24. Tammeen, S.; Solomon, T.; Leonardo, D. Seismic risk assessment of high-voltage transformers using Bayesian belief networks. *Struct. Infrastruct. Eng.* **2015**, *11*, 929–943.
25. Lin, P.; Xiong, Y.; Xu, Z.; Wang, W.; Shao, R. Risk assessment of TBM jamming based on Bayesian networks. *Bull. Eng. Geol. Environ.* **2022**, *81*, 47.1–47.15. [CrossRef]
26. An, L.; Zhang, X. Nonlinear fuzzy evaluation of construction risk of integral hoisting CFST tied arch bridge. *J. Highw. Transp. Res. Dev.* **2015**, *32*, 91–96. [CrossRef]
27. Chen, J.; Guo, Y.; Qu, C.; Yang, B. Risk assessment of heavy metal pollution to soils in chinese provincial capital cities based on triangular fuzzy numbers. *J. Xi'an Shiyou Univ. (Nat. Sci.)* **2023**, *38*, 127–135, 142.

buildings

MDPI

Article

Hybrid Intelligent Model for Estimating the Cost of Huizhou Replica Traditional Vernacular Dwellings

Jian Huang [1,2,*], Wei Huang [1], Wei Quan [1] and Yandong Xing [1]

[1] School of Architecture and Civil Engineering, Huangshan University, Huangshan 245041, China; 104037@hsu.edu.cn (W.H.); 114101@hsu.edu.cn (W.Q.); xingyandong@cumt.edu.cn (Y.X.)
[2] Anhui Institute for the Preservation and Inheritance of Hui-Style Architecture, Huangshan University, Huangshan 245041, China
* Correspondence: 114003@hsu.edu.cn; Tel.: +86-187-5595-3583

Abstract: Amidst the backdrop of rural revitalization and cultural renaissance, there is a surge in the construction demand for replica traditional vernacular dwellings. Traditional cost estimation methods struggle to meet the need for rapid and precise estimation due to the complexity inherent in their construction. To address this challenge, this study aims to enhance the accuracy and efficiency of cost estimation by innovatively developing an Adaptive Self-Explanatory Convolutional Neural Network (ASCNN) model, tailored to meet the specific cost estimation needs of replica traditional vernacular dwellings in the Huizhou region. The ASCNN model employs a Random Forest model to filter key features, inputs these into the CNN for cost estimation, and utilizes Particle Swarm Optimization (PSO) to optimize parameters, thereby improving predictive accuracy. The decision-making process of the model is thoroughly interpreted through SHAP value analysis, ensuring credibility and transparency. During the construction of the ASCNN model, this study collected and analyzed bidding control price data from 98 replica traditional vernacular dwellings. The empirical results demonstrate that the ASCNN model exhibits outstanding predictive performance on the test set, with a Root Mean Square Error (RMSE) of 9828.06 yuan, a Mean Absolute Percentage Error (MAPE) of 0.6%, and a Coefficient of Determination (R^2) as high as 0.989, confirming the model's high predictive accuracy and strong generalization capability. Through SHAP value analysis, this study further identifies key factors such as floor plan layout, roof area, and column material coefficient that are central to cost prediction. The ASCNN model proposed in this study not only significantly improves the accuracy of cost estimation for Huizhou replica traditional vernacular dwellings, but also enhances its transparency and credibility through model interpretation methods, providing a reliable basis for related investment decisions. The findings of this study also offer valuable references and insights for rapid and precise cost estimation of replica buildings in other regions worldwide.

Keywords: replica traditional vernacular dwellings; cost estimation; convolutional neural network; particle swarm optimization; SHAP value analysis

1. Introduction and Literature Review

Replicas of traditional vernacular dwellings, constructed with either traditional or modern materials that embody cultural characteristics, differ significantly from the restoration of historical buildings [1]. These replicas are not only essential carriers of "cultural confidence" for the Chinese people but also play a crucial role in promoting "rural revitalization". In recent years, driven by the dual benefits of cultural and rural revitalization policies, investments in constructing replicas of ancient architecture and gardens have surged across China.

The cost estimation of replica antique buildings is particularly challenging due to their complex designs, their intricate construction processes, and the difficulties in measurement and pricing. These factors often lead to the "three excesses" phenomenon, where the preliminary budget exceeds the estimate, the detailed budget exceeds the preliminary budget, and the final accounting exceeds the detailed budget [2]. Given that 75% to 90% of the costs in ancient replica construction projects are determined during the decision-making and design phases [3], rapid and accurate cost estimation is crucial to prevent the "three excesses" phenomenon. However, traditional cost estimation methods, such as the unit index method, expert experience method, and budget quota method, face issues like lagging index updates, insufficient experience, and low efficiency, which struggle to adapt to the complexity of cost composition in ancient replica construction [4]. These complexities are influenced by various factors such as architectural style, material selection, construction techniques, and structural types, which may involve nonlinear relationships with costs, leading to insufficient accuracy of traditional methods [5]. Therefore, there is an urgent need to develop new computational methods to improve the accuracy and efficiency of cost estimation in ancient replica construction projects to meet their specific requirements.

Artificial intelligence (AI) is gradually becoming a research hotspot in the field of engineering costs and has become a powerful tool for solving complex problems [6]. Compared with traditional methods, machine learning has a clear advantage in handling large amounts of data and discovering nonlinear relationships among data. In engineering cost estimation, AI methods are used to analyze historical cases, identify complex relationships between cost-influencing factors, and provide more accurate and reliable predictive results [7]. These methods include Backpropagation Neural Networks (BPNNs) [8–10], Support Vector Machines (SVMs) [11,12], Decision Trees (DTs) [13], Case-Based Reasoning (CBR) [14–16], and Ensemble Algorithms [17]. Studies have shown that these methods have been effectively verified in cost prediction applications for roads, tunnels, bridges, residential buildings, and public buildings, but there is little research in the field of ancient replica construction.

In the field of construction cost estimation, Deep Artificial Neural Networks (DNNs) have attracted attention for their ability to capture complex relationships among project variables, showing greater potential and value than traditional learning methods [18]. Even with limited data, DNNs can achieve highly accurate engineering valuation predictions [19–21]. For instance, Li et al. [22] introduced a construction cost estimation approach based on DNNs, which incorporates engineering characteristics and bill of quantities as inputs, while predicting the total bid price and associated taxes as outputs. This method demonstrated the DNNs model's potential to significantly improve the accuracy of construction cost forecasts, achieving a relative error of just 4.203% in predicting the total price, with relative errors for the composite unit prices V1 and V2 being 2.98% and 4.52%, respectively.

Convolutional Neural Networks (CNNs), proposed by Yann LeCun and others as a type of deep neural network [23], have achieved significant success in the field of image recognition and processing. CNNs' ability to automatically learn and extract features from raw data is particularly important in the data-intensive and diverse field of construction engineering. Xue et al. [24] utilized a CNN algorithm to perform an in-depth analysis of cost prediction for expressway projects during the conceptual design phase. The study's findings demonstrate the superior applicability and accuracy of the CNN model in handling the high-dimensional nonlinear complexities of expressway cost estimation, outperforming conventional models. Yi et al. [25] applied the Maximum Information Coefficient (MIC) to filter key indicators, combining it with CNNs to develop a cost prediction model tailored for civil engineering projects involving mountainous high-speed railways. By optimizing model parameters, the MIC-CNN model not only achieved a low average relative error of 5.476% but also exhibited a minimal prediction fluctuation of just 1.045%, outperforming the traditional CNN, BPNN, and Adaboost-SVR models in both accuracy and stability. To address the challenges of high-dimensional feature data processing and nonlinear relationships, researchers often integrate multiple predictive models within machine learning frameworks to enhance forecasting accuracy. For instance, Han et al. [26] developed an advanced construction cost estimation model, the NGO-CNN-SVM, specifically for high-standard farmland projects. This model integrates CNN with Support Vector Machines

(SVMs) and is optimized using the Northern Goshawk Optimization (NGO) algorithm. It demonstrated exceptional predictive accuracy on 120 construction cost datasets, particularly in bridge and culvert projects, with an R^2 exceeding 0.970 and a relative error below 3.548%. The NGO-CNN-SVM model outperformed traditional neural networks and other hybrid models, underscoring the effectiveness of deep learning in enhancing the precision and reliability of construction cost predictions. These studies collectively suggest that the application of deep learning technologies significantly enhances the accuracy and reliability of construction cost forecasting, providing an efficient and scientifically robust tool for construction economic analysis. Nevertheless, research on the application of CNN in cost prediction in the field of ancient replica engineering is still scarce, indicating that this area needs further exploration and development.

In AI models, feature selection is crucial for ensuring model stability and reducing computational costs [27]. In response to the complexity of factors affecting the cost of ancient replica construction, this study uses the Random Forest (RF) algorithm for preliminary feature screening to improve model efficiency [28]. The performance of CNNs in visual recognition and data processing tasks depends on the selection of hyperparameters [29]. To optimize these parameters, this study introduces the Particle Swarm Optimization (PSO) algorithm, a heuristic algorithm effective in searching for optimal solutions in high-dimensional spaces [30]. PSO, which simulates social behavior, avoids getting trapped in local minima without the need for gradient information, significantly enhancing model performance [31].

The "black box" nature of deep learning models can lead to unpredictable results and biases [32], especially when applied to assist in decision-making in the construction engineering field. Therefore, ensuring the fairness and transparency of the model is crucial for effective and responsible application [33]. SHAP (Shapley Additive exPlanations) provides a way to evaluate the marginal contribution of predictors by calculating SHAP values to measure the contribution of features, offering global and local explanations of the model [34,35]. In this study, SHAP value analysis not only enhances the credibility of the model but also provides clear decision support for stakeholders, verifying the model's ability to capture the association between covariates and cost.

This study proposes an Adaptive Self-Explanatory Convolutional Neural Network (ASCNN) model that integrates PSO, CNN, and SHAP interpretive analysis, taking replica traditional vernacular dwellings in the Huizhou region of China as a case study. The ASCNN model uses a CNN to establish a cost prediction model, a PSO to optimize network structural parameters, and SHAP to provide interpretability of model decisions. The structure of this paper is as follows: first, it introduces the construction process and working principle of the ASCNN model in detail. Then, it verifies and analyzes the ASCNN model through empirical research. Finally, it presents conclusions and discusses the potential application value and advantages of the ASCNN model in the cost estimation of replica traditional vernacular dwellings, providing references and suggestions for research and practice in related fields.

2. Methodology

2.1. Convolutional Neural Network (CNN) Algorithm

CNNs utilize convolutional layers, activation layers, pooling layers, and fully connected layers to extract spatial features from input data and perform classification or regression tasks [36]. In this paper, the CNN executes feature extraction and task execution through the following steps:

(1) Convolutional Layer: Uses convolution kernels to extract local features.
(2) Activation Layer: Introduces nonlinearity, typically employing the ReLU function to enhance the network's learning capability.
(3) Pooling Layer: Reduces the dimensionality of features and strengthens their invariance, with max pooling adopted in this paper (Equation (1)).

(4) Fully Connected Layer: Maps the extracted features to the final output (Equation (2)). The architecture and process of the CNN used in this paper are detailed in Figure 1.

$$maxpooling_{(i,j)}^{(k,l)} = max_{u,v}(max(0, (\sum_{c=1}^{C}\sum_{u=1}^{h}\sum_{v=1}^{w} \omega_{(c,u,v)}^{(k,l)} \cdot x_{(c,i\cdot s+u-1,j\cdot s+v-1)}^{(l-1)} + b^{(k,l)}))) \quad (1)$$

$$y = \varnothing(W \cdot X + b) \quad (2)$$

where x is the input data, i and j are the vertical and horizontal coordinates of the input features, k is the index of the convolutional kernel, l is the index of the convolutional group layer, C represents the number of channels, h and w represent the height and width of the input sample, ω is the weight of the convolutional kernel, u and v represent the height and width of the convolutional kernel (respectively), s denotes the pooling stride, which serves as the input to the subsequent convolutional layer, X is the one-dimensional input feature that has been flattened from the higher-dimensional feature maps, W is the weight matrix of the Fully Connected Layer, b is the bias term, and $\varnothing$ is the activation function of the output layer, which is still set as the ReLU function.

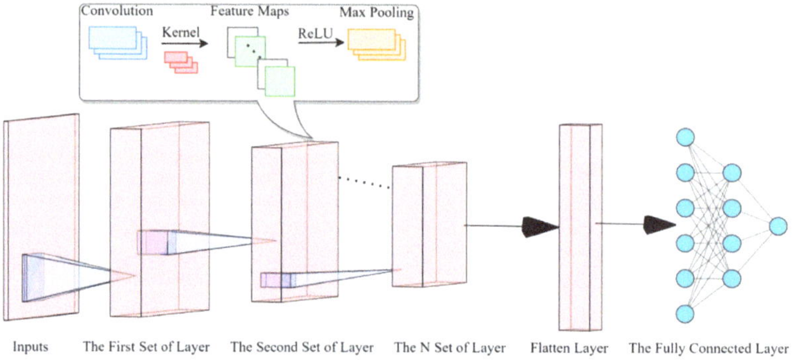

Figure 1. CNN model structure diagram.

The parameter optimization of the CNN in this paper integrates the BPNN with the Adaptive Moment Estimation (Adam) algorithm for learning rate optimization. Adam dynamically adjusts individual parameter learning rates based on first and second moment estimates, facilitating faster and more stable convergence in noisy environments or with sparse gradients [37]. For regression tasks with numerical targets, the Mean Squared Error (MSE) is chosen as the loss function (Equation (3)) to accurately measure the deviation between predicted and actual values.

$$\text{MSE} = \frac{1}{n}\sum_{i=1}^{n}(y_i - \hat{y}_i)^2 \quad (3)$$

where n is the number of samples, y_i is the actual value, and $\hat{y}_i$ is the predicted value.

2.2. Particle Swarm Optimization (PSO)

PSO is a swarm intelligence optimization method that simulates the foraging behavior of bird flocks [38]. The algorithm progressively approximates the optimal solution through collaboration and information sharing among particles within the search space. The specific process is illustrated in Figure 2, as follows:

(1) Initialize the particle swarm: Randomly generate the initial positions and velocities of particles, recording the current position of each particle and the best position in the particle swarm as the global optimum;

(2) Calculate the fitness value: Assess the fitness of each particle using the CNN prediction accuracy.

(3) Update the personal best position (*pbest*): If the current particle's fitness is better than its historical best, update the personal best position.

(4) Update the global best position (*gbest*): If the current particle's fitness is superior to the global optimum, update the global best position.

(5) Update particle velocity and position: Adjust the particle's state based on personal best and global best through velocity update and position update formulas (Equations (4) and (5)), where the inertia weight, individual learning factor, and social learning factor collectively influence the new position and velocity of the particles.

Particle position update formula:

$$x_i(t+1) = x_i(t) + v_i(t+1) \tag{4}$$

Particle velocity update formula:

$$v_i = \omega \cdot v_i(t) + c_1 \cdot r_1 \cdot (pbest_i - x_i(t)) + c_2 \cdot r_2 \cdot (gbest_i - x_i(t)) \tag{5}$$

where ω is the inertia weight, controlling the size of the particle's motion inertia, c_1 and c_2 are the individual learning factor and the social learning factor, respectively, controlling the influence of individual and group information on the particle's movement, r_1 and r_2 are random numbers, used to add randomness, $v_i(t)$ is particle velocity, and $x_i(t)$ is particle position.

(6) Assess if the stopping criteria have been met: The algorithm halts and outputs the optimal solution if the change in fitness value is below the preset threshold; otherwise, it returns to step 2 to continue the iteration process.

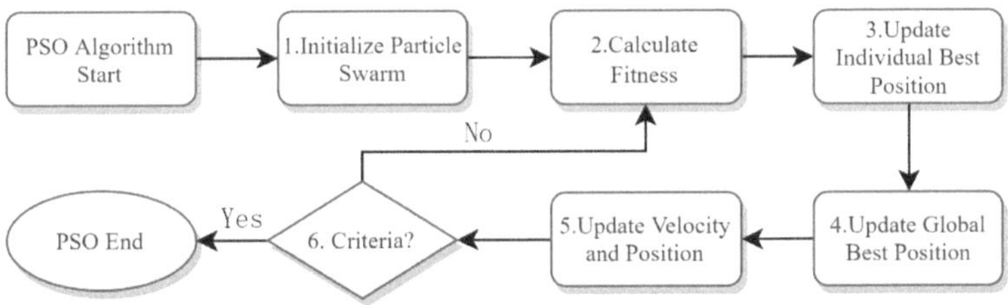

Figure 2. PSO algorithm flowchart.

2.3. SHAP Algorithm

The SHapley Additive exPlanations (SHAP) algorithm provides a method for interpreting model predictions by decomposing them into the individual and interaction contributions of features [35]. The specific computational steps are as follows: (1) Define feature subsets: considering all possible combinations of feature subsets S; (2) Calculate marginal contributions: for each subset S, compute the prediction difference before and after the feature is added to S (Equation (6)); (3) Weighted average of marginal contributions: calculate the weighted average of all marginal contributions based on the size of the subset and the total number of features (Equation (7)); (4) Sum Shapley values: sum the Shapley values of all features to obtain the contribution of each feature to the prediction.

$$M = f\left(x_{s\cup\{j\}}\right) - f(x_s) \tag{6}$$

$$SHAP_j = \sum_{\substack{S \subseteq P \\ j \notin S}} \frac{|S|}{|P|} M \tag{7}$$

where P is the set of all features, S is a subset of P excluding j, M is the marginal contribution of feature j to the prediction, $f\left(x_{s\cup\{j\}}\right)$ is the prediction with feature j included in set S,

$f(x_S)$ is the prediction with feature j excluded from set S, $|S|$ is the size of set S, and $|P|$ is the size of set P.

2.4. ASCNN for Huizhou Ancient Replica Architecture Cost Estimation

Selecting and adjusting network architecture parameters (such as the number of layers, neurons, convolutional kernels, and their sizes) and training parameters (such as learning rate, batch size, number of iterations, and regularization coefficients) are crucial for the regression accuracy and generalization capability of the CNN model [39]. This study integrates the PSO algorithm with a CNN and introduces the SHAP algorithm to enhance the model's interpretability. The PSO algorithm automatically searches for the optimal parameter combination, while SHAP values provide precise explanations for model predictions, enhancing the transparency of the model's decision-making process. The flowchart of the ASCNN algorithm is shown in Figure 3, and the steps are as follows:

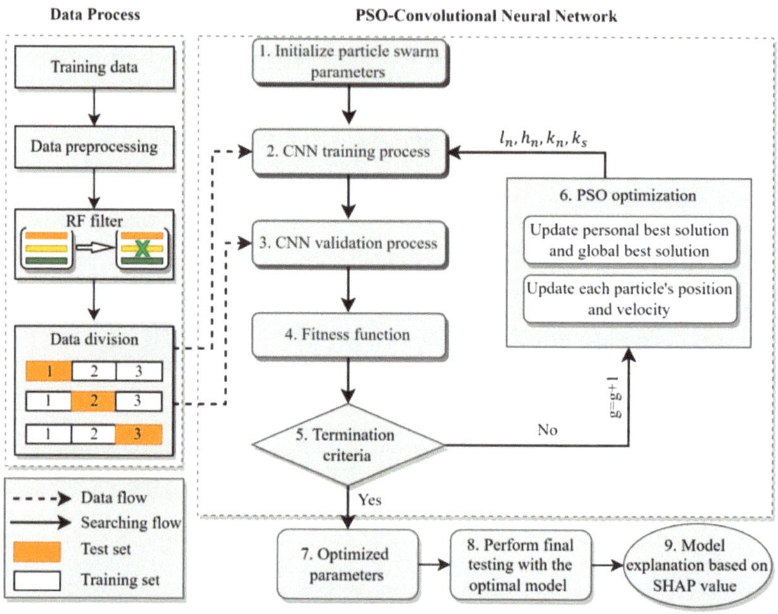

Figure 3. ASCNN structural diagram.

(1) Data Processing: The training dataset was first normalized to confine its value range to [0, 1]. After normalization, the RF algorithm was applied for feature selection, identifying and excluding factors with negligible impact on cost estimation for Huizhou replica traditional vernacular dwellings. The remaining significant factors were then used as input variables to construct the ASCNN predictive model. To meet the minimum sample size requirement for neural networks, the filtered data were randomly divided into three independent subsets, each containing at least 14 samples, ensuring adequate data for model training. During model construction, each subset was sequentially used as the validation set, while the remaining two subsets were combined to form the training set, facilitating effective training and optimization of the ASCNN predictive model.

(2) Parameter Initialization: To balance optimization effects and computational costs, this study selects five key parameters for optimization: learning rate (λ), the number of fully connected hidden layers (l_n) and their neuron counts (h_n), the number of convolutional kernels (k_n), and the size of the convolutional kernels (k_s). Considering that the learning rate is effectively adjusted in the Adam optimizer, this paper focuses on optimizing the last four parameters. Each particle's position and velocity in the particle swarm are randomly

initialized, with each particle's position representing a set of CNN model parameters (l_n, h_n, k_n, k_s). Based on experience, it is recommended to set upper and lower limits for each parameter, with the following ranges for each parameter: $l_n \in [1, 3]$, $h_n \in [1, 100]$, $k_n \in [3, 7]$, and $k_s \in [1, 5]$, all of which are positive integers.

(3) CNN Training and Validation: For each particle, a CNN model is constructed based on the current position's parameter configuration. The CNN model is trained using the training dataset, and its performance on the validation dataset is evaluated, with the evaluation results serving as the particle's fitness value.

(4) Fitness Function: To balance the model's generalization performance and complexity, Equation (8) is used as the fitness function for PSO search to identify the optimal parameter set [40].

$$Fiteness = E_{train} + E_{validation} \tag{8}$$

where E_{train} and $E_{validate}$ indicate the average training error and the validating error of three prediction models.

Extensive research has shown that integrating artificial neural networks with optimization algorithms is effective, particularly in minimizing training error and avoiding local optima during the training phase. However, overfitting can occur when a model fits too closely to the training data, indicating excessive model complexity. To prevent overfitting, it is essential to evaluate predictive accuracy using validation data when constructing the inference model. The choice of the objective function is critical in balancing the model's generalization ability with its complexity [40]. The optimization objective proposed in this paper is designed to strike an optimal balance between minimizing training error and enhancing the model's generalizability.

(5) PSO Optimization: The PSO algorithm searches for the best tuning parameters of the model by optimizing the fitness function, iterating to find the parameters that minimize the target fitness function.

(6) Stopping Condition: Once the stopping condition is met, the optimization process is terminated. The number of generations (G_{max}) or the number of function evaluations (NFE) can be used as the termination criterion; this study adopts G_{max} as the termination condition.

(7) Optimized Parameters: When the termination criterion is met, the loop is stopped, and a set of optimally adjusted parameters are used to train the entire training dataset and predict the cost of Huizhou replica traditional vernacular dwellings.

(8) Perform Final Testing with the Optimal Model: The optimized model is subjected to final testing using a test dataset that was not utilized during the training or validation phases. This step is crucial for evaluating the model's performance and ensuring its robustness in predicting unseen data.

(9) SHAP Model Interpretation: Calculate the SHAP values for each feature to evaluate the specific impact of each feature on the model's prediction.

3. Experimental Results, Analysis, and Discussion

3.1. Feature Analysis

The selection of features directly impacts the stability and accuracy of model fitting. Too many features can lead to the curse of dimensionality and feature redundancy, while too few can result in information loss and underfitting. Given the scarcity of research literature on the cost of traditional Huizhou residential construction, this paper divides the feature selection process into three steps. The first step involves using expert interviews to gather as many factors that primarily affect the cost of Huizhou residences as possible. The second step involves feature data processing, and the third step is the preliminary extraction of key features related to this task, aiming to achieve dimensionality reduction and enhance the model's generalization capability.

3.1.1. Data Collection

In this study, face-to-face expert interviews were conducted to ensure the quality and credibility of the data collected. The experts interviewed have years of experience in the design and cost estimation of Hui-style ancient architecture, contributing to a rich repository of professional knowledge. The interviewer pre-determined a semi-structured format for the interviews to elicit comprehensive responses. The interviews involved open-ended questions such as: "Could you please elaborate on the factors that you believe may influence the cost in ancient replica construction projects? Please list and describe the extent of influence for each factor". Through these expert interviews, the main factors affecting the cost of ancient replica construction were identified. Initially, eight factors were selected as the primary characteristics for the cost of Hui-style replica residential buildings, as detailed in Table 1.

Table 1. Summary of expert interviews on key factors for Huizhou replica traditional vernacular dwellings.

Main Factors	
(1) Architectural Form	(5) Brickwork Quantity and Material
(2) Woodwork Quantity and Material	(6) Stonework Quantity and Material
(3) Wood Decoration Quantity and Material	(7) Wood Frame Structure Type
(4) Tilework Quantity and Material	(8) Construction Techniques

Based on the eight influencing factors identified, this study conducted field research and consulted with experts in the field of Hui-style ancient architectural cost estimation, collecting original tender control price data for a total of 98 Huizhou replica traditional vernacular dwellings. Given the consistency in geographical location and construction timing of the projects from which the data were collected, there was no need for temporal or regional adjustments to the data.

The individual projects of Huizhou replica traditional vernacular dwellings can be categorized into two major divisions: ancient architectural sub-projects and civil engineering sub-projects. The ancient architectural sub-projects encompass key sub-projects such as large-scale timber work, wooden decoration, roofing, brickwork, and stonework, whereas the civil engineering sub-projects include sub-projects like earthwork, masonry, concrete, and wall decoration. After a meticulous analysis of the cost data for the 98 collected residential buildings, the results indicate that the cost of the ancient architectural sub-projects accounts for 88.24% of the total project cost. Within this division, the combined costs of the large-scale timber work, wooden decoration, roofing, brickwork, and stonework sub-projects constitute 80.04% of the expenses of the ancient construction division.

In accordance with the "80–20 rule", which posits that the majority of effects are often generated by a minority of critical factors [41], this study selects ancient construction components such as large-scale timber work, wooden decoration, roofing, brickwork, and stonework as the primary feature inputs. Integrating the information obtained from the preliminary expert interviews, these main factors are defined as first-level features and are further refined into 19 second-level features, as detailed in Table 2.

Table 2. Factors of Influence.

Serial Number	Primary Features	Secondary Features	Feature Attributes	Feature Encoding
1	Architectural Form	Building Area	--	m^2
2		Architectural Style	Ming Dynasty	1
			Qing Dynasty	2
3		Number of Floors	--	Floor
4		Plan Layout	"Ao"-shaped Layout	1
			"Hui"-shaped Layout	2
			"H"-shaped Layout	3
			"Ri"-shaped Layout	4

Table 2. *Cont.*

Serial Number	Primary Features	Secondary Features	Feature Attributes	Feature Encoding
5	Major Woodwork	Column Quantity	--	m^3
6		Beam Quantity	--	m^3
7		Major Woodwork Material	Chinese Fir	1
			Camphorwood	2
			Mahogany	3
8	Wood Decoration	Partition Screen Area	--	m^2
9		Partition Screen Material	Chinese Fir	1
			African Teak	2
10		Screen Door Area	--	m^2
11		Screen Door Material	Chinese Fir	1
			African Teak	2
12	Tilework	Roof Area	--	m^2
13	Brickwork	Brick Ground Area	--	m^2
14		Ground Construction	Square-Brick Ground	1
			Slate Ground	2
			Composite Soil Ground	3
15		Door Hood Construction	Arched Door Hood	1
			Inscribed Door Hood	2
			Drooping Flower Door Hood	3
			Figure-eight Door Hood	4
16		Number of Door Hood	--	Unit
17	Stonework	Patios Area	--	m^2
18	Construction	Construction Techniques	Traditional Techniques	1
			Non-traditional Techniques	2
			Combination of Traditional and Non-traditional Techniques	3
19	Wood Frame Structure	Wood Frame Type	Bracket-set Structure	1
			Combination of Bracket-set Structure and Beam-lift Structure	2
			Beam-lift Structure	3

3.1.2. Data Reduction

Data reduction is a crucial step in data analysis, aiding in reducing data dimensions, simplifying data representation, and extracting key information from the data. Data reduction is divided into two categories: quantification of qualitative features and attribute reduction.

1. Quantification of Qualitative Features

Feature factors include both qualitative and quantitative data types. Since predictive models require input features to be numerical, qualitative features must be converted into quantitative data. For example, Huizhou traditional vernacular dwellings have evolved from the traditional courtyard house form and, based on natural environmental conditions, have developed several planar types, such as "Ao", "Hui", "H", and "Ri", as shown in Figure 4 [42]. Different layouts result in varying numbers of side rooms, halls, and courtyards, significantly impacting the construction cost. Therefore, in processing qualitative data, layouts that are concave are quantified as 1, those that are "Hui"-shaped are quantified as 2, and so on, with specific quantification values detailed in Table 2.

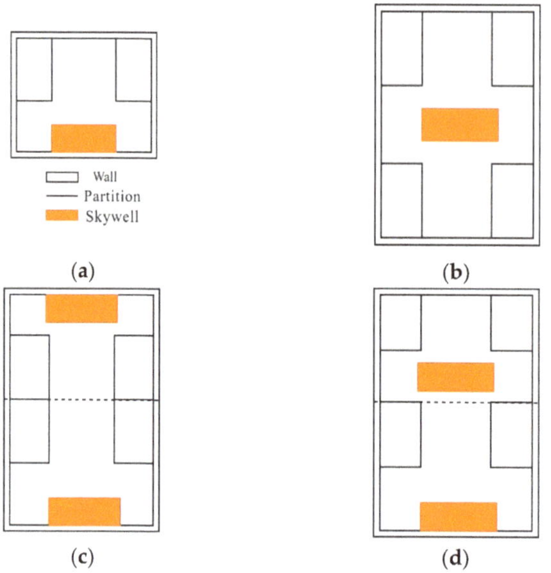

Figure 4. Floor plan of traditional Huizhou residential buildings: (**a**) "Ao"-shaped layout; (**b**) "Hui"-shaped layout; (**c**) "H"-shaped layout; (**d**) "Ri"-shaped layout.

2. Feature Aggregation

In the cost analysis of ancient replica buildings, some characteristic factors within the data exhibit high discreteness. It is necessary to integrate information scattered across multiple attribute parameters, representing the same issue in the raw data, into a single comprehensive attribute parameter. For instance, in material selection for components in Huizhou traditional vernacular dwellings, cost considerations often lead to the use of different materials for the same type of component within different size ranges. For example, columns with a diameter less than 200 mm are typically made of Chinese fir, while those with a diameter of 200 mm or greater are made of materials such as camphorwood or mahogany.

In response, this study proposes using a weighted average method to deal with the material properties of large timber works and beams. This method integrates the attribute parameters of different materials into a unified component material coefficient (η), as shown in Equation (9), thereby accurately reflecting the impact of component material on cost. This processing not only improves the consistency and analyzability of the data but also lays the foundation for establishing a more accurate cost prediction model. After data reduction, the material coefficient feature is added, increasing the number of secondary features to 20, as shown in Table 3.

$$\eta = \omega_1 \times \text{Tree Species } 1 + \omega_2 \times \text{Tree Species } 2 + \cdots \omega_n \times \text{Tree Species } n \quad (9)$$

In the formula, $\omega_1 \cdots \omega_n$ is the material weight of the tree species, determined by the ratio of the average market price of the tree species over the past five years. The proportion of tree species i is set based on the volume ratio of tree species i in similar components.

Table 3. Samples and input features.

Description	Notation	Project Number								
		1	2	3	4	5	6		97	98
Building Area	X_1	304.80	289.30	271.70	194.00	167.80	165.40	...	168.00	871.00
Architectural Style	X_2	2	2	2	2	2	2	...	2	1
Number of Floors	X_3	2	2	2	2	2	2	...	2	2
Plan Layout	X_4	2	2	2	1	2	1	...	1	4
Column Quantity	X_5	11.31	8.47	8.49	8.51	5.78	8.03	...	6.82	74.93
Beam Quantity	X_6	9.23	5.29	7.85	7.02	5.04	10.74	...	6.09	153.27
Column Material Coefficient	X_7	0.45	0.33	0.37	0.35	0.33	0.43	...	0.51	0.57
Beam Material Coefficient	X_8	0.43	0.49	0.45	0.48	0.40	0.42	...	0.67	0.71
Partition Screen Area	X_9	18.6	22.62	8.97	17.77	30.08	33.28	...	10.4	75.48
Partition Screen Material	X_{10}	2	2	2	2	2	2	...	2	2
Screen Door Area	X_{11}	114.49	165.09	83.91	64.02	62.67	176.81	...	61.30	75.48
Screen Door Material	X_{12}	1	1	1	1	1	1	...	1	1
Roof Area	X_{13}	191.09	166.68	170.38	106.67	104.54	101.52	...	92.84	188.34
Brick Ground Area	X_{14}	61.00	66.63	32.47	45.95	28.67	49.47	...	40.00	61.72
Ground Construction	X_{15}	1	1	1	1	1	1	...	1	1
Door Hood Construction	X_{16}	2	2	2	2	2	2	...	2	4
Number of Door Hoods	X_{17}	1	1	1	1	0	1	...	1	1
Patio Area	X_{18}	12.93	12.52	21.99	9.23	7.64	6.52	...	9.41	57.70
Construction Techniques	X_{19}	1	1	1	1	1	1	...	1	1
Wood Frame Type	X_{20}	1	1	1	1	1	1	...	1	3
Tender Control Price (ten thousand yuan)	Y	150.63	139.71	138.67	106.88	105.59	119.31	...	94.75	999.52

3. Min–Max Normalization for Data

The quantitative data among the input features exhibit significant differences in magnitude and order of magnitude, leading to slow and uneven model convergence, which affects the predictive results [43]. Therefore, the 98 sets of data were linearly normalized to the interval [0, 1]. When applying the trained model to the test set for cost prediction, the output data are inversely normalized according to Equation (10).

$$x = y \times (x_{max} - x_{min}) + x_{min} \tag{10}$$

3.1.3. OPTICS Clustering for Outlier Detection

The OPTICS clustering algorithm is a density-based clustering method that automatically identifies clusters and outliers without the need to preset the number of clusters, exhibiting strong robustness [44]. The algorithm generates a reachability plot that illustrates the variation in data point density, where valleys represent the presence of clusters, and their depth indicates the tightness of the connections within the clusters, as shown in Figure 5. In this study, by setting a threshold based on the 95th percentile of reachability distances (r = 0.834), data points 60 and 92 in the plot, which are above this threshold, are identified as outliers because they exhibit a greater reachability distance compared to other points in the dataset.

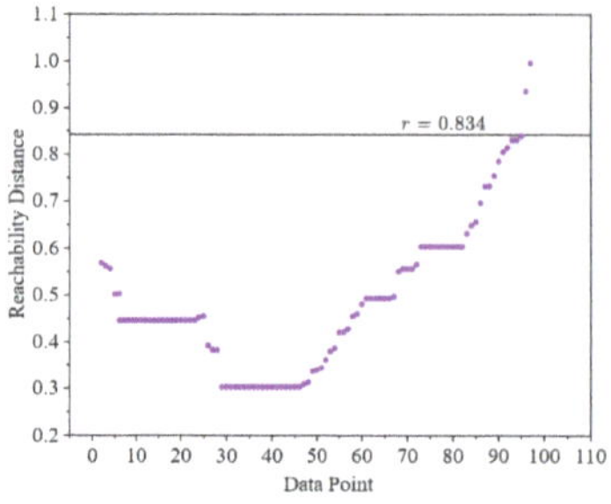

Figure 5. Optics reachability distance plot.

3.1.4. Preliminary Extraction of Principal Features

In this study, we employed the Random Forest algorithm for feature selection within the dataset to eliminate features irrelevant to the learning task. There are primarily two methods for calculating the Variable Importance Measure (VIM) in a Random Forest: one based on the reduction of the Gini index in split nodes and the other based on the Out-of-Bag (OOB) prediction error rate. Considering the dataset includes both continuous and categorical variables, we opted for an assessment method based on data permutation to evaluate feature importance, as it provides a more precise evaluation for variables whose error rates remain unchanged after data permutation.

Using a Python program, we calculated and ranked the VIM values of each feature, as shown in Figure 6. The results indicate that, among the 20 feature attributes, the VIM values for Number of Door Hoods (X_{17}), Number of Floors (X_3), Construction Techniques (X_{19}), Ground Construction (X_{15}), and Partition Screen Material (X_{10}) are all zero, rendering them ineffective as features. The relative VIM values for the other features all exceeded 20%; hence, we selected these 15 features as input features for the cost estimation of replica Huizhou traditional vernacular dwellings. This approach significantly enhanced the efficiency of data organization and model computation, reducing the time and cost for cost engineers in data preparation and model calculation.

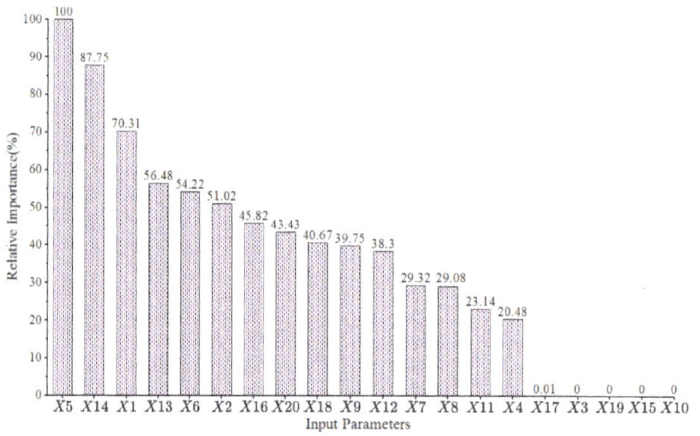

Figure 6. Features' relative importance plot.

3.2. Computational Results

A total of 96 valid datasets of bidding control prices for Huizhou replica traditional vernacular dwellings were collected in this study. The processed datasets were randomly divided into two subsets: 67 for constructing the training set of the inference model, and the remaining 29 serving as the test set for evaluating model performance. The RF algorithm was utilized to screen 20 original features, resulting in 15 key features as input variables for the ASCNN. The accuracy assessment of the model employed the Mean Squared Error (MSE), Root Mean Squared Error (RMSE), Mean Absolute Percentage Error (MAPE), and Coefficient of Determination (R^2) as evaluation metrics (Equation s (11) to (13)].

$$\mathrm{RMSE} = \sqrt{\frac{1}{n}\sum_{i=1}^{n}(y_i - \hat{y}_i)^2} \tag{11}$$

$$\mathrm{MAPE} = \frac{100\%}{n}\sum_{i=1}^{n}\left|\frac{y_i - \hat{y}_i}{y_i}\right| \tag{12}$$

$$\mathrm{R}^2 = 1 - \frac{\sum_{i=1}^{n}(y_i - \hat{y}_i)^2}{\sum_{i=1}^{n}(y_i - \overline{y})^2} \tag{13}$$

where y_i is the actual value, $\hat{y}_i$ is the predicted value, $\overline{y}$ is the average of actual values, and n is the number of samples.

Predictions were performed according to the model parameters suggested in Table 4. Figure 7 illustrates the change in normalized MSE as the number of iterations in the PSO process increases. The MSE value decreases with the increase in the number of iterations, showing a significant decline from 0 to approximately 75 iterations. During this process, the PSO algorithm effectively adjusted the model parameters, resulting in a substantial reduction in model prediction error. After 70 iterations, the value stabilizes without significant changes, indicating that the PSO algorithm has essentially converged, finding an optimal combination of model parameters. The iterative process demonstrates the effectiveness of the PSO algorithm in this optimization process. The final standard MSE value is approximately 0.015, which is a significant decrease relative to the initial value, indicating that the PSO algorithm has significantly improved the model's predictive accuracy.

Table 4. ASCNN algorithm parameter settings.

Parameter Name	Parameter Value	Parameter Name	Parameter Value
Optimizer	Adam	Learning Rate	Adaptive
Data Split	0.7	L2 Regularization	0.01
Data Shuffling	Yes	Cross-validation	3
Activation Function	ReLU	Number of Neural Network Iterations	1000
PSO Particle Number	21	PSO Maximum Iterations	100
C1, C2, W	2.0, 2.0, 0.5	Batch Size	4

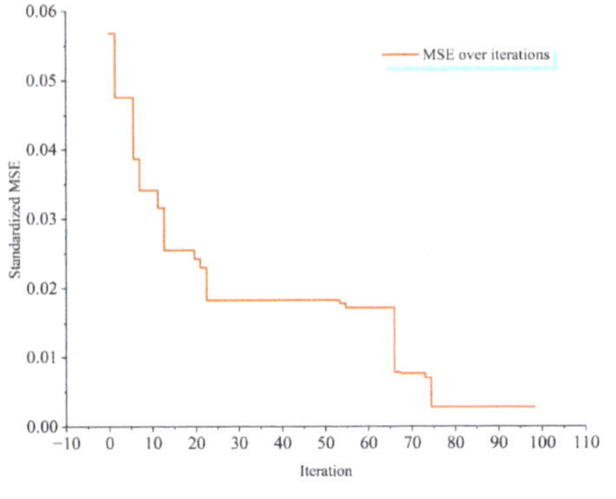

Figure 7. PSO iteration process.

The predictive results of the ASCNN model are presented in Table 5. The findings indicate that the ASCNN model demonstrated an RMSE of 9828.06 yuan and a MAPE of

0.6% on the test dataset, with an R^2 reaching 0.989, signifying highly significant predictive performance and strong generalization capabilities.

Table 5. Model results of ASCNN.

Data Set	RMSE	MAPE (%)	R^2
Training set	3018.92	0.2	0.999
Cross-validation set	8080.94	0.53	0.992
Test set	9828.06	0.6	0.989

Figure 8a illustrates the comparison between the predicted and actual values on the test dataset, showing a good match between the predicted and actual values for the majority of the data points. Notably, the overall trend of the data is well-aligned, indicating the model's proficiency in capturing the global trends of data changes. For instance, at data points 6, 9, 13, and 18, the model successfully detected peak changes in actual values, further evidencing its robust response to extreme values within the data and its ability to reflect the actual cost variations to a considerable extent.

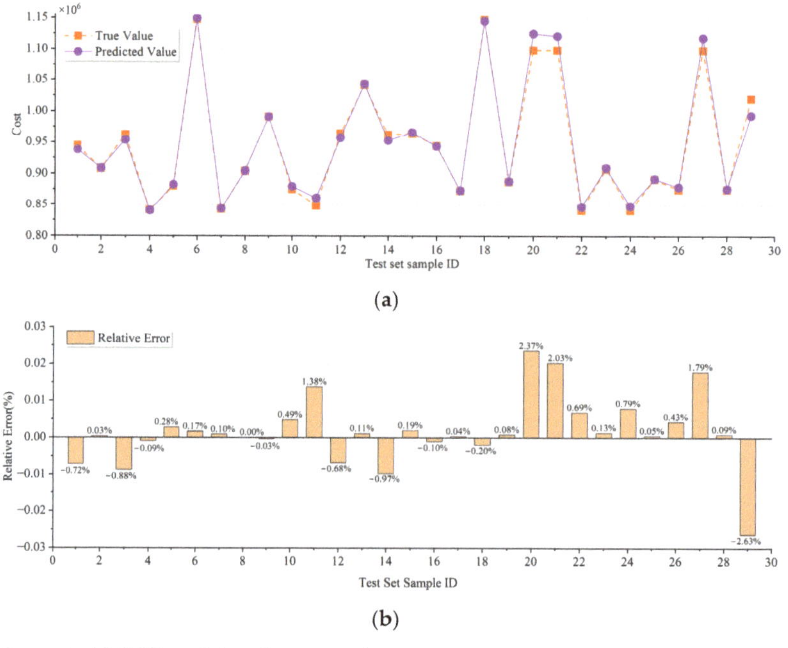

(**a**)

(**b**)

Figure 8. ASCNN model prediction results: (**a**) CNN model output results; (**b**) relative error of output results.

Figure 8b displays the distribution of relative errors for the test set samples in the cost prediction of replica traditional vernacular dwellings. The results indicate that the relative errors of the samples are all within ±3%, demonstrating that the model possesses high overall predictive accuracy and stability. Among the test samples, the maximum positive relative error is 2.37%, and the maximum negative error is −2.63%, which complies with the 3% precision range required for engineering budget preparation and is significantly lower than the requirement to control cost errors within ±10% during the detailed feasibility study phase. This indicates that the model has high predictive precision in most cases. Additionally, it is observed that the relative errors for the majority of samples are within ±2%, with a balanced error distribution and a small overall magnitude, suggesting that the model's predictions are stable and reliable, without significant fluctuations. This stability is crucial for cost budgeting and control in practical applications. The error distribution in the figure shows an alternating pattern of positive and negative values, without a noticeable bias, indicating that the model exhibits balanced performance in terms of overfitting or underfitting. It should be noted that the dataset samples collected in this study are relatively small; an increase in the quantity and quality of training samples is expected to further improve predictive accuracy.

3.3. Results Analysis

To validate the effectiveness of the ASCNN model, comparisons were made with other AI models that have achieved good results in cost estimation. These algorithms include non-optimized CNN networks, BP-ANN, Radial Basis Function Neural Networks (RBF-ANNs), and Elastic Support Vector Machines (ELSVMs). The models were assessed using RMSE, MAPE, and R^2 as metrics for evaluating model accuracy.

Figure 9 presents the performance metrics of various models on the test dataset, where the ASCNN outperforms all other models across all indicators, particularly in MAPE and R^2, demonstrating its strong predictive accuracy and fitting capability. Specifically, the RMSE of ASCNN is the lowest in both the training and test sets, indicating the smallest error between predicted and actual values. In comparison, the RMSE of the unoptimized CNN increased by 26%, and the MAPE increased by 102%. Other models show significantly higher RMSE than ASCNN, with the ELSVM model performing the worst. The MAPE of ASCNN is the lowest in both the training and test sets, with a particularly minimal value of 0.6% on the test set, indicating the smallest predictive error. Moreover, the R^2 values of ASCNN are close to 1 in both the training and test sets, indicating the best fitting effect. Other models have R^2 values lower than ASCNN, with ELSVM having the lowest R^2 value and the worst fitting effect.

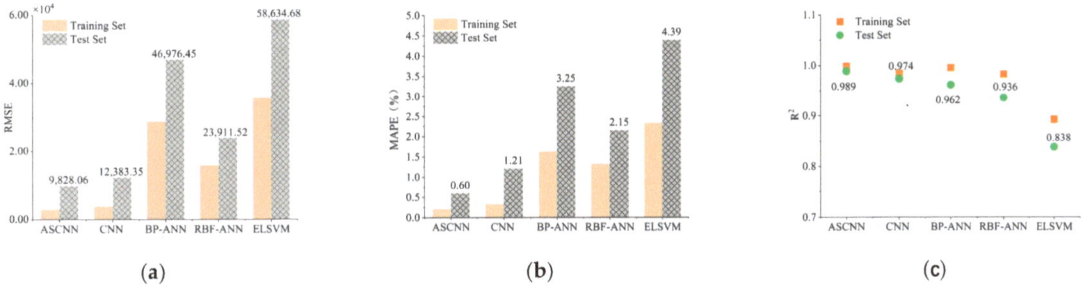

Figure 9. Results comparison of the models: (**a**) results comparison of RMSE; (**b**) results comparison of MAPE; (**c**) results comparison of MAPE.

It can be observed that ASCNN exhibits the best performance across all metrics, with its RMSE, MAPE, and R^2 superior to other models, especially with prominent performance in MAPE and R^2, indicating higher predictive accuracy and excellent fitting. Although other models such as RBF-ANN and BP-ANN show relatively good predictive performance, they are still inferior to ASCNN, particularly in MAPE and R^2. ELSVM performs the worst across all metrics, with the largest predictive error and the poorest fitting, making it unsuitable for cost prediction of replica traditional vernacular dwellings.

Therefore, based on the analysis, ASCNN is the most superior model for cost prediction of replica traditional vernacular dwellings, with high application value and reliability. In practical applications, it is recommended to prioritize the use of the ASCNN model for prediction to achieve higher predictive accuracy and better fitting effects.

3.4. Model Interpretation

This study employs SHAP to elucidate the CNN model used in the task of cost estimation for replica traditional vernacular dwellings, verifying whether the model can reasonably capture the correlation between the parameters used in the training process and the cost of these dwellings. By thoroughly analyzing the model's predictive logic, not only can the credibility of the model be enhanced, but also clearer and more persuasive decision support can be provided to relevant stakeholders.

Figure 10a presents the global feature importance plot based on SHAP values, showing the average absolute SHAP values of each feature to intuitively represent their importance. It can be observed that the average SHAP values of Plan Layout (X_4), Column Material Coefficient (X_7), Roof Area (X_{13}), Beam Material Coefficient (X_8), Screen Door Area (X_{11}), Partition Screen Area (X_9), and Building Area (X_1) rank in the top seven, indicating their greatest impact on the model's prediction and further indicating that the Plan Layout, Tilework, Major Woodwork, Wood Decoration, and Building Area have the most significant influence on the cost of these dwellings.

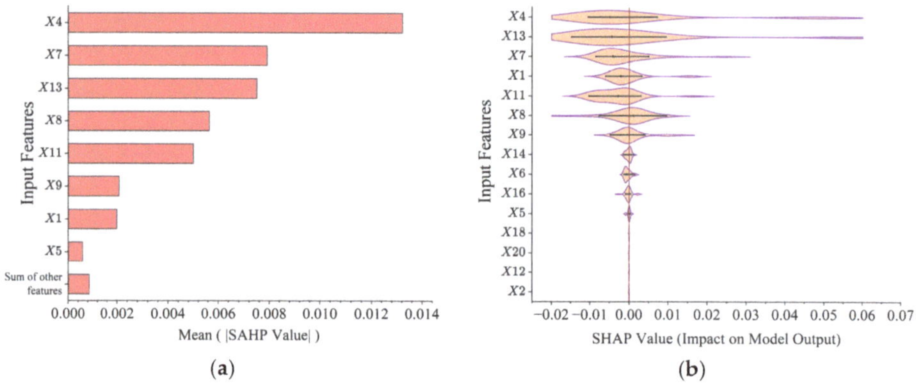

Figure 10. SHAP value: (**a**) feature importance ranking plot; (**b**) SHAP value violin plot.

The SHAP value distribution illustrated in Figure 10b offers an in-depth insight into the contribution of each feature in the model's prediction. The results show that the SHAP values of Plan Layout (X_4) and Roof Area (X_{13}) cover a wide range with significant differences, indicating that these two features play a notable role in the cost prediction of ancient dwellings. In particular, the broad distribution of SHAP values for Plan Layout (X_4) and Roof Area (X_{13}) suggests a potentially complex nonlinear relationship between these factors and the predicted outcomes.

The Column Material Coefficient (X_7), Building Area (X_1), Screen Door Area (X_{11}), Beam Material Coefficient (X_8), and Partition Screen Area (X_9) exhibit a moderate range of SHAP values, indicating a stable but significant impact on cost prediction. This finding aligns with common knowledge in the field of architecture that the material of wooden structures and wood decoration are key factors affecting the construction cost of these dwellings.

On the other hand, features such as Brick Ground Area (X_{14}) and Patio Area (X_{18}) have a more concentrated distribution of SHAP values, indicating their relatively minor contribution to the model's prediction. This suggests that the Brickwork and Stonework parts of Huizhou replica traditional vernacular dwellings have a smaller impact on the cost. The impact of the other six features on the model's prediction is relatively limited, indicating a degree of redundancy.

To further validate the robustness and interpretability of the ASCNN model, in-depth interviews were conducted with eight experts, each possessing extensive experience in cost estimation of traditional construction projects. The interview findings revealed a high degree of consistency between the SHAP value analysis and the experts' professional insights, thereby significantly enhancing the credibility of the model's predictive outcomes.

Delving into the contributions of features to model predictions and their mechanisms of impact is crucial for precisely controlling the construction costs of replica traditional vernacular dwellings. This analysis not only identifies the key factors that dominate cost predictions, but also reveals the nonlinear dynamics in cost influences, providing a solid foundation for accurate budget preparation. Through this process, cost engineers can gain profound insights into the mechanisms of ancient dwelling cost formation, quantify the specific impacts of various features on costs with precision, and thus comprehensively understand how architectural elements and material choices collectively affect construction costs. This in-depth understanding is of significant practical importance for formulating reasonable cost budgets, optimizing resource allocation, and enhancing the overall efficiency of construction projects.

4. Conclusions

Huizhou-style residences are celebrated for their exquisite architectural structures and intricate decorative craftsmanship. These characteristics involve the use of numerous materials and artisanal skills, establishing a complex nonlinear relationship with construction costs. This complexity significantly increases the difficulty of cost estimation, causing traditional methods to often fall short of the expected accuracy and leading to frequent budget overruns. In the current context, where the Engineering, Procurement, and Construction (EPC) model is increasingly popular and the profit margins of construction projects are gradually decreasing, accurately predicting construction costs is particularly crucial. To address this challenge, this study proposes a model that fully utilizes historical data, combining PSO with a CNN model, referred to as the ASCNN, to predict the costs of replica traditional vernacular dwellings in the Huizhou area. This model not only provides more precise cost estimation but also enhances the transparency and credibility of the model through interpretation of SHAP values, offering reliable support for investment decisions in replica traditional vernacular dwellings.

A total of 96 bidding control price datasets for newly constructed replica traditional vernacular dwellings were collected to assess the performance of the ASCNN. The experimental results indicate that the ASCNN model can accurately predict the construction costs of Huizhou replica traditional vernacular dwellings, achieving the best results in RMSE, MAPE, and R^2, with RMSE and MAPE being 26% and 102% higher in accuracy than the second most accurate CNN algorithm, respectively. Additionally, by removing five redundant input parameters, the ASCNN model saves more time and effort in data updating and collection compared to other algorithms. Based on SHAP value calculations, this study identified Plan Layout (X_4), Roof Area (X_{13}), Column Material Coefficient (X_7), Screen Door Area (X_{11}), Beam Material Coefficient (X_8), Building Area (X_1), and Partition Screen Area (X_9) as the most important factors affecting the accuracy of cost prediction for Huizhou replica traditional vernacular dwellings, while finding that brickwork has a relatively minor impact on cost estimation.

Cost engineers can use this model to accurately identify and quantify the key factors affecting the cost of replica traditional vernacular dwellings, providing precise cost estimates. Through in-depth SHAP value analysis, this study reveals elements that significantly impact costs, enabling engineers to understand the internal mechanisms of cost composition more deeply and make precise adjustments accordingly.

This study has successfully developed an efficient artificial intelligence model aimed at providing a powerful tool for auxiliary decision-making and cost control in the field of cost estimation for replica Huizhou traditional dwellings. The proposed ASCNN model integrates the advantages of various artificial intelligence technologies, including: (1) adaptively adjusting the input parameters of the CNN to optimize performance; (2) providing high reliability and relatively accurate cost predictions for Huizhou replica traditional dwellings, enhancing the accuracy of estimation; (3) improving the objectivity and efficiency of the model by reducing manual operations; and (4) reducing the workload and time required for data updating and collection, improving the efficiency of data processing.

Despite the significant achievements of this study in the cost estimation of replica Huizhou traditional dwellings, some limitations need to be addressed in future research. (1) Geographical Limitations: The current study's samples are mainly concentrated in specific areas of Huizhou traditional architecture. Future work should expand the geographical scope of the samples to verify the model's universality and applicability in different regions. (2) Sample Diversity: The collected samples mainly come from a few single buildings of local cultural and tourism projects, and the construction time is relatively short, which limits the diversity of the samples and does not fully consider the impact of time factors on cost. Follow-up studies should consider the impact of time changes on the cost of replica traditional vernacular dwellings to improve the model's dynamic adaptability and the accuracy of long-term predictions.

Author Contributions: Conceptualization, J.H.; resources, J.H.; formal analysis, J.H.; writing—original draft preparation, J.H.; writing—review and editing, J.H. and W.H.; validation, W.H.; funding acquisition, J.H. and W.Q.; supervision, W.Q. and Y.X. All authors have read and agreed to the published version of the manuscript.

Funding: The research was supported by the Natural Science Research Key Project of Anhui Educational Committee (Grant No. 2022AH051956), the Natural Science Foundation of Huangshan University (2021xkjq003), and the Huizhou Culture Research Project of Huangshan University, China (Grant No. 2019xhwh005).

Data Availability Statement: The raw data supporting the conclusions of this article will be made available by the authors on request.

Conflicts of Interest: The authors declare no conflicts of interest.

References

1. Zhang, M.; Zhang, J.; Liu, Q.; Li, T.; Wang, J. Research on the Strategies of Living Conservation and Cultural Inheritance of Vernacular Dwellings—Taking Five Vernacular Dwellings in China's Northern Jiangsu as an Example. *Sustainability* **2022**, *14*, 12503. [CrossRef]
2. Yan, Y. Cost Management of Ancient Building Relocation Repair Projects. *J. Fujian Univ. Technol.* **2023**, *21*, 481–486. [CrossRef]
3. Vitasek, S.; Macek, D. A Study of Factors Influencing the Compliance of Design Estimates at the Construction Stage of Residential Buildings. *Buildings* **2024**, *14*, 2010. [CrossRef]
4. Liu, Y. Study on the Cost of Ancient Buildings Repair Project in South China. Master's Thesis, Guangdong University of Technology, Guangzhou, China, 2019. [CrossRef]
5. Yang, L. Research on Influencing Factors and Countermeasures of Renovation Cost of Historical Buildings. Master's Thesis, Tianjin University of Science and Technology, Tianjin, China, 2022. [CrossRef]
6. Elmousalami, H.H. Artificial Intelligence and Parametric Construction Cost Estimate Modeling: State-of-the-Art Review. *J. Constr. Eng. Manag.* **2020**, *146*, 03119008. [CrossRef]
7. Seidu, R.; Young, B.; Clack, J.; Adamu, Z.; Robinson, H. Innovative Changes in Quantity Surveying Practice through BIM, Big Data, Artificial Intelligence and Machine Learning. *Appl. Sci. Univ. J. Nat. Sci.* **2020**, *4*, 37–47.
8. Emsley, M.W.; Lowe, D.J.; Duff, A.R.; Harding, A.; Hickson, A. Data Modelling and The Application of a Neural Network Approach to The Prediction of Total Construction Costs. *Constr. Manag. Econ.* **2002**, *20*, 465–472. [CrossRef]
9. Lowe, D.J.; Emsley, M.W.; Harding, A. Predicting Construction Cost Using Multiple Regression Techniques. *J. Constr. Eng. Manag.* **2006**, *132*, 750–758. [CrossRef]
10. Wilmot, C.G.; Mei, B. Neural Network Modeling of Highway Construction Costs. *J. Constr. Eng. Manag.* **2005**, *131*, 765–771. [CrossRef]
11. An, S.H.; Park, U.Y.; Kang, K.I.; Cho, M.Y.; Cho, H.H. Application of Support Vector Machines in Assessing Conceptual Cost Estimates. *J. Comput. Civ. Eng.* **2007**, *21*, 259–264. [CrossRef]
12. Cheng, M.Y.; Hoang, N.D.; Wu, Y.W. Hybrid Intelligence Approach Based on LS-SVM and Differential Evolution for Construction Cost Index Estimation: A Taiwan Case Study. *Autom. Constr.* **2013**, *35*, 306–313. [CrossRef]
13. Moussa, M.; Ruwanpura, J.; Jergeas, G. Decision Tree Modeling Using Integrated Multilevel Stochastic Networks. *J. Constr. Eng. Manag.* **2006**, *132*, 1254–1266. [CrossRef]
14. Doğan, S.Z.; Arditi, D.; Murat Günaydin, H. Using Decision Trees for Determining Attribute Weights in a Case-Based Model of Early Cost Prediction. *J. Constr. Eng. Manag.* **2008**, *134*, 146–152. [CrossRef]

15. Ji, S.H.; Ahn, J.; Lee, E.B.; Kim, Y. Learning Method for Knowledge Retention in CBR Cost Models. *Autom. Constr.* **2018**, *96*, 65–74. [CrossRef]
16. Kim, S. Hybrid Forecasting System Based on Case-based Reasoning and Analytic Hierarchy Process for Cost Estimation. *J. Civ. Eng. Manag.* **2013**, *19*, 86–96. [CrossRef]
17. Cao, Y.; Ashuri, B.; Baek, M. Prediction of Unit Price Bids of Resurfacing Highway Projects through Ensemble Machine Learning. *J. Comput. Civ. Eng.* **2018**, *32*, 04018043. [CrossRef]
18. Pan, Y.; Zhou, S.; Guan, J.; Wang, Q.; Ding, Y. Limited Field Images Concrete Crack Identification Framework Using PCA and Optimized Deep Learning Model. *Buildings* **2024**, *14*, 2054. [CrossRef]
19. Wang, Y.; Ning, X.; Zhen, D.; Yong, W.; Zhang, H. Research on Construction Project Cost Prediction Model Based on Recurrent Neural Network. In *SHS Web of Conferences, Proceedings of the 2023 International Conference on Digital Economy and Management Science, Kai Feng, China, 21–23 April 2023*; EDP Sciences: Les Ulis, France, 2023; Volume 170, p. 02009.
20. Al-tawal, D.R.; Arafah, M.; Sweis, G.J. A Model Utilizing the Artificial Neural Network in Cost Estimation of Construction Projects in Jordan. *Eng. Constr. Archit. Manag.* **2021**, *28*, 2466–2488. [CrossRef]
21. Saeidlou, S.; Ghadiminia, N. A Construction Cost Estimation Framework Using DNN and Validation Unit. *Build. Res. Inf.* **2024**, *52*, 38–48. [CrossRef]
22. Li, B.; Xin, Q.; Zhang, L. Engineering Cost Prediction Model Based on DNN. *Sci. Program.* **2022**, *2022*, 3257856. [CrossRef]
23. Lecun, Y.; Bengio, Y.; Hinton, G. Deep Learning. *Nature* **2015**, *521*, 436–444. [CrossRef]
24. Xue, X.; Jia, Y.; Tang, Y. Expressway Project Cost Estimation with a Convolutional Neural Network Model. *IEEE Access* **2020**, *8*, 217848–217866. [CrossRef]
25. Yi, M.; Zeng, Y.; Qin, Z.; Xia, Z.; He, Y. Cost Prediction Model of Mountain High-speed Railway Civil Engineering Based on MIC-CNN. *Railw. Stand. Des.* **2023**, *67*, 44. [CrossRef]
26. Han, K.; Wang, W.; Huang, X. Predicting the Construction Cost of High Standard Farmland Irrigation Projects using NGO-CNN-SVM. *Trans. Chin. Soc. Agric. Eng.* **2024**, *40*, 62–72. [CrossRef]
27. Jović, A.; Brkić, K.; Bogunović, N. A Review of Feature Selection Methods with Applications. In Proceedings of the 2015 38th International Convention on Information and Communication Technology, Electronics and Microelectronics (MIPRO), Opatija, Croatia, 25–29 May 2015.
28. Jaiswal, J.K.; Samikannu, R. Application of Random Forest Algorithm on Feature Subset Selection and Classification and Regression. In Proceedings of the 2017 World Congress on Computing and Communication Technologies (WCCCT), Tiruchirappalli, India, 2–4 February 2017.
29. Gaspar, A.; Oliva, D.; Cuevas, E.; Zaldívar, D.; Pérez, M.; Pajares, G. *Hyperparameter Optimization in a Convolutional Neural Network Using Metaheuristic Algorithms. Metaheuristics in Machine Learning: Theory and Applications*; Springer International Publishing: Cham, Switzerland, 2021; pp. 37–59.
30. Kennedy, J.; Eberhart, R. Particle swarm optimization. In Proceedings of the ICNN'95-International Conference on Neural Networks, Perth, WA, Australia, 27 November 1995.
31. Jiyue, E.; Liu, J.; Wan, Z. A Novel Adaptive Algorithm of Particle Swarm Optimization Based on the Human Social Learning Intelligence. *Swarm Evol. Comput.* **2023**, *80*, 101336.
32. Montavon, G.; Samek, W.; Müller, K.R. Methods for Interpreting and Understanding Deep Neural Networks. *Digital Signal Process.* **2018**, *73*, 1–15. [CrossRef]
33. Wieland, R.; Lakes, T.; Nendel, C. Using Shapley Additive Explanations to Interpret Extreme Gradient Boosting Predictions of Grassland Degradation in Xilingol, China. *Geosci. Model Dev.* **2021**, *14*, 1493–1510.
34. Zhu, P.; Cao, W.; Zhang, L.; Zhou, Y.; Wu, Y.; Ma, Z.J. Interpretable Machine Learning Models for Prediction of UHPC Creep Behavior. *Buildings* **2024**, *14*, 2080. [CrossRef]
35. Lundberg, S.M.; Lee, S.I. A Unified Approach to Interpreting Model Predictions. *Adv. Neural Inf. Process. Systems.* **2017**, *30*, 4768–4777.
36. Dhillon, A.; Verma, G.K. Convolutional Neural Network: A Review of Models, Methodologies and Applications to Object Detection. *Prog. Artif. Intell.* **2020**, *9*, 85–112. [CrossRef]
37. Kinga, D.; Adam, J.B. A Method for Stochastic Optimization. In Proceedings of the International Conference on Learning Representations (ICLR), Boston, MA, USA, 6 July 2015.
38. Wang, D.; Tan, D.; Liu, L. Particle Swarm Optimization Algorithm: An Overview. *Soft Comput.* **2018**, *22*, 387–408. [CrossRef]
39. Shafi, S.; Assad, A. *Exploring the Relationship between Learning Rate, Batch Size, and Epochs in Deep Learning: An Experimental Study*; Springer Nature Singapore: Singapore, 2023; pp. 201–209.
40. Cao, M.T.; Cheng, M.Y.; Wu, Y.W. Hybrid Computational Model for Forecasting Taiwan Construction Cost Index. *J. Constr. Eng. Manag.* **2015**, *141*, 04014089. [CrossRef]
41. Le, C.; Jeong, H.D.; Damnjanovic, I.; Bukkapatnam, S. Pareto Principle in Scoping-Phase Cost Estimating: A Multiobjective Optimization Approach for Selecting and Applying Optimal Major Work Items. *J. Constr. Eng. Manag.* **2022**, *148*, 04022076. [CrossRef]
42. Shan, D.Q. *Chinese Vernacular Dwellings: People's Daily Life with Their Houses*; China Intercontinental Press: Beijing, China, 2010; pp. 35–52.

43. Liang, K.Y.; Zeger, S.L. Regression Analysis for Correlated Data. *Annu. Rev. Public Health* **1993**, *14*, 43–68. [CrossRef] [PubMed]
44. Zhao, Y.; Li, H.; Yu, X.; Ma, N.; Yang, T.; Zhou, J. An independent central point OPTICS clustering algorithm for semi-supervised outlier detection of continuous glucose measurements. *Biomed. Signal Process. Control* **2022**, *71*, 103196. [CrossRef]

buildings

MDPI

Article

Study on Deformation of New Tunnels Overcrossing Existing Tunnels Underneath Operating Railways

Can Xie, Yuhang Qu , Haiyan Lu and Shuguang Song *

School of Traffic Engineering, Shandong Jianzhu University, Jinan 250101, China; xiecansdu@163.com (C.X.); 15244567062@163.com (Y.Q.); 13385316864@163.com (H.L.)
* Correspondence: twilightsong@126.com

Abstract: With the increasing utilization of urban underground space, new tunnels frequently intersect with existing tunnels and operational railways. However, sometimes the excavation and unloading of new tunnels can cause deformation of adjacent existing tunnels and railways, significantly affecting their normal operation. We used finite element software to predict the influence of new tunnel construction on overcrossing existing tunnels and down-traversing operational railways by a dynamic tunneling model based on a connection channel project of the east and west squares of a railway station. This article is not only control the distance between the two tunnels, but the new tunnel and the existing tunnel, as well as the new tunnel and the operation of the railway, the positional relationship between the three, the deformation laws of existing tunnels and operational railways during the construction of new tunnels with different buried depths are analyzed. The results show that the deformation curves of existing tunnels and operational railways present a normal distribution. The maximum deformation position is at the intersection with the new tunnel upon completion of the new tunnel excavation construction. Moreover, an increase in the buried depth of the new tunnel increases the deformation of the operational railway and the existing tunnel. The influence of the depth change of the new tunnel on the settlement of the operational railway is greater than that of the existing tunnel.

Keywords: excavation construction; existing tunnels; overcrossing and down-traversing; operating railways

1. Introduction

In the course of urban development, given the limited space above ground, the development of underground transportation has emerged as an essential means for metropolitan progress. Nevertheless, as the development of underground transportation lags behind the process of spatial urbanization, and the construction of urban facilities is imperfect, underground transportation projects constructed in the later stage often intersect with existing above-ground projects and other underground ones. However, the construction of a new underground shield tunnel will lead to alteration of ground stress, which will inevitably exert a series of adverse effects on the adjacent above-ground and underground projects.

Many researchers have analyzed the impact of new tunnel construction on adjacent existing tunnels or operating railways. Different methods and models have been used to predict settlement during tunnel excavation [1,2]. Jin et al. [3] discuss several key factors influencing the settlement of existing tunnels, such as spatial position, support pressure, and tunnel stiffness, and propose an empirical equation for estimating the settlement of existing tunnels caused by the excavation of new shield tunnels. Feng et al. [4,5] derived the stress added under the action of shield tunneling based on a modified Gaussian formula and proposed a method for closely predicting stress-induced deformation of newly constructed tunnels under existing tunnels. He et al. and Li et al. [6,7] investigated the impact of shield tunnel excavation on rectangular pipe jacking and the surface, respectively.

Zhang et al. [8] developed an analytical solution by utilizing Timoshenko beams placed on Kerr foundations, investigated the response of existing tunnels to the excavation of new tunnels underground, and examined the applicability of the proposed analytical solution through centrifuge laboratory tests and field measurements collected at a construction site. Ma et al. [9] conducted a three-dimensional numerical simulation to investigate the influence of overburden thickness and examined the distribution characteristics and variation of land subsidence. Within the range of 0.5 times the width of the central axis of the large pipe tunnel, cumulative ground settlement decreases linearly as the thickness of the casing increases. Cumulative settlement of the foundation rises with the increase in casing thickness. A simplified analysis method was proposed by Liang et al. [10], which considers the tunnel merely as a continuous Euler–Bernoulli beam with a certain equivalent flexural stiffness. When the existing tunnel is disregarded, the unloading stress caused by crossing the tunnel is calculated by the Mindlin solution. Fei et al. [11–15] used theoretical and numerical simulation methods to reveal the deformation patterns of existing shield tunnels deformed by newly constructed overpass tunnels. Furthermore, Zhang [16] and Pan [17] considered different construction parameters for sandy layer and water-rich strongly weathered sandstone layers to investigate deformation effects on operating railways during the construction of new tunnels under the railways and possible deformation-control measures. Shield tunnels also have different effects on railway tracks and roadbeds under different working conditions [18–20]. Mroueh H. and Shahrour I. [21] carried out a numerical investigation on the interaction between tunnels and adjacent structures in soft soil. The numerical simulation was performed through full three-dimensional computations, considering the existence of structures during tunneling. The analysis indicates that the force induced by the tunnel significantly depends on the presence of adjacent structures. The structural stiffness was disregarded in the tunnel-structure analysis, resulting in a serious overestimation of the internal force of the structural members. Hu [22] studied the axial force, bending moment, and pore water pressure of shield tunnel segments in soft and hard uneven strata, clay layers, and weathered granite strata of overlying buildings by establishing a rectangular element mechanical model based on the field test method. Zhang [23–25] studied the impact of a new tunnel on an existing tunnel through numerical analysis and verified the numerical analysis results of horizontal convergence displacement of the side wall of the new tunnel and longitudinal cracks of the existing tunnel through real-time monitoring. Zhang et al. [26–28] studied vertical deformation of the surface, arch foot, and ballast bed of the existing tunnel and analyzed in detail the effect of technical measures to control deformation of the existing shield tunnel. Therefore, it is of certain significance to study the influence of shield tunnel excavation on surrounding buildings [29,30]. Cheng's [31] study shows that the force on the tunnel segments and the shield excavation speed are increasing. Too fast shield tunneling speed will lead to an increase of tunnel segment load and aggravate uneven distribution of grouting pressure.

There have been many studies focused on the deformation of newly built tunnels crossing existing tunnels or operating railways. However, there are only a few studies investigating construction of shield tunnels in complex environments. Therefore, this paper takes the underground connecting passage of the east and west squares of a railway station as the engineering background. The newly built shield tunnel is in a complex environment, spanning the subway tunnel and beneath the railway. Aiming at the new shield tunnel project in a complex environment, this paper utilizes finite element software to establish six distinct numerical models with buried depths of 5 m, 7 m, 9 m, 11 m, 13 m, and 15 m, which are employed to simulate and calculate the influence of underground connection channels with different buried depths on adjacent subway tunnels and operational railways in a complex environment. The deformation patterns of the existing tunnel and operating railway during the dynamic tunneling of the new shield tunnel under different buried depths are analyzed. This study offers the theoretical basis for determining the location of the new shield tunnel and the rational utilization of underground space and provides a reference for on-site construction and similar projects.

2. Project Overview

The project is the construction of a shield tunnel that serves as an underground pedestrian exchange channel between the east and west squares of a railway station. The channel crosses Metro Line 1 at an angle of 34.2° and crosses four tracks of the operating railway. The positions of the newly constructed tunnel, the overlying subway tunnel, and the operating railway are shown in Figure 1.

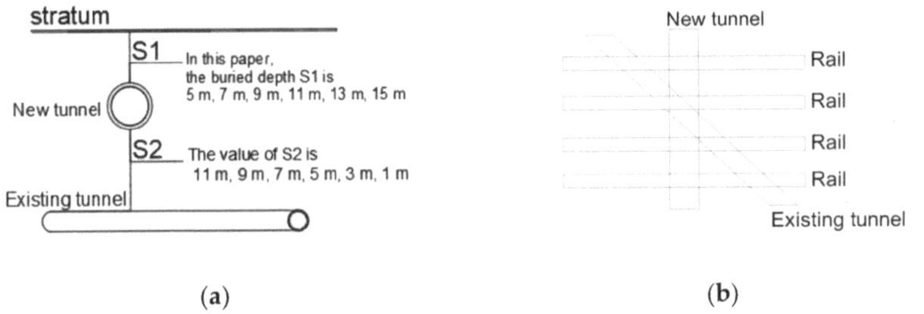

Figure 1. The positions of the new tunnel, the upper subway tunnel, and the operating railway. (**a**) Front view; (**b**) top view.

The outer diameter and inner diameter of the existing subway tunnel segments are 6 m and 5.5 m, respectively, with a width of 2 m and a thickness of 0.25 m. The outer and inner diameters of the newly constructed shield tunnel segments are 6 m and 5.5 m, with a width of 2 m and a thickness of 0.25 m. According to the geological survey, the surface layer is composed of artificial miscellaneous fill soil, and below it is silt formed by recent quaternary alluvial deposits consisting of silty clay. The groundwater level of the site is approximately 10.5–11.7 m below the natural ground level (with an elevation of approximately 91.0 m), corresponding to the quaternary loose rock pore water. The main sources of groundwater recharge are atmospheric precipitation infiltration recharge and groundwater runoff recharge, with the main discharge methods being artificial extraction and groundwater runoff. The annual variation in the groundwater level at the site is approximately 1.0–2.0 m, and the highest water level at the site within 3–5 years was approximately 8.0 m (elevation of approximately 94.0 m). The historically highest groundwater level is at a depth of approximately 6.0 m underground (elevation 96.0 m).

3. Numerical Simulation

3.1. Model Establishment

MIDAS GTS NX 2021 software was used to create a comprehensive three-dimensional model based on the spatial relationships among the new shield tunnel, existing tunnel, and operating railway. The model establishes boundary constraints around and beneath the soil mass, considering the soil mass as infinite and keeping the top surface of the soil mass free at the same time. The distance from the lateral boundary of the model and the distance between the lower bound of the model from top should be taken as sufficient, so that the effects of the boundaries in the numerical model was minimized. The displacement and the stress contours in the finite element software indicate that this distance is sufficient. The model is divided into 531,700 units within an 80 m × 40 m × 60 m space, with a simulated excavation depth of 2 m (Figure 2). Different vertical clearances between the new shield tunnel and both the operating railway (S1) and existing tunnel (S2) are analyzed (5, 7, 9, 11, 13, and 15 m) to simulate and analyze the deformation of the existing tunnels and operational railways during the excavation of new underpass tunnels. The specific working conditions are detailed in Table 1, and the grid calibration is automatically inspected and calibrated by the software.

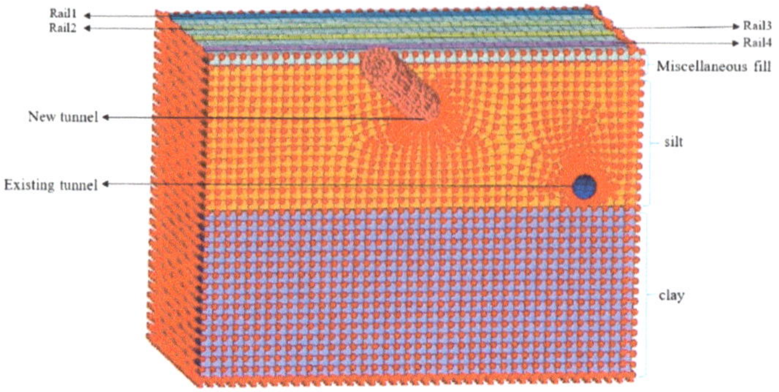

Figure 2. Calculation model.

Table 1. Different buried depths of new tunnels.

Model	S1	S2
1	5 m	11 m
2	7 m	9 m
3	9 m	7 m
4	11 m	5 m
5	13 m	3 m
6	15 m	1 m

3.2. Calculation Parameters and Operating Conditions

Solid elements are used to simulate the segments, existing tunnels, grouting, shield shells, and strata. The segments, existing tunnels, and grouting layers are based on elastic constitutive models, while others are based on Mohr–Coulomb elastoplastic constitutive models. The calculation parameters are listed in Tables 2 and 3.

Table 2. Formation parameters.

Stratum	Modulus of Elasticity KN/m^2	Poisson's Ratio	Unit Weight KN/m^3	Cohesion KN/m^2	Internal Friction Angles °
Miscellaneous fill	15,000	0.42	18	5	15
silt	20,000	0.4	18	20	25
clay	40,000	0.35	20	25	36

Table 3. Supporting parameters.

Type	Modulus of Elasticity KN/m^2	Poisson's Ratio	Unit Weight KN/m^3
railway	31.5	0.2	20
lining segment	31.5	0.25	25
shield shell	34.5	0.25	25
slip casting	28.0	0.25	25

The surface of the model is free, while all other surfaces constrain normal displacement. A built-in function of the simulation software is used to automatically divide the solid mesh into tetrahedral elements with four nodes. According to Code for Design of Railway Bridges and Culverts [32], the calculation model must apply ZKH live load, uniformly distributed load, concentrated load, and concentrated load at the center of the road surface.

The softening modulus method is used to simulate the release of stress during excavation. For simulating shield tunnel excavation, MIDAS GTS NX is used to analyze the activation and passivation of elements. Firstly, the element is "activated," then the excavated soil element is "passivated," and some node forces are released. Finally, the lining element of the pipe segment is "changed in attributes," and all other node forces are released. The simulation of the shield tunnel excavation is presented in Figure 3, and the calculation conditions are listed in Table 4.

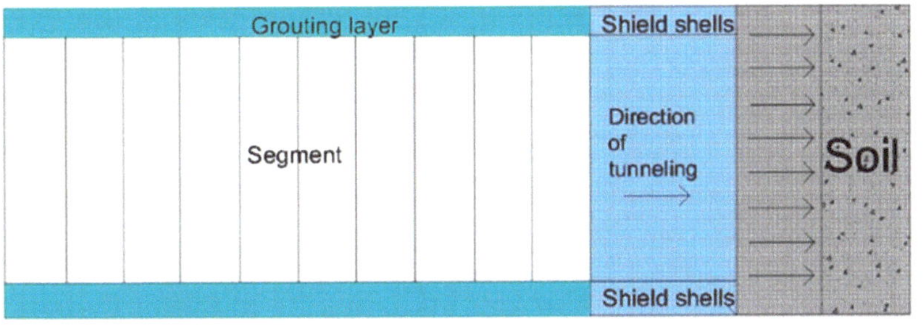

Figure 3. Schematic diagram of simulated excavation of shield tunnel.

Table 4. Construction step.

Step	Step Content
1	Initial stress balance
2	Displacement clearing
3	Subway operation
4	Displacement clearing
5–23	Inner and outer diameter excavation, shield application
7–26	Segment construction and grouting simulation

4. Analysis of Calculation Results

4.1. Impact of New Tunnels on Operating Railways

The cloud map of railway track settlement after completion of the new tunnel excavation is shown in Figure 4. The maximum settlement position of each track is at the intersection of the newly built tunnel and the operating railway. The self-intersection is symmetrically distributed along both sides of the tracks with an impact range of 30 m on both sides of the track, which conforms to a normal distribution.

The maximum settlement curve of the operating railway track corresponding to different steps in the construction of the new tunnel is depicted in Figure 5. As shown in Figure 5, under different burial depths, with the excavation of the new tunnel, the settlement of rail 1, rail 2, rail 3, and rail 4 increases continuously. After completion of construction, maximum settlement of rail 1 is achieved, and as the burial depth increases, rail 1's maximum settlement continuously increases; the maximum settlement value of rail 1 is 12.2 mm (when S1 = 15 m). During the process of construction, new tunnels must not be buried too deep, and corresponding reinforcement measures should be adopted for operating railways.

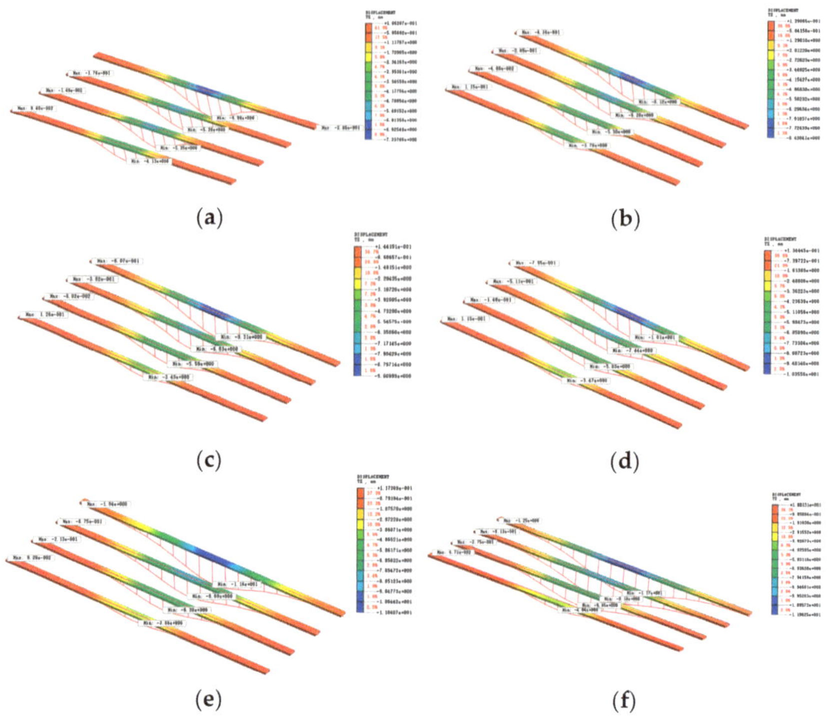

Figure 4. Settlement cloud map of the operational railway track after completion of new tunnel construction. (**a**) S1 = 5 m; (**b**) S1 = 7 m; (**c**) S1 = 9 m; (**d**) S1 = 11 m; (**e**) S1 = 13 m; (**f**) S1 = 15 m.

Taking railway track 2, for example, Figure 6 shows that in the range of 1D to 3D buried depth, the maximum settlement of the operating railway increases with the increase of the buried depth S1. This is because the self-bearing capacity of the surrounding rock cannot be fully utilized when the buried depth of the tunnel is shallow. With the increase of tunnel buried depth, the self-weight stress and self-bearing capacity of surrounding rock will increase, too, while the growth rate of the self-bearing capacity of surrounding rock is relatively less. Due to the limitation of the actual distance between the operating railway and the existing tunnel, this paper does not study the situation in which the buried depth of the tunnel is greater than 3D. Combined with the research results of the literature [33], when the buried depth of the tunnel is greater than 3D, the surface settlement decreases with the increase of the buried depth of the tunnel. As the excavation of the new tunnel steps proceed, the rate of settlement also undergoes certain changes, and these changes can be divided into the following stages: the advanced-deformation stage, the intensified-deformation stage, and the slow-deformation stage. Tunnel excavation causes disturbances to the soil around the operating railway, but the rate of settlement change at this time is low; this is the advanced deformation stage. During the construction step, it is the highest when the location of the construction is directly below the operating railway. The rate of settlement change is the highest in the stage of intensified deformation. Owing to the application of shield tail compensation grouting and lining, the settlement deformation of the operating railway slows down in the slow-deformation stage.

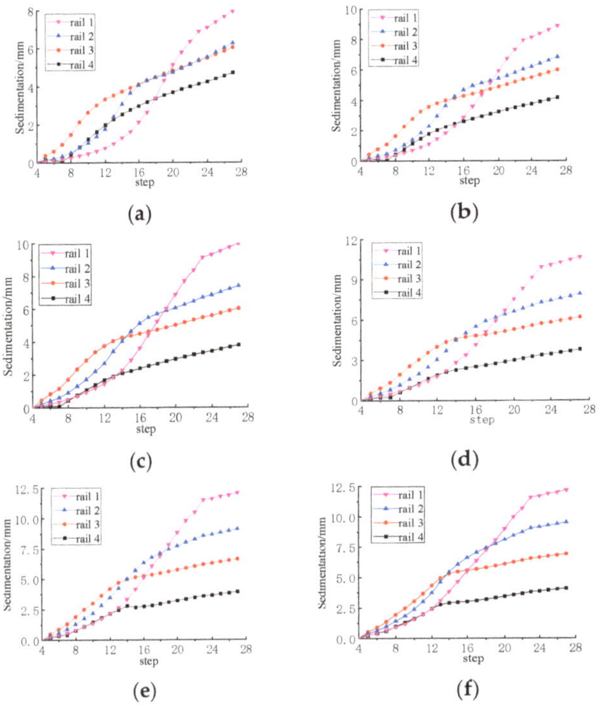

Figure 5. Maximum settlement curve of railway tracks under different construction steps of the new tunnel (**a**) S1 = 5 m; (**b**) S1 = 7 m; (**c**) S1 = 9 m; (**d**) S1 = 11 m; (**e**) S1 = 13 m; (**f**) S1 = 15 m.

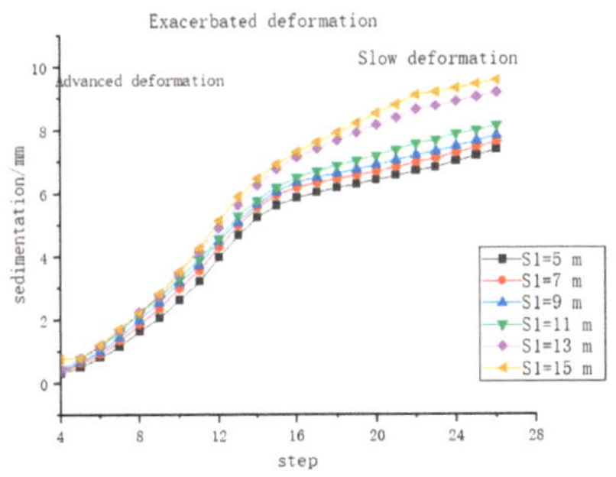

Figure 6. Variation curve of track 2 settlement with construction steps under different S1 values.

4.2. Deformation of Existing Tunnels

The total displacement cloud map of the arch of the existing tunnel is shown in Figure 7 for the condition in which the deformation of the existing tunnel is S2 = 11 m. From the figure, it can be seen that, after construction of the new tunnel, the arch of the existing tunnel undergoes uneven deformation. The closer the existing tunnel is to the intersection of the new and old tunnels, the greater the total deformation, with a maximum value of 1.19 mm. Moreover, the extent of overall deformation of the existing tunnel is substantial in the middle and small on both sides, which conforms to a normal distribution. When the S2 value is 9, 7, 5, 3, and 1 m, the deformation patterns are consistent with that at S2 = 11 m. Due to space limitations, we will not elaborate further.

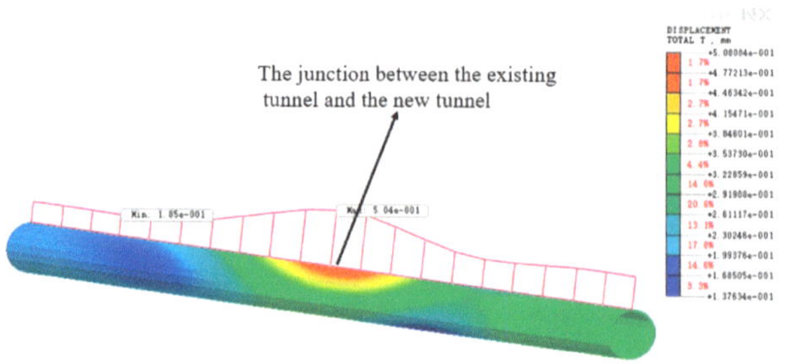

Figure 7. Cloud image of the total displacement of the arch roof of the existing tunnel when S2 = 11 m.

The total displacement curve of the existing tunnel arch during the excavation of new tunnels at different burial depths is shown in Figure 8. As the new tunnel excavation progresses, the maximum total displacement of the arch continuously increases and eventually remains unchanged. During excavation and construction of a new tunnel, the total displacement rate of the existing tunnel arch increases with the decrease in the S2 value. When the S2 value is between 7 m and 11 m (i.e., the vertical clear distance is 1–2D, where D is the diameter of the new tunnel), the total displacement rate of the existing tunnel arch is relatively low. When S2 is between 1 m and 5 m (i.e., the vertical clear distance is 0–1D), the total displacement of the existing arch crown changes rapidly and tends to stabilize at the 14th construction step. The maximum total displacement of the arch of the existing subway tunnel gradually increases with the decrease in the value of S2. When S2 = 1 m, the maximum total displacement of the arch is 6.66 mm. In actual construction, the location of the new tunnel needs to be reasonably selected by considering the potential deformation of the underpass operating railway and by taking corresponding settlement-control measures.

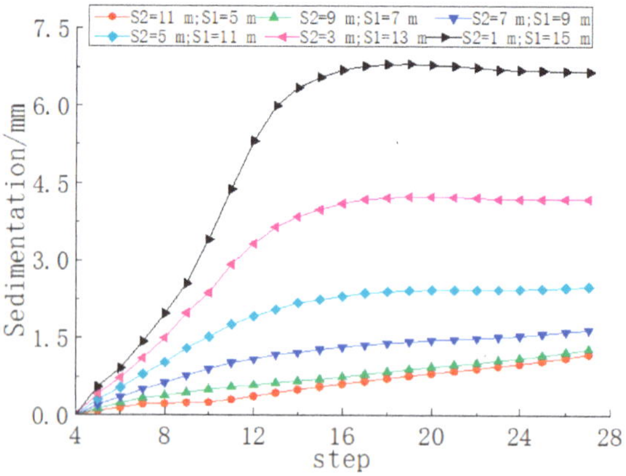

Figure 8. Variation curve of existing tunnel settlement with construction steps under different burial depths.

5. Conclusions

During shield construction of urban subway tunnels, it is inevitable that existing underground pipelines, building foundations, subway stations, existing subway tunnels, and other related underground structures will be traversed and overridden underground. The construction of tunnel shields causes soil disturbance, which affects the stability of related structures and even leads to functional damage to existing structures and accidents. Based on the east–west square connecting passage project of a railway station, this paper establishes a dynamic tunnel connection model using finite element software and discusses the influence of the construction of the new tunnel on the existing tunnel crossing and down-running railway. Moreover, it also analyzes the deformation of the existing tunnel and the operating railway under the construction condition of the new tunnel at different buried depths by controlling the distance between the existing tunnel and the new tunnel.

(1) The excavation of new shield tunnels can cause uneven deformation of adjacent existing tunnels and operating railways. In the burial depth range of 1 to 3D, the impact of the construction of a new shield tunnel on the deformation of existing tunnels and operating railways increases with the burial depth of the new tunnel.
(2) After completion of new shield tunneling construction, both the operating railway and the existing tunnel exhibit maximum deformation at the intersection with the new tunnel, gradually decreasing from the center point toward both sides and following a normal distribution.
(3) As construction of the new tunnel progresses, the settlement changes of the operating railways can be divided into three stages: the advanced-deformation stage, intensified-deformation stage, and slow-deformation stage. During excavation and construction of a new tunnel, the rate of change in the total displacement of the existing tunnel arch increases with the decrease in the value of S2.
(4) For the existing tunnel, when the buried depth is shallow, the deformation of the second stage (strengthening deformation stage) constitutes 42.8% of the principal deformation. However, with the progressive increase in buried depth, the proportion of the first stage gradually rises and eventually reaches 70%. The principal deformation is concentrated in the first (advanced-deformation) and the second (strengthening deformation) stages, each accounting for approximately 40%, and the third stage merely accounts for approximately 13%.

Author Contributions: Conceptualization, C.X. and Y.Q.; Methodology, C.X. and Y.Q.; Software, C.X., Y.Q. and H.L.; Data curation, C.X., Y.Q., H.L. and S.S.; Writing—original draft, C.X. and Y.Q.; Writing—review and editing, C.X., Y.Q., S.S. and H.L. All authors have read and agreed to the published version of the manuscript.

Funding: This research was funded by Shandong Province Higher Education Youth Innovation Technology Support Program Project, grant number 2021KJ058, and the Shandong Jianzhu University College Student Innovation and Entrepreneurship Competition Project (GCX23019501).

Data Availability Statement: The raw data supporting the conclusions of this article will be made available by the authors on request.

Conflicts of Interest: The authors declare no conflicts of interest.

References

1. Jin, D.L.; Shen, X.; Yuan, D.J. Theoretical Analysis of Three-dimensional Ground is Placements Induced by Shield Tunneling. *Appl. Math. Model.* **2020**, *79*, 85–105. [CrossRef]
2. Liang, R.Z. Simplified Analytical Method for Evaluating the Effects of Overcrossing Tunnelling on Existing Shield Tunnels Using the Nonlinear Pasternak foundation model. *Soils Found.* **2019**, *59*, 1711–1727. [CrossRef]
3. Jin, D.L.; Yuan, D.J.; Li, X.G.; Zheng, H.T. Analysis of the Settlement of an Existing Tunnel Induced by Shield Tunneling Underneath. *Tunn. Undergr. Space Technol.* **2018**, *81*, 209–220. [CrossRef]
4. Feng, G.H.; Xu, C.J.; Zheng, M.W. Study of Tunnel-soil Interaction Induced by Tunneling Underlying. *Eng. Mech.* **2023**, *40*, 59–68.
5. Feng, G.H. Deformation Response Analysis on Existing Tunnel Caused by Shield Tunneling Underlying Considering Axial Internal Force. *J. Southeast Univ. (Nat. Sci. Ed.)* **2022**, *53*, 523–529. (In Chinese)
6. He, J.; Yang, Z.; Zhang, X. Effect of Undercrossing Shield Tunnels Excavation on Existing Rectangular Pipe-Jacking Tunnels. *Appl. Sci.* **2023**, *13*, 12235. [CrossRef]
7. Li, Y.; Bian, X.; Peng, H.; Zhu, B.; Zhou, Y. Additional Stress of Soil and Surface Settlement during Tunnel Shield Construction. *Buildings* **2023**, *13*, 1437. [CrossRef]
8. Zhang, D.M.; Huang, Z.K.; Li, Z.L.; Zong, X.; Zhang, D.M. Analytical Solution for the Response of an Existing Tunnel to a New Tunnel Excavation Underneath. *Comput. Geotech.* **2018**, *108*, 197–211. [CrossRef]
9. Ma, S.J.; Li, M.Y.; Jin, J.W.; Bai, K. The Influence of Shallow Buried Double-line Parallel Rectangular Pipe Jacking Construction on Ground Settlement Deformation. *Alex. Eng. J.* **2021**, *60*, 1911–1916. [CrossRef]
10. Liang, R.Z.; Xia, T.D.; Hong, Y.; Yu, F. Effects of above-crossing Tunnelling on the Existing Shield Tunnels. *Tunn. Undergr. Space Technol.* **2016**, *58*, 159–176. [CrossRef]

11. Fei, R.; Peng, L.; Zhang, C.; Zhang, J.; Zhang, P. Experimental and Numerical Studies of a Shield Twin Tunnel Undercrossing the Existing High-Speed Railway Tunnel. *Geotech. Geol. Eng.* **2024**, *42*, 1871–1886. [CrossRef]
12. Liu, H.Y.; Small, J.C.; Carter, J.P.; Williams, D.J. Effects of Tunnelling on Existing Support Systems of Perpendicularly crossing Tunnels. *Comput. Geotech.* **2009**, *36*, 880–894. [CrossRef]
13. Chen, H.D.; Shen, W.M.; Zhang, W.J. Viscoelastic Analysis of Deformation of the Existing Tunnel Due to the Above-Crossing Tunneling. *Chin. Q. Mech.* **2022**, *43*, 958–969. (In Chinese)
14. Liu, H.Y.; Small, J.C.; Carter, J.P. Full 3D Modelling for Effects of Tunnelling on Existing Support Systems in the Sydney Region. *Tunn. Undergr. Space Technol.* **2008**, *23*, 399–420. [CrossRef]
15. Luo, C.S.; Cheng, Y.H.; Bai, Z.; Shen, T.; Wu, X.Y.; Wang, Q.G. Study on Deformation Law of Subway Construction under Passing Existing Line in Short Distance. *Adv. Civ. Eng.* **2021**, *2021*, 6626927. [CrossRef]
16. Zhang, R.; Xiong, H.; Su, D.; Han, W.; Li, Q.; Chen, X. Micromechanical Investigation into the Mechanism of Slurry Fracturing and the Effect of Covering Depth Concerningshield Tunneling in a Sandy Stratum. *Tunn. Undergr. Space Technol.* **2024**, *144*, 105538. [CrossRef]
17. Pan, P.; Chen, W.; Zheng, J.C. Construction Technology of Pipe-jacking Tunnel in Water-rich Intensely Weathered Sandstone Formation crossing underneath Existing Railway. *Sci. Technol. Eng.* **2023**, *23*, 12714–12720. (In Chinese)
18. Fang, M.; Zhou, C.Y.; Liu, Z. Influence of Construction Parameters and Cross Angle on Existing Tunnel Settlement During undercrossing Shield Tunneling Construction. *Eng. Mech.* **2011**, *28*, 133–138. (In Chinese)
19. Li, S. Monitoring and Early Warning of Subgrade Settlement for High Speed Railway Undercrossed by Shield Tunnel. *Railw. Eng.* **2023**, *63*, 127–130. (In Chinese)
20. Liu, X.; Fang, H.C.; Jiang, A.N.; Zhang, D.L.; Fang, Q.; Lu, T.; Bai, J.R. Mechanical Behaviours of Existing Tunnels Due to Multiple-tunnel Excavations Consideringconstruction Sequence. *Tunn. Undergr. Space Technol.* **2024**, *152*, 105870. [CrossRef]
21. Mroueh, H.; Shahrour, I. A full 3-D Finite Element Analysis of Tunneling-adjacent Structures Interaction. *Comput. Geotech.* **2003**, *30*, 245–253. [CrossRef]
22. Hu, H.; Xue, T.; Li, J.; Liu, P.; Wang, B.; Liu, Y. Mechanical Performance of Concrete Segment Lining Structure of Shield Tunneling in Different Strata. *Buildings* **2023**, *13*, 3118. [CrossRef]
23. Xiang, P.; Wei, G.; Jiang, H.; Ding, Y.; Guo, H.; Zhang, R. Experimental Study and Theoretical Analysis of the Effect of Surcharge/Unloading at Both Sides of the Ground Surface on an Existing Shield Tunnel at Different Burial Depths. *Buildings* **2023**, *13*, 1911. [CrossRef]
24. Wang, C.W.; Liu, X.L.; Song, D.Q.; Wang, E.Z.; He, Z.H.; Tan, R.S. Structural response of former Tunnel in the Construction of Closely-Spaced cross-river Twin Tunnels. *Tunn. Undergr. Space Technol.* **2024**, *147*, 105652. [CrossRef]
25. Ding, Z.; Zhang, X.; He, S.H.; Qi, J.Y.; Lin, G.C. Experimental and Theoretical Study on Longitudinal Deformation and Internal Force of Shield Tunnel under Surcharge. *Tunn. Undergr. Space Technol.* **2024**, *144*, 105506. [CrossRef]
26. Zhang, Y.; Tao, L.; Liu, J.; Zhao, X.; Guo, F.; Tan, L.; Wang, Z. Construction Techniques and Mechanical Behavior of Newly-built Large-Span Tunnel Ultra-short Distance up-crossing the Existing Shield Tunnel with Oblique Angle. *Tunn. Undergr. Space Technol.* **2023**, *138*, 105162. [CrossRef]
27. Yao, J.; Wang, Z.; Li, Z.; Xiao, X.; Wu, H.; Ma, W.; Huo, Z.; Pu, Y. Research on Structural Deformation Law and Allowable Deformation of Shield Tunnel Underneath Passing High-Speed Railway Tunnel. *Proc. SPIE-Int. Soc. Opt. Eng.* **2021**, *12058*, 758–764. [CrossRef]
28. Shi, Y.; Lin, L.; Xu, J.; Qing, J. Analysis of Construction Mechanics of Expansive Soil Tunnel Passing Through Existing subway station in short distance Under humidification Condition. *J. Railw. Sci. Eng.* **2020**, *18*, 200–210. [CrossRef]
29. Zhu, Q.L.; Ding, Y.P. Impact of New undercrossing Tunnel Excavation on the Stability of the Existing Tunnel. *Front. Earth Sci.* **2022**, *10*, 915882. [CrossRef]
30. Wu, K.; Zheng, Y.; Li, S.; Sun, J.; Han, Y.; Hao, D. Vibration Response Law of Existing Buildings Affected by Subway Tunnel Boring Machine Excavation. *Tunn. Undergr. Space Technol.* **2021**, *120*, 104318. [CrossRef]
31. Cheng, Y.; Liu, X. Research on the Pressure Distribution Law of Synchronous Grouting in ShieldTunnels and the Force on Segments. *Buildings* **2024**, *14*, 1099. [CrossRef]
32. *TB 10002-2017*; Code for Design on Railway Bridge and Culvert. National Railway Administration: Beijing, China, 2017.
33. Yan, P.F. Three-dimensional Finite element Analysis of Influence by Excavation of different buried Tunnels on Ground Subsidence. *J. Guangxi Univ.* **2021**, *46*, 60–66.

buildings

MDPI

Article

Intelligent Identification of Surrounding Rock Grades Based on a Self-Developed Rock Drilling Test System

Quanwei Liu [1], Junlong Yan [2], Hongzhao Li [2,3], Peiyuan Zhang [1], Yankai Liu [2,*], Linsheng Liu [1], Shoujie Ye [1] and Haitao Liu [4]

1 Qingdao Metro Line 6 Co., Ltd., Qingdao 266427, China; anboil@126.com (Q.L.); 18553402127@163.com (P.Z.); liulsh2011@126.com (L.L.); yeshoujie813@163.com (S.Y.)
2 Geotechnical and Structural Engineering Research Center, Shandong University, Jinan 250061, China; 18769746170@163.com (J.Y.); lhz13515319837@163.com (H.L.)
3 Shandong Transportation Institute, Jinan 250031, China
4 China Railway First Bureau Group Fifth Engineering Co., Ltd., Baoji 721000, China; lyystu@126.com
* Correspondence: yankai_liu@sdu.edu.cn

Abstract: The classification of surrounding rock is crucial for formulating safe tunnel construction plans and support measures. However, the complex geological environment of tunnels presents a challenge in obtaining accurate drilling parameters for rock mass classification. This paper presents the development of a rock drilling testing system, which includes a propulsion speed acquisition system, oil pressure acquisition system, air pressure acquisition system, and an automatic data acquisition system. This system enables real-time, high-precision automatic collection and storage of parameters such as propulsion speed, with data collected twice per second for each parameter. Leveraging the Qingdao Metro Line 6 as a case study, we conducted rock mass drilling and constructed a rock mass classification database. By employing kernel density estimation and Pearson correlation analysis, we quantified the correlation between rock mass classification and the drilling parameters. The results indicated that relying on a single drilling parameter is insufficient for accurately determining rock mass classification. Both impact pressure and rotational pressure showed the strongest correlation with rock mass classification, each with a correlation coefficient below −0.8 (indicating a strong negative correlation). Outlier values of drilling parameters were excluded using the interval method. Based on the remaining data, we established an intelligent rock mass classification model using the random forest algorithm. This model demonstrated good accuracy and generalization performance, with an average accuracy exceeding 0.9. The proposed rock drilling testing system, combined with the intelligent rock mass classification model, forms an integrated system for the intelligent identification of rock mass grades. This system has significant implications for the intelligent and safe construction of drill-and-blast tunnels.

Keywords: rock drilling test system; drilling and blasting method; drilling parameters; intelligent classification of surrounding rocks; random forest algorithm

1. Introduction

Drilling and blasting method has high flexibility in tunnel construction, can adapt to complex rock types and geological structures, and is widely used [1,2]. The classification of surrounding rocks is essential for tunnel excavation using the drilling and blasting method, as it provides insights into the rock's strength, deformation characteristics, and stability at the tunnel face. Accurate classification can significantly mitigate the risk of geological hazards such as water and mud inflow, collapses, and rock bursts during excavation. Currently, the most common classification methods for surrounding rocks are the Q system [3,4], Rock Mass Rating (RMR) [5–7], and Basic Quality (BQ) systems [8–10]. These methods generally require obtaining rock cores through drilling and assessing rock quality indicators like uniaxial compressive strength through laboratory tests. However, accurate assessments depend heavily on expertise and engineering experience; without this, evaluations might be biased.

With advances in artificial intelligence, using machine learning for rapid classification has become a new direction for rock classification [11,12]. Neural networks have performed well in predicting RMR values, taking physical and mechanical parameters such as density, compressive strength, and RQD as inputs. Zheng et al. [13] utilized the least squares support vector machine (LSSVM) to express the implicit relationship between classification indicators and rock mass grades. They established a rock mass classification model using geological forecasting and rock mass strength rebound results as classification indicators. Jalalifar et al. [14] utilized parameters like uniaxial compressive strength, RQD, joints, and groundwater conditions as inputs to develop an intelligent classification model based on a neuro-fuzzy method. Hasegawa et al. [15] explored the applicability of artificial neural networks for classifying rock masses in mountainous tunnels, achieving improved classification accuracy using geophysical datasets (seismic velocity and resistivity). Bressan et al. [16] constructed various intelligent classification models, including multilayer perceptron and random forest models, based on geophysical data and compared their performance differences.

When there is an intrinsic relationship between input and output indicators and a sufficient number of samples are available, machine learning algorithms can continuously optimize classification boundaries through automated analysis and iterative learning. These algorithms have proven to be highly effective in rock mass classification [17,18]. They tend to outperform traditional statistical methods in highly nonlinear problems, with better performance and higher computational efficiency. However, the input parameters for these models are typically obtained through field or laboratory testing, which can be inefficient.

Since drill rigs are in direct contact with rocks during core extraction, parameters such as thrust and torque are indicative of rock quality. Numerous scholars have investigated this relationship to enhance our understanding and application of these parameters in rock quality assessment. Tan et al. [19] utilized the discrete element method (DEM) to simulate the rock-breaking process of drilling rigs. Their findings revealed a strong correlation between the average drilling rate, vibration acceleration, vibration frequency, and the surrounding rock grade. These results were validated through experimental tests conducted at a tunnel site. Torno et al. [20] studied the feasibility of using drilling parameters to predict RMR values and established a fuzzy logic model for predicting geomechanical properties. Mostofi et al. [21] developed a rock strength prediction model based on drilling rate, rotation speed, bit weight, and torque, finding that worn drill bits often overestimate rock strength. Kalantari et al. [22] used limit equilibrium principles, considering factors like contact friction, fracture zones, and drill bit geometry during rotary drilling, to create a theoretical model for estimating rock strength parameters, later validating its accuracy through standard tests. Kalantari et al. [23] found that strength parameters like internal friction angle and cohesion can be estimated from drilling data alone, unaffected by drill bit wear. Lakshminarayana et al. [24] built a mathematical model for estimating rock mechanical properties using drilling variables (thrust, torque, vibration parameters) and acoustic parameters from rotary drilling. They estimated the uniaxial compressive and tensile strengths of sedimentary rocks, noting an estimated error of about 10% through experiments.

This study analyzed the operating principles of drill rigs and identified four drilling parameters closely related to rock quality. Based on this analysis, a rock drilling testing system was developed to efficiently collect these parameters in real time. Using data from Qingdao Metro Line 6, a surrounding rock grade database was constructed, and the correlation between rock grades and the four parameters was quantitatively analyzed. Finally, an intelligent classification model based on the random forest algorithm was built, showing excellent performance. These findings provide a solid foundation for intelligent tunneling and information-based monitoring and management.

2. Development of the Rock Drilling Testing System

2.1. Working Performance of the Down-the-Hole Drill Rig

This study uses the SK150 crawler-mounted surface down-the-hole (DTH) drill rig manufactured by Hubei Shoukai Machinery Co., Ltd. (The company is located in Huangshi City, Hubei Province, China) for drilling operations, as shown in Figure 1. The primary task of the SK150 DTH drill rig is impact drilling, which is accomplished by a hammer. In this study, a threaded bit designed for hydraulic jumbos is used, and the drilling holes are 95 mm in diameter.

Figure 1. The SK150 crawler-mounted down-the-hole (DTH) drill rig.

The main technical specifications of the SK150 crawler-mounted surface down-the-hole drill rig are shown in Table 1.

Table 1. Key technical specifications of the SK150 down-the-hole drill rig.

Parameter	Value
Rock Hardness	f = 6~18
Borehole Diameter	90~146 mm
Borehole Depth	30 m
Lowest Horizontal Hole	500 mm
Highest Horizontal Hole	3200 mm
Travel Speed	2.5 km/h
Climbing Gradient	30°
Gradient for Trailing Compressor	15°
Rotation Speed	0~170 r/min
Rotation Torque	2960 Nm
Compensation Length	900 mm
Operating Pressure	0.7~1.7 MPa
Air Consumption	9~14 m^3/min
Minimum Ground Clearance	472 mm
Propulsion Beam Tilt	Tilt down 105°, up 13°
Propulsion Beam Swing	Right 5°, Left 90° (or Right 90°, Left 5°)
Boom Tilt Angle	Up 49°, Down 25°
Boom Swing Angle (Horizontal Boom)	Left 27°, Right 29°
Dimensions (Transport State) L × W × H	6500 × 2110 × 2400 mm
Diesel Tank Capacity	85 L
Hydraulic Oil Tank Capacity	188 L
Total Weight	5.95 t

By analyzing existing studies [12] and the principles and processes of surrounding rock drilling, it was determined that the four drilling parameters most closely correlated with the surrounding rock grade are propulsion speed, impact pressure, rotary pressure, and propulsion pressure. A detailed explanation of each parameter is as follows:

(1) Propulsion speed: This measures the drilling speed of the drill bit under various pressures. While multiple factors affect the propulsion speed, it generally reflects the rock mass's strength and integrity. In general, the speed decreases with increasing rock hardness or integrity.

(2) Impact pressure: This is the air pressure transferred from the compressor to the hammer in the down-the-hole drilling rig during impact drilling. As the main factor in breaking surrounding rock, impact pressure increases with rock hardness. Additionally, more impact pressure is required for intact rock masses of the same hardness, as they need more force to fracture. Therefore, as rock integrity improves, the average impact pressure required also increases.

(3) Rotary pressure: This is the oil pressure in the hydraulic cylinder during rotary motion. After the impact, some of the rock mass may remain partially fractured. Under the rotary mechanism, the drill bit rotates, changing its angle to fully crush the rock. Due to the positive correlation between shear and uniaxial compressive strength, rotary pressure rises with increased rock hardness. Also, more intact rock masses require higher rotary pressure, as more rock needs to be cut during rotation.

(4) Propulsion pressure: This is the oil pressure in the hydraulic cylinder during forward movement. Propulsion pressure ensures the drill bit remains in close contact with the rock. Higher propulsion pressure is needed to maintain this contact when rock strength is higher. More intact rock masses also require higher propulsion pressure because they generate stronger reverse impact pressures, increasing the propulsion force required. Therefore, as rock hardness or integrity increases, so too does the propulsion pressure.

2.2. Development of the Rock Drilling Testing System

The SK150 crawler-mounted surface down-the-hole drill rig used in this study does not have an automatic data recording system for drilling parameters. Instead, the rotary, propulsion, and impact pressures are displayed separately on the hydraulic and air pressure gauges, as shown in Figure 2.

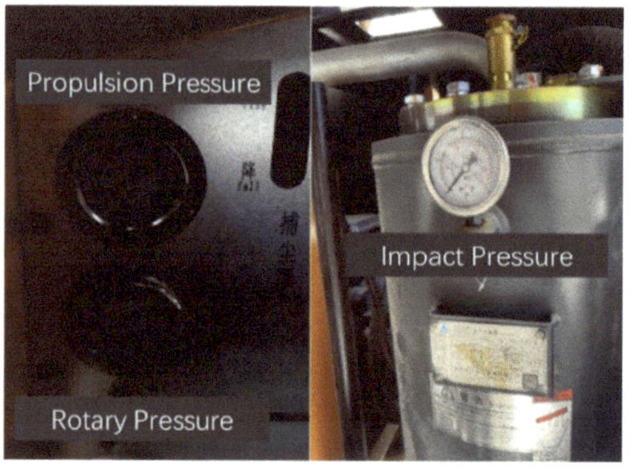

Figure 2. Hydraulic and air pressure gauges of the down-the-hole drill rig.

To automatically collect a large volume of drilling parameters, sensors are strategically placed at various locations on the rig. They can record multiple parameters, including the four main drilling parameters: impact pressure, rotary pressure, propulsion pressure, and propulsion speed. Table 2 lists the sensors and their respective measurement methods for each of these parameters.

Table 2. Measurement methods for drilling parameters.

No.	Drilling Parameter	Sensor	Measurement Method
1	Propulsion Speed	Displacement Sensor	After drilling begins, data are collected twice per second. Every 512 bytes of data are stored once the buffer is full. The displacement measured by the sensor is divided by time to obtain the propulsion speed.
2	Impact Pressure	Air Pressure Sensor	Using a 141.5H-141.5A tee fitting, the air pressure sensor is connected to the tee and data are transferred through a high-pressure oil pipe to a custom data acquisition and storage system.
3	Rotary Pressure	Hydraulic Pressure Sensor	Using a 141.5H-141.5A tee fitting, the hydraulic pressure sensor is connected to the tee, and data are transferred through a high-pressure oil pipe to the custom data acquisition and storage system.
4	Propulsion Pressure	Hydraulic Pressure Sensor	Using a 141.5H-141.5A tee fitting, the hydraulic pressure sensor is connected to the tee, and data are transferred through a high-pressure oil pipe to the custom data acquisition and storage system.

2.2.1. Propulsion Speed Acquisition System

The schematic diagram of the propulsion speed acquisition system is shown in Figure 3.

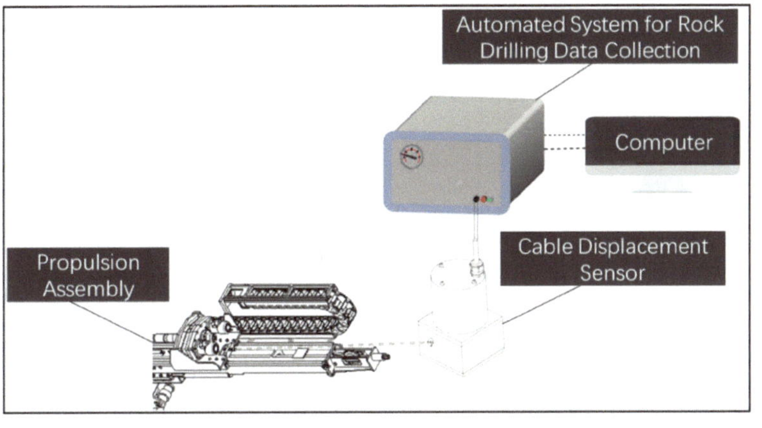

Figure 3. Schematic diagram of the propulsion speed acquisition system.

The actual setup of the propulsion speed acquisition system is shown in Figure 4. The displacement sensor's draw wire is fixed to the rear of the propulsion cylinder in the down-the-hole drill rig assembly. During drilling operations, the drill bit is positioned against the rock surface, and the hammer begins to move forward. The draw-wire displacement sensor records the relative displacement of the propulsion assembly every 0.5 s, storing the data in the automatic rock drilling data collection system. This process allows us to determine the drilling speed of the down-the-hole drill rig.

In this study, an MT200-3500 draw-wire displacement sensor was used to collect displacement data. Its fully enclosed design ensures strong anti-interference capabilities, making it effective in harsh environments like underground spaces with moisture, dripping water, and high vibration. Table 3 provides the detailed specifications of this sensor.

Figure 4. Actual setup of the propulsion speed acquisition system.

Table 3. Specifications of the displacement sensor.

Parameter	Value
Signal Output	Digital (pulse) signal
Measurement Range	0–3500 mm
Signal Type	Differential, open circuit, push–pull, voltage, RS485/232
Resolution	0.01/0.02/0.05 (selectable)
Linearity Accuracy	0.05% FS
Repeatability	0.01%
Maximum Speed	1000 mm/s
Wire Rope	High-flex imported plastic-coated steel wire
Pull Force at Cable Outlet	5 N
Service Life	2 million cycles
Operating Voltage	5 V, 5–24 V, 10–30 V
Operating Temperature	−25–75 °C
Protection Rating	IP54 (standard), IP65 (optional)
Housing Material	Imported aluminum alloy, anti-static, nonconductive

2.2.2. Hydraulic Pressure Acquisition System

The SK150 down-the-hole drill rig is equipped with a dashboard displaying propulsion and rotary pressures, but it lacks automatic real-time recording functionality. To address this, the rig was modified internally. The schematic diagram of the hydraulic pressure acquisition system is shown in Figure 5, and the actual setup is illustrated in Figure 6. The YC-131 pressure transmitter is installed in the hydraulic pipeline behind the pressure gauge. The transmitter outputs two lines: one connects to the dashboard displaying propulsion and rotary pressures, and the other connects to the automatic rock drilling data acquisition system for real-time pressure recording.

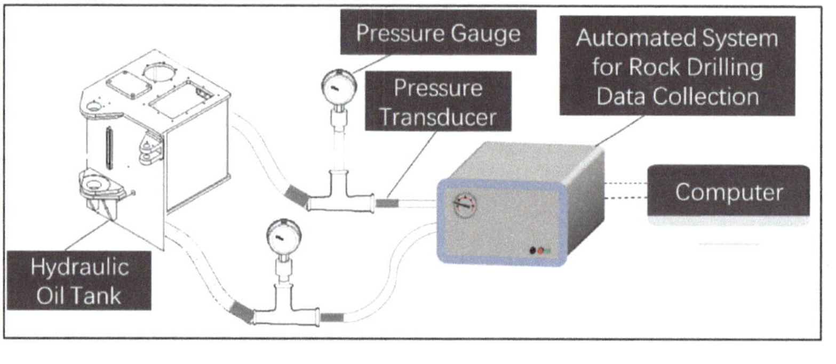

Figure 5. Schematic diagram of the hydraulic pressure acquisition system.

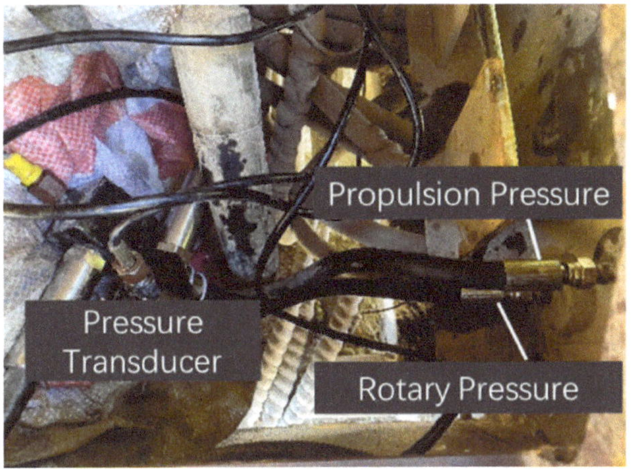

Figure 6. Actual setup of the hydraulic pressure acquisition system.

In this study, the specifications of the YC-131 pressure transmitter are outlined in Table 4. This pressure transmitter uses imported circuitry based on American BB integrated chips. It features an advanced diaphragm isolation technology with imported diffused silicon pressure-sensitive elements. Its compact structure, convenient installation, lightning protection, anti-interference, vibration resistance, high stability, rapid response, and high accuracy make it an excellent choice.

Table 4. Specifications of the pressure transmitter.

Parameter	Specification
Pressure Type	Gauge pressure, negative pressure, absolute pressure
Material	Core: 316 stainless steel; Shell: 316 stainless steel
Output Signal	4–20 mA, 0–5 VDC, 0–10 VDC, 1–5 VDC
Ambient Temperature	−40 to 85 °C
Overload Capacity	200% FS
Vibration Resistance	25 g (20–2000 Hz)
Response Frequency	≤500 Hz
Measuring Range	0–30 MPa
Accuracy	0.5%
Power Supply	24 V DC
Output	4–20 mA

2.2.3. Air Pressure Acquisition System

During operation, the SK150 drill rig receives its impact pressure from a Kaishan brand screw compressor (KSDY-15/17), which has a dashboard displaying impact pressure but lacks real-time automatic recording capabilities. To address this, the compressor was modified, and the schematic diagram of the air pressure acquisition system is shown in Figure 7, with the actual setup in Figure 8. The YC-131 pressure transmitter was installed at the connection between the pressure gauge and air pipe. The transmitter outputs two lines: one to the impact pressure display dashboard, and the other to the automatic rock drilling data collection system for real-time impact pressure recording.

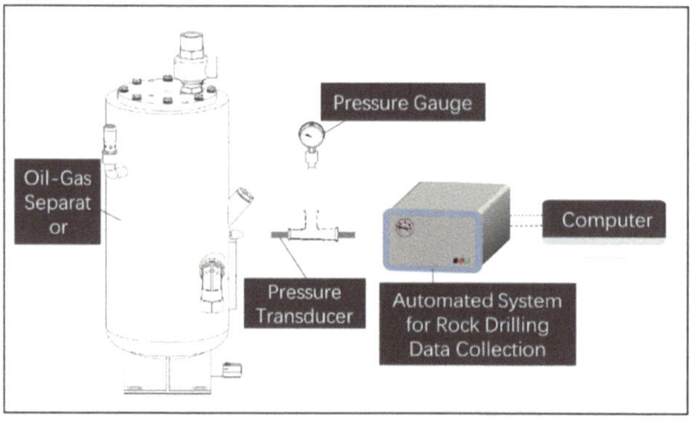

Figure 7. Schematic diagram of the air pressure acquisition system.

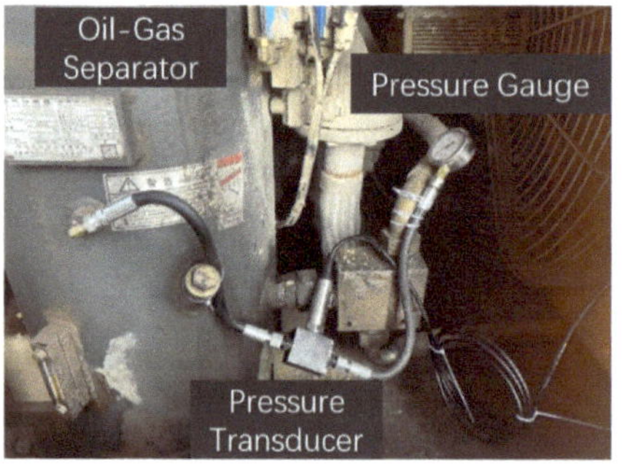

Figure 8. Actual setup of the air pressure acquisition system.

2.2.4. Automatic Data Acquisition System

The automatic rock drilling data acquisition system is based on the STM32F103RCT6 microcontroller. The ADC chip used for the hydraulic data collection is CS1237, a high-precision, low-power analog-to-digital conversion chip. For storage, the system uses a W25Q256 flash chip, providing 32 MB of storage space. The encoder section uses the AM26C32 chip, and data are collected twice per second. Once the buffer reaches 512 bytes, the data are saved.

The data acquisition and storage system includes the following components: ① and ② are ADC collection interfaces, ③ is the encoder interface, and ④ is the USB interface, as shown in Figure 9.

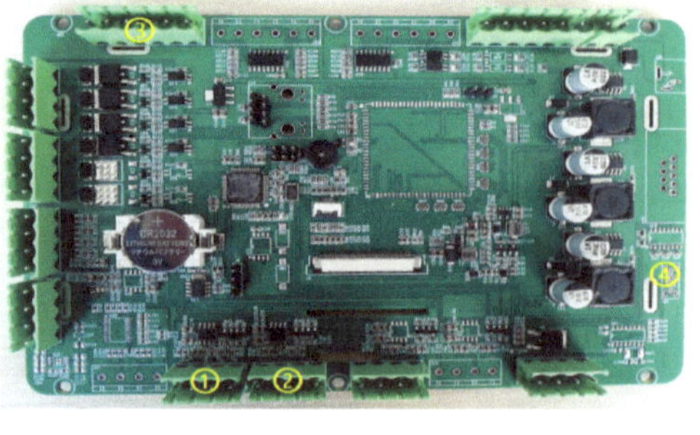

Figure 9. Data acquisition and storage system.

The primary function of the ADC (analog-to-digital converter) is to convert digital signals to analog signals, and it is mainly used for data conversion in data acquisition. The ADC chip used in this study, CS1237, is a high-precision, low-power ADC with a differential input channel, built-in temperature sensor, and high-precision oscillator. The ADC data output rate in normal mode can be selected at 10 Hz, 40 Hz, 640 Hz, or 1.28 kHz, with 10 Hz being the default. The CS1237 includes an internal crystal oscillator, integrated temperature sensor, power-down function, and a 2-wire SPI interface with a maximum speed of 1.1 MHz.

The encoder section of this study uses the AM26C32 chip, which collects data twice per second. When the buffer reaches 512 bytes, data are saved. The AM26C32 is a quadruple differential line receiver designed for balanced or unbalanced digital data transmission. The enable function is common to all four receivers and provides a choice of active-high or active-low input. The status output allows direct connection to bus-organized systems. Its failsafe design ensures that if the input is open, the output is always high. The main specifications of the AM26C32 are presented in Table 5.

Table 5. Specifications of the AM26C32.

Technical Parameter	Parameter Value
DC Voltage	4.50 V (min)
Supply Current	15 mA
Operating Temperature (Max/Min)	70 °C/0 °C
Supply Voltage (Max/Min)	5.5 V/4.5 V

In this study, the data storage uses the W25Q256 flash chip, providing 32 MB of storage. Compared to conventional NAND flash memory, it has slower erase speeds and smaller capacity, but it offers faster read speeds, a lower probability of bad blocks, and improved security. These characteristics make it suitable for this research environment, which involves vibrations. The main technical specifications are presented in Table 6.

Table 6. Technical specifications of the W25Q256 chip.

Parameter Name	Specification
Mounting Style	SMD/SMT
Package	WSON-8
Maximum Frequency	133 MHz
Interface Type	SPI
Data Bus Width	8-bit
Supply Voltage (Max/Min)	1.95 V/1.7 V
Operating Temperature (Max/Min)	85 °C/−40 °C

3. Data Collection and Correlation Analysis

3.1. Data Collection

At the Qingdao Metro Line 6 underground excavation station near the Qingdao Medical West Campus, drilling parameters were collected using the rock drilling testing system. This paper employs the Basic Quality (BQ) classification method to categorize the surrounding rock masses. The BQ method, known for its high accuracy, classifies the rock masses through a comprehensive analysis of the saturated uniaxial compressive strength and the rock mass integrity index. The BQ method comprises two steps:

(1) Calculate the basic quality index (BQ value) using Equation (1).

$$BQ = 100 + 3R_c + 250K_v \quad (1)$$

In the formula, BQ represents the Basic Quality Index of the rock mass, R_c is the uniaxial saturated compressive strength of the rock, and K_v is the rock mass integrity index, which is the square of the ratio of the elastic longitudinal wave velocities in the rock mass and rock.

When using Equation (1), the following constraints should be observed:

1. When $R_c > 90\ K_v + 30$, R_c should be set to $90\ K_v + 30$ and K_v should be substituted into the equation to calculate the BQ value.
2. When $K_v > 0.04\ R_c + 0.4$, K_v should be set to $0.04\ R_c + 0.4$ and R_c should be substituted into the equation to calculate the BQ value.

(2) Based on the BQ value and the qualitative characteristics of the basic quality of the rock mass, the rock mass grade is determined according to Table 7.

Table 7. BQ method rock mass classification table.

Rock Mass Grade	Qualitative Characteristics of Rock Mass Quality	BQ Value
I	Hard rock, intact rock mass.	>550
II	Hard rock, relatively intact rock mass; Moderately hard rock, intact rock mass.	550~451
III	Hard rock, moderately fractured rock mass; Moderately hard rock, relatively intact rock mass; Moderately soft rock, intact rock mass.	450~351
IV	Hard rock, fractured rock mass; Moderately hard rock, moderately fractured to fractured rock mass; Moderately soft rock, relatively intact to moderately fractured rock mass; Soft rock, intact to relatively intact rock mass.	350~251
V	Moderately soft rock, fractured rock mass; Soft rock, moderately fractured to fractured rock mass; All extremely soft rocks and all extremely fractured rocks.	≤250

This study conducted uniaxial saturated compressive strength tests and rock mass wave velocity tests, as shown in Figures 10 and 11. The final results indicate that the rock mass grades in the drilling test area range from II to V.

Figure 10. Uniaxial saturated compressive strength test of rock.

To ensure the accuracy of machine learning training results, we selected approximately the same amount of data for each of the three surrounding rock grades from the samples collected to build the database relating drilling parameters to surrounding rock grades. The database contains a total of 436 rock mass classification samples. The drilling parameters and sample sizes for each rock mass grade are shown in Table 8.

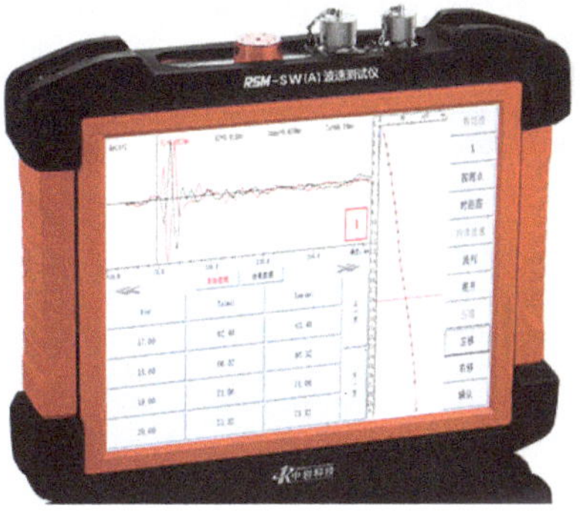

Figure 11. Rock mass wave velocity test.

Table 8. Surrounding rock classification database.

Propulsion Speed (m/min)	Impact Pressure (Pa)	Propulsion Pressure (Pa)	Rotary Pressure (Pa)	Rock Classification	Sample Size
0.91	135.08	87.11	99.37		
1.04	137.95	76.33	91.12		
1.37	133.88	77.02	94.15	Grade 2	76
0.96	153.25	74.26	95.04		
...	...	...	...		
1.41	144.34	63.38	84.12		
1.62	143.36	62.15	81.56		
1.89	147.86	70.12	95.14	Grade 3	119
1.64	151.77	70.03	96.58		
...	...	...	...		
2.32	143.12	77.33	83.22		
2.14	138.86	72.12	79.86		
2.35	129.43	67.72	78.58	Grade 4	129
2.18	138.89	72.23	79.81		
...	...	...	...		
3.28	124.29	49.39	65.11		
3.31	127.09	54.15	75.93		
2.61	122.55	49.61	68.11	Grade 5	112
2.88	122.92	58.13	73.42		
...	...	...	...		

3.2. Correlation Analysis

To investigate the relationship between drilling parameters and surrounding rock grades, we studied the direction (positive or negative) and magnitude of the correlation coefficients for each drilling parameter. Based on the drilling parameter and surrounding rock grade database, kernel density estimation (KDE) plots were drawn to visualize the relationships between propulsion speed, impact pressure, rotary pressure, propulsion pressure, and surrounding rock grades, as shown in Figure 12.

From Figure 12, it is evident that:

(1) For any drilling parameter, each surrounding rock classification corresponds to a specific distribution range, indicating a correlation between drilling parameters and surrounding rock grades. However, there is significant overlap in the distribution ranges of drilling parameters for different surrounding rock classifications, making it challenging to accurately identify the grade using a single parameter alone.

(2) Comparing the density distribution of different drilling parameters with surrounding rock classifications reveals a monotonic change in the distribution ranges of propulsion

speed, impact pressure, rotary pressure, and propulsion pressure as the surrounding rock grade increases.

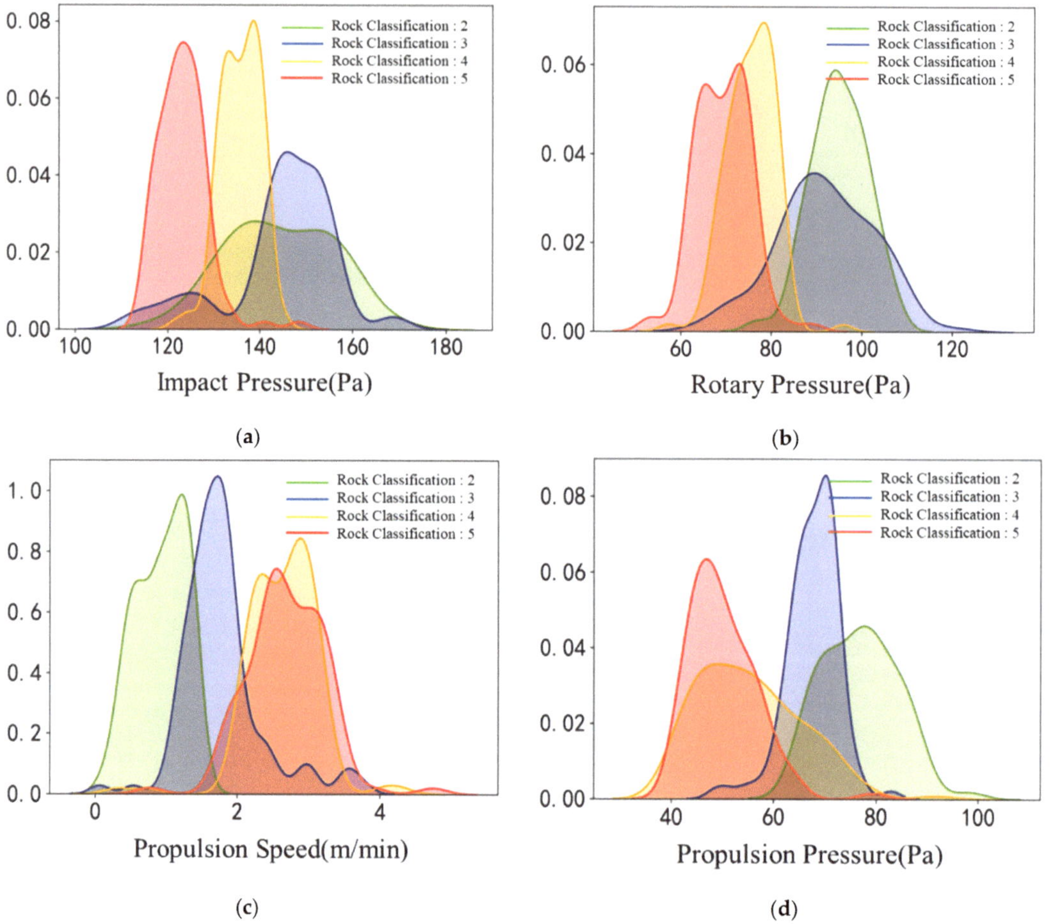

Figure 12. Core density map of drilling parameters and surrounding rock classification: (**a**) Impact pressure; (**b**) rotary pressure; (**c**) propulsion speed; (**d**) propulsion pressure.

According to the empirical method for normal distribution testing, if the ratio of the sample median to the arithmetic mean is between 0.9 and 1.1, and the arithmetic mean is greater than three times the standard deviation, the sample can be considered to follow a normal distribution. Based on the rock mass grade sample database, calculations show that all four drilling parameters meet the empirical method criteria, indicating that the samples follow a normal distribution. Pearson correlation analysis was used to quantify the correlation between the drilling parameters and rock mass grades, with the rock mass grade represented by the BQ value. The formula for calculating the Pearson correlation coefficient is shown in Equation (2). The correlation coefficient ranges between -1 and 1, with -1 and 1 indicating complete negative correlation and complete positive correlation, respectively. The greater the absolute value of the correlation coefficient, the higher the linear correlation between the variables.

$$\rho = \frac{\mathrm{Cov}(x,\ y)}{\sigma_x \sigma_y} = \frac{\sum_{i=1}^{n} (x_i - \overline{x})(y_i - \overline{y})}{\sqrt{\sum_{i=1}^{n} (x_i - \overline{x})^2} \sqrt{\sum_{i=1}^{n} (y_i - \overline{y})^2}} \quad (2)$$

In the formula, ρ represents the correlation coefficient, Cov denotes the covariance, and σ signifies the standard deviation.

The heat map showing the correlation between drilling parameters and surrounding rock classification is presented in Figure 13. Key observations include the following:

(1) The absolute values of the correlation coefficients (ρ) for the four drilling parameters with the surrounding rock classification are all greater than 0.6, indicating a strong correlation between these parameters and rock classification.

(2) Propulsion speed shows a positive correlation with surrounding rock classification, whereas impact pressure, propulsion pressure, and rotary pressure all display a negative correlation.

(3) The absolute value of the correlation coefficient is highest between impact pressure and surrounding rock classification, suggesting the strongest correlation. It is followed by rotary pressure (−0.83), propulsion speed (0.74), and propulsion pressure (−0.69).

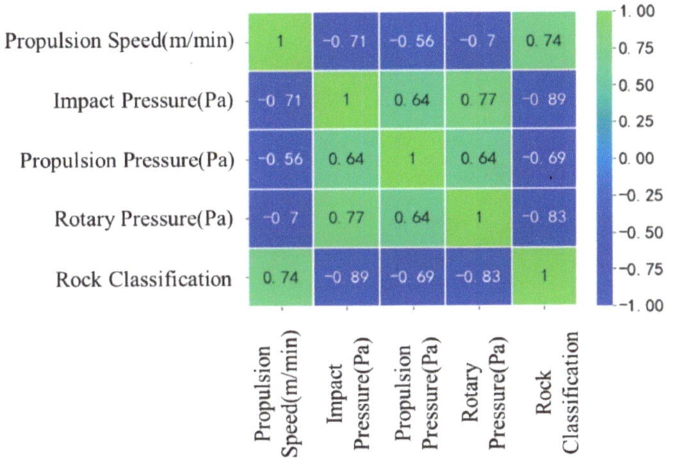

Figure 13. Heat map of drilling parameters and surrounding rock classification.

4. Data Preprocessing

4.1. Removing Outliers

Due to human and other factors, the collected drilling parameters often contain some outliers. The presence of outliers hinders building a model that reflects general patterns, so the standard deviation and mean interval method are used to eliminate these data anomalies.

From the parameter correlation analysis, impact pressure and rotary pressure show the strongest correlation with surrounding rock classification, so the samples are primarily filtered based on these two parameters.

The data for different rock classifications are processed in batches. Let $\alpha 2$ and $\beta 2$ represent the mean and standard deviation of impact pressure for Grade 2 surrounding rock data, respectively. Similarly, for Grade 3 surrounding rock data, let $\alpha 3$ and $\beta 3$ denote the mean and standard deviation of impact pressure. The corresponding values for Grade 4 and Grade 5 surrounding rocks are $\alpha 4$, $\beta 4$, and $\alpha 5$, $\beta 5$, respectively.

For rotary pressure, let $\gamma 2$ and $\delta 2$ denote the mean and standard deviation for Grade 2 surrounding rock data, $\gamma 3$ and $\delta 3$ for Grade 3, $\gamma 4$ and $\delta 4$ for Grade 4, and $\gamma 5$ and $\delta 5$ for Grade 5.

Samples are filtered based on these metrics, with the sample ranges shown in Table 9.

Table 9. Sample selection range table.

Rock Classification	Drilling Parameter	Selection Range
Grade 2	Impact Pressure	$[\alpha 2 - \beta 2, \alpha 2 + \beta 2]$
	Rotary Pressure	$[\gamma 2 - \delta 2, \gamma 2 + \delta 2]$
Grade 3	Impact Pressure	$[\alpha 3 - \beta 3, \alpha 3 + \beta 3]$
	Rotary Pressure	$[\gamma 3 - \delta 3, \gamma 3 + \delta 3]$
Grade 4	Impact Pressure	$[\alpha 4 - \beta 4, \alpha 4 + \beta 4]$
	Rotary Pressure	$[\gamma 4 - \delta 4, \gamma 4 + \delta 4]$
Grade 5	Impact Pressure	$[\alpha 5 - \beta 5, \alpha 5 + \beta 5]$
	Rotary Pressure	$[\gamma 5 - \delta 5, \gamma 5 + \delta 5]$

After removing outliers, the statistical patterns of the four drilling parameters—propulsion speed, impact pressure, propulsion pressure, and rotary pressure—are presented for each rock classification in the sample library. The data are summarized in Table 10.

Table 10. Statistical table of drilling parameters database before and after filtering.

Rock Classification	Time Point	Statistic	Impact Pressure (Pa)	Rotary Pressure (Pa)	Sample Size
Grade 2	Before Filtering	Mean	145.14	97.32	76
		Standard Deviation	14.28	7.36	
	After Filtering	Mean	144.43	94.95	64
		Standard Deviation	11.37	6.09	
Grade 3	Before Filtering	Mean	144.07	92.08	119
		Standard Deviation	12.75	11.66	
	After Filtering	Mean	147.85	95.38	104
		Standard Deviation	6.17	8.44	
Grade 4	Before Filtering	Mean	136.44	75.16	129
		Standard Deviation	8.92	6.50	
	After Filtering	Mean	135.90	75.52	111
		Standard Deviation	4.16	4.53	
Grade 5	Before Filtering	Mean	123.33	69.18	112
		Standard Deviation	7.34	6.39	
	After Filtering	Mean	123.02	69.48	92
		Standard Deviation	4.61	4.97	

A total of 371 ideal samples were obtained, including 64 samples for Grade 2 surrounding rock, 104 samples for Grade 3, 111 samples for Grade 4, and 92 samples for Grade 5. Figure 14 illustrates that after removing redundant data, the degree of data dispersion across different surrounding rock grades has significantly reduced. The samples in each rock classification are more concentrated and less dispersed than before filtering.

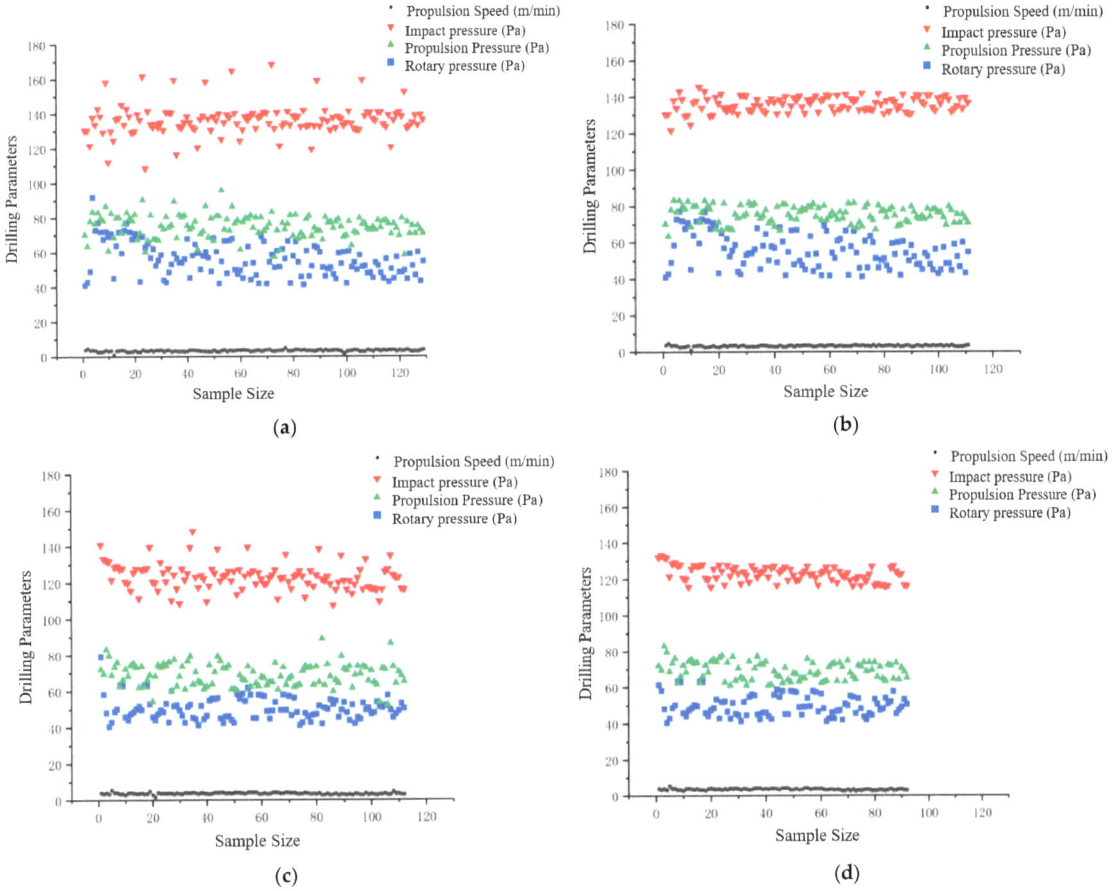

Figure 14. Scatter plot of drilling parameters database before and after filtering: (**a**) Grade 4 surrounding rock before filtering; (**b**) Grade 4 surrounding rock after filtering; (**c**) Grade 5 surrounding rock before filtering; (**d**) Grade 5 surrounding rock after filtering.

4.2. Data Standardization and Dataset Division

In the original dataset, the features have different units and scales. Discrepancies in these scales and initial values can cause certain features to overshadow others in their influence on surrounding rock classification. Therefore, standardization is employed to process the drilling parameters. The standardization formula is shown in Equation (3). Subsequently, the standardized samples are divided into training and testing datasets with a 7:3 ratio.

$$X = \frac{x - mean}{\sigma} \tag{3}$$

In the formula, X represents the standardized parameter, x is the parameter before standardization, *mean* is the mean value, and σ denotes the standard deviation.

5. Intelligent Rock Classification

5.1. Random Forest

The random forest algorithm model was first proposed in 2001 by Leo Breiman and Adele Cutler. This ensemble algorithm is based on decision trees and bagging techniques. Random forest offers good generalization performance and fast processing speeds. The core principles are as follows:

(1) Random sampling: Using the bootstrap method, it repeatedly samples *n* samples with replacement from the original training data to form a bootstrap sample set.
(2) Decision tree construction: For each bootstrap sample set, a subset of features is randomly selected to build a decision tree. This random selection ensures that each tree is different and aims to improve the model's generalization ability.
(3) Decision tree integration: Once a sufficient number of decision trees are built, each tree provides a classification result for new input data. The random forest algorithm aggregates these results through a voting mechanism to determine the final classification.

In the random forest model, a subset of attributes is randomly selected from the attribute set of each base decision tree node. The best attribute is then chosen to perform the split. Its core algorithm, the decision tree, minimizes entropy using conditional entropy and information gain to achieve the optimal structure. Given the engineering context of this paper, the workflow for intelligent surrounding rock classification using the random forest model is illustrated in Figure 15.

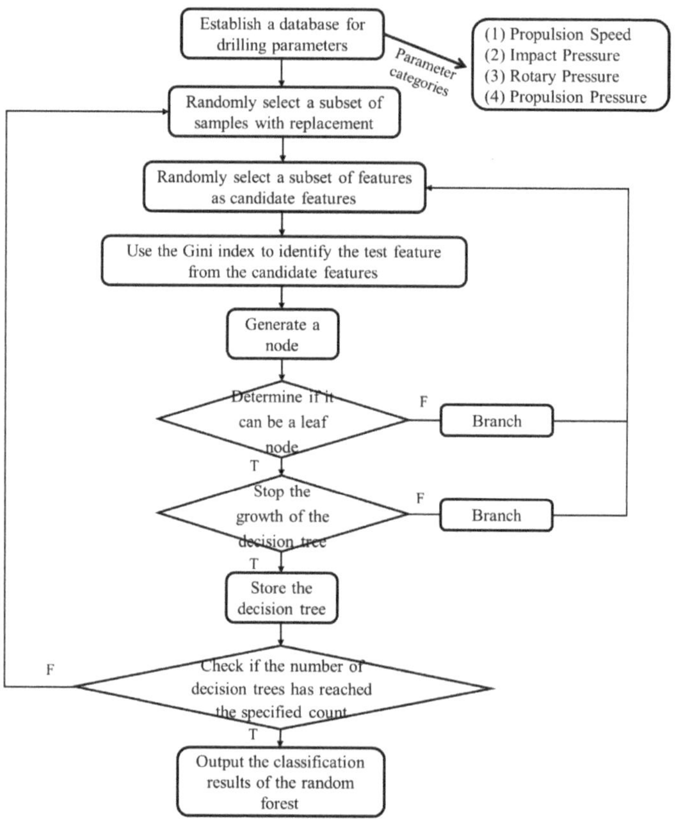

Figure 15. Flowchart of intelligent rock mass classification based on random forest.

5.2. Rock Mass Classification Results

The confusion matrix for the random forest on the test set samples is shown in Figure 16. There are 19 samples of secondary-grade rock mass in the test set, of which 15 were correctly predicted, while 4 were misclassified as tertiary grade. For the tertiary-grade rock mass, 31 samples were tested, of which 30 were correctly predicted, while 1 was classified as secondary grade. For the quaternary-grade rock mass, 34 samples were tested, with 32 correctly classified, 1 misclassified as tertiary grade, and 1 as quinary grade. Lastly, the quinary-grade rock mass had 28 samples, of which 26 were correctly classified, while 1 was misclassified as tertiary grade and another as quaternary grade.

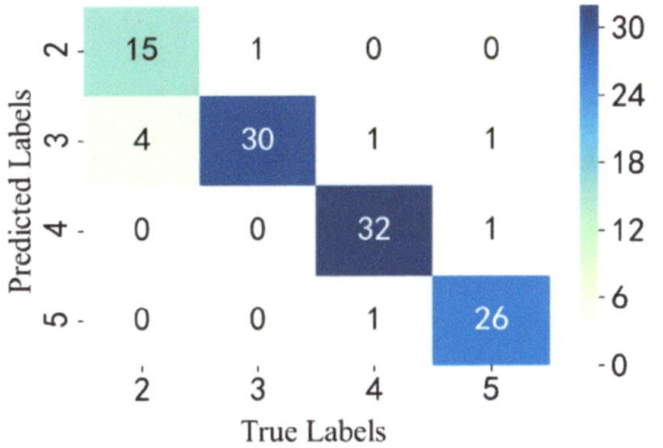

Figure 16. Confusion matrix of the random forest.

In this paper, precision, recall, and F1 score are used to evaluate the random forest model for rock mass classification. The formula for precision is shown in Equation (4), the formula for recall is given in Equation (5), and the formula for the F1 score is presented in Equation (6).

$$Precision = \frac{TP}{TP + FP} \quad (4)$$

TP (true positive): The sample is positive, and the prediction is also positive, which means that the positive class is correctly predicted. FP (false positive): The sample is negative, but the prediction is positive, indicating a misclassification of the negative class. FN (false negative): The sample is positive, but the prediction is negative, indicating a misclassification of the positive class.

$$\text{Recall} = \frac{TP}{TP + FN} \quad (5)$$

$$F1 = 2\frac{Precision \times Recall}{Precision + Recall} \quad (6)$$

The evaluation results are shown in Table 11. The average precision for rock mass classification is 0.93, with each grade having a precision exceeding 0.8. This indicates that the random forest model not only has high accuracy in identifying positive classes but also maintains high stability. The average recall for rock mass classification is 0.91, demonstrating that the model has a high coverage rate for positive samples, meaning relatively few false negatives. An average F1 score of 0.92 indicates that the model achieves a good balance between precision and completeness. Overall, the random forest classification model performs well in classifying rock mass grades.

Table 11. Results of random forest evaluation.

Rock Mass Grade	Precision	Recall	F1 Score
Grade 2	0.94	0.79	0.86
Grade 3	0.83	0.97	0.90
Grade 4	0.97	0.94	0.96
Grade 5	0.96	0.93	0.95
Average Accuracy	0.93	0.91	0.92

6. Conclusions

This paper focuses on the efficient classification of rock mass grades during tunnel excavation using the drill and blast method. Based on the development of a down-the-hole drilling rig, a rock drilling testing system capable of real-time acquisition of drilling parameters was developed. Using the obtained drilling parameters, an intelligent rock mass classification model was established. This study contributes to the efficient interoperability between data acquisition technologies and model methodologies in rock mass classification. The specific results are as follows:

(1) With displacement sensors and pressure transmitters as core components, a feed speed acquisition system, an oil pressure acquisition system, and an air pressure acquisition system were developed. Using an ADC (analog-to-digital converter), an encoder, and a flash storage chip, an automatic data acquisition system was developed. This system can collect data twice per second and has the advantages of high precision, low power consumption, and high stability. Together, these four systems form the rock drilling testing system, enabling the down-the-hole drilling rig to automatically collect and store impact pressure, rotation pressure, feed pressure, and feed speed in real time during the drilling process. This greatly improves the efficiency and accuracy of data acquisition.

(2) By analyzing kernel density plots and Pearson correlation heatmaps between drilling parameters and rock mass grades, it was found that a single drilling parameter is insufficient for determining rock mass grades. The rock mass grade has a strong positive correlation with feed speed and a strong negative correlation with impact pressure, feed pressure, and rotation pressure. Specifically, the correlation with impact pressure and rotation pressure is strongly negative, with correlation coefficients both exceeding 0.8.

(3) Outliers were removed using an interval method, and the dataset was standardized, effectively improving model robustness. The random forest model used in this study has excellent rock mass classification performance, with precision, recall, and F1 scores all exceeding 0.9. Classification performance across rock mass grades is fairly balanced, and the model demonstrates good generalizability.

(4) The rock drilling test system efficiently acquires drilling parameters, and the random forest model can accurately classify rock mass grades based on these parameters. Combining the two forms an intelligent tunnel rock mass classification system, which plays a critical role in determining tunnel excavation plans and quality evaluation. This system can effectively promote the transition and upgrade of tunnel construction processes from traditional operations to intelligent ones.

7. Discussion

The geological environment at engineering sites is often complex, and it may not always be feasible to obtain a comprehensive dataset. Enhancing the applicability of rock mass classification models to other engineering projects remains a significant challenge. This paper suggests the following approaches to provide insights for developing more generalized models.

The first is to develop comprehensive models with stronger generalization ability, such as the voting method. Ensemble voting methods offer a significant advantage for enhancing the practical application and universality of our classification algorithm. By combining the predictions of multiple models, ensemble voting can improve overall classification accuracy and robustness. This method leverages the strengths of different models, mitigating the impact of individual model biases and errors. Furthermore, ensemble voting is particularly effective in adapting to diverse geological conditions, ensuring reliable performance across various environments.

The second is transfer learning. Transfer learning presents a valuable approach for enhancing the practical application and universality of our classification algorithm. By leveraging pretrained models on large, diverse geological datasets, transfer learning can significantly improve the adaptability and performance of the algorithm in various real-world conditions. The process involves pretraining on a comprehensive source dataset and fine-tuning the model with specific target datasets, enabling efficient and robust classification across different geological environments.

With the rapid development of artificial intelligence, new classification algorithms and techniques will continue to emerge. Improving the engineering practicality of rock mass classification models will remain a key development trend in the future.

Author Contributions: Q.L.: Methodology, writing—original draft. J.Y.: Data curation, supervision. H.L. (Hongzhao Li): Resources, investigation. P.Z.: Supervision, investigation. Y.L.: Validation, data curation. L.L.: Project administration, software. S.Y.: Formal analysis, writing—review and editing. H.L. (Haitao Liu): Supervision, project administration. All authors have read and agreed to the published version of the manuscript.

Funding: This research was funded by the National Key R&D Program of China (grant numbers: 2021YFB2600800).

Data Availability Statement: The data that support the findings of this study are available from the corresponding author upon reasonable request.

Conflicts of Interest: Author Quanwei Liu, Peiyuan Zhang, Linsheng Liu and Shoujie Ye are employed by the Qingdao Metro Line 6 Co., Ltd. Author Haitao Liu is employed by the China Railway First Bureau Group Fifth Engineering Co., Ltd. The remaining authors declare that the research was conducted in the absence of any commercial or financial relationships that could be construed as a potential conflict of interest.

References

1. Kumar, P.; Mohammadi, H.; Chopra, R.; Kumar Tyagi, S.; Kumar Pandey, S. A Newly Developed Blasting Cut in Tunnels; Application of "Combined Method" in Small to Medium-Sized Tunnels. *Tunn. Undergr. Space Technol.* **2023**, *142*, 105426. [CrossRef]
2. Wang, M.; Zhao, S.; Tong, J.; Wang, Z.; Yao, M.; Li, J.; Yi, W. Intelligent Classification Model of Surrounding Rock of Tunnel Using Drilling and Blasting Method. *Undergr. Space* **2021**, *6*, 539–550. [CrossRef]
3. Barton, N. Some New Q-Value Correlations to Assist in Site Characterisation and Tunnel Design. *Int. J. Rock Mech. Min. Sci.* **2002**, *39*, 185–216. [CrossRef]
4. Jiang, F.; He, P.; Wang, G.; Zheng, C.; Xiao, Z.; Wu, Y.; Lv, Z. Q-Method Optimization of Tunnel Surrounding Rock Classification by Fuzzy Reasoning Model and Support Vector Machine. *Soft Comput.* **2022**, *26*, 7545–7558. [CrossRef]
5. Somodi, G.; Bar, N.; Kovács, L.; Arrieta, M.; Török, Á.; Vásárhelyi, B. Study of Rock Mass Rating (RMR) and Geological Strength Index (GSI) Correlations in Granite, Siltstone, Sandstone and Quartzite Rock Masses. *Appl. Sci.* **2021**, *11*, 3351. [CrossRef]
6. Zhang, Q.; Huang, X.; Zhu, H.; Li, J. Quantitative Assessments of the Correlations between Rock Mass Rating (RMR) and Geological Strength Index (GSI). *Tunn. Undergr. Space Technol.* **2019**, *83*, 73–81. [CrossRef]
7. Jiang, W.; Wang, Y.; Yang, J.; Zhang, Z. Surrounding Rock Quality Evaluation and Application Development for Highway Tunnel Based on Engineering Applicability. *Bull. Eng. Geol. Environ.* **2023**, *82*, 115. [CrossRef]
8. Qiu, D.; Fu, K.; Xue, Y.; Tao, Y.; Kong, F.; Bai, C. TBM Tunnel Surrounding Rock Classification Method and Real-Time Identification Model Based on Tunneling Performance. *Int. J. Geomech.* **2022**, *22*, 04022070. [CrossRef]
9. Liu, Q.; Liu, J.; Pan, Y.; Kong, X.; Hong, K. A Case Study of TBM Performance Prediction Using a Chinese Rock Mass Classification System—Hydropower Classification (HC) Method. *Tunn. Undergr. Space Technol.* **2017**, *65*, 140–154. [CrossRef]
10. Xie, H.T. Bayesian Network Based Expert System for Tunnel Surrounding Rockmass Classification. *Appl. Mech. Mater.* **2013**, *482*, 248–251. [CrossRef]
11. Niu, G.; He, X.; Xu, H.; Dai, S. Development of Rock Classification Systems: A Comprehensive Review with Emphasis on Artificial Intelligence Techniques. *Eng* **2024**, *5*, 217–245. [CrossRef]

12. Zhao, S.; Wang, M.; Yi, W.; Yang, D.; Tong, J. Intelligent Classification of Surrounding Rock of Tunnel Based on 10 Machine Learning Algorithms. *Appl. Sci.* **2022**, *12*, 2656. [CrossRef]
13. Zheng, S.; Jiang, A.N.; Yang, X.R.; Luo, G.C. A New Reliability Rock Mass Classification Method Based on Least Squares Support Vector Machine Optimized by Bacterial Foraging Optimization Algorithm. *Adv. Civ. Eng.* **2020**, *2020*, 1–13. [CrossRef]
14. Jalalifar, H.; Mojedifar, S.; Sahebi, A.A.; Nezamabadi-pour, H. Application of the Adaptive Neuro-Fuzzy Inference System for Prediction of a Rock Engineering Classification System. *Comput. Geotech.* **2011**, *38*, 783–790. [CrossRef]
15. Hasegawa, N.; Hasegawa, S.; Kitaoka, T.; Ohtsu, H. Applicability of Neural Network in Rock Classification of Mountain Tunnel. *Mater. Trans.* **2019**, *60*, 758–764. [CrossRef]
16. Bressan, T.S.; Kehl De Souza, M.; Girelli, T.J.; Junior, F.C. Evaluation of Machine Learning Methods for Lithology Classification Using Geophysical Data. *Comput. Geosci.* **2020**, *139*, 104475. [CrossRef]
17. Liu, X.; Wang, H.; Jing, H.; Shao, A.; Wang, L. Research on Intelligent Identification of Rock Types Based on Faster R-CNN Method. *IEEE Access* **2020**, *8*, 21804–21812. [CrossRef]
18. Song, S.; Xu, G.; Bao, L.; Xie, Y.; Lu, W.; Liu, H.; Wang, W. Classifying the Surrounding Rock of Tunnel Face Using Machine Learning. *Front. Earth Sci.* **2023**, *10*, 1052117. [CrossRef]
19. Tan, F.; You, M.; Zuo, C.; Jiao, Y.-Y.; Tian, H. Simulation of Rock-Breaking Process by Drilling Machine and Dynamic Classification of Surrounding Rocks. *Rock Mech. Rock Eng.* **2022**, *55*, 423–437. [CrossRef]
20. Torno, S.; Toraño, J.; Menéndez, M.; Gent, M.; Velasco, J. Mathematical and fuzzy logic models in prediction of geological and geomechanical properties of rock mass by excavation data on underground works. *J. Civ. Eng. Manag.* **2011**, *17*, 197–206. [CrossRef]
21. Mostofi, M.; Rasouli, V.; Mawuli, E. An Estimation of Rock Strength Using a Drilling Performance Model: A Case Study in Blacktip Field, Australia. *Rock Mech. Rock Eng.* **2011**, *44*, 305–316. [CrossRef]
22. Kalantari, S.; Hashemolhosseini, H.; Baghbanan, A. Estimating Rock Strength Parameters Using Drilling Data. *Int. J. Rock Mech. Min. Sci.* **2018**, *104*, 45–52. [CrossRef]
23. Kalantari, S.; Baghbanan, A.; Hashemalhosseini, H. An Analytical Model for Estimating Rock Strength Parameters from Small-Scale Drilling Data. *J. Rock Mech. Geotech. Eng.* **2019**, *11*, 135–145. [CrossRef]
24. Lakshminarayana, C.R.; Tripathi, A.K.; Pal, S.K. Prediction of Mechanical Properties of Sedimentary Type Rocks Using Rotary Drilling Parameters. *Geotech. Geol. Eng.* **2020**, *38*, 4863–4876. [CrossRef]

buildings

MDPI

Article

Adoption of Fourth Industrial Revolution Technologies in the Construction Sector: Evidence from a Questionnaire Survey

Julia Menegon Lopes * and Luiz Carlos Pinto da Silva Filho

Programa de Pós-Graduação em Engenharia Civil: Construção e Infraestrutura (PPGCI), Universidade Federal do Rio Grande do Sul (UFRGS), Porto Alegre 90035-190, Brazil; lcarlos66@gmail.com
* Correspondence: menegonjulia@gmail.com

Abstract: The fourth industrial revolution (4IR) can significantly benefit the construction sector, improving productivity, efficiency, collaborative efforts, and product quality while promoting safety and sustainability. However, research on applying 4IR technologies in construction is scarce in developing countries. It is crucial to understand the ability of construction companies to adopt new technologies and identify factors influencing the success of technology implementation. In this study, a questionnaire-based survey was conducted with construction professionals to evaluate the level of technological development of the construction market in an emerging economy, assess the potential for innovation implementation, and identify factors that might influence technological development. The results showed that most innovations are in the early stages of implementation in the construction sector, and their adoption tends to occur differently, depending on the size of the company and the stage of the construction lifecycle in which they operate. Furthermore, technologies tend to be progressively adopted and driven by virtualization technologies. This article presents a framework to assist in decision-making regarding the adoption of 4IR technologies at different phases of the lifecycle of construction projects and identifies the potential barriers and promoters of this adoption in the analyzed context.

Keywords: Industry 4.0; Construction 4.0; new technology; digital transformation; developing country

1. Introduction

Contemporary society is in the midst of the fourth industrial revolution (4IR), opening new avenues for industrial relations [1]. Based on the digital revolution and the Internet of Things (IoT), 4IR, which started at the beginning of the 21st century in the manufacturing industry under the name of Industry 4.0, has since spread to several other fields [2,3]. Focused on creating smart processes, procedures, and products [3], 4IR, once consolidated, has the potential to generate diverse economic, environmental, and social benefits [4]. Like many other fields, the construction industry can benefit significantly from 4IR technologies, which allow process integration and automation along the entire value chain, increasing productivity, efficiency, collaboration, and final product quality while improving key parameters of the sector, such as safety and sustainability [5].

Given the benefits associated with 4IR technologies, researchers have expressed a growing interest in the topic. Most studies, however, were conducted in developed countries [6], whose socioeconomic panorama differs significantly from that of developing countries. Findings [6–8] indicate that research efforts dedicated to understanding 4IR technology adoption are scarce in Brazil, particularly in construction. A report from the National Confederation of Industry (CNI) [9] revealed that, although considered a subject of great importance, adopting innovative technologies is scant in most Brazilian industries. PwC Consulting [10] pointed out that Brazil is lagging behind the global industrial scenario regarding digitization and integration. In agreement with these observations, Brazilian studies [11,12] concluded that the national industry is still transitioning from the second to the third IR, suggesting that there is still a large technological gap to be bridged. Technologies that are well consolidated or under consolidation in developed countries are likely incipient in the Brazilian market. Thus, the Brazilian manufacturing industry is missing the opportunity to harness the advantages of current technologies [13].

Understanding how prepared construction companies are to adopt existing technologies and which factors influence technology implementation success is crucial for advancing research on the topic [5]. In view of this and of the need to achieve the practical consolidation of emerging concepts in developing countries, this study sought to examine the capacity of the local industry to absorb 4IR technologies and understand the perspectives and objectives of those involved in innovation adoption. The following two-part question was raised: What is the current state of the Brazilian construction market in terms of 4IR technologies, and what factors influence technology adoption? To answer this question, we conducted a quantitative exploratory study using structured questionnaires. The aim was to describe the current reality of the Brazilian construction market to identify the implementation potential of 4.0 solutions and examine intervening factors that might influence the future scenario. Responses of professionals working in the Brazilian construction industry were analyzed using descriptive and inferential statistical techniques.

First, this article presents and discusses a literature review exploring 4IR concepts and technologies, as well as their potential benefits, barriers, and impacts on the construction sector. The next section provides an analysis of the construction market's technological maturity, construction professionals' knowledge of 4IR, and the expected benefits associated with the adoption of novel technologies. The potential for technology adoption was assessed based on participants' responses about current technology use, interest in technology adoption, and perceptions of time, cost, and market readiness for technology absorption. Cluster analyses were applied to identify patterns in technology use. Finally, the barriers cited by construction professionals were analyzed, and inferences were made about critical factors for the adoption of 4IR technology in the local construction industry.

This study provides an overview of the main expectations, concerns, and challenges for implementing 4IR technologies in the construction sector within the studied context. We also identify the factors that may influence technology adoption. We categorized the concepts into four groups: undeveloped, incipient, under development, and consolidated, based on their average level of adoption. The results indicate that none of the technologies have fully consolidated within the sample. We identified the lack of qualified personnel, resistance to change, and perceptions of time and cost as significant barriers to innovation in the sector. On the other hand, the company's high technological maturity, market readiness, widespread 4IR knowledge, and early career professionals tend to be critical drivers of technological advancement.

Our findings reveal that the adoption of innovative tools occurs progressively among the respondents. This means those who have already initiated the digitalization process are more likely to adopt other technologies. Furthermore, we have identified three groups of technologies, namely, virtualization, automation, and manufacturing technologies, with distinct usage patterns among respondents. We propose a framework to guide the adoption of these concepts based on the construction lifecycle phase in which they will be applied, indicating which group of technologies is most suitable for each application.

This greater understanding of the current scenario can support the development of further studies and strategies to foster and disseminate current innovations so that construction industries in developing countries can benefit from the ongoing IR.

2. Industry and Construction 4.0

2.1. Industry 4.0

Coined in Germany, the term Industry 4.0 alludes to a new version of industry and the ensuing changes in industrial production. This paradigm shift emerged from combining the so-called futuristic technologies and the widespread use of the internet, empowering common physical objects with autonomy and "intelligence" [1]. Many of the technologies that form the pillars of the new industrial paradigm have been evolving since the creation of the first computers and have long been used, in isolation, in manufacturing [4,14]. However, in the interaction between physical and digital media lies the transformative power of Industry 4.0 [4,14], as it allows various devices to interact through an internet network [7], collect information, and assist in decision-making. Thus, existing technologies form an integrated system that has the potential to revolutionize relations between suppliers, manufacturers, and customers and improve the efficiency of the production chain [2].

In addition to technology-driven changes, the new industrial scenario also brings forth a multitude of social, economic, and organizational changes, such as greater focus on consumers, information integration, decentralized decision-making, and new business possibilities. Consumers are the greatest beneficiaries of this paradigm shift, as they are provided with new internet-based products and services that promote efficiency in everyday activities [2]. Additionally, 4IR is modifying the role of human beings within productive systems, requiring workers to develop new skills to perform tasks of greater complexity assisted by novel technologies [15].

2.1.1. Fourth Industrial Revolution Technologies and Principles

According to the Boston Consulting Group (BCG), Industry 4.0 is supported by the following nine pillars [14]: additive manufacturing, augmented reality, autonomous robotics, big data and analytics, cloud computing, cybersecurity, vertical and horizontal integration, IoT, and simulation. Additive manufacturing, exemplified by three-dimensional (3D) printing, is the opposite of subtractive manufacturing. Whereas subtractive manufacturing removes surplus materials to shape parts and objects, additive manufacturing builds objects layer-by-layer according to a pre-existing 3D model [7]. Its main advantage lies in the ability to produce small batches of customized products rapidly and efficiently [13,14].

As for augmented reality, one of its major benefits is the possibility of assisting workers in their activities. The technology integrates information from computer models into the real environment, representing a valuable tool to guide teams during the execution of familiar and unfamiliar tasks. Data, graphics, and virtual images can be reproduced in a user's field of view, allowing them to interact with information projected onto their surroundings [7,16]. A wide variety of activities can be facilitated by augmented reality, such as maintenance services, wherein workers can receive instructions and remote support during task execution and stock selection, and virtual training for emergencies, wherein practitioners can receive specific instructions in a controlled environment [14].

Robotics has long been used in manufacturing, having evolved and becoming increasingly useful over the years [14]. Autonomy, flexibility, and cooperation are some characteristics attributed to the new generation of autonomous robots [14]. Given their ability to self-configure and negotiate with each other to adjust to changing needs [4], autonomous robots are expected to play a key role in smart manufacturing. Collaborative robots have been developed to interact with humans and provide support during work activities [17]. Safe and collaborative human–machine interactions have been envisioned and are expected to become widespread when such equipment costs decrease [14].

As a result of the various data acquisition and storage technologies that emerged with the ubiquitous use of the internet, big data and analytics became important pillars of Industry 4.0. This comes as no surprise, given that the capacity to analyze large amounts of data is essential for the digital transformation of companies [7]. Big data technologies can be used to process and select data quickly and efficiently, separating relevant from less important information [18] amidst the gigantic realm of available data—a task beyond the capacity of any other method, especially human processing. Algorithms based on correlations and probabilities can mine the data, evaluate patterns, and generate information for knowledge building [4]. Big data-derived knowledge has the potential to improve production quality, reduce energy consumption, assist in rapid decision-making, and improve equipment operation [14].

Most of the analyzed data are stored in the cloud, which represents another pillar of 4IR. Possibly one of the most widespread tools nowadays, cloud computing allows the creation of a network connecting people, data, services, and objects through the internet [7]. With the ability to store data in remote databases [19], cloud services provide easy access to information [17] and make it financially affordable to store the exponential amount of data generated over time [4].

This plethora of data-sharing and connectivity technologies explains the importance of cybersecurity for the diffusion of Industry 4.0. The need to protect industrial systems and information from cyberattacks is fundamental and expanding [14]. Malicious softwares can spread through interconnected machines to modify processes, destroy data [7], or steal inside information. Therefore, technologies that reduce concerns about cyberattacks have a strong appeal in the new industrial reality. Security requirements vary according to the needs of each networked system. It should be recognized, however, that the complex reality of interconnected environments makes it unfeasible to attain complete security [4]. Nevertheless, it is possible to create means for the real-time detection of atypical behaviors and generate quick responses to keep network-connected equipment and users safe [4].

Another principle of the new IR is production chain integration, both horizontally and vertically. Vertical integration is defined as the integration of information systems along the hierarchical levels of a company [3], which results in more flexible and faster communication between levels [20]. Such an integration model encompasses product development and purchase, manufacturing, logistics, and services [10]. Horizontal integration, on the other hand, refers to the connection between different phases of production and design processes that involve the exchange of materials, energy, or information between the different companies participating in a value chain [3]. The purpose of integration is to connect both ends of the value chain. This represents an important innovation in that it fully interconnects information technologies, culminating in an extraordinary level of association between companies, suppliers, and clients, as well as between departments within companies [14].

At the heart of information exchange and storage lies another key concept of 4IR—IoT. Objects enriched with sensors and actuators are able to communicate in real-time at high speeds with each other and with controllers, creating an intelligent and interconnected environment [4,14]. Ultimately, products will be able to communicate with other products and systems in a manner that amplifies their performance and offers novel and improved solutions before and after sales [13], altering the course of business strategies [19]. IoT-based solutions play a key role in increasing efficiency in the field of logistics and mobility, as they allow the real-time monitoring of objects and goods in transport and urban mobility services [4,18]. Three characteristics make IoT a revolutionary technology [18]: (i) context, whereby objects can provide information on location, weather, and physical conditions; (ii) ubiquity, i.e., capacity for large-scale communication between objects; and (iii) optimization, whereby objects can acquire multiple functionalities. The smart ecosystem formed by interconnected objects supports decentralized decision-making and real-time responsivity to changes and needs [14].

Finally, the last pillar is simulation, considered the cornerstone of Industry 4.0 by BCG. Although its use was common in modeling before the current IR, simulation technology has gained new uses and applications. Current models are able to mirror the real environment, including not only geometric but also behavioral characteristics in real time [3,14]. Simulation tests and optimizations carried out using virtual models improve the quality of final products and the rate of introduction of new products into the market [9]. Logistics and transport alternatives can be tested, relevant risks associated with production processes can be assessed, and costs and environmental impacts can be compared between suppliers through simulations [3].

The different technologies of Industry 4.0 can be classified into two types: frontend and base technologies [17]. Technologies that connect and smarten existing technologies are called base and form the foundation upon which Industry 4.0 resides. Examples include IoT, cloud computing, and big data and analytics. Frontend technologies, on the other hand, are linked to operational activities and market needs and can be divided into four dimensions: smart manufacturing, smart products, smart working, and smart supply chain. Smart production technologies are at the core of research on Industry 4.0, whereas smart working has received less attention [21]. However, it is the implementation of base technologies that sets apart the new paradigm from previous stages of industrial development, ultimately transforming a conventional company into a smart one [21].

From a theoretical point of view, the implementation of 4IR technologies can be conducted in one, a few, or all four dimensions, depending on the objectives of digitization. Nevertheless, it should be noted that, in practice, 4IR technologies are considered complementary and tend to be implemented progressively, with new technologies being added as the maturity of the company increases [17]. As stated by Schwab [2], innovations "build on and amplify each other", and integration between different dimensions leverages the benefits of Industry 4.0 [21].

Consumers' decision to adopt or not innovations was shown to be influenced by the following five factors [22]: (i) the perception of economic advantage, social prestige, convenience, or increased satisfaction in comparison with the current state; (ii) the perception of compatible values, experiences, and needs; (iii) the level of complexity of technology use; (iv) the ability to test or experiment technologies for a period of time; and (v) the observation of the results of peers who used the innovation. As for organizations, a cautious attitude and a lack of trained professionals represent structural challenges that may delay technology adoption in medium-sized and small companies [3]. There are also concerns related to the high initial financial investment required to implement technologies, which can be intimidating for smaller companies, especially on a return basis [23]. In line with these observations, studies conducted in the manufacturing industry indicate that larger organizations tend to be at more advanced stages of Industry 4.0 implementation [17].

2.1.2. Industry 4.0 Trends in the Construction Sector

Compared to other industrial sectors, the construction industry lags significantly behind in adopting 4IR technologies, potentially because of its conservative nature [24,25]. Despite this, there is great potential for the digitization of the sector, which can provide cost and time savings to projects [24], among other benefits. Oesterreich and Teuteberg [5] identified several Industry 4.0 technologies and concepts that are key to the construction sector and enable process digitization, automation, and integration. A previous study [6] classified some of these concepts into five groups according to their similarity of application in the construction sector, as shown in Table 1.

Table 1. Industry 4.0 principles and technologies with prevalent applications in construction [6].

Cluster	Concept/Technology
Data intelligence	Cloud computing Big data Product lifecycle management
Robotics and automation	Robots/drones Automation
Virtual environments	Building information modeling Simulation/modeling Virtual and augmented reality
Smart technologies and objects	Internet of Things Mobile devices Embedded sensors/cyber–physical systems Digitization
Advanced manufacturing	Additive manufacturing Prefabrication and modularization

Fourth Industrial Revolution technologies can play significant roles in different phases of the lifecycle of a construction project [25], often serving different purposes in each of them. Because of the fragmented and dynamic nature of the construction industry, innovation needs differ between phases. There is a tendency toward a more organic approach to innovation in the initial phases of a project (e.g., planning and design) and toward a more systematic approach during subsequent phases, which typically require greater discipline as a result of stricter deadlines [26]. Such differences in approach may indicate the need for different technologies. Industry 4.0 concepts have been most explored in the planning and management phases, during which the main focus of technologies lies on task execution, smart manufacturing, and smart working [6], that is, dimensions related to internal processes of companies [21]. Technologies applied to external processes (smart products and smart supply chain), as well as those based on Industry 4.0, remain little explored in the construction sector [6].

The potential applications of Industry 4.0 principles and technologies in construction are summarized in Table 2, which was constructed based on a previous literature review [6]. IoT, sensors, and cyber–physical system (CPS) technologies were grouped under a single concept, given their similarity and interrelatedness.

Table 2. Application of Industry 4.0 principles and technologies in construction.

Technology	Applications
Cloud computing	A large amount of data can be stored and accessed from the cloud, facilitating information sharing between design team members and assisting in the development of designs collaboratively and simultaneously between individuals in different geographical locations.
Big data	This technology assists in the collection and selection of relevant information from the universe of available data. It has the potential to simplify database searches and assist in choosing between different alternatives of engineering designs and evaluating parameters, such as cost and energy efficiency, for each design alternative in a rapid and automated way.
Product lifecycle management	Data collected and stored are used to integrate and manage product information from the design to the manufacture and use phases until the end of a product's useful life.
Robots and drones	This technology has the potential to replace human labor in everyday tasks. Drones can capture aerial images that enable and facilitate services such as construction and asset management, inspection, and maintenance.
Automation	Potential applications encompass several areas, such as the quality monitoring of concrete trucks, soil compaction, parameter control during concreting, design automation, building monitoring in the use phase, and so on.
Building information modeling	Tool for the centralization of the information generated and accumulated at each stage of the construction process.
Simulation and modeling	Modeling and simulation of reality to foresee behaviors and characteristics of the final product and production stages. It can be used for the simulation of construction processes, conflict identification, resource allocation, and assessment of energy efficiency and flows of people, among others.
Virtual reality and augmented reality	Virtual environments that mimic reality and allow the interaction and visualization of situations in real dimensions.
Internet of Things, sensors, and cyber–physical systems	Common physical systems equipped with sensors and devices that interact and exchange information among themselves and/or with an operator. They can be used to automate processes, control inventory, machinery, and human resources, track material transportation, and monitor the behavior of existing buildings and their facilities.
Mobile devices	Use of smartphones, tablets, and applications as tools to support communication and collaboration throughout the production cycle.
Three-dimensional printing	Printing of objects in three dimensions, comprising either entire buildings or individual parts for subsequent assemblage.
Prefabrication and modularization	Construction industrialization, mass production, and off-site part production for later installation at the final destination.

Source: adapted from Menegon and Silva Filho [6].

2.1.3. Impact of New Technologies

The ongoing technological revolution is expected to have a prominent impact on the economic, social, and cultural spheres of societies worldwide, particularly on economic development and the labor market [2]. Current innovations may dramatically affect skill profiles and workplace activities [3], potentially exerting some negative effects in the short term owing to the rapid replacement of human labor by computers [2]. Schwab [2] argued that, in the long term, however, new demands for services and products are likely to catalyze the emergence of new professions, which eventually absorb the available workforce. With the new IR, workers tend to be more focused on creative and added-value activities and dedicate less time to routine and repetitive activities [3], as the latter can be easily replaced by machines. There is a prospect that there will be an increased supply of high-salary positions with high cognitive and creative demands, just as there could be a reduced need for low-paid, fundamentally manual occupations [2].

For the construction sector, connected sensors, smart construction equipment, mobile devices, and the use of applications can improve productivity, manage complexity, reduce delays and cost increase, and also ensure quality, safety, and collaboration [6], changing the way we build and maintain our assets.

The planning and management phase is poised for transformation through new technologies in the construction lifecycle. Industry 4.0 advancements in data acquisition, storage, and processing will significantly enhance the planning phase. Nearly all Industry 4.0 technologies are relevant to this stage [6]. They can enhance the production process by decentralizing decision-making, emphasizing pre-construction phases, and enabling real-time progress monitoring through integrated information. Furthermore, the construction stage is significantly influenced by the emergence of automated construction processes, gradually replacing conventional manual labor. Indeed, the industry can enormously benefit from data acquisition, storage, and processing [6].

The expected benefits of 4IR can be classified into three main categories [13]: (i) product-related benefits, including those directly linked to the performance, quality, and release timing of final products; (ii) operation-related benefits, which refer to improvements in internal production activities, such as increased yields and reduced operating costs; and (iii) side benefits, which are not directly linked to products or productivity but can be equally advantageous to companies [13]. Table 3 describes some of the benefits that new technologies can provide to the construction industry, stratified into categories.

Table 3. Expected benefits of the adoption of Construction 4.0 technologies.

Product Benefits	Operational Benefits	Side Benefits
Improved final product quality [27–37] Reduced release time [34,38–41] Preventive maintenance support [31,42–45]	Increased productivity [31,39,43,44] Reduced rework [32] Reduced cost [33,41,43,45,46] Improved communication and information exchange [27,28,32,47–50] Reduction in repetitive work [29,32,39,40,46] Reduction in manual labor and physical exertion [36,37,39,44–46,51]	New business opportunities [28,31,43] Labor reallocation [38,39] Increased employee safety [31,41,46]

In the context of Brazilian manufacturing, certain 4IR technologies can be associated with different benefits; that is, by adopting a certain technology, there is a greater probability of achieving the benefits related to it [13]. Such an association, if well established, could allow users to direct technology adoption efforts according to the desired goals.

Despite the countless gains that can be achieved with the Industry 4.0 model, many difficulties still have to be addressed for the full development of this industrial age. Some of the barriers that may hamper the progress of 4IR include a lack of regulations and standards [9,12,52–55], job cuts [2,12], information security risk [9,12,23], insufficient infrastructure [9,12], a lack of customer demand [23,53,55], a limited clarity of returns and benefits [9,12,53], the difficulty and lack of time for implementation [9,12,53], a lack of knowledge or insufficient information [12,53,55,56], a lack of trained professionals [9,12,53], resistance to change [9,12,52,53,55], and high implementation costs [9,12,23,53,54].

Consideration should also be given to the structural challenges that emerging economies need to overcome to achieve a satisfactory level of technological implementation [13]. Many emerging countries differ greatly from developed countries in terms of technological, scientific, and social barriers and market peculiarities that interfere with the acceptance of innovations [57]. For this reason, in order for adaptation to occur in a satisfactory manner, it is "important to understand the results of the 4IR in the context of each specific industry and country" [2]. It is expected that, with the satisfactory implementation of Industry 4.0, Brazilian companies will experience renewed growth, increased efficiency, and reduced costs [10]. Such results may stem from new products and services that generate additional revenue, as well as from the improvement of operational factors, such as process digitization, real-time quality control, inventory management, and production flexibility [10].

3. Methods

3.1. Hypothesis Formulation

In view of the foregoing and based on a literature review, we developed the following hypotheses to be tested by this study:

H1. *Larger companies make greater use of 4IR technologies.*

H2. *There is a difference in expected benefits and observed barriers to 4IR implementation between companies of different sizes.*

H3. *Professionals working at different stages of the construction lifecycle have a preference for different technologies.*

H4. *The type of expected benefits influences the choice of technologies.*

To test the above-mentioned hypotheses and understand the perceptions of professionals on promising technologies, structured questionnaires were administered to engineers and architects working at any stage of the construction project lifecycle or developing academic research in the field of construction. This type of questionnaire, with a strict sequence of questions and predetermined response options, facilitates the statistical analysis of the results [58]. The aim was to understand the current state of the Brazilian construction sector, assess future expectations regarding the adoption of innovations, and evaluate the potential of the sector to adhere to new technologies. Additionally, critical factors and barriers to the success of technology adoption were identified. Questionnaire data were analyzed using descriptive and inferential statistical techniques to identify behavioral patterns among respondent groups.

3.2. Data Collection Instrument

The structured questionnaire was designed following the recommendations of Manzato and Santos [59]. The instrument contained 24 items divided into five sections. Most of the questions were closed-ended to facilitate data analysis [58,59]. In the first section, questions assessed the characteristics of construction professionals. Respondents were classified based on their professional training, experience, and field of expertise within the construction industry. The second session aimed to characterize the companies where professionals worked based on size, technological maturity degree, and geographical location.

After the characterization of respondents and companies, we sought to evaluate the perceptions of actors in the construction industry about the level of technological advancement of the construction sector and the anticipated benefits of innovations. So, the third section comprised four 5-point Likert-scale questions, one multiple-choice question, and one single-choice question. The first one was related to the role of new technologies in fostering the development of construction activities. The second and third questions in this section assessed the current level of technological development of the construction industry and the perspective for innovation in the next five years, respectively. Responses were rated on a scale ranging from 1 to 5. We assessed user expectations regarding the adoption of innovative technologies in construction by asking respondents to select five benefits they would expect from technology adoption. Additionally, at the end of this section, we inquired about participants' knowledge of terms such as Construction 4.0, Fourth Industrial Revolution, and Industry 4.0. This helped us understand the permeability of 4IR ideas, concepts, and technologies among stakeholders in the construction industry.

In section four, we sought to investigate the development potential of concepts and technologies considered promising in the construction industry by understanding the level of current use and future interest, and assessed the perceptions about the cost and time involved in their implementation. The concepts described in Table 2 were presented to the respondents, aiming to provide a brief elucidation of their application in the sector. Respondents were presented with a list of different concepts and asked to rate the level of application of each concept within organizations on a 5-point scale (very low/absent, scarce, reasonable, good, and high). These levels were assigned integer values ranging from 1 to 5 for quantification. The same scale and items were used to measure the respondents' degree of interest in adopting the concepts over a 5-year period. Academic professionals were invited to answer these questions about the use and interest in technologies, considering the research they carry out in the construction field. With the aim of analyzing the perception of construction professionals about the cost and time required for adopting technologies, participants were asked to rank the cost and time needed to apply technologies/concepts on a 5-point scale (very low, low, reasonable, high, and very high). The items were assigned scores ranging from 1 to 5 for analytics. The participants' perception of the preparedness of companies to adopt technologies was assessed using a 5-point numerical scale, with 1 representing unpreparedness to adopt technologies and 5 representing complete preparedness to adopt technologies.

Finally, the fifth section investigated preferences for innovation adoption and the main barriers to technology diffusion in construction. In this phase of the research, participants were asked to choose three technologies to be adopted in the day-to-day of the company/organization in which they work. As the final part of the survey, we sought to identify factors hindering the adoption of Industry 4.0 technologies in the construction sector according to the opinions of participants. To this end, participants were invited to report the three main barriers that prevent or hinder technology adoption in order of importance. There was also a final open-ended question for additional comments.

After creating the survey instrument, it was pre-tested by 20 volunteers to identify any issues or ambiguities [59]. The instrument was developed using Google Forms. A link was sent to participants through social and professional networks and through institutions such as the Rio Grande do Sul Construction Industry Union (SINDUSCON-RS) and the Santa Catarina Association of Technology (ACATE). A total of 104 valid responses were obtained, representing a relevant sample for this exploratory study.

3.3. Data Analysis

The first step of data analysis involved a descriptive statistical approach, which was used to summarize and describe data both graphically and numerically in an informative manner [60]. Descriptive analysis examines events and understands trends, identifying patterns and drawing conclusions based on them. This approach does not intend to predict outcomes. It focuses on correlations rather than causality, and data interpretation depends on the studied context. Descriptive analysis was applied to all sections of the questionnaire using tables and graphs. These visual representations primarily described the frequency of answers for each alternative. Additionally, we compared data from professionals working at different company sizes to assess differences in perceptions among the groups. In the fourth part of the survey, which evaluated the use, interest, and perception of cost, time, and preparedness for adopting innovations, the graphs displayed mean values based on participant responses. These analytics were conducted using Microsoft Excel tools. We also analyzed information about the current use of technologies to assess the level of development of 4IR innovations in the investigated context. Technologies were classified according to the following criterion: low technology use ($1 \leq$ mean < 2), undeveloped; intermediate technology use ($2 \leq$ average use < 3), incipient; high technology use ($3 \leq$ average use < 4), undergoing development; and very high technology use ($4 \leq$ average use < 5), consolidated.

Section four comprised, in addition to descriptive ones, some inferential analysis conducted using IBM SPSS Statistics 18 software. Firstly, we conducted a cluster analysis to group respondents with similar use patterns. For these analytics, we employed a two-step cluster analysis aiming to identify different groups within our sample, following a previous research approach [17]. Hair et al. [61] suggest the possibility of combining a hierarchical approach to select the number and characterize cluster centers, and a non-hierarchical method, which aggregates all observations using seed points to provide more accurate allocations.

So, at first, an agglomerative hierarchical method, that combined respondents into clusters based on the similarity of their technology use responses, was applied to determine the appropriate number of clusters. This method is capable of generating a tree-like structure, called dendrogram, that captures various consistent partitions at different levels [61]. As suggested by the dendrogram generated in this step, it was inferred that respondents could be classified into three groups, thereby avoiding the dispersion of the sample across several groups with little representativeness or the concentration of heterogeneous respondents in the same group [61]. Subsequently, the sample was divided into these three groups by using non-hierarchical K-means clustering. The objective of this approach is to divide the sample into K (in our case, three) distinct groups, aiming to maximize similarity among members of the same cluster while identifying dissimilarity between different cluster members [61]. This implies that respondents within the same group exhibit similar technology usage patterns among themselves and differ from the use expressed by respondents of other groups. Then, we assessed the demographics of group members to understand the distribution of company sizes within each cluster.

After that, the statistical technique of Principal Component Analysis (PCA) was employed to reduce the number of variables of the dataset [13], aiming to group technologies with similar patterns of use. PCA is a statistical technique that helps us understand relationships among many variables. Its goal is to summarize the information from multiple original variables into a smaller set of factors. These factors represent the common inherent dimensions in the data [61]. The analysis afforded the categorization of the technologies into three groups.

To identify the group where the variable will be placed, the factor loading matrix is analyzed, which indicates the correlation between the variable and the factors. The greater this factor loading, the more strongly the variable is related to that group. However, to facilitate the interpretation of values, the rotated factor loading matrix is often used [62]. In this study, Varimax orthogonal rotation was applied to facilitate data interpretation. Values greater than 0.50 were considered to place a technology in one of the three groups [61]. The technologies that did not exhibit sufficiently high factor loadings were considered indeterminate and were not included in any grouping [62]

The adequacy of the sample was tested by the Kaiser–Meyer–Olkin (KMO) test, Bartlett's sphericity test, and measures of sample adequacy (MSA) [61]. The results showed that the sample was adequate, with a KMO value of 0.852, a significant Bartlett's test result ($p < 0.001$), and an MSA greater than 0.50. The internal consistency of the clusters was measured by Cronbach's alpha [61].

Finally, by averaging technology use, interest, and preparedness, we assessed the absorption potential of Industry 4.0 technologies in the construction sector within our study context. Subsequently, we conducted multiple regression analysis to identify factors influencing the adoption of technology. Regression is a technique in which a variable is dependent or can be explained by other independent variables (predictors). The objective is to predict changes in the dependent variable in response to changes in the independent ones. Two variables are considered correlated when changes in one variable are associated with changes in the other. This correlation is reflected in the regression coefficients. To achieve reliable results, multivariate analysis should be conducted using a dataset that adheres to specific criteria, including normality, homoscedasticity, and linearity. It is crucial to avoid multicollinearity, which occurs when independent variables are highly correlated with each other. Multicollinearity can weaken the predictive power of the analysis [61]. In our study, we assessed these factors using the aforementioned software. The significance of the relationships was evaluated using ANOVA analysis, which provides a statistical test for the overall model fit in terms of the F ratio [61].

Each cluster of technologies underwent regression using two models. The first model included professionals' characteristics (field of activity, years of experience, and knowledge of 4IR) as predictors, while the second model also considered company characteristics (size and technological maturity). Both models yielded significant results in explaining adoption patterns ($p < 0.05$) [61], with significance increasing when company characteristics were included. This analysis helped consolidate the framework linking the 4IR technologies to different construction lifecycle phases.

Additionally, we applied Poisson regression with robust error variance [63,64] for each technology to identify factors influencing use and interest in that technology. This method is suitable for counting data, especially when large counts are rare events [64]. For the analysis, a binary variable was created for low use and interest and another for high use and interest (1 = high; 0 = low). We then assessed the proportion of respondents within each option for potential influencing factors. For instance, within the group exhibiting high technology use, we evaluated the proportion of professionals working in smaller companies compared to those in larger companies. Similar analyses were conducted for "low use", "high interest", and "low interest", considering other factors such as company maturity and sector, as well as perceptions related to time, cost, and market readiness for technology adoption. To address the potential overestimation of relative risk errors in Poisson regression due to binomial data, we employed a robust error variance procedure [63]. Results with p-values below 0.05 were considered significant, indicating that a factor influences technology use or interest.

Additionally, to explore the relationship between anticipated benefits and technology preferences, the standardized Pearson chi-squared and Fisher's exact tests were employed. These tests evaluate two variables' independence when comparing independent and uncorrelated groups—benefits and preferences. The first method is suitable for a large sample, while the second is more appropriate for a small sample [61,65]. When fewer than five respondents chose a specific technology, we applied Fisher's test.

In the final sections of this paper, we summarized the results of both descriptive and inferential analyses, ultimately identifying critical factors for innovation in the studied context.

3.4. Sample Characterization

Different phases of the construction industry can be impacted by the use of new technologies. Thus, this study sought to include individuals with different professional profiles who participate in different phases of the construction lifecycle. A description of the sample is shown in Table 4.

Table 4. Characterization of the sample of construction professionals.

Variable	Description	Absolute Frequency	Relative Frequency
Academic degree	Architecture/Urban Planning	12	11.5%
	Civil Engineering	88	84.6%
	Other	4	3.8%
Level of education	Doctoral degree	5	4.8%
	Undergraduate degree	32	30.8%
	Master's degree	26	25.0%
	Specialization (postgraduate degree lato sensu)	41	39.4%
Field of expertise	Academic research/teaching	7	6.7%
	Project management	33	31.7%
	Budget/planning	22	21.2%
	Inspection	14	13.5%
	Supervision	20	19.2%
	Technical evaluation	2	1.9%
	Other	6	5.8%
Professional experience	1 to 3 years	28	26.9%
	4 to 6 years	21	20.2%
	7 to 10 years	15	14.4%
	11 to 15 years	11	10.6%
	16 to 20 years	5	4.8%
	More than 20 years	24	23.1%
Sector	Private	77	74%
	Public	27	26%

Most participants (85%) had a degree in Civil Engineering, and a smaller proportion (11%) had a degree in Architecture and Urbanism. Only 4% of respondents had training in other fields of engineering. The question about continuing education showed that 69% of participants had some graduate degree, either stricto or lato sensu. As for professional experience, it was found that 53% of respondents had more than seven years of experience in construction, whereas the other 47% had worked in the field for a shorter time.

Most of the research participants stated that they worked in private companies (74%). The area with the highest proportion of respondents was that of project management (32%). The sample also included professionals who worked in academic research and/or teaching (7%).

Finally, 32% of respondents reported that they worked in areas directly related to the construction phase, such as inspection and supervision, and 53% of respondents worked in pre-construction phases, such as project management, budget, and planning.

3.5. Characterization of Construction Companies

The second section of the questionnaire assessed the characteristics of the companies where participants worked. Most responses were concentrated in one of two extremes: companies were either micro (37%) or large (40%) in size (Figure 1). This distribution revealed an interesting comparison of the technology implementation profile of companies of different sizes. Given the lower frequency of respondents in medium-sized companies, the results were grouped into two groups for statistical analysis purposes (micro and small companies were grouped into smaller companies, and medium-sized and large companies were grouped into larger companies).

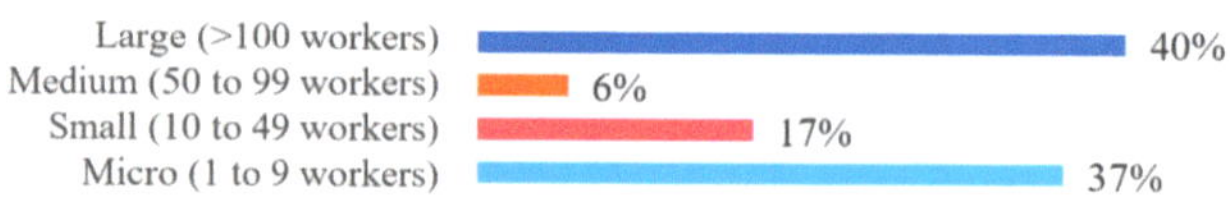

Figure 1. Size characterization of construction companies.

In terms of geographical location, respondents from companies across thirteen states in Brazil participated in the questionnaire. However, half of the answers were from workers based in Rio Grande do Sul State, the southernmost region of the country. Therefore, it is important to highlight that, given the large territorial extension of Brazil, the results may not reflect the reality of the country as a whole.

The last question related to company characterization assessed the degree of technological maturity in terms of digital transformation. The majority of respondents (56%) rated the maturity of their organization as "undergoing development", indicating that they were at the beginning of the digital journey, with leaders showing an understanding of the importance of digital transformation but without a clear implementation strategy. It should be noted that 27% of respondents classified their organization as "sophisticated". This means that just over a quarter of companies already reap the benefits of digital transformation with a clear implementation strategy and a high level of employee engagement. On the other hand, only 5% of companies were classified as "innovative", which represents the maximum degree of technological maturity. Furthermore, 12% of companies were described as "traditional", with no strategy or plan for digital transformation. Overall, the results suggest that the construction industry is moving toward digitization, but implementation strategies have not yet been consolidated (Figure 2).

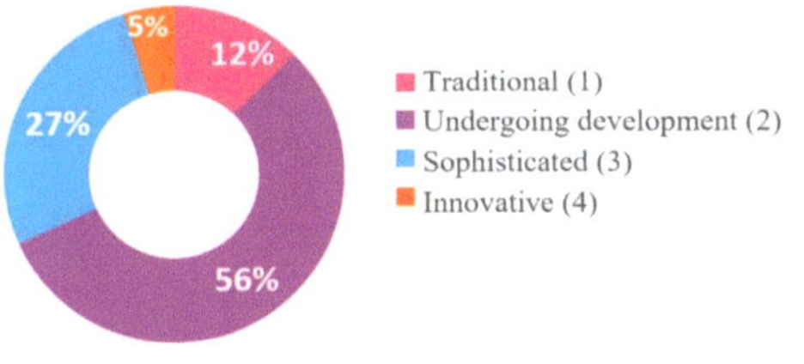

Figure 2. Level of technological maturity of construction companies. (1) There is no defined digital strategy. Leaders and teams are not prepared for the required transformations, and most services are not digital. (2) The institution has embarked on its digital journey by acquiring software and/or technological equipment, but there is no clear implementation strategy, team training, or evaluation of results. Nevertheless, leaders understand that digital transformation is essential for the company. (3) The institution is reaping the fruits of digital transformation, having a clear strategy and employee engagement. Most processes have been digitized, and there is information integration between some of them. (4) The institution has achieved an advanced level of digitization, has a well-defined and structured information management strategy, and performs impact assessment and continuous improvement. There is the use of disruptive technologies associated with the fourth industrial revolution, such as the Internet of Things and artificial intelligence.

When comparing technological maturity between company sizes, it was found that larger companies have a greater level of innovation adoption. Large companies were more frequently evaluated as sophisticated than micro and small companies and were considered innovative more than twice as frequently as micro enterprises (Figure 3). It should be noted, however, that smaller companies are moving toward technological maturity, even if slowly. Given the low number of respondents from medium-sized companies, the results of this group were considered unrepresentative.

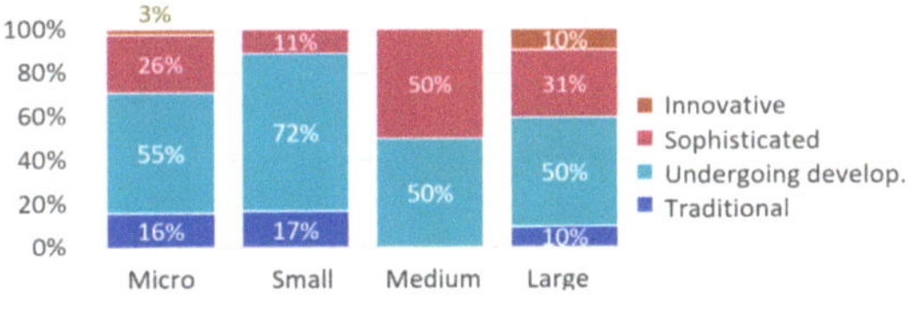

Figure 3. Technological maturity according to company size.

4. Results

4.1. Characterization of Technological Advancement in Construction

By collecting data from five out of the six questions in Section 3 of the questionnaire, we were able to characterize technological advancements in the construction industry according to respondents' perceptions. The majority of respondents (79%) regarded technological innovation as extremely important for the construction sector, assigning the maximum score to the question, and no respondents considered it unimportant.

Although the value of technological innovation was widely recognized by the study sample, the pace of innovation adoption over the previous five years was perceived as insufficient, with 39% of respondents reporting a moderate level of innovation adoption (score 3 out of 5) and 36% reporting a below average level (<3). Only 3% of construction professionals observed a significant evolution in the construction sector, attributing it a score of 5. These findings show that there is a growing movement in search of innovation, but adoption is still slow.

When asked about the level of development expected for the sector in the following five years, respondents were reasonably optimistic about the future, with 46% of valuations above the intermediate value, tending toward a "rapid and significant advancement". As for the current level of technological development, almost half of the participants (49%) considered it below average, indicating that the industry is still more focused on traditional methods (Figure 4).

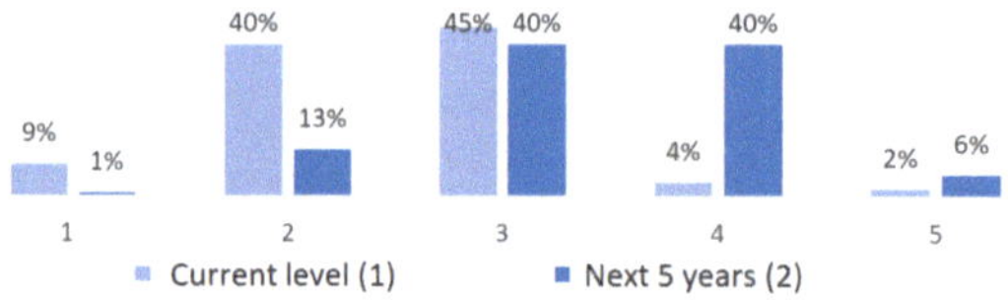

Figure 4. Level of technological development of the construction industry. (1) Responses range from 1 (traditional) to 5 (innovative). (2) Responses range from 1 (no advance) to 5 (rapid and significant advance).

A comparison of the expectations of respondents from companies of different sizes (Figure 5) revealed a greater degree of optimism about the near future in smaller companies, with 54% of respondents expecting an above-average pace of advance, compared to 37% of respondents from larger companies. This expectation may indicate a more optimistic perspective and greater interest in small and micro enterprises in adopting new technologies in the coming years.

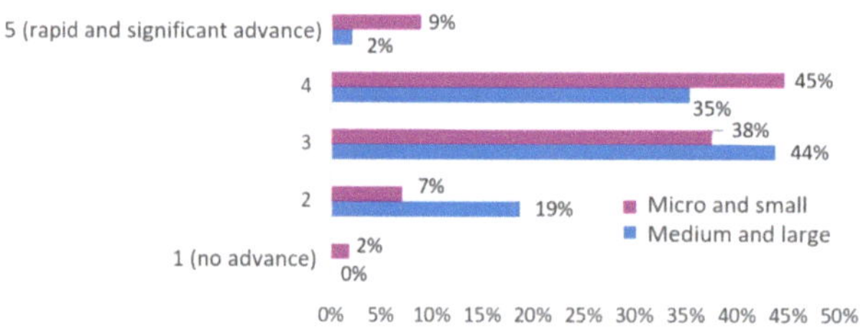

Figure 5. Level of technological development expected over the next 5 years according to company size.

As shown in Figure 6, 57% of respondents said that they had no knowledge about these topics and, of these, 13% reported not even having heard these terms before. Those with advanced knowledge comprised 6% of the sample, and 37% considered they had only basic knowledge. These results evidence the need to disseminate knowledge about 4IR among construction professionals and raise awareness about the current movements of the industry in general.

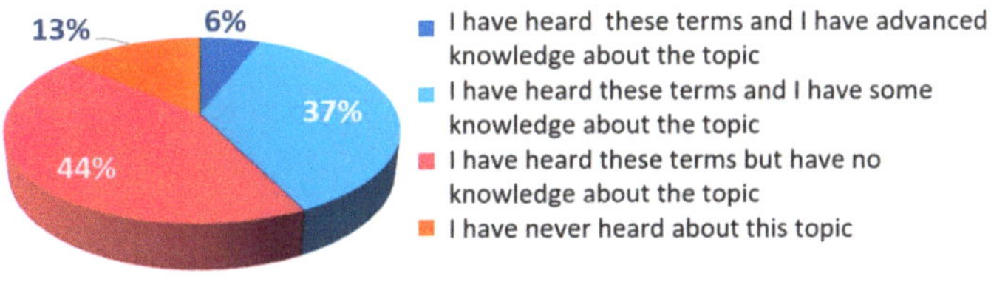

Figure 6. Knowledge of terms related to Industry 4.0.

The analysis of knowledge level according to company size (Table 5) showed, as expected, that professionals with advanced knowledge about Industry 4.0 terms are more frequent in large companies (12%) than in microenterprises (3%). Surprisingly, however, individuals who had never heard of the terms were also more common in large companies. This may indicate that knowledge in larger companies might be concentrated in some agents without vertical dissemination. Additionally, it can be inferred that smaller companies are aware of the new IR but still lag behind in terms of learning and knowledge consolidation. For medium-sized companies, the number of respondents was not significant to identify general behavioral trends.

Table 5. Knowledge of Industry 4.0 terms according to company size.

Response	Company Size			
	Micro	Small	Medium	Large
I have never heard about this topic	11%	11%	0%	19%
I have heard these terms but have no knowledge about the topic	50%	28%	33%	45%
I have heard these terms, and I have some knowledge about the topic	37%	56%	67%	24%
I have heard these terms, and I have advanced knowledge of the topic	3%	6%	0%	12%
Total number of responses	38	18	6	42

4.2. Expected Benefits of Technology Use

The results obtained from the last question in Section 3 of the questionnaire revealed that gains in productivity (74%) and final product quality (68%) were the most anticipated benefits, followed by reduced rework (60%) and production costs (59%) (Figure 7). The least frequent expectation indicated by the respondents was an increase in employee safety. This perception may be related to the large number of participants who work in pre-construction phases, where safety concerns are not as evident.

Figure 7. Expected benefits from using new technologies.

It is surprising that preventive maintenance support was one of the least expected benefits. One of the great advantages of using emerging technologies is the possibility of the end-to-end integration of production chain information and asset monitoring through sensors and models, which would greatly benefit the use and maintenance phases. This finding indicates that the potential of new technologies in construction may not be sufficiently clear and that professionals still have little interest in the lifecycle management of buildings. It underscores, therefore, the need for professionals to be trained and raised aware of the applications of new technologies.

In comparing the perceptions of professionals from large and small companies about the benefits of technologies, we found that a reduction in release time was significantly more expected among workers from small companies ($p < 0.05$ in the Poisson regression test). A similar result was observed for private company workers as compared with public employees ($p = 0.05$), explained by the fact that release times are usually less rigid in the public sector. On the other hand, increased productivity and a better allocation of resources were more important for larger organizations. Cost reduction was more relevant for large companies than for small ones. Small organizations valued more rework reduction and final product quality. Small companies also emphasized the reduction in repetitive work.

These differences in perceptions between companies of different sizes are expected, given that they have different objectives. The results support and validate the second hypothesis of this study (H2) about the expected benefits of using emerging technologies. Whereas larger companies seek to optimize processes to reduce production costs and increase productivity, smaller companies are more focused on growth and market gain. To do this, smaller companies need to overcome obstacles that, in general, have already been overcome by larger companies, such as those related to repetitive work, rework, and release time. Larger companies seek to improve resource allocation, productivity, and the exchange of information between stakeholders to remain competitive.

Operational benefits, those associated with productivity and efficiency, were the most expected. The expectation of benefits related to products was, however, somewhat lower. Although there is strong potential for increasing quality with the introduction of innovations, the reduction in release time and support for preventive maintenance remained in the background. On the other hand, side benefits were the least expected among respondents, which might be related to a lack of interest in these benefits or a limited understanding of the potential of technologies.

Hierarchical linear regression was performed to identify factors influencing the importance given to the different types of benefits. It was inferred that benefits related to products increase in importance with increasing experience in the activity. On the other hand, workers with less experience tend to value the operational and side benefits of emerging technologies, particularly the latter.

4.3. Potential of Industry 4.0 Technologies in Construction

Drawing on the results obtained from Section 4 of the questionnaire and combining them with data from previous questions, this section aims to explore the adoption patterns of concepts and technologies within the construction sector in the investigated context.

4.3.1. Use and Interest

The mean level of use of Industry 4.0 technologies by construction professionals is shown in Figure 8. Cloud computing was the technology with the highest degree of implementation. However, in line with what was observed among manufacturing companies [19], cloud computing is mostly applied in the form of remote servers for data storage or online software. IoT, sensors, and CPS, which depend on the use of clouds, were second-to-last in terms of current application. This finding demonstrates that construction equipment and products are not yet connected to the cloud; this would enable communication between objects and servers or controllers. In addition to cloud computing, only mobile devices exceeded the average usage level. Even building information modeling (BIM), which has shown great impact potential and is a topic of wide visibility in discussions concerning the construction industry, is not yet consolidated in emerging markets.

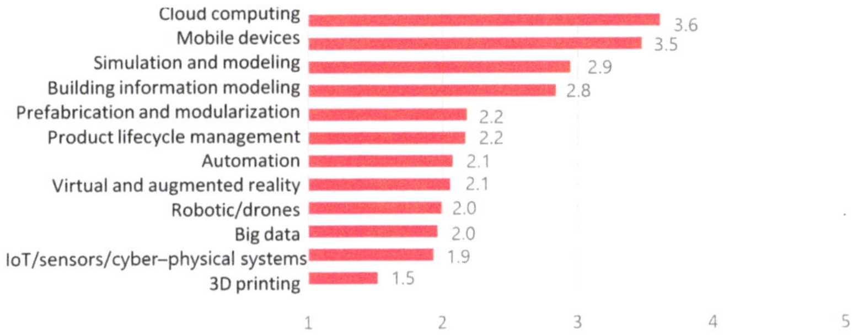

Figure 8. Use of Industry 4.0 technologies in the construction sector.

Additive manufacturing, represented by 3D printing, had the lowest level of adoption among respondents, reflecting its incipient development in the construction sector. This reality contrasts with the international literature showing vast interest in additive manufacturing [6]. In Brazil, this technology remains at the academic level without practical application.

No significant differences in the degree of technology use were observed between small and large companies. Therefore, it was not possible to confirm the first hypothesis of this study (H1) through descriptive analysis. Among small companies, we observed a slight trend toward the use of innovations related to virtual environments (BIM, simulation and modeling, and virtual and augmented reality) and mobile devices. Large companies, on the other hand, made more use of concepts related to advanced manufacturing, data intelligence, and automation. The identified pattern might be related to the focus of action of small companies, that is, pre-construction phases, such as design, planning, and budgeting.

In light of the current level of technology use, we investigated the degree of development of innovations among the respondents. As explained in Section 3.3, we classified the technologies into four categories: undeveloped, incipient, under development, and consolidated, based on their mean level of use (Figure 9). Although Oesterreich and Teutberg [5] stated that "several digitization and automation technologies for construction have reached market maturity and thus are currently available", the vast majority of technologies are not in an advanced state of use. Furthermore, those that are fundamental to the consolidation of emerging technologies in construction (base technologies) have not yet been developed. Our findings indicate that no technology has been fully consolidated in the studied context.

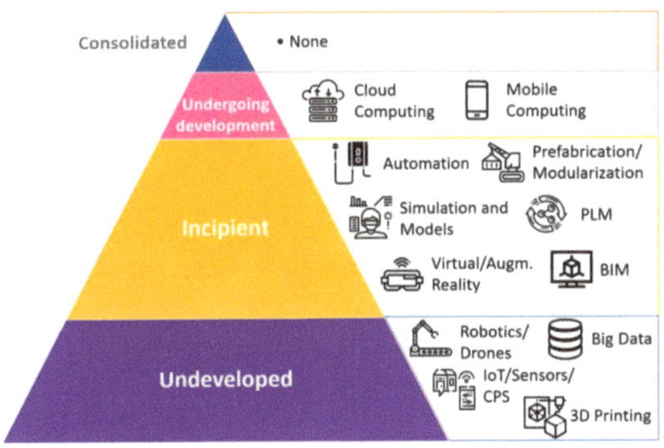

Figure 9. Level of development of 4IR technologies in the construction industry of the studied context.

Upon comparing the level of current usage with the expressed interest in adopting technology, respondents showed a willingness to expand the use of all concepts in the coming years (Figure 10). The highest levels of interest were in mobile devices, cloud computing, and BIM. The industry is closer to achieving the desired development in the first two technologies, as shown by our results. The other technologies are still far from reaching the desired level of use expressed by the respondents, particularly virtual and augmented reality, BIM, and automation.

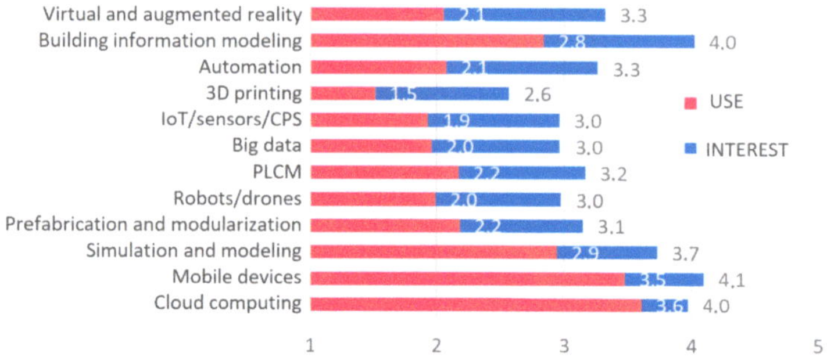

Figure 10. Current use and interest in adopting Industry 4.0 technologies in the construction sector.

Applying the Poisson regression, it was inferred again that the technological maturity of companies is a fundamental factor for technology use. For all innovations, there was a significant increase in the level of use with increasing technological maturity. The same was observed for interest: professionals working in companies that had already begun the transition to digital technologies showed more interest in new technologies, even if such technologies were not yet used. The only technologies for which this increase in interest was not significant were IoT/sensors/CPS and product lifecycle management. These results indicate that traditional companies, in addition to not having yet started the digital transformation process, have less interest in adopting technologies in the next five years.

There was not enough statistical power to identify differences in the use of technologies between public and private companies. However, the interest of professionals in the public sector was significantly lower than that of private sector employees. This result denotes greater disinterest in digitization among public workers.

Regarding the influence of respondents' areas of expertise, Poisson regression indicates that the use of robots and drones proved to be superior among professionals working in phases related to construction. On the other hand, simulation and modeling were more predominant in the preconstruction phases. The intention to adopt BIM in the coming years was higher among participants working in project design, planning, and budgeting. These observations are in line with previous results. This result supports our third hypothesis (H3).

Cluster Analysis

Cluster analysis was performed to identify similarities between respondents and technologies. First, the sample was grouped according to similarities in technology use by hierarchical and non-hierarchical clustering [59]. As a result, three different usage profiles were obtained, namely, (i) high technology use, composed of professionals who use practically all emerging technologies more frequently than the other groups, except cloud computing; (ii) moderate technology use, comprising professionals whose average use of technologies is higher than that of the low group and who frequently use mobile devices and cloud computing; and (iii) low technology use, comprising professionals who have a low degree of technology use (mean < 1.5).

Table 6 shows the technology use results (mean and standard deviation) for the identified groups. It also includes the percentage of respondents in each group and the representativeness of companies of different sizes. The use of all concepts has statistical significance in determining the group in which the respondents were classified; that is, members of each group use all concepts differently from members of the others.

Table 6. Grouping of construction professionals according to the degree of technology use.

Technology	High Use		Moderate Use		Low Use		*F*-Value
	Mean	SD	Mean	SD	Mean	SD	
Mobile devices	4.19	0.75	4.02	1.06	2.35	1.30	30.64 ***
Cloud computing	3.96	0.96	**4.39**	0.83	2.49	1.39	31.52 ***
Simulation and modeling	3.96	1.00	3.56	1.21	1.54	1.07	46.91 ***
Building information modeling	3.92	0.80	3.27	1.18	1.59	0.64	55.50 ***
Virtual and augmented reality	3.35	1.06	2.02	0.91	1.22	0.53	49.27 ***
Product lifecycle management	3.27	1.34	2.32	1.06	1.27	0.65	29.85 ***
Automation	3.15	0.73	2.24	1.20	1.16	0.50	39.31 ***
Robots and drones	3.04	1.04	1.85	1.01	1.41	0.76	23.82 ***
Big data	2.73	1.15	2.20	1.03	1.19	0.62	22.46 ***
IoT, sensors, and CPS	3.31	1.05	1.85	0.96	1.08	0.28	57.11 ***
Prefabrication/modularization	3.42	1.03	1.68	0.96	**1.86**	1.32	21.71 ***
3D printing	2.65	1.02	1.24	0.54	1.05	0.23	57.22 ***
% of respondents in each group	**25%**		**39%**		**36%**		
Large and medium companies	42%		44%		51%		
Micro and small companies	58%		56%		49%		

IoT, Internet of Things; CPS, cyber–physical system; *** $p < 0.001$.

In agreement with what has been observed in manufacturing [19], construction professionals tended to increase technology use in a homogeneous and progressive way. In other words, professionals who used one technology more frequently tended to adopt other technologies over time. Therefore, technology use increases with the increase in the technological maturity of the company. This inference is corroborated by the degree of technological maturity of professionals from different groups: in the high technology use group, 54% of professionals reported working in a company with sophisticated or innovative characteristics, whereas, in the low technology use group, this number dropped to 14%. Furthermore, the group of professionals with high technology use showed greater interest in the future, demonstrating that those who already adopted innovative technologies intend to further expand their use in the coming years.

Another interesting result is that there was no proportional relationship between company size and the adoption of technologies, which is different from that observed in the manufacturing industry [17]. The cited study found that there were more small and medium-sized companies in the high technology use group than large and medium-sized companies. Again, the first hypothesis of the current study (H1) could not be confirmed.

The categorization of technologies into groups, performed through PCA to facilitate further analysis, is represented in Table 7. As explained earlier, the analysis grouped variables with similar use patterns among the survey participants and revealed three clusters, which were labeled as virtualization, automation, and manufacture, based on the technologies they encompass. IoT/sensors/CPS was not statistically included in any of the three groups. In Table 7, factor loadings greater than 0.50, considered significant [61], are highlighted in bold, showing the items that make up each group. The internal consistency of the first two clusters, as measured by Cronbach's alpha, was high. The consistency of the third group was low, but the cluster was maintained, given the exploratory nature of the study and the low number of variables involved.

The first cluster (C1, virtualization) comprised mobile devices, cloud computing, and virtual environment technologies (BIM, simulation and modeling, and virtual and augmented reality). This cluster had the highest mean utilization score among respondents. It is understood, therefore, that this cluster represents the first innovations absorbed by the market, either because they are more consolidated or because they are perceived to provide more benefits. The second cluster (C2, automation) had intermediate adoption among respondents. This cluster includes technologies related to data intelligence, such as big data and product lifecycle management, as well as automation technologies, robots, and drones.

Table 7. Grouping of similar technologies.

Technology	Factor			Commonality
	Virtualization	Automation	Manufacture	
Cloud computing	**0.629**	0.475	−0.361	0.751
Big data	0.143	**0.756**	0.243	0.651
PLM	0.312	**0.543**	0.402	0.554
Robots and drones	0.053	**0.721**	0.217	0.570
Automation	0.383	**0.608**	0.199	0.556
BIM	**0.861**	0.137	0.234	0.815
Simulation and modeling	**0.843**	0.114	0.221	0.772
VR and AR	**0.642**	0.210	0.491	0.697
IoT, sensors, CPS	0.492	0.446	0.354	0.566
Mobile devices	**0.540**	0.489	−0.117	0.544
3D printing	0.205	0.279	**0.686**	0.591
Prefabrication and modularization	0.065	0.173	**0.761**	0.613
Eigenvalue	5.29	1.37	1.02	
Cumulative variance (%)	44.09	55.52	64.00	
Cronbach's alpha	0.76	0.86	0.54	

PLM, product lifecycle management; IoT, Internet of Things; CPS, cyber–physical system.

Cluster 3 (C3, manufacture) includes prefabrication/modularization and 3D printing, both related to the advancement of construction techniques. This cluster had the lowest degree of use among research participants. However, prefabrication was more applied than 3D printing, particularly among high technology use professionals, who have greater technological maturity. This result shows that most companies still adopt traditional construction techniques, although prefabrication has been gaining ground in companies with greater technological maturity.

4.3.2. Perception of Cost, Time, and Preparedness of Companies to Adopt Emerging Technologies

In this section, we analyze construction professionals' perceptions regarding the costs and time associated with technology adoption, as well as the market's readiness to embrace it. As shown in Figure 11, robots and drones were perceived to have the highest implementation costs, followed by automation and 3D printing. The perception of cost seems to be strongly linked to the acquisition of technologies, not to the actual implementation process. It should be noted that practically all technologies exceeded the average value in terms of cost, with the exception of mobile devices and cloud computing, explaining their high degree of use.

The same pattern was observed for implementation time: mobile devices and cloud computing were perceived to have the shortest time of implementation. Automation and IoT/sensors/CPS, by contrast, were perceived to require more time for implementation, probably because these technologies are believed to have greater complexity. In general, the perception of implementation time was lower than that of implementation costs, indicating that the latter might be more important in the decision to adopt technologies.

Nevertheless, the Poisson regression inferred that a perception of longer implementation time is strongly associated with reduced interest in adopting technologies. In several cases, respondents who perceived a technology as more time-consuming to implement also showed a low use intention in the coming years.

Also, according to the respondents, the industry is unprepared to adopt the vast majority of technologies. The technologies with scores higher than the average were mobile devices and cloud computing, followed by prefabrication and modularization (intermediate scores). This finding suggests a market that is still insecure and has a low capacity to modernize itself. In this scenario, there is a greater need for qualified professionals to assist in the transition to modernization and technological maturity.

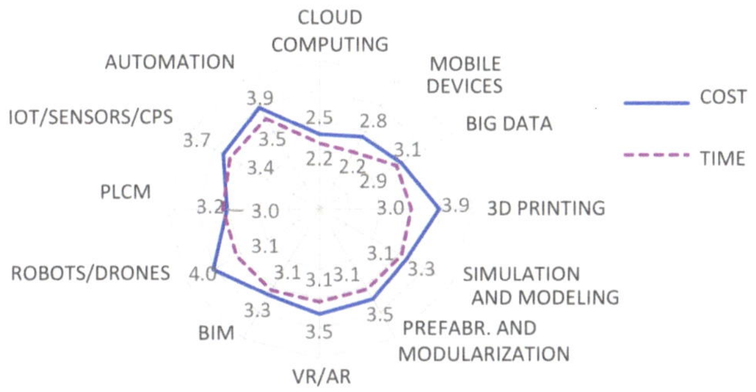

Figure 11. Perception about the cost and time to adopt Industry 4.0 technologies.

The perception of market preparedness was directly related to interest in adopting technologies, as indicated by the Poisson regression analysis. Professionals who perceived the market as well prepared to receive innovations tended to express greater interest in adopting new technologies. Such a factor might be related to confidence in technologies that have already been tested and approved by peers, as noted by Rogers [22].

4.3.3. Technology Absorption Potential

By analyzing the results for technology use, interest, and preparedness, it can be concluded that the technologies with the highest potential for absorption in the studied context are mobile devices and cloud computing, followed by BIM, simulation and modeling, and prefabrication and modularization (Figure 12).

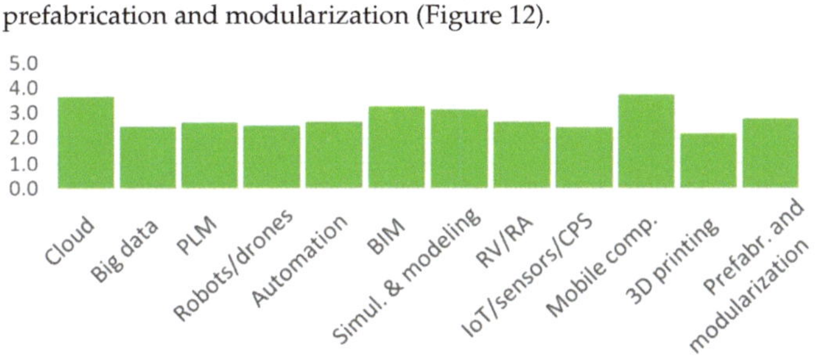

Figure 12. Absorption potential of Industry 4.0 technologies in the studied context.

Table 8 presents the results of multiple regression performed to identify what can influence the absorption potential of 4IR technologies. The values of the table indicate the correlation between dependent and independent variables. The negative sign indicates a negative influence. In regard to the area of expertise, greater values represent more advanced phases in the construction lifecycle. For manufacturing technologies and IoT/sensors/CPS, significance was only observed for the model that included company characteristics—maturity and size ($p < 0.001$). It is clear that the adoption of these two concepts is primarily dependent on company maturity.

The adoption potential of virtualization (C1) technologies, besides being greater than the other clusters, was also statistically higher among smaller companies than among larger companies. C3, composed of smart manufacturing technologies, had higher adoption potential among large companies, but this difference was not statistically significant. The adoption potential of automation (C2) technologies was significantly higher among professionals working in the most advanced lifecycle phases, such as supervision, inspection, and the preparation of technical and building evaluation reports, as well as among professionals with a greater knowledge of Industry 4.0.

Table 8. Factors influencing the adoption of Industry 4.0 technologies in the construction sector.

Factor	Virtualization		Automation		Manufacture		IoT	
Area	0.075	0.072	0.285 ***	0.302 *	0.088	0.050	0.109	0.119
Experience	−0.104 **	−0.092 **	−0.046	−0.061 *	−0.044	−0.037	−0.004	−0.012
Knowledge	0.138	0.109	0.152 *	0.113	0.056	0.061	−0.050	−0.091
Maturity		0.538 ***		0.392 ***		0.283 ***		0.440 ***
Size		−0.151 ***		−0.070		0.037		−0.099
F-value	2.711 **	8.518 ***	4.069 ***	5.919 ***	0.659	2.142 *	0.328	2.547 **

* $p < 0.10$; ** $p < 0.05$; *** $p < 0.001$.

For all technology clusters, there was a positive correlation between absorption potential and the technological maturity of companies, further corroborating that companies that are more mature are more likely to absorb any Industry 4.0 concept. The experience of professionals was inversely proportional to technology absorption potential, with significant differences in C1 and C2. This finding suggests that more experienced professionals have greater skepticism toward emerging technologies.

On the basis of the results, it can be inferred again that there is a relationship between the lifecycle phase of construction projects and the choice of certain technologies, confirming the third hypothesis (H3) of this study. The framework presented in Figure 13 illustrates which concepts are more interesting for each lifecycle stage and the likely order of adoption by practitioners in the construction industry. The design phase can make use of technologies related to virtualization. Inspection and technical reports and building evaluation phases may benefit from technologies related to both virtualization and automation. Manufacture technologies are applied mainly in the construction phase, which may also benefit from automation concepts and the use of IoT/sensors/CPS. The budget and planning phase could benefit from all concepts analyzed here. IoT, sensors, and CPS, which were not included in any cluster, are depicted to lie across all three groups, as they provide integration between emerging technologies.

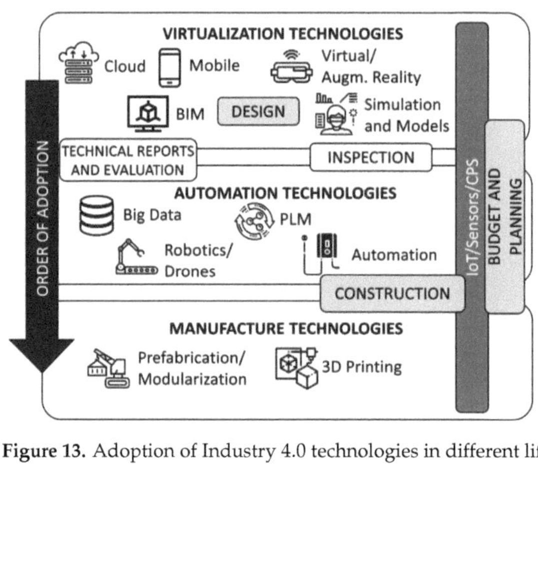

Figure 13. Adoption of Industry 4.0 technologies in different lifecycle phases of construction.

4.4. Preference of Respondents

In the fifth section of the questionnaire, participants were asked to express their preferences among the presented technologies. The most chosen technology was BIM, reported by 63% of respondents (Figure 14). The second most chosen technology was cloud computing (52%). Simulation and modeling were chosen by 27% of professionals. Less than a quarter of the participants selected the remaining technologies.

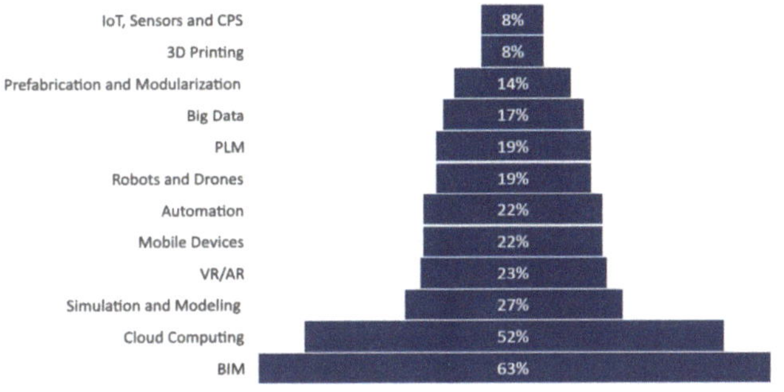

Figure 14. Choice of Industry 4.0 technologies among construction professionals.

It is noteworthy that mobile devices, which had great adoption interest in the previous survey, were chosen by only 22% of respondents. This finding shows that, although there is interest in this technology, it is not considered a priority or that respondents already feel satisfied with the current level of use of the technology, in agreement with the results presented in Figure 10. IoT/sensors/CPS and 3D printing ranked last, being selected by only 8% of participants. The dissociation between IoT and cloud computing, as previously mentioned, corroborates the focus on file storage and use of cloud software and not on integration between objects via devices and sensors. The finding also highlights the distance between construction and 4IR, given that the basis of the revolution is the connectivity promoted by sensors.

We did not identify statistically significant relationships between respondents' choice of technologies and the objectives they indicated. Thus, the fourth hypothesis (H4) of this research was rejected, and it can be inferred that professionals are not yet able to associate the available innovations with the objectives they are aiming for.

4.5. Barriers to Technology Adoption

Finally, we assessed the potential barriers to the adoption of 4IR technologies in the studied context. Figure 15 shows that the most chosen barrier, regardless of priority, was the high cost of implementation (56%), followed by resistance to change (53%), which is characteristic of the sector. The third most cited factor (45%) was the lack of qualified professionals to lead the necessary changes, underscoring the need for training. The factors considered less important were a lack of regulation and standards, a reduction in jobs, and risks to information security.

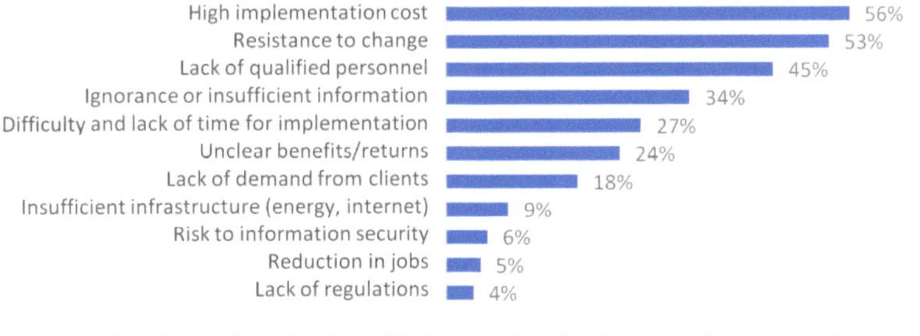

Figure 15. Barriers to the adoption of Industry 4.0 technologies in the construction sector.

We stratified results according to company size. Implementation cost was cited as a major barrier by medium and large companies. For smaller companies, however, the biggest difficulty to overcome was the culture of resistance to change. High costs were also perceived as a barrier. Larger companies showed more concern about data security, available infrastructure, and a lack of regulations, although these factors were not considered of great relevance. Professionals from small and micro companies have lower demands from customers and, therefore, may feel less inclined to adopt new technologies.

The results indicate that organizations of different sizes encounter different barriers to the adoption of emerging technologies, supporting the second hypothesis of this study (H2). Larger companies are more concerned about operational factors, such as costs and benefits, data security, infrastructure, and regulation, whereas smaller companies are concerned about initial barriers, such as resistance to change, the difficulty of implementation, and low customer demand.

Barriers differed according to the degree of technology use, grouped in Table 6 (Figure 16). A lack of clarity of benefits and the difficulty of implementation were less important for professionals who already made use of innovative technologies, whereas resistance to change and a lack of customer demand gained prominence. For respondents who were in the low technology use group, ignorance or lack of information about innovations was evident, appearing as the third most important barrier. This finding indicates that knowledge dissemination may contribute to introducing innovations in traditional companies.

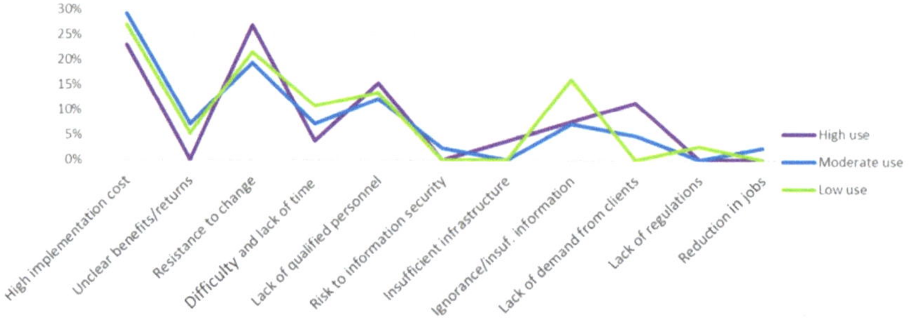

Figure 16. Important barriers according to the degree of innovation use.

Understanding the obstacles perceived by construction professionals is crucial for guiding the development of actions in research and development.

4.6. Overview

The questionnaire administered to engineers and architects working in the Brazilian construction industry allowed us to test the hypotheses formulated and gain insights into the adoption of Industry 4.0 technologies in this context.

Section 4.1 of the present study provided an overview of respondents' perceptions regarding the importance of technologies for the construction sector's evolution, as well as perceptions about technological advancements observed in this context. It also assessed participants' knowledge of the topic. In turn, Section 4.2 addressed the expected benefits associated with the introduction of technologies in construction activities and the factors influencing those expectations. In this section, we tested hypothesis H2 concerning the differences in expected benefits for companies of different sizes.

Section 4.3 presented data on technology usage and interest among participants, highlighting factors that influence these aspects. These data allowed us to classify technologies into different development levels within the studied context and test the first and third hypotheses (H1 and H3). Additionally, we conducted clustering analyses, grouping respondents with similar usage profiles and identifying technologies with similar adoption patterns. We identified three respondent groups with varying technology usage levels (high, medium, and low) and three technology groups (virtualization, automation, and manufacturing) used similarly by participants. In Section 4.3, we also evaluated perceptions related to time, cost, and market readiness for technology adoption, identifying the potential for technology absorption among respondents and the factors affecting that potential. The analyses led to the creation of a framework identifying the likely order of technology adoption by professionals and the lifecycle phases to which these concepts best apply.

Section 4.4 analyzed the respondents' hypothetical choices among various observed technologies, assessing participant preferences and testing the fourth hypothesis (H4). Finally, in Section 4.5, we evaluated barriers to technology adoption within the studied context. We also investigated potential differences in the barriers identified by professionals from larger versus smaller companies, as well as variations among professionals with different technology usage levels. This analysis concluded the testing of hypothesis H2. Figure 17 summarizes significant results, listing factors identified as barriers or promoters of innovation in the construction industry of the studied context.

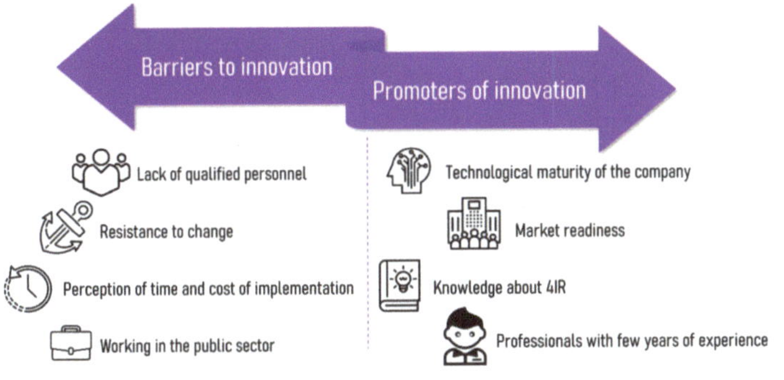

Figure 17. Critical factors for innovation in the construction sector of studied context.

Table 9 summarizes the tests conducted for hypothesis validation, the bibliographic sources from which they originated, the data collection source in the applied questionnaire, the location where the hypothesis results were presented in the text, and the test outcomes. Hypotheses H2 and H3 were supported by the analyses performed, while hypotheses H1 and H4 could not be confirmed in this study.

Table 9. Summary of the hypothesis tests and results.

Hypothesis	Reference	Statistical Test	Source of Data (Questionnaire)	Evidence in the Text	Result
H1	[3,17,23]	Descriptive analysis Linear regression	Section 4	Section 4.3.1 Table 6 Table 8	Rejected
H2	[3,23]	Poisson regression Descriptive analysis	Section 3 Section 5	Section 4.2 Section 4.5	Supported
H3	[6,25,26]	Poisson regression Linear regression	Section 4	Section 4.3.1 Table 8	Supported
H4	[13]	Descriptive analysis Chi-square Fisher test	Section 3 Section 5	Section 4.4	Rejected

5. Discussion

The results of this study allowed for the identification of trends in the current interest of adopting innovative solutions by the construction sector. First, it was found that construction professionals perceive the importance of emerging technologies and show interest in expanding their use, acknowledging the value of digital transformation in the sector.

However, construction companies still have little knowledge about the characteristics, applicability, and potential of technological solutions and have a low degree of technological maturity. These factors, combined with a lack of qualified workers and high implementation costs, hinder, delay, or prevent the incorporation and use of innovative technologies.

A large part of respondents, when asked about the degree of technological maturity of their organizations, stated that companies are evolving, with a growing perception of the need for transformation. Nevertheless, professionals stated that companies do not have clear strategies for the acquisition, training, or implementation of new technologies and innovative solutions. The majority of respondents claimed to have no knowledge about 4IR concepts.

These observations indicate that the most important step in promoting the adoption of innovations in the construction sector involves the dissemination of knowledge and the development of strategies to support acculturation and the incorporation of new technologies and processes. Companies with greater technological maturity showed greater potential for the adoption of new technologies. This finding was corroborated by the participants' perceptions of barriers to technological development, mainly a culture of resistance to change, a lack of knowledge, and a lack of qualification. Companies and associations should seek to train construction professionals, for instance, in partnerships with academic institutions, to reduce resistance to change, facilitate implementation, and accelerate the consolidation of innovative technologies. The exchange of experiences between peers can also foster the adoption of innovations, as the decision may be influenced by communication and discussions with companies and professionals in the sector who have already opted for the adoption or rejection of innovations [22].

Our findings show that the gains in productivity and final product quality are the most anticipated benefits among respondents. The least frequent expectation was an increase in employee safety. Furthermore, we observed differences in the expected benefits between companies of different sizes, given their different goals and needs. The outcomes also indicate that companies of different sizes face different challenges. These findings support our second hypothesis.

Technologies considered the base for the implementation of Industry 4.0 by Frank et al. [17], namely, cloud computing, big data, and IoT, seem to be not yet consolidated in the construction industry, similar to what occurs in other sectors of the Brazilian industry [13,17]. Cloud computing was the most used technology by the professionals interviewed, but IoT, sensors, and CPS had low applications. This indicates that cloud computing is likely directed toward data storage only, without real-time analysis for decision-making support. In other words, the technology is adopted in its most basic form, constituting an alternative form of remote data archiving. The low maturity of these base technologies places the construction industry of the studied context still far from 4IR. Another technology with a low level of adoption among the respondents was 3D printing, which seems to be very incipient in the studied context. Although the interest and importance of Construction 4.0 is growing, the level of maturity and ability to incorporate and promote changes in this direction is still limited.

Larger companies usually take the lead in digitization initiatives, given their greater investment capacity. The findings of this study, however, do not corroborate this assumption in the context of the construction industry. So, our first hypothesis could not be supported by our results. Within the studied context, larger companies did not tend to adopt new technologies more than smaller companies or, at least, not in such a way that engineers and architects were aware of the strategy. Technologies with greater application in the sector (e.g., mobile devices, cloud computing, simulation and modeling, BIM, and virtual and augmented reality) were more applied in smaller companies.

To categorize technologies according to their average adoption, we classified them into four categories: undeveloped, incipient, under development, and consolidated. The result indicates that none of the technologies are fully consolidated in the studied context. Based on statistical clustering, we identified three groups of respondents: those with high technology use, medium technology use, and low technology use. With this analysis, we inferred that the respondents tend to adopt technology in a homogeneous and progressive way. Additionally, three technology groups were established based on the respondents' patterns of use: virtualization, automation, and manufacture. The first seems to be adopted earlier than the other two.

Statistical tests reveal a significant relationship between respondents' technology usage levels and the construction lifecycle phases they engage in, supporting our third hypothesis. In this way, to assist decision-making, we propose a framework that outlines the preferred order of adoption of the different groups of technologies. It also illustrates which concepts can be more suitable for each lifecycle phase of the construction industry. In the context of design, virtualization technologies can be utilized. Inspection and technical reports and building evaluation may benefit from both virtualization and automation technologies. Manufacturing technologies primarily find application during the construction phase, which can also benefit from automation concepts and the use of IoT/sensors/CPS. Additionally, the budget and planning phase could benefit from all the concepts analyzed. Notably, IoT, sensors, and CPS are positioned to bridge across all three technology clusters, facilitating integration between emerging technologies.

One important factor possibly hindering the adoption of innovative solutions in the construction sector is the perception of implementation and training costs, and the acquisition of technologies seems to be the main concern among respondents when it comes to cost. The lack of clarity about the returns associated with new technologies hinders cost–benefit assessment, contributing to a sense of risk in investing in innovations. Furthermore, investments are often made errantly, given the lack of a clear strategy for the adoption and use of new technologies and solutions. Many companies, for example, invest first in the acquisition of software and equipment. However, without adequate training and acculturation, the return of isolated investments in technological acquisition will be scarce and limited, and the overall results of changes may not meet expectations [66]. Investment in technology must, therefore, take place after objectives have been defined and strategies outlined.

In the present study, no statistical relationship was found between professionals' interest in certain technologies and the expected benefits, so the fourth hypothesis was rejected. This finding may indicate that professionals are not yet able to identify which of the innovations can best help them achieve the results they are seeking. It would be a good practice to conduct initial pilot projects with reduced investments and expectations to help understand the results, implications, demands, and costs associated with the adoption of Industry 4.0 solutions. This strategy could also promote experimentation with technologies, which tends to facilitate their adoption [22]. The practical visualization of the benefits and costs of technologies could reduce uncertainties, allowing the adjustment of plans and expectations and encouraging a more directed and effective financial investment, both internally and by third parties. PwC Consulting [10] suggested starting discreet pilot projects and selecting a specific and accurate scope. In this way, the first positive results obtained can generate confidence for larger and more complex projects. Technologies with greater consolidation and maturity in the academic environment might represent safer starting points.

It is crucial for the construction industry to maintain a broad and holistic view of the possibilities emerging from new technologies and understand that innovation may impact not only production activities but also relationships and ways of working, the integration of the supply chain, and the products offered by the sector [17]. Otherwise, it is possible that the focus of technological development in the construction industry remains centered on smart manufacturing, leaving aside the other equally important dimensions of 4IR, as has occurred in other sectors [21]. It is precisely innovations in products and services that may be the key for companies to remain competitive in the face of new industrial paradigms [2].

For the rapid development of technologies, a sectoral effort and the involvement of several actors are necessary. This is a fundamental issue that requires attention to prevent companies from falling behind in the rapidly advancing technological landscape. The acculturation, adoption, and use of new solutions within the field of construction 4.0 are fundamental to ensure the competitiveness of companies. Those who adapt more quickly can make use of this competitive advantage in important ways in the coming years.

6. Conclusions

This study aimed to evaluate the implementation potential of Industry 4.0 solutions in the construction market of an emerging country by mapping the current reality of the sector. The factors influencing the adoption of innovations were identified. Construction professionals recognized the value of emerging technologies and express interest in their adoption. However, limited knowledge, low technological maturity, a scarcity of skilled workers, and high costs hinder the widespread use of innovative solutions in the industry. Many respondents recognized their organizations' evolving technological maturity and the need for transformation. However, results highlight a lack of clear strategies for technology implementation. Surprisingly, most respondents were unfamiliar with 4IR concepts. To promote innovation adoption in construction, knowledge dissemination and strategic support are crucial, mainly because companies with higher technological maturity exhibit greater potential for adopting new technologies.

When asked about the anticipated benefits, respondents prioritized productivity gains and final product quality, with less emphasis on employee safety. Also, it was possible to observe that most Industry 4.0 innovations are poorly developed in the construction sector of emerging countries, especially some technologies that are fundamental for the consolidation of 4IR in the sector.

Based on clustering analysis, three groups of technologies with a similar level of use among professionals were observed: virtualization, automation, and manufacture. Virtualization technologies showed a higher level of use among professionals, and, therefore, there is a tendency for these to be adopted first. Manufacture technologies, on the other hand, had lower use and will likely take longer to be absorbed by the industry. Furthermore, it was observed that professionals and companies tend to absorb innovations progressively; that is, the adoption of one innovation leads to the adoption of others over time.

Two of our hypotheses were confirmed, and two were rejected. The benefits and barriers to technology use differed according to company size, demonstrating that companies face distinct challenges in innovation. These factors should be taken into account during implementation. However, it was not possible to associate company size with the level of technology use. Thus, smaller companies are also attentive to the industrial transformation process. It was not possible to identify a relationship between the benefits expected by professionals and the technologies they chose either, which may indicate that there are still considerable uncertainties among agents of the sector regarding the benefits of innovations.

Professionals working at different stages of the building lifecycle had different preferences for technologies. Some innovations are more useful at certain phases than others. We developed a framework relating the different stages of the construction lifecycle to the types of technologies available, which may assist in decision-making. Finally, we identified the critical barriers and promoters of 4IR technology adoption in developing countries. This analysis is important to enhance the absorption capacity of these technologies by the local industry, as it allows actions to be conducted in a targeted manner. A key obstacle to adopting innovative solutions in construction is the perceived implementation and training costs. Respondents expressed significant concern about technology acquisition expenses. The lack of clarity regarding returns from new technologies fosters a sense of risk in innovation investments.

It should be noted that the results of the current study are limited by the sample, which comprises a restricted portion of agents working in the Brazilian construction sector, especially in the southernmost region of the country. The extrapolation of results to other developing countries should be performed with caution, given the peculiarities of local markets. Moreover, there must be other aspects, benefits, concepts, and factors involved in the adoption of new technologies that were not addressed by the research. Additional studies are needed to validate the findings and the proposed framework. Nevertheless, in line with its exploratory purpose, this study was able to elucidate several relevant points for the progress of research in the area, representing a starting point to expand the knowledge of the implementation of Industry 4.0 technologies in the construction sector of emerging countries.

Author Contributions: Conceptualization, J.M.L.; Formal analysis, J.M.L.; Investigation, J.M.L.; Methodology, L.C.P.d.S.F.; Supervision, L.C.P.d.S.F.; Writing—original draft, J.M.L.; Writing—review and editing, L.C.P.d.S.F. All authors have read and agreed to the published version of the manuscript.

Funding: This research received no external funding.

Data Availability Statement: The data presented in this study are available on request from the corresponding author.

Conflicts of Interest: The authors declare no conflicts of interest.

References

1. Lasi, H.; Fettke, P.; Kemper, H.-G.; Feld, T.; Hoffmann, M. Industry 4.0. *Bus. Inf. Syst. Eng.* **2014**, *6*, 239–242. [CrossRef]
2. Schwab, K. *A Quarta Revolução Industrial*, 1st ed.; Edipro: Providencia, Chile, 2016.
3. Kagermann, H.; Wahlster, W.; Helbig, J. *Recommendations for Implementing the Strategic Initiative INDUSTRIE 4.0—Final Report of the Industrie 4.0 Working Group*; Forschungsunion, Acatech: Frankfurt, Germany, 2013. Available online: https://en.acatech.de/publication/recommendations-for-implementing-the-strategic-initiative-industrie-4-0-final-report-of-the-industrie-4-0-working-group/ (accessed on 2 July 2024).
4. Kagermann, H. Change through Digitization—Value Creation in the Age of Industry 4.0. In *Management of Permanent Change*; Springer: Berlin/Heidelberg, Germany, 2015; pp. 23–45. [CrossRef]
5. Oesterreich, T.D.; Teuteberg, F. Understanding the implications of digitisation and automation in the context of Industry 4.0: A triangulation approach and elements of a research agenda for the construction industry. *Comput. Ind.* **2016**, *83*, 121–139. [CrossRef]
6. Menegon, J.; da Silva Filho, L.C.P. The Impact of Industry 4.0 Concepts and Technologies on Different Phases of Construction Project Lifecycle: A Literature Review. *Iran. J. Sci. Technol. Trans. Civ. Eng.* **2022**, *47*, 1293–1319. [CrossRef]
7. Kamble, S.S.; Gunasekaran, A.; Gawankar, S.A. Sustainable Industry 4.0 framework: A systematic literature review identifying the current trends and future perspectives. *Process Saf. Environ. Prot.* **2018**, *117*, 408–425. [CrossRef]
8. Porto, G.D.B.P.; Kadlec, T.M.D.M. Mapeamento de Estudos Prospectivos de Tecnología na Revolução 4.0: Um Olhar Para a Indústria da Construção Civil. **2018**, *70*. Available online: http://repositorio.utfpr.edu.br/jspui/handle/1/8115 (accessed on 2 July 2024).
9. CNI, SPECIAL SURVEY: Industry 4.0. 2016. Available online: https://static.portaldaindustria.com.br/media/filer_public/13/e7/13e7e7bd-9b1d-4c16-8099-99b6d844d04e/special_survey_industry_40.pdf (accessed on 2 July 2024).
10. PwC, P.B.L. Indústria 4.0: Digitização como Vantagem Competitiva no Brasi. 2016. Available online: https://www.pwc.com.br/pt/publicacoes/servicos/assets/consultoria-negocios/2016/pwc-industry-4-survey-16.pdf (accessed on 2 July 2024).
11. Federação das Indústrias do Estado do Rio de Janeiro. *Panorama da Inovação: Indústria 4.0*; FIRJAN, SENAI: Rio de Janeiro, Brazil, 2016.
12. Firmino, A.S.; Perles, G.X.; Mendes, J.V.; da Silva, J.E.A.R.; Silva, D.A.L. Towards Industry 4.0: A SWOT-based analysis for companies located in the Sorocaba Metropolitan Region (São Paulo State, Brazil). *GestãoProdução* **2020**, *27*, 1–21. [CrossRef]
13. Dalenogare, L.S.; Benitez, G.B.; Ayala, N.F.; Frank, A.G. The expected contribution of Industry 4.0 technologies for industrial performance. *Int. J. Prod. Econ.* **2018**, *204*, 383–394. [CrossRef]
14. Rubmann, M.; Lorenz, M.; Gerbert, P.; Waldner, M.; Justus, J.; Engel, P.; Harnisch, M. Industry 4.0: The Future of Productivity and Growth in Manufacturing Industries. *Boston Consult. Group* **2015**, *9*, 54–89.
15. Longo, F.; Nicoletti, L.; Padovano, A. Smart operators in industry 4.0: A human-centered approach to enhance operators' capabilities and competencies within the new smart factory context. *Comput. Ind. Eng.* **2017**, *113*, 144–159. [CrossRef]
16. Paelke, V. Augmented Reality in the Smart Factory: Supporting Workers in an Industry 4.0. Environment. In Proceedings of the IEEE Emerging Technology and Factory Automation (ETFA), Barcelona, Spain, 16–19 September 2014.
17. Frank, A.G.; Dalenogare, L.S.; Ayala, N.F. Industry 4.0 technologies: Implementation patterns in manufacturing companies. *Int. J. Prod. Econ.* **2019**, *210*, 15–26. [CrossRef]
18. Witkowski, K. Internet of Things, Big Data, Industry 4.0—Innovative Solutions in Logistics and Supply Chains Management. *Procedia Eng.* **2017**, *182*, 763–769. [CrossRef]
19. ABIMAQ, NEO, Indústria 4.0: Mapeamento das Tecnologias. Relatório Geral. 2018. Available online: https://www.ufrgs.br/neo/wp-content/uploads/2018/10/Relat%C3%B3rio-Geral-Completo.pdf (accessed on 2 July 2024).

20. Gerber, T.; Bosch, H.; Johnsson, C. *Service Orientation in Holonic and Multi Agent Manufacturing and Robotics*; Springer: Berlin/Heidelberg, Germany, 2013. [CrossRef]
21. Meindl, B.; Ayala, N.F.; Mendonça, J.; Frank, A.G. The four smarts of Industry 4.0: Evolution of ten years of research and future perspectives. *Technol. Forecast. Soc. Chang.* **2021**, *168*, 120784. [CrossRef]
22. Rogers, E.M. *Diffusion of Innovations*; The Free Press: New York, NY, USA, 1983.
23. Müller, J.M.; Buliga, O.; Voigt, K.I. Fortune favors the prepared: How SMEs approach business model innovations in Industry 4.0. *Technol. Forecast. Soc. Chang.* **2018**, *132*, 2–17. [CrossRef]
24. Barrow, K.; Verzat, P. Construction 4.0: Delivering an infrastructure revolution. *Int. Railw. J.* **2018**, 23–24. Available online: https://www.railjournal.com/in_depth/construction-40-delivering-an-infrastructure-revolution/ (accessed on 2 July 2024).
25. Karmakar, A.; Delhi, V.S.K. Construction 4.0: What we know and where we are headed? *J. Inf. Technol. Constr.* **2021**, *26*, 526–545. [CrossRef]
26. Loosemore, M. Construction Innovation: Fifth Generation Perspective. *J. Manag. Eng.* **2015**, *31*, 04015012. [CrossRef]
27. Merschbrock, C.; Munkvold, B.E. Effective digital collaboration in the construction industry—A case study of BIM deployment in a hospital construction project. *Comput. Ind.* **2015**, *73*, 1–7. [CrossRef]
28. Dounas, T.; Lombardi, D.; Jabi, W. Framework for decentralised architectural design BIM and Blockchain integration. *Int. J. Archit. Comput.* **2021**, *19*, 157–173. [CrossRef]
29. Bianconi, F.; Filippucci, M.; Buffi, A. Automated design and modeling for mass-customized housing. A web-based design space catalog for timber structures. *Autom Constr.* **2019**, *103*, 13–25. [CrossRef]
30. Perera, S.; Nanayakkara, S.; Rodrigo, M.N.N.; Senaratne, S.; Weinand, R. Blockchain Technology: Is it Hype or Real in the Construction Industry? *J. Ind. Inf. Integr.* **2020**, *17*, 100125. [CrossRef]
31. Meng, Q.; Zhang, Y.; Li, Z.; Shi, W.; Wang, J.; Sun, Y.; Xu, L.; Wang, X. A review of integrated applications of BIM and related technologies in whole building life cycle. *Eng. Constr. Archit. Manag.* **2020**, *27*, 1647–1677. [CrossRef]
32. Ma, Z.; Ma, J. Formulating the application functional requirements of a BIM-based collaboration platform to support IPD projects. *KSCE J. Civ. Eng.* **2017**, *21*, 2011–2026. [CrossRef]
33. Hassan, S.M.; Azab, M.; Mokhtar, A. Smart concrete transportation in semi-automated construction sites. In Proceedings of the 2019 IEEE 10th Annual Information Technology, Electronics and Mobile Communication Conference, IEMCON 2019, Vancouver, BC, Canada, 17–19 October 2019; pp. 661–667. [CrossRef]
34. Feucht, T.; Lange, J.; Erven, M.; Costanzi, C.B.; Knaack, U.; Waldschmitt, B. Additive manufacturing by means of parametric robot programming. *Constr. Robot.* **2020**, *4*, 31–48. [CrossRef]
35. Martinez, P.; Al-Hussein, M.; Ahmad, R. Intelligent vision-based online inspection system of screw-fastening operations in light-gauge steel frame manufacturing. *Int. J. Adv. Manuf. Technol.* **2020**, *109*, 645–657. [CrossRef]
36. Jensen, M.B.; Foged, I.W.; Andersen, H.J. A framework for interactive human–robot design exploration. *Int. J. Archit. Comput.* **2020**, *18*, 235–253. [CrossRef]
37. Dai, R.; Kerber, E.; Reuter, F.; Stumm, S.; Brell-Cokcan, S. The digitization of the automated steel construction through the application of microcontrollers and MQTT. *Constr. Robot.* **2020**, *4*, 251–259. [CrossRef]
38. Schimanski, C.P.; Monizza, G.P.; Marcher, C.; Matt, D.T. Pushing digital automation of configure-to-order services in small and medium enterprises of the construction equipment industry: A design science research approach. *Appl. Sci.* **2019**, *9*, 3780. [CrossRef]
39. Moon, S.; Ham, N.; Kim, S.; Hou, L.; Kim, J.H.; Kim, J.J. Fourth industrialization-oriented offsite construction: Case study of an application to an irregular commercial building. *Eng. Constr. Archit. Manag.* **2020**, *27*, 2271–2286. [CrossRef]
40. Thomsen, M.R.; Nicholas, P.; Tamke, M.; Gatz, S.; Sinke, Y.; Rossi, G. Towards machine learning for architectural fabrication in the age of industry 4.0. *Int. J. Archit. Comput.* **2020**, *18*, 335–352. [CrossRef]
41. Bogue, R. What are the prospects for robots in the construction industry? *Ind. Robot* **2018**, *45*, 1–6. [CrossRef]
42. Siwiec, J. Comparison of Airborne Laser Scanning of Low and High Above Ground Level for Selected Infrastructure Objects. *J. Appl. Eng. Sci.* **2019**, *8*, 89–96. [CrossRef]
43. Chan, B.; Guan, H.; Hou, L.; Jo, J.; Blumenstein, M.; Wang, J. Defining a conceptual framework for the integration of modelling and advanced imaging for improving the reliability and efficiency of bridge assessments. *J. Civ. Struct. Health Monit.* **2016**, *6*, 703–714. [CrossRef]
44. Zhou, Z.; Irizarry, J.; Lu, Y. A Multidimensional Framework for Unmanned Aerial System Applications in Construction Project Management. *J. Manag. Eng.* **2018**, *34*, 04018004. [CrossRef]
45. Heesom, D.; Boden, P.; Hatfield, A.; de Los Santos Melo, A.; Czarska-Chukwurah, F. Implementing a HBIM approach to manage the translocation of heritage buildings. *Eng. Constr. Archit. Manag.* **2020**, *28*, 2948–2966. [CrossRef]
46. Reinhardt, D.; Haeusler, M.H.; London, K.; Loke, L.; Feng, Y.; de Oliveira Barata, E.; Firth, C.; Dunn, K.; Khean, N.; Fabbri, A.; et al. CoBuilt 4.0: Investigating the potential of collaborative robotics for subject matter experts. *Int. J. Archit. Comput.* **2020**, *18*, 353–370. [CrossRef]
47. Sun, S.; Zheng, X.; Villalba-Díez, J.; Ordieres-Meré, J. Data handling in industry 4.0: Interoperability based on distributed ledger technology. *Sensors* **2020**, *20*, 3046. [CrossRef]
48. Tezel, A.; Papadonikolaki, E.; Yitmen, I.; Hilletofth, P. Preparing construction supply chains for blockchain technology: An investigation of its potential and future directions. *Front. Eng. Manag.* **2020**, *7*, 547–563. [CrossRef]

49. Elghaish, F.; Abrishami, S.; Hosseini, M.R. Integrated project delivery with blockchain: An automated financial system. *Autom Constr.* **2020**, *114*, 103182. [CrossRef]
50. Hilal, M.; Maqsood, T.; Abdekhodaee, A. A hybrid conceptual model for BIM in FM. *Constr. Innov.* **2019**, *19*, 531–549. [CrossRef]
51. Gao, W.; Su, Q.; Zhang, J.; Xie, H.; Wen, F.; Li, F.; Liu, J. Steel Bridge Construction of Hong Kong–Zhuhai–Macao Bridge. *Int. J. Steel Struct.* **2020**, *20*, 1498–1508. [CrossRef]
52. Chan, D.W.M.; Olawumi, T.O.; Ho, A.M.L. Critical success factors for building information modelling (BIM) implementation in Hong Kong. *Eng. Constr. Archit. Manag.* **2019**, *26*, 1838–1854. [CrossRef]
53. Marefat, A.; Toosi, H.; Hasankhanlo, R.M. A BIM approach for construction safety: Applications, barriers and solutions. *Eng. Constr. Archit. Manag.* **2019**, *26*, 1855–1877. [CrossRef]
54. Banawi, A. Barriers to Implement Building Information Modeling (BIM) in Public Projects in Saudi Arabia. In *Advances in Intelligent Systems and Computing*; Springer: Cham, Switzerland, 2018; pp. 119–125. [CrossRef]
55. Suprun, E.V.; Stewart, R.A. Construction innovation diffusion in the Russian Federation barriers, drivers and coping strategies. *Constr. Innov.* **2015**, *15*, 278–312. [CrossRef]
56. Silverio-Fernandez, M.A.; Renukappa, S.; Suresh, S. Evaluating critical success factors for implementing smart devices in the construction industry: An empirical study in the Dominican Republic. *Eng. Constr. Archit. Manag.* **2019**, *26*, 1625–1640. [CrossRef]
57. Ramani, S.V.; Thutupalli, A.; Urias, E. High-value hi-tech product introduction in emerging countries: The role and construction of legitimacy. *Qual. Mark. Res.* **2017**, *20*, 208–225. [CrossRef]
58. Bechhofer, F.; Peterson, L. To interview or not to interview. In *Principles of Research Design in the Social Sciences*, 1st ed.; Routledge: London, UK, 2012; p. 192.
59. Manzato, A.; Santos, A. *A Elaboração de Questionários na Pesquisa Quantitativa*; UNESP: São Paulo, Brazil, 2012; Available online: https://www.inf.ufsc.br/~vera.carmo/Ensino_2012_1/ELABORACAO_QUESTIONARIOS_PESQUISA_QUANTITATIVA.pdf (accessed on 2 July 2024).
60. Green, J.L.; Manski, S.E.; Hansen, T.A.; Broatch, J.E. Descriptive statistics. In *International Encyclopedia of Education*, 4th ed.; Tierney, R.J., Rizvi, F., Ercikan, K., Eds.; Elsevier: Amsterdam, The Netherlands, 2023; pp. 723–733. [CrossRef]
61. Hair, J.F.; Black, W.C.; Babin, B.J.; Anderson, R.E.; Tatham, R.L. *Análise Multivariada de Dados*, 6th ed.; Bookman Companhia Editora Ltda: Porto Alegre, Brazil, 2009.
62. Malhotra, N.K. *Pesquisa de Marketing-: Uma Orientação Aplicada*; Bookman Editora: Porto Alegre, Brazil, 2019.
63. Zou, G. A Modified Poisson Regression Approach to Prospective Studies with Binary Data. *Am. J. Epidemiol.* **2004**, *159*, 702–706. [CrossRef]
64. Kutner, M.H.; Nachtsheim, C.J.; Neter, J.; Li, W. *Applied Linear Statistical Models*, 5th ed.; McGraw-Hill Irwin Series Operations and Decision Sciences: New York, NY, USA, 2005.
65. Kim, H.Y. Statistical notes for clinical researchers: Chi-squared test and Fisher's exact test. *Restor. Dent. Endod.* **2017**, *42*, 152–155. [CrossRef]
66. Frank, A.G.; Cortimiglia, M.N.; Ribeiro, J.L.D.; de Oliveira, L.S. The effect of innovation activities on innovation outputs in the Brazilian industry: Market-orientation vs. technology-acquisition strategies. *Res. Policy* **2016**, *45*, 577–592. [CrossRef]

buildings

MDPI

Article

Investigations on the Environmental Characteristics and Cracking Control of Plateau Concrete

Xiaochuan Hu [1], Manping Liao [1], Ming Li [2], Fuqiang Wang [3], Xiang Lyu [4] and Mei-Ling Zhuang [5,6,7,*]

1 The Civil Engineering Group Corporation of China Second Engineering Bureau Ltd., Beijing 101100, China; huxiaochuan@cscec.com (X.H.); liaomanping@cscec.com (M.L.)
2 Jiangsu Sobute New Materials Co., Ltd., Nanjing 211103, China; liming@cnjsjk.cn
3 School of New Energy, Harbin Institute of Technology at Weihai, Weihai 264209, China; wangfuqiang@hitwh.edu.cn
4 China Construction Second Engineering Bureau Ltd., Beijing 100000, China; lvxiang18@mails.jlu.edu.cn
5 Water Resources Research Institute of Shandong Province, Jinan 250013, China
6 School of Civil Engineering, Shandong University, Jinan 250061, China
7 School of Transportation and Civil Engineering, Nantong University, Nantong 226019, China
* Correspondence: ml_zhuang99@163.com or ml_zhuang@ntu.edu.cn

Abstract: In the present study, first, the environmental challenges and cracking characteristics during the construction of plateau concrete on the Sichuan–Tibet route were revealed. Then, using a multi-field coupled shrinkage model with hydration temperature humidity constraints, the early and long-term cracking risks in the core of plateau pier bodies were investigated. Later, the effects of tensile strength, pouring interval age and adiabatic temperature rise on the cracking risk were analyzed. Finally, various control measures for high-altitude concrete cracking were proposed. The results indicated that the complex environment of the plateau led to different forms of cracks in the pier body, especially vertical cracks in the straight sections. The long-term risk of core cracking in the plateau pier body is significantly greater than the risk of early cracking. This risk was strongly influenced by factors such as the concrete tensile strength, pouring interval age and adiabatic temperature rise, which should be given more attention. Deformation compensation can significantly enhance the peak and residual deformation capacities of plateau concrete, with peak values greater than 900 $\mu\varepsilon$ and residual deformation greater than 200 $\mu\varepsilon$ at day 60, as well as its resistance to cracking. Strategies such as adopting radiant cooling techniques, improving construction techniques and implementing effective management measures can all play a vital role in improving the cracking resistance of highland concrete.

Keywords: plateau concrete; environmental characteristics; cracking risk; cracking control

1. Introduction

The average altitude of the Qinghai–Tibet plateau is 3000 m above sea level, which is significantly higher than that in neighboring regions at the same latitude. Due to its distinctive geographical structure and location, the Qinghai–Tibet plateau presents extreme environmental conditions such as large temperature variations, low air pressure, dryness, intense ultraviolet radiation and high wind speeds. The serviceability of concrete engineering in the plateau region faces significant challenges [1,2], among which concrete cracking is the most prominent problem. The causes of concrete cracking are multifaceted, including factors such as the climate, environment, construction technology and raw material quality. Among these factors, the environment plays a crucial role in the occurrence of concrete cracking. In response to the unique environmental conditions of the plateau, several studies have investigated the behavior of concrete in high-altitude environments. Hu and Cao [3] examined the dynamic modulus and porosity of concrete surface layers in plateau regions. The results indicated that the surface layer of concrete in plateau regions was susceptible to environmental erosion and deterioration, resulting in a 10% decrease in the dynamic modulus and an increase in the proportion of harmful porosity. Li et al. [4] explored the effect of air pressure on the air content of concrete and found that when the ambient air pressure was decreased to 50 kPa, the air content of concrete decreased by about 20% to 49%, and the air content performed a linear decreasing pattern as the air pressure was further reduced. He et al. [1] investigated how the environment affects the strength and permeability of concrete in Lhasa, Tibet, and observed that concrete performed poorly under natural outdoor curing conditions at the plateau. The 28-day compressive strength decreased by about 21.9%, and the relative coefficient of permeability increased significantly. Ge et al. [5,6] also demonstrated the effect of low pressure on the strength and chemical resistance of concrete, showing that the effect of air pressure on these two properties was decreasing. It can be seen that with the development of railroad and highway projects on the Qinghai–Tibet plateau, the effect of a high-altitude environment on the deterioration of concrete mechanical properties has attracted more and more attention [7–12]. Improving the service performance of concrete in high-altitude environments has become a hot and difficult research topic in recent years.

In high-altitude regions such as the Qinghai–Tibet railway and the Sichuan–Tibet railway in China, great efforts have been made to improve the performance and cracking resistance of concrete [13,14], and some positive results have been achieved. However, some degree of cracking still exists in concrete structures in the plateau region, indicating that the risk of cracking and mitigation measures in the environment are not fully understood. In particular, little research has been conducted on the environmental conditions and risks of concrete use in the middle section between Sichuan and Tibet. Therefore, an in-depth study of the challenges facing concrete in plateau regions and the development of effective crack control strategies are imperative.

The challenges and in situ cracking characteristics in plateau concrete projects on the Sichuan–Tibet route were firstly investigated in the present study. Then, using a multi-field coupling shrinkage model with hydration–temperature–humidity constraints, the early and long-term cracking risks of plateau piers located at an altitude of more than 3500 m above sea level were quantitatively evaluated. Finally, crack control measures such as deformation compensation and radiation cooling technology for plateau concrete were proposed to provide insights and guidance for improving crack control and serviceability in plateau concrete projects.

2. Environmental Characteristics

2.1. Temperature and Wind Speed

The distribution of the average annual temperature in China in 2022 is depicted in Figure 1, showing that the temperature on the Sichuan–Tibet route is obviously low, and the temperature is below zero degrees in some areas. Meanwhile, due to the large temperature difference in the plateau region, it was necessary to shorten the construction period in order to ensure the quality of the concrete. For instance, in the high-altitude Xindu Bridge area in Kangding, Sichuan, which is 3500 m above sea level, the temperature can soar to 30 °C in summer and plummet to −15 °C in winter, and the difference in the temperature between day and night can be as much as 35 °C within a year. Thunderstorms and hailstorms occur frequently, and even in the summer months, there can be sudden icing, snow and sudden drops in the temperature. The harsh temperature conditions pose substantial risks to the use of concrete. According to the data from the micro meteorological station, the average wind speed in the Xindu Bridge area in 2020 ranged from 8 to 10 m/s, with a maximum speed of 20 m/s. Wind speeds are usually low from July to October. Strong winds can cause a significant drop in concrete slump, worsen drying shrinkage and trigger cracking. In addition, high wind speeds require more stringent, early concrete insulation measures due to increased heat exchange between the formwork and concrete surfaces. These environmental challenges are less pronounced in the plains. Therefore, controlling concrete cracking becomes particularly difficult in high wind-speed environments.

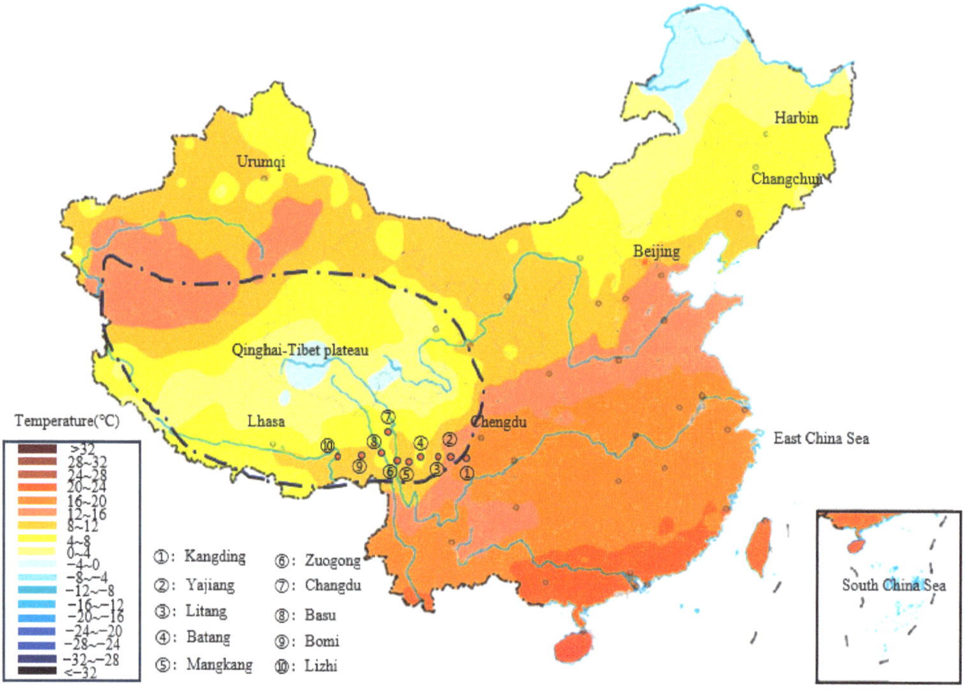

Figure 1. Statistics of average temperature of China in 2022. Note that the data are from the National Climate Centre, China (http://www.ncc-cma.net/, accessed on 20 October 2023).

2.2. Severe Freeze–Thaw Cycles

The annual freeze–thaw cycle refers to the number of times in a year that the temperature drops from +3 °C to below −3 °C and then rises back to +3 °C again. The freeze–thaw characteristics of the Kangding–Bomi section in the plateau region were statistically analyzed based on the monitoring data from meteorological stations. The frequency of freeze–thaw cycles varies greatly from year to year, with the most severe freeze–thaw cycles generally occurring from November of one year to March of the next, accounting for more than 80% of the total number of freeze–thaw cycles in the year. Figure 2 shows the average annual number of freeze–thaw cycles in some highland areas along the Sichuan–Tibet route from 2020 to 2022. Litang, Mangkang, Zuogong and Bangda experienced the most significant freeze–thaw cycles, with more than 110 cycles per year, and even more than 150 cycles in Mangkang. The average annual number of freeze–thaw cycles in Bomi and Yajiang was relatively low, at 48 and 71 cycles, respectively. Although the annual freeze–thaw occurrences are minimized, severe freeze–thaw conditions are still encountered in the plateau regions, highlighting the need for greater attention to be paid to the consideration of freeze–thaw effects in the design of concrete durability.

The freeze–thaw failure of concrete is a gradual process in which the internal microparticles fail due to expansion pressure and permeation pressure [15,16]. This process is accompanied by mortar detachment on the concrete surface, and the internal loosening, cracking and filling of cracks with frozen bodies, all of which can significantly reduce the durability of concrete engineering. Figure 3 illustrates the failure mode of C20 test concrete after one year of storage at a high altitude of 3500 m, where water is abundant, and driving loads and freeze–thaw conditions exist. Water diffuses and freezes in the gaps and cracks of the concrete particles, leading to the severe deterioration of the concrete structure and strength. This highlights the increased risk of freeze–thaw damage to highland concrete.

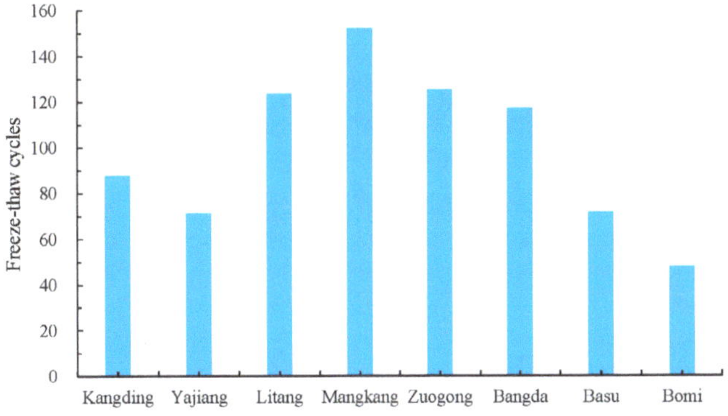

Figure 2. Average annual freeze–thaw cycles in certain plateau areas of Sichuan–Tibet route.

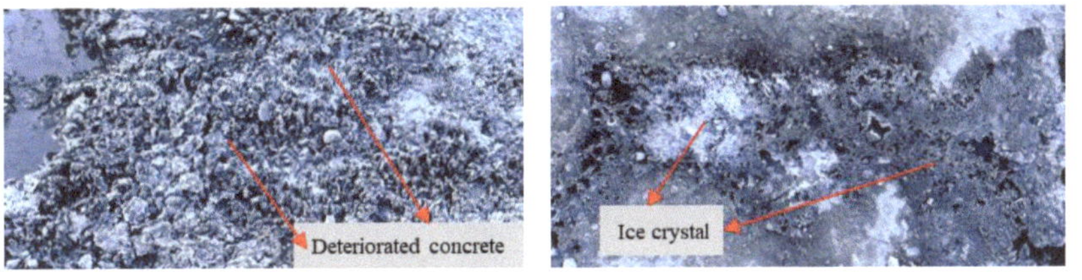

Figure 3. Concrete freeze–thaw disaster.

2.3. Low Air Pressure

The air pressure distribution in high-altitude areas on the Sichuan–Tibet route is shown in Figure 4. The air pressure level in these areas is about 30–40% lower than that in the plains, with Zuogong, Bangda and Basu having particularly low air pressure values, below 60 kPa. Other areas have slightly higher pressures, mostly around 60 kPa, with Bomi close to 70 kPa. A lower air pressure led to reduced air solubility and increased the surface tension of bubbles and decreased the foam volume upon bubble burst [17–19], affecting the effectiveness of air-entraining agents. In addition, under low air pressure conditions, the concrete hydration products decreased, and number of concrete micro-cracks increased, which seriously affected the mechanical and durability properties of concrete. This usually leads to a loss of slump in concrete mixtures and poses challenges to field workers in placing concrete, thus affecting concrete quality control in engineering practices.

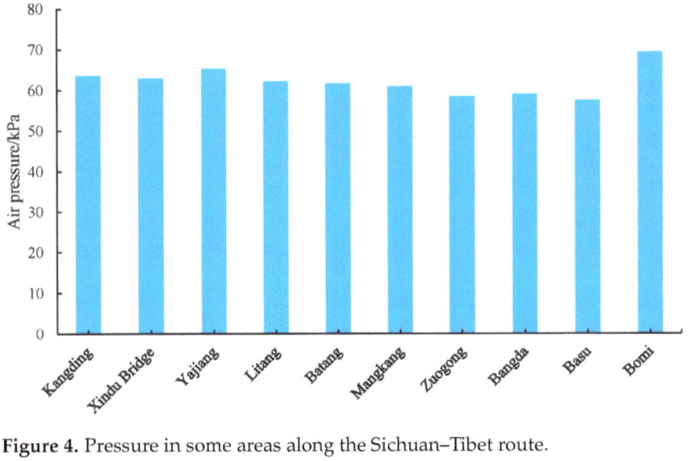

Figure 4. Pressure in some areas along the Sichuan–Tibet route.

2.4. Light Intensity and Temperature Differences

Figure 5 shows the average annual light intensity in various high-altitude regions on the Sichuan–Tibet route. The data revealed that Mangkang and Basu had the highest light intensity of more than 2500 h/year, while Kangding and Bomi had a light intensity of about 1250 h/year, which was significantly higher than that in many plains. Adequate sunlight can lead to moisture evaporation during concrete curing, accelerating drying shrinkage and potentially causing cracking. Moreover, light intensity is related to radiation intensity, and intense ultraviolet radiation at high altitudes may cause internal cracks in the concrete, reducing its performance and ultimately leading to concrete failure.

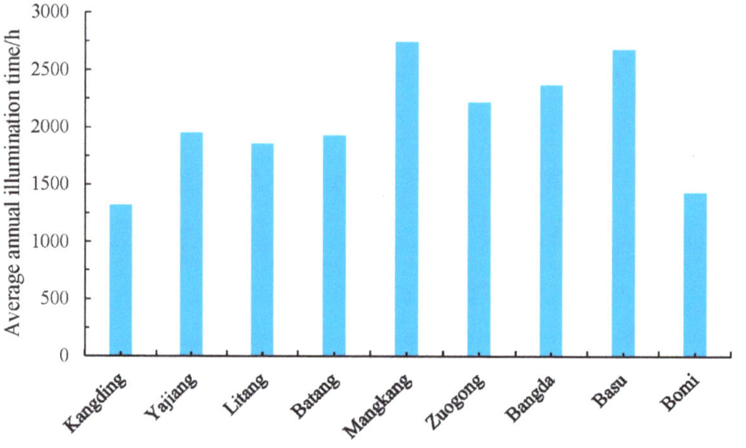

Figure 5. Average annual illumination time.

In fact, the temperature difference between the sunny side and the shaded side of a bridge abutment on the plateau has the potential to cause concrete damage. On-site monitoring has shown that the temperatures on the sunny side can reach 30–40 °C, while temperatures on the shaded side are still close to ambient temperature. The temperature difference between the two sides can affect the hydration strength of the poured concrete, resulting in a 10% reduction in the rebound strength [20]. Additionally, a significant temperature difference can lead to high tensile stresses from temperature gradients, which can be detrimental to preventing concrete cracking. Figure 6 presents the simulation results of the surface temperature stresses on the pier body, considering the sunny surface temperatures of 30 °C and 40 °C and the cloudy surface temperature of 10 °C. The results indicated that the tensile stress on the concrete surface was 3.08 MPa (positive for tension and negative for compression) at the sunny temperature of 30 °C, while at the sunny temperature of 40 °C, the tensile stress was about 4.63 MPa. Conventional concrete typically has a tensile strength of 2–4 MPa, suggesting that the temperature-induced tensile stress resulting from the temperature difference between the sunny and shady sides of a pier could potentially lead to micro-cracks or even outright cracking. Over time, this will bring a great risk of damage to the concrete structure. Based on observations of the post-construction concrete of a large number of bridge piers on the Qinghai–Tibet plateau, many bridge piers and abutments in a specific plateau project have shown extensive cracking in the years of construction, with more severe damage on the sunny side. Therefore, it is critical to investigate methods to regulate the temperature distribution on both sides of the pier body in a high-altitude environment.

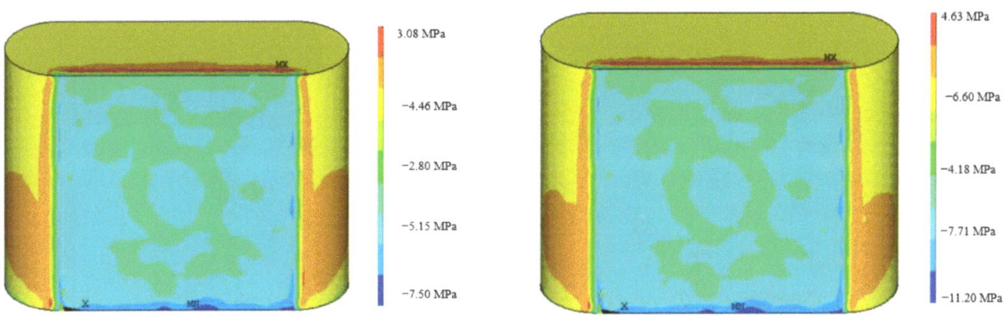

(**a**) 30–10 °C (**b**) 40–10 °C

Figure 6. Temperature stress caused by temperature difference between sunny and shaded sides of pier.

2.5. Dynamic Disturbance

The faults and other discontinuous tectonic surfaces along the Sichuan–Tibet route are developed, with strong tectonic movements and frequent strong earthquakes. When rupture occurs, the stress drops instantaneously, and a large amount of energy is released in the form of dynamic stress waves. This causes significant additional dynamic loads on concrete structures. In addition, the opening of certain railway projects with speeds of up to 200 km/h will also cause additional dynamic stresses on concrete structures (1–300 kPa, acceleration of 0.5–20 m/s^2). It has been shown that dynamic loading can accelerate crack initiation, coalescence or nucleation and causes intricate changes in the stress field near the crack tip, resulting in fatigue and degradation effects [21,22]. It has been observed that 80% to 90% of static stresses and micro-dynamic disturbances can initiate material failure: (1) under dynamic disturbance loading and material properties, crack propagation is time dependent, meaning that crack propagation accelerates as the dynamic disturbance time increases, and the material properties degrade over time; (2) dynamic stress waves can be reflected and refracted inside the concrete structure, which leads to a localized concentration of dynamic stresses, initiating adjacent micro-cracks, thus exacerbating concrete damage.

2.6. Other Factors

Precipitation is generally low in high-altitude areas such as the Qinghai–Tibet plateau. Areas such as Sichuan and Tibet receive less than 600 mm of rainfall, which is significantly lower than in central and eastern China. In addition, the relative humidity in high-altitude regions is also lower than that in the low-altitude eastern areas. Failure to timely maintain concrete in high altitudes can lead to a significant decrease in the rate of the hydration reaction and strength development. With the proliferation of construction projects in these regions in recent years, the challenges of transportation difficulties and the urgent need for raw material have led to problems of pile breakage and segregation in concrete structures. The key to solving these problems is to improve the frequency and standard of raw material quality monitoring to prevent such problems from occurring.

3. Investigation of Concrete Cracking

3.1. On-Site Cracking Investigation

The cracking characteristics of bridge piers in high altitudes are summarized from the on-site tests, as listed in Table 1. Cracks were mainly found on the rectangular surfaces and circular arc segments of the piers, with most of the cracks occurring on the rectangular surfaces. Cracks were found in a variety of forms including vertical, horizontal, diagonal and irregular, with vertical cracks being the most common and horizontal cracks being less common. Cracks usually appeared about 1.2 m from the bottom of the pier. Cracks were formed by a variety of factors, including temperature shrinkage, dry shrinkage, chemical shrinkage and high-temperature stresses.

Table 1. Cracking of on-site piers.

Distribution	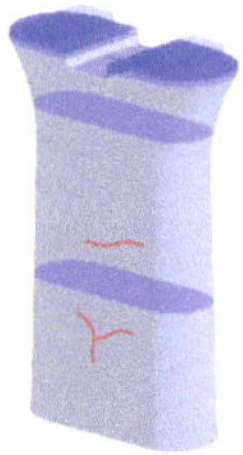	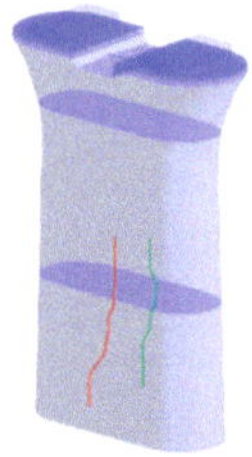	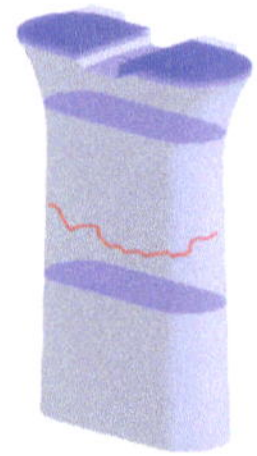	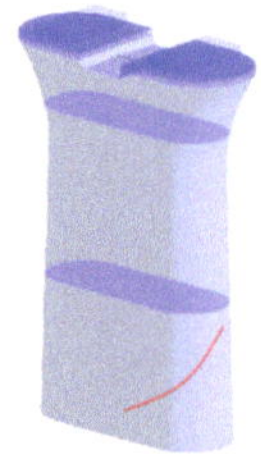	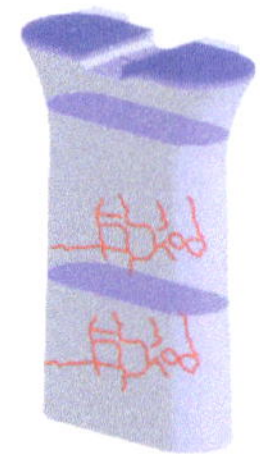
Cases		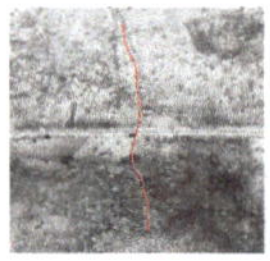	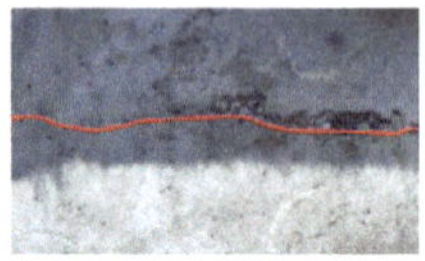	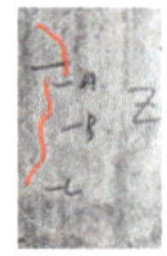	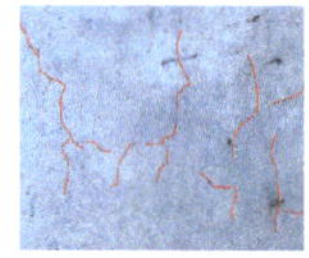
Descriptions	Irregular cracks	Vertical cracks	Horizontal cracks	Arc segment cracking	Honeycomb cracks

3.2. *Risk Analysis of Cracking*

Cracking in plateau concrete is a process that develops from continuous development to stabilization. Factors such as the temperature, humidity, wind speed and other environmental variables change from year to year at high altitudes. When assessing cracking, these factors must be taken into account. Engineers often assess the crack resistance of concrete by observing the early cracking conditions, especially after demolding. However, the long-term cracking risk of concrete structures is often overlooked, especially in high-altitude environments. To remedy this deficiency, the early cracking risk and long-term cracking risk of concrete in high-altitude environments have been systematically evaluated. Early cracking time is defined as within 60 days after pouring completion, while long-term cracking time is defined as at least one minimum winter temperature cycle. The risk of cracking was investigated using a plateau bridge pier as a case study. A numerical model (21,532 solid 65 units) was developed for the pier with C45 concrete for the lower 1.0 m of the pier and C35 concrete for the upper pier (Figure 7). The bottom of the numerical model was fixed in all directions, and the displacement of the node at $x = 0$ in the x-direction was fixed. A multi-field coupled shrinkage model with hydration, temperature and humidity constraints [23,24] was used in the numerical simulation process, taking into account the time interval between the pier body and pier cap casting (denoted by the interval age $\triangle t$), the varying tensile strengths of the concrete σ_t, adiabatic temperature rise (ATR), construction and environment. The risk coefficient for concrete cracking is defined as

$$\eta = \sigma_t^T / f_t^T \tag{1}$$

where σ_t and f_t denote the tensile stress and tensile strength of concrete at time T, both of which depend on the degree of hydration. When $\eta \geq 1.0$, the risk of concrete cracking is extremely high. When $0.7 < \eta < 1.0$, the risk is high. When $\eta \leq 0.7$, the concrete is unlikely to crack, and the crack-free guarantee is not less than 95% [24].

The early elastic modulus E and tensile strength f_t of concrete are important parameters for determining the shrinkage stress and cracking risk and dependent on the hydration degree α [23,25,26]. The calculation formula is

$$E(\alpha) = E^{\infty}(\frac{\alpha - \alpha_0}{\alpha^{\infty} - \alpha_0})^p \tag{2}$$

$$f_t(\alpha) = f_t^{\infty}(\frac{\alpha - \alpha_0}{\alpha^{\infty} - \alpha_0})^q \tag{3}$$

where E^{∞} and f_t^{∞} denote the final elastic modulus and final tensile strength of concrete; p and q are exponential constants with values of 0.5 and 1, respectively; and α_0 and α^{∞} denote the initial and final values of the degree of hydration α, respectively.

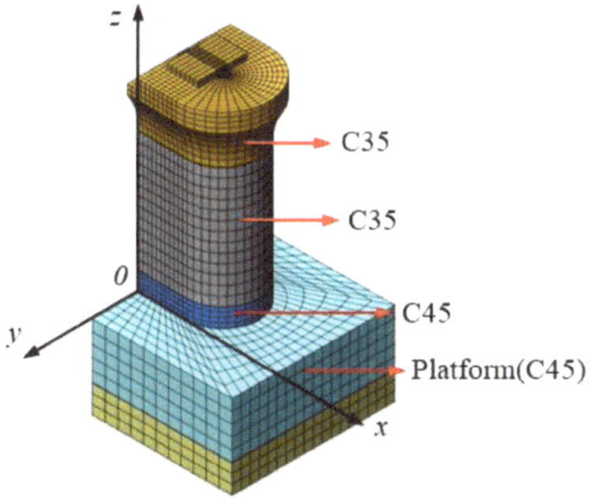

Figure 7. Numerical calculation model.

To simulate the maintenance measures after the pouring of a pier, the surface heat dissipation coefficient of the concrete was set to 20 kJ/(m^2·h·K). The thermal conductivity coefficient of concrete was set to 1.5 W/(m·K). In assessing the risk of early cracking, the ambient temperature was set at 8 °C, and the concrete pouring temperature was set at 24 °C. In assessing the risk of long-term cracking, the annual temperature variation was taken into account and defined as

$$T = T_0 + (A/2)cos(\pi/6(t - t_0)) \quad (4)$$

where T_0 is the average annual temperature (8.6 °C); A is the annual variation in the temperature (26 °C); t_0 is the highest temperature of the year in a month, usually in mid-July, with a value of 6.5. The annual variation in the temperature of the plateau project site can be seen in Figure 8. More calculation parameters are detailed in Table 2.

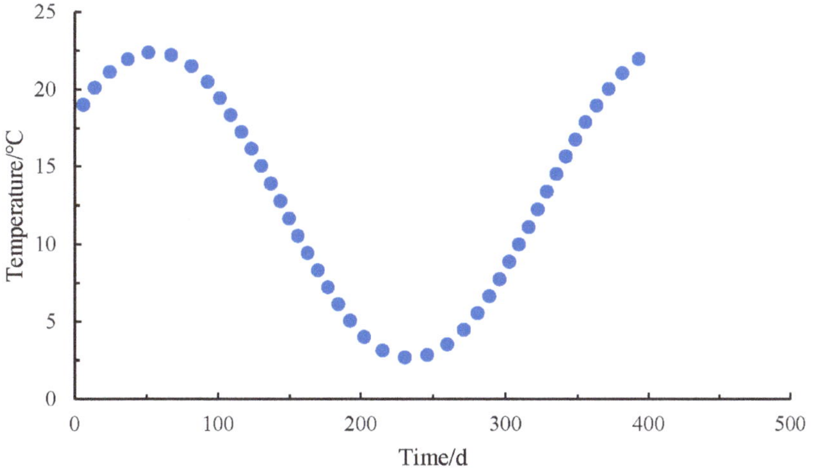

Figure 8. Annual temperature change.

Table 2. Model parameters.

Parameters	C35	C45
Final elastic modulus E^{∞}	32 GPa	36 GPa
Final tensile strength f_t^{∞}	2.3 MPa	2.6 MPa
Poisson's ratio	0.23	0.23
Pouring interval age	7/15/30 d	7/15/30 d
7-day adiabatic temperature rise	43 °C	47 °C
28-day shrink	150 με	200 με

Figure 9 displays the stress distribution of the pier body and pier cap at day 63 (positive values represent tensile stress; negative values represent compressive stress). It can be seen that the core tensile stresses of the pier body and pier cap are relatively high at day 63, posing a risk of cracking. However, the surface tensile stress is relatively low, reducing the risk of cracking, since the surface temperature matches the ambient temperature. Therefore, the risk of core cracking in the pier body and pier cap may exist for a long time. Frequently, the core cracking of the pier is often overlooked by engineers. Therefore, a comprehensive analysis of the early and long-term cracking risks of the core concrete of a pier is provided in this study.

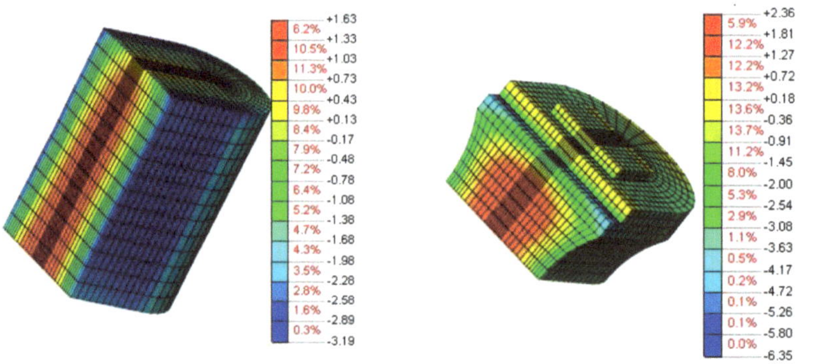

(**a**) Pier body (**b**) Pier cap

Figure 9. Stress of the pier (unit: MPa).

Tables 3–5 lists the cracking risk in the core of the pier body and pier cap, with the higher of the two values indicated. The data indicated that at a tensile strength of 2.3 MPa and an age of 7 days, with an adiabatic temperature rise of 43 °C, the early cracking risk coefficient was less than 0.7. Conversely, with a tensile strength of 2.0 MPa, an adiabatic temperature rise of 50 °C or a pouring interval age of the pier body and pier cap of 14 days or more, the early cracking risk coefficient was greater than 0.7, indicating that the risk of cracking is higher. For an age of 30 days after pouring, the early cracking risk coefficient was 1.24, indicating a very high risk of cracking (Table 5). Additionally, the long-term risk of cracking in the core was higher than in the short term under different working conditions, which was especially noticeable when the pouring interval for the pier body and pier cap was 14 days or 30 days, where the cracking risk coefficients exceeded 1.0 (Table 5).

Table 3. Effect of tensile strength on the maximum risk of core cracking (ATR = 43 °C, $\triangle t$ = 7 d).

Concrete	σ_t/MPa	Early Cracking	Long Term
C35	2.0	0.92	1.1
	2.3	0.69 (↓33%)	0.91 (↓21%)

Table 4. Effect of adiabatic temperature rise on the maximum risk of core cracking (σ_t = 2.3 MPa, $\triangle t$ = 7 d).

Concrete	Adiabatic Temperature Rise/°C	Early Cracking	Long Term
C35	43	0.69	0.91
	50	0.87 (↑26%)	1 (↑10%)

Table 5. Effect of pouring interval age on the maximum risk of core cracking (σt = 2.3 MPa, ATR = 43 °C).

Concrete	Pouring Interval Age/d	Early Cracking	Long Term
C35	7	0.69	0.91
	14	0.93 (↑35%)	1.15 (↑26%)
	30	1.24 (↑79%)	1.35 (↑48%)

Increasing the tensile strength from 2 to 2.3 MPa can reduce the risk of early cracking by about 33% and long-term cracking by about 21% (Table 3). There is a positive correlation between adiabatic temperature rise and the risk of cracking, where the lower the temperature rise, the lower the cracking risk (Table 4). Extending the interval age between pier body and pier cap pouring from 7 to 30 days increased the risk of early cracking by 79% and long-term cracking by 48% (Table 5), suggesting that a decrease in the age of this interval was recommended to reduce the risk of concrete cracking. To maintain the early cracking risk coefficient of the concrete core below 0.7 and the long-term cracking risk coefficient below 1.0, it is crucial to improve the concrete mix ratio, control the adiabatic temperature rise and manage the pouring interval age.

4. Cracking Control Measures

More than 80% of cracks in concrete are caused by shrinkage [24,27]. Early shrinkage cracking poses a significant challenge in modern cement concrete engineering. There are two approaches to mitigate shrinkage-induced cracking in cement-based materials: one is to enhance the crack resistance of the material itself, e.g., by adding fibers to increase the tensile strength, and the other is to enhance the volume stability of the materials, e.g., by optimizing the raw materials, using shrinkage-reducing agents or curing agents and incorporating expansion components to reduce various shrinkage factors. In practical application, the crack resistance of concrete can be greatly improved by incorporating admixtures with deformation compensation, together with advanced curing construction techniques; thus, the cracking resistance of concrete can be significantly improved.

4.1. Increase Deformation Compensation

Concrete cracking is closely related to deformation shrinkage. Deformation compensation is an important approach to enhance crack resistance and minimize the risk of cracking [24,27]. Therefore, an anti-cracking agent (HME®-V) with deformation compensation properties has also been used to develop an advanced high-altitude deformation compensation and crack control strategy, aiming at minimizing deformation shrinkage and potentially achieving zero shrinkage during the temperature drop stage after concrete pouring. A comparison was made between two site piers: pier #1 without an anti-cracking agent at a height of 7.5 m and pier #2 at a height of 7 m with an added anti-cracking agent. The cement used was Jiahua Low Heat P LH 42.5 with Class II fly ash, with three types of single aggregates with particle sizes of 5–10 mm, 10–20 mm and 16–31.5 mm and a continuous graded crushed stone size of 5–31.5 mm. The Shanxi Jinkeqi high-performance water-reducing agent was also used. The concrete had a slump of about 195 mm in the laboratory and showed good workability during on-site pouring. Details of the mix ratio are shown in Table 6. Geotextiles and membranes were used to cover the pier body to provide insulation and moisture retention. Additionally, intermittent water sprinkling was conducted for maintenance purposes. A strain monitoring point was placed at each pier body and pier cap core. Two mutually perpendicular rigid chord DY-YT-500B sensors were installed along the length direction (x-direction) and thickness direction (y-direction) to monitor the strain in the pier body and pier cap core. The sensor layout and on-site testing are shown in Figure 10.

Table 6. Concrete mix ratio (kg/m^3).

	Cement	Fly Ash	HME®-V	Sand	Aggregate	Water
Pier 1#	320	80	0	774	1070	150
Pier 2#	300	68	32	774	1070	150

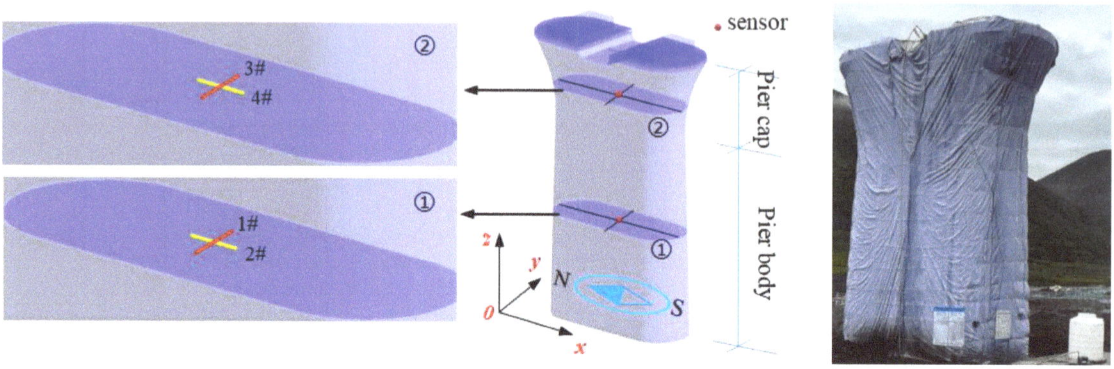

Figure 10. Sensor layout and on-site testing. Note that 1#~4# refer to stain sensors.

Figure 11 shows the deformation monitoring results for the benchmark comparison pier (#1) and the test pier (#1). It is worth noting that the positive strains indicate expansion and compression, while negative strains indicate contraction and tension. The peak deformation of the test pier ranged from 600 to 900 με, indicating that the concrete was in compression (Figure 11b). The peak strain of the reference pier was about 400 με, indicating a significant deformation compensation effect of the anti-cracking agent before the peak strain was reached (Figure 11a). The deformation of the test pier stabilized around 50 days and remained in expansion and compression. Even after 60 days, the deformation still exceeded 200 με. The deformation of the benchmark pier gradually stabilized after about 65 days and transitioned to contraction and tension, with a deformation of less than −100 με. This highlighted the compensatory effect of the anti-cracking agent in the post-peak strain stage. The deformation curves of the test piers performed a relatively smooth pattern without abrupt fluctuations in the data (Figure 11b). The deformation curve of strain sensor #2 in benchmark pier showed a significant local variation around 40 days (Figure 11a), indicating that local cracks may exist in the pier. Concrete has limited tensile stress but can withstand significant compressive stresses. Therefore, anti-cracking agents help to compensate for the deformation and improve the crack resistance by keeping the concrete in a slightly expanded state internally.

The early and long-term cracking risks of core concrete with the addition of anti-cracking agents (HME®-V) were analyzed. The results are listed in Tables 7–9. The mix proportion was consistent with that of pier #2. Compared with the concrete without adding anti-cracking agents, the risk coefficient for both early and long-term cracking was significantly lower for the concrete with anti-cracking agents. For instance, at an adiabatic temperature rise of 43 °C, pouring interval of 7 days and tensile strength of 2.0 MPa, the risk coefficient of early cracking decreased from 0.92 to 0.73. In terms of long-term cracking, keeping the pouring interval between the pier body and pier cap below 14 days led to a risk of cracking lower than 1.0. Practical experience has demonstrated that maintaining an early cracking risk coefficient below 0.7 and a long-term cracking risk coefficient below 1.0 can effectively control concrete cracking. Therefore, the use of deformation compensation technology in plateau concrete can greatly reduce the risk of cracking.

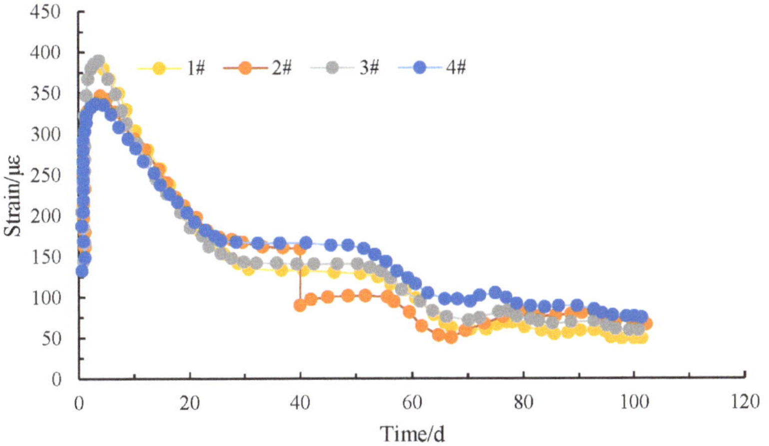

(**a**) Pier 1 #

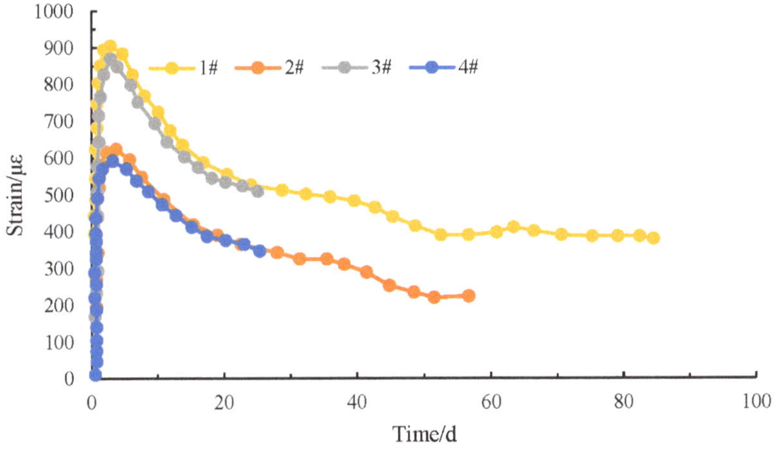

(**b**) Pier 2 #

Figure 11. Concrete deformation of on-site piers.

Table 7. Effect of tensile strength on the maximum risk of core cracking after adding HME®-V (ATR = 43°C, △t = 7 d).

Concrete	σ_t/MPa	Early Cracking	Long Term
C35-HME	2.0 2.3	0.73 0.56 (↓30%)	0.92 0.77 (19%)

Table 8. Effect of adiabatic temperature rise on the maximum risk of core cracking after adding HME®-V (σ_t = 2.3 MPa, $\triangle t$ = 7 d).

Concrete	Adiabatic Temperature Rise/°C	Early Cracking	Long Term
C35-HME	43	0.56	0.77
	50	0.72 (↑28%)	0.88 (↑14%)

Table 9. Effect of pouring interval age on the maximum risk of core cracking after adding HME®-V (σ_t = 2.3 MPa, *ATR* = 43 °C).

Concrete	Pouring Interval Age/d	Early Cracking	Long Term
C35-HME	7	0.56	0.77
	14	0.77 (↑37%)	1.0 (↑30%)
	30	1.11(↑98%)	1.25(↑62%)

4.2. Radiation Refrigeration Technology

Using radiation refrigeration technology, the temperature difference between the sunny and shaded sides of the bridge pier body can be controlled, thus mitigating the effect of thermal radiation on the temperature field of the concrete and ultimately improving the durability of the concrete. To verify the technology, concrete specimens with dimensions of 100 mm × 100 mm × 100 mm were prepared and sprayed with radiatively cooled polymer batch on their surrounding surfaces. The specimens without any spraying measures were also prepared. Both specimens were placed in the outdoor environment at the same time. The temperature field variations of the specimens under the two different working conditions were monitored, as shown in Figure 12.

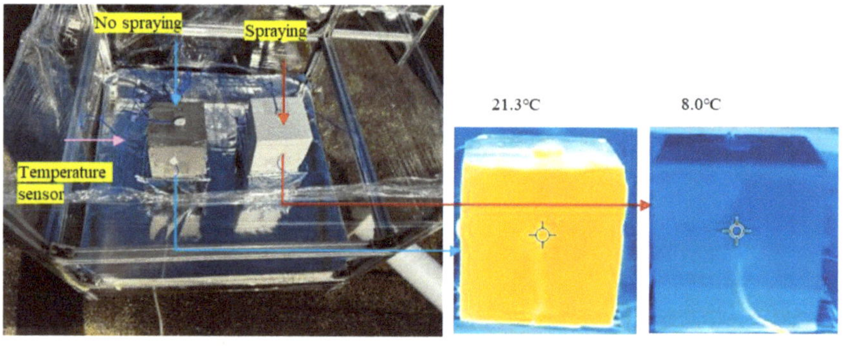

Figure 12. Radiation refrigeration test.

Figure 13 shows the temperature field variations and temperature difference between the sunny and shaded sides of concrete specimens under different working conditions. The maximum temperature of the specimen was decreased by 12 °C after spraying with the radiation-cooled polymer materials compared with the specimen without the radiation-cooled polymer material. The temperature difference between the sunny side and shady side was significantly decreased, with a temperature difference of 5.4 °C for the unsprayed specimens and only 1.7 °C for the specimen the radiation-cooled polymer material. Therefore, the spraying of radiation-cooled polymer materials can effectively reduce the overall temperature of the concrete structure and minimize the temperature difference in different directions, thus reducing the risk of concrete cracking.

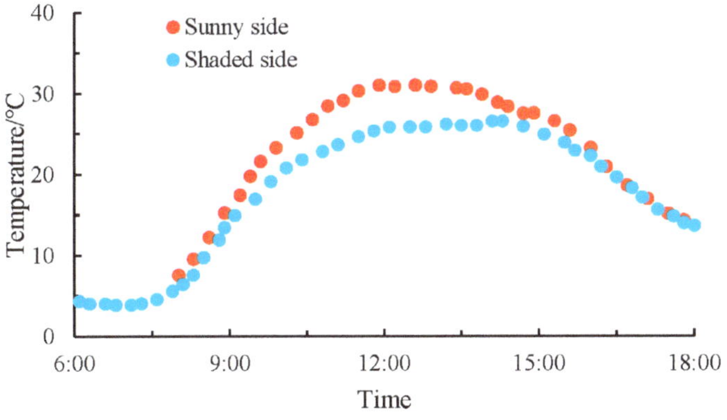

(**a**) Without spraying.

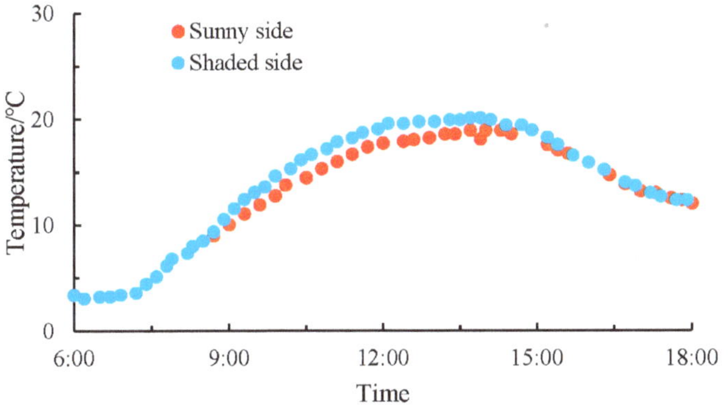

(**b**) With radiation-cooled polymer materials.

Figure 13. Temperature development of sunny and shaded sides of the test specimen.

4.3. Improve Construction Technology

Figure 14 shows the shrinkage deformation of concrete under different plateau curing methods. During joint curing with insulation and film, the shrinkage strain of the concrete was 150 $\mu\varepsilon$ and remained relatively constant over time. The shrinkage strain of concrete after 40 days in natural conditions exceeded 400 $\mu\varepsilon$, an increase of 167%. As mentioned earlier, shrinkage was a significant factor contributing to concrete cracking. Effective maintenance practices can greatly reduce this risk. Delayed curing of freshly poured concrete can lead to the rapid evaporation of water, preventing the full hydration of cement particles and formation of stable crystals. This inadequate hydration led to insufficient cohesion, resulting in spalling of the concrete surface. By controlling the proportion of cement clinker components, continuously optimizing the concrete mix proportion and implementing strategies such as increasing cooling water pipes, lowering the molding temperature and mixing mineral powder and silicon powder, the release of the hydration heat was controlled, and the serviceability of concrete was improved. These measures ultimately improved the quality and crack resistance of the concrete. The specific construction process control measures are outlined in Table 10.

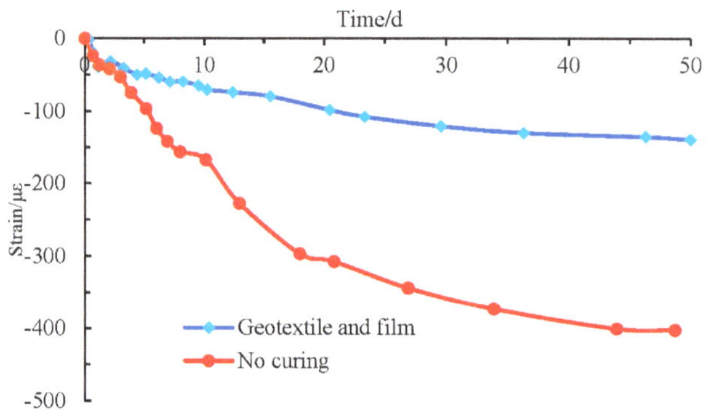

Figure 14. The influence of concrete curing methods on deformation.

Table 10. Construction process control measures for concrete.

No.	Strategy	Stands
1	Water	Low-temperature water prepared using groundwater or cold-water machine.
2	Cement	With 3 d hydration heat ≤ 230 kJ/kg, 7 d hydration heat ≤ 260 kJ/kg, $C_2S \geq 40\%$ and $C_3A \leq 6\%$ [28].
3	Mixing time	The mixing time of each concrete mixer should not be less than 120 s.
4	Transportation	Concrete transport vehicles should have insulation or thermal insulation measures; the transportation time should not exceed 60 min.
5	Pumping	① The concrete molding temperature falls within the range of 5–25 °C. ② The pumping temperature should be lowered, or construction should be conducted during low temperature seasons. This can also mitigate the risk of long-term concrete cracking.
6	Pouring and vibration	① In cases where there is a substantial loss of slump that fails to meet the requirements, it can be rectified by adding a suitable amount of water-reducing agent twice, with direct water addition being strictly prohibited. ② Concrete should be poured continuously in layers, with each layer having a thickness of 30–50 cm. The time between layers should not exceed the initial setting time of the concrete. ③ The thickness of vibration pouring ranged from 30–50 cm, and the vibration time was determined based on the factors of concrete surface bleeding, absence of significant sinking and no appearance of bubbles.
7	Demolding and maintenance	① Concrete formwork should be removed during periods of high temperatures. After the removal of concrete formwork, a water energy film should be applied and covered with a minimum of two layers of geotextiles with plastic lining. ② In conditions of strong winds, rainy days and sudden temperature drops, it was recommended to extend the demolding time to prevent cold cracking. ③ The cooling rate of concrete should be ≤2.0 °C/d, and the temperature difference between the inner and outer surfaces should be maintained at ≤ 20 °C. ④ When mold curing was adopted, it was recommended to embed insulation materials such as a rubber sponge and XPS board in the steel formwork, and to wrap windproof tarpaulin around the exterior of the formwork. Additionally, post-dismantling of the formwork, it was recommended to continue long-term insulation and moisturizing curing. ⑤ The temperature difference between the curing water and the concrete surface should be less than 15 °C.

4.4. Improve Management Measures

Careful management is necessary to improve the crack resistance of concrete. Essential elements such as well-equipped laboratories, efficient mixing plants and effective site management are essential to maintain quality control of concrete. Moreover, proper secondary finishing techniques can help to minimize the formation of concrete micro-cracks. Usually, the use of an intelligent control system (MES) has proved beneficial in the production of concrete for highland projects. The system allows for precise adherence to proportions during automatic discharging, controls the water–cement ratio and ensures the overall quality of the concrete, such as the strength, slump and workability. Therefore, the workability of the concrete placed on site has been significantly improved, improving the crack resistance and overall quality of the piers after placement.

5. Conclusions

The challenges faced in concrete construction in plateau areas on the Sichuan–Tibet route were investigated. The cracking characteristics and early and long-term cracking risks of the plateau pier using a multi-field coupled shrinkage model were analyzed. Furthermore, crack control technologies such as deformation compensation and radiation cooling for plateau concrete were discussed. The main conclusions are outlined as follows.

(1) Various factors such as the temperature, wind speed, freeze–thaw cycles, air pressure, lighting conditions, temperature differentials between sunny and shady sides and dynamic disturbances significantly affected the durability and service performance of the plateau concrete. These factors usually led to different types of cracking in the vertical, longitudinal and transverse directions of the bridge piers. Vertical cracks in the straight section of the pier body are the most common type of crack.

(2) The risk of concrete cracking in the core of a plateau pier body was influenced by factors such as the tensile strength of the concrete, pouring interval age and insulation temperature rise. The long-term cracking risk of the core of the plateau pier was significantly higher compared with the early cracking risk. To mitigate this risk, it is necessary to optimize the concrete mix ratio, regulate the insulation temperature rise and carefully manage the pouring interval age. Through these measures, it is possible to ensure that the early cracking risk coefficient for pier concrete remains below 0.7 and that the long-term cracking risk coefficient does not exceed 1.0.

(3) Deformation compensation can significantly enhance the peak and residual deformation capacities of plateau concrete, with peak values greater than 900 $\mu\varepsilon$ and residual deformation greater than 200 $\mu\varepsilon$ at day 60, which greatly reduced the possibility of early and long-term cracking of plateau concrete.

(4) The implementation of radiation cooling technologies, advances in construction techniques and effective management practices can help to mitigate cracking in high-altitude concrete. However, further research is needed to fully understand the effects of air pressure, high-frequency freeze–thaw cycles and other factors on the degradation of plateau concrete.

Author Contributions: Conceptualization, X.H.; Formal analysis, X.H. and M.L. (Manping Liao); Investigation, X.H. and M.L. (Ming Li); Supervision, M.-L.Z.; Writing—original draft, X.H.; Writing—review and editing, X.H., M.L. (Manping Liao), M.-L.Z., F.W. and X.L. All authors have read and agreed to the published version of the manuscript.

Funding: This work was supported financially by the CSCEC Technology R&D Program Funding Projects (CSCEC-2021-S-1) and the China Construction Science and Technology Innovation Platform Grant (CSCEC-PT-017).

Data Availability Statement: The original contributions presented in the study are included in the article; further inquiries can be directed to the corresponding author.

Acknowledgments: The authors would like to express their gratitude for the support received.

Conflicts of Interest: Authors Xiaochuan Hu and Manping Liao were employed by the company The Civil Engineering Group Corporation of China Second Engineering Bureau Ltd. Author Ming Li was employed by the company Jiangsu Sobute New Materials Co., Ltd. Author Xiang Lyu was employed by the company China Construction Second Engineering Bureau Ltd. The remaining authors declare that the research was conducted in the absence of any commercial or financial relationships that could be construed as a potential conflict of interest.

References

1. He, R.; Wang, T.; Cheng, X.H.; Xue, C.; Bai, H.Y. Impact of Qinghai-Tibet plateau's climate on strength and permeability of concrete. *China J. Highw. Transp.* **2020**, *33*, 29–41.
2. Li, Y.; Wang, Z.D.; Xue, C.; Bai, Y.H.; Wang, L. Influence of low air pressure on the performance of concrete in road engineering. *China J. Highw. Transp.* **2021**, *34*, 194–202.
3. Hu, Y.B.; Cao, R.P. Differences of internal and external structural properties of concrete in the plateau area. *Bull. Chin. Ceram. Soc.* 2017; 36, 213–218. (In Chinese)
4. Li, X.F.; Fu, Z.; Luo, Z.; Ke, G.J. Effect of low-pressure in plateau environment on air content of fresh concrete. *J. Southeast Univ. (Nat. Sci. Ed.)* **2014**, *44*, 1046–1051. (In Chinese)
5. Ge, X.; Ge, Y.; Du, Y. Effect of low air pressure on mechanical properties and shrinkage of concrete. *Mag. Concrete Res.* **2018**, *70*, 919–927. [CrossRef]
6. Ge, X.; Ge, Y.; Li, Q.F. Effect of low air pressure on the durability of concrete. *Constr. Build. Mater.* **2018**, *187*, 830–838. [CrossRef]
7. Shi, Y.; Yang, H.Q.; Zhou, S.H.; Wang, A.G.; Lv, X.D. Effect of atmospheric pressure on performance of AEA and air entraining concrete. *Adv. Mater. Sci. Eng.* **2018**, 1–7. [CrossRef]
8. Huo, J.Y.; Wang, Z.J.; Chen, H.X.; He, R. Impacts of low atmospheric pressure on properties of cement concrete in plateau areas: A literature review. *Materials* **2019**, *12*, 1384. [CrossRef]
9. Zeng, X.H.; Lan, X.L.; Zhu, H.S.; Long, G.C.; Xie, Y.J. Investigation on air-voids structure and compressive strength of concrete at low atmospheric pressure. *Cement Concrete Comp.* **2021**, *122*, 104139. [CrossRef]
10. Mi, L.; Mao, J.H.; Li, L.; Shi, Q.; Fang, K.; Li, S.L.; Deng, X.; Li, G. Microstructural damage and durability of plateau concrete: Insights into freeze-thaw resistance and improvement strategies. *Structures* **2024**, *60*, 105888. [CrossRef]
11. Qin, Y.H.; Ma, H.Y.; Zhang, L.L.; Yin, J.S.; Zheng, X.H.; Li, S. Quantification of the concrete freeze–thaw environment across the Qinghai–Tibet Plateau based on machine learning algorithms. *J. Mt. Sci.* **2024**, *21*, 322–334. [CrossRef]
12. Zhang, R.L.; Long, Z.F.; Long, G.C.; Guo, H.Z.; Zhang, F.J.; Zhang, X.P. Investigating concrete performance: Compressive strength and pore-structure evolution in simulated permafrost conditions of the Qinghai–Tibet Plateau zone. *J. Mater. Civ. Eng.* **2024**, *36*, 04024063. [CrossRef]
13. Lu, C.F.; Cai, C.X. Challenges and countermeasures for construction safety during the Sichuan–Tibet railway project. *Engineering* **2019**, *5*, 833–838. [CrossRef]
14. Chen, Z.; Chen, B.; Zheng, J.L.; Liu, J.P.; Ye, Z.X.; Wu, C.J.; Xu, W. Methodology on evaluating the compactness of core concrete in CFST serving under low atmospheric pressure over the Qinghai-Tibet plateau. *China Civ. Eng. J.* **2021**, *54*, 1–13. (In Chinese)
15. Powers, T.C. A working hypothesis for further studies of frost resistance of concrete. *J. Am. Concr. Inst.* **1945**, *16*, 245–272.
16. Qin, X.C.; Meng, S.P.; Cao, D.F.; Tu, Y.M.; Sabourova, N.; Grip, N.; Ohlsson, U.; Blanksvärd, T.; Sas, G.; Elfgren, L. Evaluation of freeze-thaw damage on concrete material and prestressed concrete specimens. *Construct Build. Mater.* **2016**, *125*, 892–904. [CrossRef]
17. Li, X.F.; Fu, Z. Effect of low-pressure of environment on air content and bubble stability of concrete. *J. Chin. Ceram. Soc.* **2015**, *43*, 1076–1082.
18. Lan, X.L.; Zeng, X.H.; Zhu, H.S.; Long, G.C.; Xie, Y.J. Experimental investigation on fractal characterstics of pores in air-entrined concrete at low atmospheric pressure. *Cem. Concr. Compos.* **2022**, *130*, 104509. [CrossRef]
19. Li, Y.L.; Duo, Z.S.; Li, Y.; Liu, Y.H.; Sun, X.J.; Guo, H.Y. Mechanical properties and damage characteristics of concrete under different atmospheric pressures. *J. Build. Eng.* **2024**, *82*, 108196. [CrossRef]
20. Chen, H.X.; Wang, T.; He, R.; Zhang, J.S.; Fang, J.H.; Bai, Y.H. Effect of complex climatic environment on pore structure and mechanical properties of concrete. *J. Chang'an Univ. (Nat. Sci. Ed.)* **2020**, *40*, 30–37. (In Chinese)
21. Lu, J.Z.; Zhu, K.F.; Tian, L.Z.; Guo, L. Dynamic compressive strength of concrete damaged by fatigue loading and freeze-thaw cycling. *Construct Build. Mater.* **2017**, *152*, 847–855. [CrossRef]
22. Simon, K.M.; Kishen, J.M.C. A multiscale approach for modeling fatigue crack growth in concrete. *Int. J. Fatigue* **2017**, *98*, 1–13. [CrossRef]
23. Li, H.; Liu, J.P.; Wang, Y.J.; Yao, T.; Tian, Q.; Li, S.C. Deformation and cracking modeling for early-age sidewall concrete based on the multi-field coupling mechanism. *Constr. Build. Mater.* **2015**, *88*, 83–94. [CrossRef]
24. Liu, J.P.; Tian, Q.; Wang, Y.J.; Li, H.; Xu, W. Evaluation method and mitigation strategies for shrinkage cracking of modern concrete. *Engineering* **2021**, *7*, 348–357. [CrossRef]
25. Ulm, F.J.; Coussy, O. Modeling of thermo-chemo-mechanical couplings of concrete at early ages. *J. Eng. Mech.* **1995**, *121*, 785–794. [CrossRef]
26. Cervera, M.; Oliver, J.; Prato, T. Thermo-chemo-mechanical model for concrete I: Hydration and Aging. *J. Eng. Mech.* **1999**, *125*, 1018–1027. [CrossRef]

27. Liu, J.P.; Tian, Q. *Early Deformation and Crack Control of Modern Concrete*; China Communications Press: Beijing, China, 2020.
28. *GB/T 200-2017*; National Standards of the People's Republic of China. Moderate-Heat Portland Cement, Low-Heat Portland Cement. China Standard Press: Beijing, China, 2017.

buildings

MDPI

Article

Displacement and Internal Force Response of Mechanically Connected Precast Piles Subjected to Horizontal Load Based on the m-Method

Li Gao [1,2], Mei-Ling Zhuang [3,4,*], Qunqun Zhang [5], Guangdong Bao [4], Xiaoyang Yu [4], Jiahao Du [4], Shengbo Zhou [1,2] and Mingsen Wang [3]

1 School of Civil Engineering and Architecture, Suqian University, Suqian 223800, China; gaoli@squ.edu.cn (L.G.); zhoushengbo2005@163.com (S.Z.)
2 Jiangsu Province Engineering Research Center of Prefabricated Building and Intelligent Construction, Suqian University, Suqian 223800, China
3 Water Resources Research Institute of Shandong Province, Jinan 250013, China; wang_ms88@163.com
4 School of Transportation and Civil Engineering, Nantong University, Nantong 226019, China; 18051133608@163.com (G.B.); 15050658281@163.com (X.Y.); 15093016704@163.com (J.D.)
5 School of Transportation and Civil Engineering, Anhui Jianzhu University, Hefei 23061, China; 15532395385@163.com
* Correspondence: ml_zhuang@ntu.edu.cn or ml_zhuang99@163.com

Abstract: Mechanically connected precast piles are a type of precast piles that utilise snap-type mechanical connectors to restrain the pile ends of two identical or different precast piles at the top and bottom so as to quickly realise the purpose of the connection. However, the gap problem in the connectors of mechanically connected piles can lead to uneven and uniform deformation of the piles under horizontal loading, resulting in additional displacements and rotation angles of the piles at the connection. Solving the problem of calculating the internal force response of discontinuous deformed piles is a prerequisite for promoting and applying mechanically connected precast piles. Firstly, the theoretical derivation of mechanically connected piles with fixed constraints at the pile bottom is carried out. Secondly, the pile response equations of mechanically connected piles are established, and the theoretical solutions of pile displacement and internal force response of mechanically connected piles under horizontal loading are derived. Thirdly, the pile-soil model of the test pile is established using ABAQUS software (ABAQUS 2016) in combination with the design data of the test pile. The numerical simulation displacements and angles of rotation are compared with the test results. Finally, the theoretical and numerical simulation displacements and internal forces of the ordinary pile and the mechanically connected pile are compared. The relative errors of the displacements and angles of rotation of the established pile-soil model are less than 10%, indicating that the established model has good accuracy. The relative errors of the theoretical and numerical simulation displacements and internal forces of the mechanically connected pile are less than 10%, proving the correctness of the theoretical calculation by the m-method. This study can provide effective theoretical support and methodological guidance for the displacement and internal force response of discontinuous piles.

Keywords: prestressed solid square piles; mechanically connected piles; m-method; displacement curve equations; soil-pile model

1. Introduction

A pile foundation is one of the most widely used forms of upper load bearing in various deep foundation categories in actual projects. It can adapt to various sizes of loads, various directions of distribution, and various complex geological conditions. It also has the advantages of large bearing capacity, high stability, relatively small settlement value, and good durability under the condition of well-treated foundations. Therefore, pile foundations have been widely used in construction fields such as roads and bridges, super-high-rise buildings, and major projects such as harbour terminals [1–3].

Among the commonly used pile foundations, bored piles are the most widely used [4], but they require on-site pouring construction, which has the disadvantages of high construction noise, serious environmental pollution and a long construction period, which is no longer in line with the development prospect of industrialisation, information technology, and greening of China's construction industry, and thus has limitations in the process of its use. While precast structures [5,6] are produced in factory workshops, the environmental pollution of on-site construction can be effectively controlled, which has a high application prospect. At present, welding is still the main method for connecting prefabricated piles, such as square piles and tubular piles. However, the welding process, though simple, is demanding and often cannot be effectively implemented at construction sites, resulting in inconsistent welding quality. During the piling process, any defects in the joint welds are magnified, making the joints the weak points of the piles. Among prefabricated pile foundation project quality accidents, the accidents caused by the quality of joint welding accounted for a large proportion.

Mechanically connected precast piles are an improved type of precast piles using snap-in mechanical connectors (see Figure 1a), which can confine the pile ends of two identical or different precast piles (see Figure 1b) at the top and bottom to quickly realize the purpose of connection [7–9]. Compared with ordinary precast square piles, the connection process of mechanically connected piles does not require welding, and the connection effect can be achieved by extruding and butt jointing, which have the advantages of high efficiency, environmental protection and reliability [10,11]. In addition, due to the replaceability of mechanically connected piles, the precast piles have a wider range of combinations of pile lengths, cross-sections, and reinforcement ratios, which improves the fault tolerance of precast pile production.

(**a**) (**b**)

Figure 1. Photos of prestressed solid square piles and a nap-type mechanical connector. (**a**) Prestressed solid square piles. (**b**) A snap-in mechanical connector.

Under the action of horizontal external load, the precast pile will undergo bending deformation, and the soil body also squeezes reaction force on the pile body [12]. Therefore, the response of the pile under horizontal loading mainly involves the joint bearing of the pile body and soil body to the external load. Different soil conditions and depths of entry can greatly affect the internal force response generated by the pile. Different theoretical methods have been proposed for the displacement and internal force response of pile foundations under horizontal action [13]. Among the widely used theoretical methods, the main ones are the ultimate foundation reaction method, the elastic foundation reaction method and the elastoplastic foundation reaction method. The elastic foundation reaction force method is a method based on Winkler's elastic foundation beam model [14]. It regards the pile as an elastic beam and considers the magnitude of the reaction force per unit length of the foundation beam to be proportional to the settlement of the foundation beam of that length. Then, according to the force relationship between the pile and foundation soil when they act together, the differential equations are established to be solved by the m-method [15,16]. For the horizontal load response of ordinary piles, researchers have jointly discussed the feasibility of the m-method in the horizontal load response of piles through theoretical analysis, numerical simulations and experimental studies and put forward a series of problems and suggestions [17,18]. Our latest public bridge regulations issued in 2019 [19] provided a detailed and systematic introduction to the calculation of horizontal displacement and action effects of elastic piles by the m-method, proposing a weight conversion method for the equivalent m-value of the scale factor for multi-storey foundations and a correction method for the maximum bending moment of the pile. García et al. [20] performed horizontal load tests on near-true scale foundation models of defective and intact three-pile systems. On the basis of the experimental data, a preliminary calibration of the 3D finite element analysis was carried out, which was subsequently used to investigate the effect of defective elements on the pile load distribution and the horizontal base force at the pile shaft. The results indicated that the presence of a defective pile increased the raft tilting, which affected vertical and horizontal load distributions among the raft and the piles, as well as among tailing and leading piles.

The problem of gaps in the connectors of mechanically connected piles can lead to uneven and uniform deformation of the pile under horizontal loading, which in turn leads to additional displacement and rotation angles of the piles at the connection. Solving the problem of calculating the internal force response of discontinuous deformed piles is a prerequisite for the popularisation and application of mechanically connected precast piles. Firstly, the theoretical derivation of mechanically connected piles with fixed constraints at the pile bottom is carried out. Secondly, the pile response equations of mechanically connected piles are established, and the theoretical solutions of pile displacement and internal force response of mechanically connected piles under horizontal load are derived. Thirdly, the pile-soil model of the test pile from reference [21] is established using ABAQUS software. The numerical simulation results of the pile-soil model are compared with the test results to verify the correctness of the pile-soil model. Finally, the theoretical and numerical simulation displacement and internal forces of the ordinary pile and mechanically connected pile are compared to verify the correctness of the theoretical calculation by the m-method.

2. Theoretical Calculation of Displacement and Internal Force Response of Mechanically Connected Piles

2.1. Description of Stress Phases and Basic Assumptions

Under horizontal loading, the deformation of the pile body after the mechanically connected pile is subjected to force is not as uniform as that of ordinary tubular piles, and an additional structural angle of rotation will be generated at the connection due to the existence of a gap in the mechanical connection. The physical drawing and conceptual drawing of the angle of rotation diagram of the mechanically connected pile structure are shown in Figure 2a,b.

The following assumptions are made based on the stressing process and state of mechanically connected piles. The assumptions are made as follows.

(1) Upper pile stress stage.

When the load is small and acts on the top of the upper pile, the top of the pile first undergoes a displacement and an angle of rotation when the horizontal external load acts on the top of the connected pile. Due to the structural angle of rotation φ, it is assumed that the bottom connection of the upper pile undergoes only rotation without horizontal displacement, i.e., the connection is considered as an equivalent plastic hinge, while the lower pile is considered as fixed and unresponsive.

(2) Critical state

The horizontal load continues to increase until the structural rotation angle reaches the limit value of the structural angle of rotation φ_0. The structural angle of rotation reaches the limit value, meaning that the simple upper pile force reaches the critical state, and the combined effect of the external loads in the critical state is solved by the boundary condition of the state, which is used as the basis for judgement of the connecting pile force state.

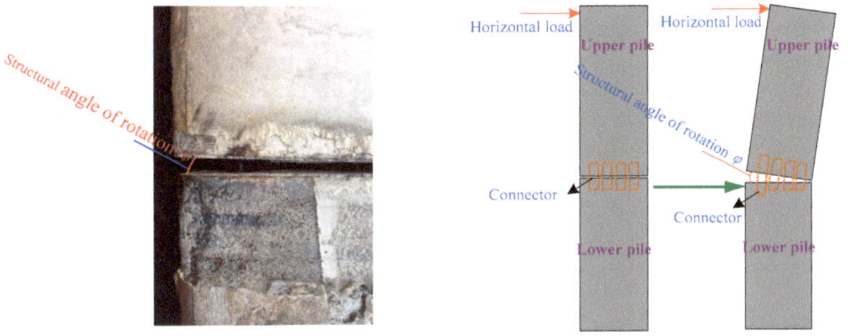

(a) (b)

Figure 2. The angle of rotation diagram of mechanically connected pile structure. (**a**) Physical drawing. (**b**) Conceptual drawing.

(3) Synergistic force stage of upper and lower piles

The horizontal load continues to be applied, and when the structural angle of rotation exceeds this limit value, this plastic hinge begins to transmit bending moment and shear force, and the lower piles begin to participate in the force work. The displacement curve equations of the upper and lower piles are established separately and brought into the respective boundary conditions, and the system solution of the overall pile response is obtained by coupling the geometrical and mechanical relationship equations of the members.

Under the joint action of the pile and the soil of the mechanically connected piles, with the increase of external forces Q^0 and M^0 at the top of the upper pile, the rotation angle of the connectors gradually increases until it reaches the limit angle of rotation φ_0. Prior to this time, it is assumed that the lower piles are not subject to the transfer of moments and shear forces from the upper piles, that no displacements are generated, and that the action of the pile-soil at the top meets the assumption of the Winkel's elastic foundation beams as shown in Figure 3. In Figure 3, "0" denotes Section 0; "1" denotes Section 1; "2" denotes Section 2; "3" denotes Section 3.

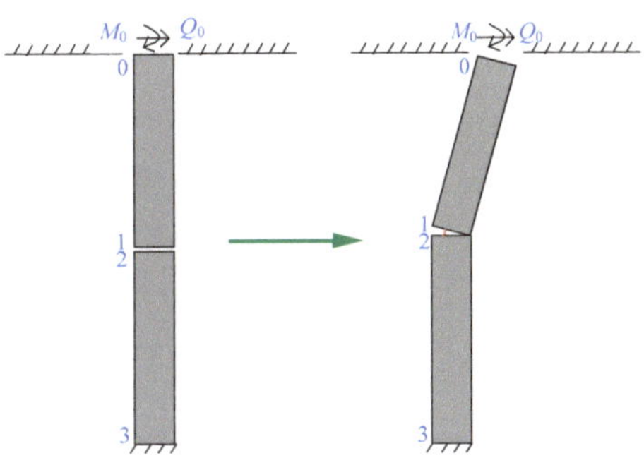

Figure 3. Deformation form of embedded pile body.

2.2. Dual-Pile Cooperative Work

When φ_1 reaches the limit value of the gap φ_0, which is determined by the construction of the joint, the angle of rotation reaches the limit state value. If the external load continues to be applied, the connection begins to transmit internal forces, and the lower pile begins to displace horizontally. Then, the combination of load effects at the limit state is expressed as Equation (1). It is a guideline for determining at what stage the pile is stressed.

$$\overline{\varphi_1} = \overline{\varphi^0} = \delta_{\overline{\varphi_1}\overline{M_0}} \cdot \overline{M_0} + \delta_{\overline{\varphi_1}\overline{Q_0}} \cdot \overline{Q_0} \tag{1}$$

After the external load reaches the ultimate load and the load continues to be applied, the lower pile will be horizontally displaced under the driving of the connection; the upper and lower piles are driven by the connection, constituting a stressed whole. The deflection equations were established for the upper and lower piles, respectively. According to the force transmission characteristics of the mechanical connections, the deflection equations of the upper and lower piles are related to obtain the overall force state. The deflection equation of the upper pile is established based on the m-method. The second stage is basically the same as the first stage of the force condition, only the setting of boundary conditions is different; the deflection differential equation of the upper pile is

$$E_1 I_1 \frac{d^4 y}{dz^4} + m b_0 z y = 0,\ 0 < z < L_1 \tag{2}$$

In Equation (2), the calculated formulas for b_0 can be obtained from the specifications for the design of the foundation of highway bridges and culverts [19].

Let $\alpha_1 = \sqrt[5]{\frac{mb_0}{E_1 I_1}}$, then Equation (2) can be rewritten as

$$\frac{d^4 y}{dz^4} + {\alpha_1}^5 z y = 0 \tag{3}$$

Equation (3) can be solved by the power series method and dimensionless simplification:

$$\begin{cases} \overline{y(z)} = \overline{y_0} A_1 + \overline{\varphi_0} B_1 + \overline{M_0} C_1 + \overline{Q_0} D_1 \\ \overline{\varphi(z)} = \overline{y_0} A_2 + \overline{\varphi_0} B_2 + \overline{M_0} C_2 + \overline{Q_0} D_2 \\ \overline{M(z)} = \overline{y_0} A_3 + \overline{\varphi_0} B_3 + \overline{M_0} C_3 + \overline{Q_0} D_3 \\ \overline{Q(z)} = \overline{y_0} A_4 + \overline{\varphi_0} B_4 + \overline{M_0} C_4 + \overline{Q_0} D_4 \end{cases} \tag{4}$$

When the external loads on the upper pile are M^0 and Q^0, the horizontal displacement at its lower end is no longer 0 in the second stage of deformation, so the intermediate variables are introduced as the horizontal displacement and the angle of rotation angle in Section 1. Then, the boundary condition of the upper pile can be established as

$$\begin{cases} y(z = L_1) = y_1 \\ \varphi(z = L_1) = \varphi_1 \end{cases} \tag{5}$$

Bringing Equation (5) into Equations (1) and (2), it can obtain

$$\begin{cases} \overline{y_0} A_{11} + \overline{\varphi_0} B_{11} + \overline{M_0} C_{11} + \overline{Q_0} D_{11} = \overline{y_1} \\ \overline{y_0} A_{21} + \overline{\varphi_0} B_{21} + \overline{M_0} C_{21} + \overline{Q_0} D_{21} = \overline{\varphi_1} \end{cases} \tag{6}$$

Then, the solution is

$$\begin{cases} \overline{y_0} = \delta_{\overline{y_0}\ \overline{y_1}} \overline{y_1} + \delta_{\overline{y_0}\ \overline{\varphi_1}} \overline{\varphi_1} + \delta_{\overline{y_0}\overline{M_0}} \overline{M_0} + \delta_{\overline{y_0}\overline{Q_0}} \overline{Q_0} \\ \overline{\varphi_0} = \delta_{\overline{\varphi_0}\ \overline{y_1}} \overline{y_1} + \delta_{\overline{\varphi_0}\ \overline{\varphi_1}} \overline{\varphi_1} + \delta_{\overline{\varphi_0}\overline{M_0}} \overline{M_0} + \delta_{\overline{\varphi_0}\overline{Q_0}} \overline{Q_0} \end{cases} \tag{7}$$

where

$$\begin{cases} \delta_{\overline{y_0 y_1}} = -\frac{B_{21}}{A_{21}B_{11}-A_{11}B_{21}} \\ \delta_{\overline{y_0 \varphi_1}} = \frac{B_{11}}{A_{21}B_{11}-A_{11}B_{21}} \\ \delta_{\overline{y_0 M_0}} = \frac{B_{21}C_{11}-B_{11}C_{21}}{A_{21}B_{11}-A_{11}B_{21}} \\ \delta_{\overline{y_0 Q_0}} = \frac{B_{21}D_{11}-B_{11}D_{21}}{A_{21}B_{11}-A_{11}B_{21}} \\ \delta_{\overline{\varphi_0 y_1}} = \frac{A_{21}}{A_{21}B_{11}-A_{11}B_{21}} \\ \delta_{\overline{\varphi_0 \varphi_1}} = \frac{A_{11}}{A_{21}B_{11}-A_{11}B_{21}} \\ \delta_{\overline{\varphi_0 M_0}} = -\frac{A_{21}C_{11}-A_{11}C_{21}}{A_{21}B_{11}-A_{11}B_{21}} \\ \delta_{\overline{\varphi_0 Q_0}} = -\frac{A_{21}D_{11}-A_{11}D_{21}}{A_{21}B_{11}-A_{11}B_{21}} \end{cases} \quad (8)$$

These eight coefficients in Equation (8) are determined from the converted depth $\alpha_1 L_1$ in Section 1 and are given in Table 1.

Table 1. Coefficient of influence of upper pile boundary on pile top displacement response.

$\alpha_1 L_1$	$\delta_{\overline{y_0 y_1}}$	$\delta_{\overline{y_0 \varphi_1}}$	$\delta_{\overline{y_0 M_0}}$	$\delta_{\overline{y_0 Q_0}}$	$\delta_{\overline{\varphi_0 y_1}}$	$\delta_{\overline{\varphi_0 \varphi_1}}$	$\delta_{\overline{\varphi_0 M_0}}$	$\delta_{\overline{\varphi_0 Q_0}}$
0.0	1.00000	0.00000	0.00000	0.00000	0.00000	1.00000	0.00000	0.00000
0.1	1.00000	−0.10000	0.00500	0.00033	0.00000	1.00000	−0.10000	−0.00500
0.2	0.99999	−0.20000	0.02000	0.00267	0.00007	0.99999	−0.20000	−0.02000
0.3	0.99992	−0.29999	0.04500	0.00900	0.00034	0.99994	−0.30000	−0.04500
0.4	0.99966	−0.39992	0.07999	0.02133	0.00107	0.99974	−0.39996	−0.07999
0.5	0.99896	−0.49970	0.12495	0.04165	0.00260	0.99922	−0.49988	−0.12495
0.6	0.99741	−0.59910	0.17983	0.07192	0.00539	0.99806	−0.59962	−0.17983
0.7	0.99442	−0.69772	0.24448	0.11406	0.00997	0.99581	−0.69902	−0.24448
0.8	0.98913	−0.79492	0.31867	0.16985	0.01698	0.99185	−0.79783	−0.31867
0.9	0.98051	−0.88976	0.40199	0.24092	0.02706	0.98538	−0.89562	−0.40199
1.0	0.96718	−0.98085	0.49373	0.32854	0.04099	0.97539	−0.99179	−0.49375
1.1	0.94771	−1.06643	0.59294	0.43351	0.05937	0.96077	−1.08560	−0.59294
1.2	0.92031	−1.14417	0.69811	0.55589	0.08288	0.94021	−1.17605	−0.69811
1.3	0.88336	−1.21143	0.80737	0.69488	0.11189	0.91246	−1.26199	−0.80737
1.4	0.83530	−1.26524	0.91831	0.84855	0.14654	0.87636	−1.34213	−0.91831
1.5	0.77503	−1.30263	1.02816	1.01382	0.18656	0.83106	−1.41518	−1.02815
1.6	0.70202	−1.32086	1.13380	1.18632	0.23122	0.77612	−1.47990	−1.13379
1.7	0.61669	−1.31797	1.23219	1.36088	0.27927	0.71183	−1.53540	−1.23218
1.8	0.52033	−1.29295	1.32058	1.53179	0.32906	0.63909	−1.58115	−1.32059
1.9	0.41522	−1.24616	1.39688	1.69343	0.37861	0.55957	−1.61718	−1.39688
2.0	0.30444	−1.17928	1.45979	1.84091	0.42581	0.47549	−1.64405	−1.45979
2.2	0.07998	−0.99776	1.54549	2.08041	0.50555	0.30397	−1.67490	−1.54548
2.4	−0.12589	−0.77951	1.58566	2.23974	0.55677	0.14426	−1.68520	−1.58567
2.6	−0.29362	−0.55639	1.59617	2.32965	0.57544	0.01051	−1.68665	−1.59619
2.8	−0.41438	−0.35193	1.59262	2.37119	0.56380	−0.09103	−1.68717	−1.59265
3.0	−0.48805	−0.17917	1.58606	2.38548	0.52764	−0.16039	−1.69051	−1.58609
3.5	−0.50409	0.09557	1.58435	2.38891	0.37428	−0.22003	−1.71100	−1.58437
4.0	−0.37581	0.18363	1.59979	2.40075	0.20194	−0.18093	−1.73218	−1.59983

The expression for the displacement at the pile top for a pile with a fully consolidated bottom is

$$\overline{y_0} = \delta_{\overline{y_0 M_0}} \overline{M_0} + \delta_{\overline{y_0 Q_0}} \overline{Q_0} \quad (9)$$

$$\overline{y_0} = \delta_{\overline{y_0 M_0}} \overline{M_0} + \delta_{\overline{y_0 Q_0}} \overline{Q_0} + \delta_{\overline{y_0}\ \overline{\varphi_1}} \overline{\varphi_1} + \delta_{\overline{y_0}\ \overline{y_1}} \overline{y_1} \quad (10)$$

where $\delta_{\overline{y_0}\ \overline{y_1}}$ is negative, the horizontal displacement $\overline{y_1}$ at the pile bottom acts against the displacement $\overline{y_0}$ at the pile top (in the same direction).

Compared Equation (9) with Equation (10), it can be found that the larger the displacement at the bottom of the pile, the smaller the displacement at the pile top. When the pile bottom is not fully restrained, the external load is transferred from the pile body to the pile bottom to produce a positive horizontal displacement, and the horizontal displacement of the pile bottom causes the soil body to produce part of the soil resistance to help balance the external load, which results in a decrease in the displacement of the pile body. During the loading of the mechanically connected piles, the pile top displacement will briefly stagnate at the beginning of the second stage, and the overall internal forces in the piles will be similarly redistributed. However, in general, the displacement generated at the bottom of the pile is not in the same direction as that at the top of the pile; in the case of consolidation and articulation, the displacement at the bottom of the pile is in the direction of 0, while in the case of the free boundary, the displacement at the bottom of the pile is in the opposite direction to that at the top of the pile. In this study, the positive direction displacement of the pile bottom is actually the displacement of the whole pile at the connection, and it is very reasonable to divide the pile body as a whole and consider the positive displacement here as the displacement at the bottom of the pile, and then explain it with the theory and further speculate the change of the pile body internal force in the next step. The displacement at the top of the pile can be further reduced if a boundary condition is added at the bottom of the pile to produce a positive displacement at the bottom of the pile. Rigid and flexible piles are temporarily unable to generate positive horizontal displacements under any embedment conditions.

Bringing Equation (7) into Equation (4), the internal forces in the upper pile can be expressed by the unknown $\overline{\varphi_1}$ and $\overline{y_1}$:

$$\begin{cases} \overline{y(z)} = \delta_{\overline{y}\overline{y_1}}\overline{y_1} + \delta_{\overline{y}\overline{\varphi_1}}\overline{\varphi_1} + \delta_{\overline{y}\overline{M_0}}\overline{M_0} + \delta_{\overline{y}\overline{Q_0}}\overline{Q_0} \\ \overline{\varphi(z)} = \delta_{\overline{\varphi}\overline{y_1}}\overline{y_1} + \delta_{\overline{\varphi}\overline{\varphi_1}}\overline{\varphi_1} + \delta_{\overline{\varphi}\overline{M_0}}\overline{M_0} + \delta_{\overline{\varphi}\overline{Q_0}}\overline{Q_0} \\ \overline{M(z)} = \delta_{\overline{M}\overline{y_1}}\overline{y_1} + \delta_{\overline{M}\overline{\varphi_1}}\overline{\varphi_1} + \delta_{\overline{M}\overline{M_0}}\overline{M_0} + \delta_{\overline{M}\overline{Q_0}}\overline{Q_0} \\ \overline{Q(z)} = \delta_{\overline{Q}\overline{y_1}}\overline{y_1} + \delta_{\overline{Q}\overline{\varphi_1}}\overline{\varphi_1} + \delta_{\overline{Q}\overline{M_0}}\overline{M_0} + \delta_{\overline{Q}\overline{Q_0}}\overline{Q_0} \end{cases}, 0 \le L \le L_1 \tag{11}$$

where

$$\begin{cases} \delta_{\overline{y}\overline{y_1}} = \frac{B_1 A_{21} - A_1 B_{21}}{A_{21} B_{11} - A_{11} B_{21}} \\ \delta_{\overline{y}\overline{\varphi_1}} = \frac{A_1 B_{11} - B_1 A_{11}}{A_{21} B_{11} - A_{11} B_{21}} \\ \delta_{\overline{y}\overline{M_0}} = \frac{A_1(B_{21}C_{11} - B_{11}C_{21}) - B_1(A_{21}C_{11} - A_{11}C_{21})}{A_{21} B_{11} - A_{11} B_{21}} + C_1 \\ \delta_{\overline{y}\overline{Q_0}} = \frac{A_1(B_{21}C_{11} - B_{11}C_{21}) - B_1(A_{21}C_{11} - A_{11}C_{21})}{A_{21} B_{11} - A_{11} B_{21}} + D_1 \\ \delta_{\overline{\varphi}\overline{y_1}} = \frac{B_2 A_{21} - A_2 B_{21}}{A_{21} B_{11} - A_{11} B_{21}} \\ \delta_{\overline{\varphi}\overline{\varphi_1}} = \frac{A_2 B_{11} - B_2 A_{11}}{A_{21} B_{11} - A_{11} B_{21}} \\ \delta_{\overline{\varphi}\overline{M_0}} = \frac{A_1(B_{21}C_{11} - B_{11}C_{21}) - B_1(A_{21}C_{11} - A_{11}C_{21})}{A_{21} B_{11} - A_{11} B_{21}} + C_2 \\ \delta_{\overline{\varphi}\overline{Q_0}} = \frac{A_1(B_{21}C_{11} - B_{11}C_{21}) - B_1(A_{21}C_{11} - A_{11}C_{21})}{A_{21} B_{11} - A_{11} B_{21}} + D_2 \\ \delta_{\overline{M}\overline{y_1}} = \frac{B_3 A_{21} - A_2 B_{21}}{A_{21} B_{11} - A_{11} B_{21}} \\ \delta_{\overline{M}\overline{\varphi_1}} = \frac{A_3 B_{11} - B_2 A_{11}}{A_{21} B_{11} - A_{11} B_{21}} \\ \delta_{\overline{M}\overline{M_0}} = \frac{A_1(B_{21}C_{11} - B_{11}C_{21}) - B_1(A_{21}C_{11} - A_{11}C_{21})}{A_{21} B_{11} - A_{11} B_{21}} + C_3 \\ \delta_{\overline{M}\overline{Q_0}} = \frac{A_1(B_{21}C_{11} - B_{11}C_{21}) - B_1(A_{21}C_{11} - A_{11}C_{21})}{A_{21} B_{11} - A_{11} B_{21}} + D_3 \\ \delta_{\overline{Q}\overline{y_1}} = \frac{B_4 A_{21} - A_2 B_{21}}{A_{21} B_{11} - A_{11} B_{21}} \\ \delta_{\overline{Q}\overline{\varphi_1}} = \frac{A_4 B_{11} - B_2 A_{11}}{A_{21} B_{11} - A_{11} B_{21}} \\ \delta_{\overline{Q}\overline{M_0}} = \frac{A_1(B_{21}C_{11} - B_{11}C_{21}) - B_1(A_{21}C_{11} - A_{11}C_{21})}{A_{21} B_{11} - A_{11} B_{21}} + C_4 \\ \delta_{\overline{Q}\overline{Q_0}} = \frac{A_1(B_{21}C_{11} - B_{11}C_{21}) - B_1(A_{21}C_{11} - A_{11}C_{21})}{A_{21} B_{11} - A_{11} B_{21}} + D_4 \end{cases} \tag{12}$$

The 16 coefficients $\delta_{\overline{y}\,\overline{y_1}}$_$\delta_{\overline{QQ_0}}$ in Equation (12) represent the effects of the four boundary conditions $\overline{y_1}$, $\overline{\varphi_1}$, $\overline{M_0}$, and $\overline{Q_0}$ on the internal forces $\overline{y}$, $\overline{\varphi}$, $\overline{M}$, and $\overline{Q}$ in any cross-section, and the combination of their effects is the value of the internal force in that cross-section, which is used for calculating the response of the pile body in the upper pile. When L = 0, the values of $A_1 - D_4$ at $\alpha_1 L = 0$ reveals that the expression is exactly the same as that of $\delta_{\overline{y_0 y_1}}$ – $\delta_{\overline{QQ_0}}$ in Equation (8), thus enabling the validation of the correctness of Equation (12).

When L = L1, the internal force in Section 1 at the bottom of the upper pile can be obtained as Equation (13).

$$\begin{cases} \overline{M}(z=L_1) = \delta_{\overline{M_1 y_1}}\overline{y_1} + \delta_{\overline{M_1 \varphi_1}}\overline{\varphi_1} + \delta_{\overline{M_1 M_0}}\overline{M_0} + \delta_{\overline{M_1 Q_0}}\overline{Q_0} \\ \overline{Q}(z=L_1) = \delta_{\overline{Q_1 y_1}}\overline{y_1} + \delta_{\overline{Q_1 \varphi_1}}\overline{\varphi_1} + \delta_{\overline{Q_1 M_0}}\overline{M_0} + \delta_{\overline{Q_1 Q_0}}\overline{Q_0} \end{cases} \quad (13)$$

where

$$\begin{cases} \delta_{\overline{M_1 y_1}} = \frac{B_{31}A_{21} - A_{31}B_{21}}{A_{21}B_{11} - A_{11}B_{21}} \\ \delta_{\overline{M_1 \varphi_1}} = \frac{A_{31}B_{11} - B_{31}A_{11}}{A_{21}B_{11} - A_{11}B_{21}} \\ \delta_{\overline{M_1 M_0}} = \frac{A_{31}(B_{21}C_{11} - B_{11}C_{21}) - B_{31}(A_{21}C_{11} - A_{11}C_{21})}{A_{21}B_{11} - A_{11}B_{21}} + C_{31} \\ \delta_{\overline{M_1 Q_0}} = \frac{A_{31}(B_{21}D_{11} - B_{11}D_{21}) - B_{31}(A_{21}D_{11} - A_{11}D_{21})}{A_{21}B_{11} - A_{11}B_{21}} + D_{31} \\ \delta_{\overline{Q_1 y_1}} = \frac{B_{41}A_{21} - A_{41}B_{21}}{A_{21}B_{11} - A_{11}B_{21}} \\ \delta_{\overline{Q_1 \varphi_1}} = \frac{A_{41}B_{11} - B_{41}A_{11}}{A_{21}B_{11} - A_{11}B_{21}} \\ \delta_{\overline{Q_1 M_0}} = \frac{A_{41}(B_{21}C_{11} - B_{11}C_{21}) - B_{41}(A_{21}C_{11} - A_{11}C_{21})}{A_{21}B_{11} - A_{11}B_{21}} + C_{41} \\ \delta_{\overline{Q_1 Q_0}} = \frac{A_{41}(B_{21}D_{11} - B_{11}D_{21}) - B_{41}(A_{21}D_{11} - A_{11}D_{21})}{A_{21}B_{11} - A_{11}B_{21}} + D_{41} \end{cases} \quad (14)$$

These eight coefficients in Equation (14) are determined solely from the converted depth $\alpha_1 L_1$ in Section 1 and are given in Table 2.

Table 2. Coefficients of influence of the upper pile boundary on the internal force response in Section 1.

$\alpha_1 L_1$	$\delta_{\overline{M_1 y_1}}$	$\delta_{\overline{M_1 \varphi_1}}$	$\delta_{\overline{M_1 M_0}}$	$\delta_{\overline{M_1 Q_0}}$	$\delta_{\overline{Q_1 y_1}}$	$\delta_{\overline{Q_1 \varphi_1}}$	$\delta_{\overline{Q_1 M_0}}$	$\delta_{\overline{Q_1 Q_0}}$
0.0	0.00000	0.00000	1.00000	0.00000	0.00000	0.00000	0.00000	1.00000
0.1	−0.00017	0.00001	1.00000	0.10000	−0.00500	0.00017	0.00000	1.00000
0.2	−0.00133	0.00014	0.99999	0.20000	−0.02000	0.00133	−0.00007	0.99999
0.3	−0.00450	0.00068	0.99994	0.29999	−0.04500	0.00450	−0.00034	0.99992
0.4	−0.01067	0.00214	0.99974	0.39992	−0.08000	0.01067	−0.00107	0.99966
0.5	−0.02082	0.00520	0.99922	0.49969	−0.12497	0.02082	−0.00260	0.99896
0.6	−0.03597	0.01079	0.99806	0.59909	−0.17989	0.03597	−0.00540	0.99741
0.7	−0.05704	0.01996	0.99581	0.69772	−0.24467	0.05702	−0.00996	0.99442
0.8	−0.08497	0.03398	0.99184	0.79492	−0.31917	0.08497	−0.01699	0.98914
0.9	−0.12055	0.05419	0.98538	0.88977	−0.40312	0.12056	−0.02706	0.98050
1.0	−0.16447	0.08209	0.97540	0.98087	−0.49609	0.16447	−0.04096	0.96720
1.1	−0.21717	0.11910	0.96076	1.06642	−0.59746	0.21717	−0.05937	0.94770
1.2	−0.27877	0.16652	0.94020	1.14417	−0.70631	0.27877	−0.08288	0.92032
1.3	−0.34898	0.22532	0.91246	1.21143	−0.82146	0.34897	−0.11188	0.88337
1.4	−0.42698	0.29602	0.87634	1.26524	−0.94111	0.42873	−0.14923	0.83349
1.5	−0.51144	0.37841	0.83103	1.30260	−1.06438	0.51146	−0.18659	0.77501
1.6	−0.60044	0.47147	0.77611	1.32084	−1.18843	0.60044	−0.23125	0.70203
1.7	−0.69168	0.57337	0.71181	1.31794	−1.31162	0.69173	−0.27937	0.61661
1.8	−0.78259	0.68141	0.63906	1.29292	−1.43221	0.78260	−0.32910	0.52030
1.9	−0.87067	0.79242	0.55953	1.24615	−1.54875	0.87067	−0.37866	0.41524
2.0	−0.95371	0.90297	0.47545	1.17925	−1.66040	0.95371	−0.42584	0.30446
2.2	−1.09885	1.11028	0.30389	0.99772	−1.86835	1.09885	−0.50559	0.08003
2.4	−1.21330	1.28484	0.14416	0.77944	−2.05964	1.21332	−0.55690	−0.12590
2.6	−1.30183	1.42061	0.01039	0.55624	−2.24232	1.30187	−0.57544	−0.29352
2.8	−1.37369	1.52161	−0.09126	0.35182	−2.42376	1.37374	−0.56387	−0.41420
3.0	−1.43774	1.59667	−0.16074	0.17894	−2.60815	1.43780	−0.52806	−0.48816
3.5	−1.59636	1.72536	−0.22055	−0.09593	−3.08241	1.59642	−0.37477	−0.50400
4.0	−1.76223	1.82594	−0.18144	−0.18388	−3.54605	1.76217	−0.20311	−0.37654

Note: The two unknown variables $\overline{y_1}$ and $\overline{\varphi_1}$ in Equation (13) will be eliminated by the internal force case of the lower pile.

2.3. Solving for Internal Forces in the Lower Pile

The force form of the lower pile in the stage of double-pile cooperative work is shown in Figure 4. It is first assumed that the lower pile is through the length and its top is flush with the ground. The lower deflection curve of the through-length pile is ensured to be the same as the actual deflection curve by ensuring that the deflection of the through-length pile in Section 1 is equal to the actual deflection. For a virtual through-length pile with length $(L_1 + L_2)$ and flexural stiffness E_2I_2, the soil resistance on the pile side satisfies Winkle's elastic foundation model; the basic deflection equation is expressed as Equation (15).

$$E_2I_2\frac{d^4y}{dz^4} + mb_0zy = 0,\ (0 \le z \le L_1 + L_2) \tag{15}$$

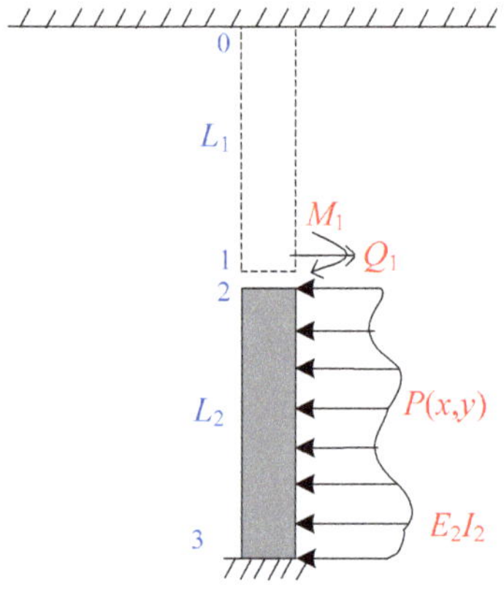

Figure 4. The force form of the lower pile in the stage of double-pile cooperative work.

Let $\alpha_2 = \sqrt[5]{\frac{mb_0}{E_2I_2}}$, then it can obtain

$$\frac{d^4y}{dz^4} + {\alpha_2}^5zy = 0 \tag{16}$$

The solution of the through-length pile deflection equation is obtained by solving the power series method and simplifying it by dimensionless simplification:

$$\begin{cases} \overline{y(z)}^{\Delta} = \overline{y_0}^{\Delta}A_1 + \overline{\varphi_0}^{\Delta}B_1 + \overline{M_0}^{\Delta}C_1 + \overline{Q_0}^{\Delta}D_1 \\ \overline{\varphi(z)}^{\Delta} = \overline{y_0}^{\Delta}A_2 + \overline{\varphi_0}^{\Delta}B_2 + \overline{M_0}^{\Delta}C_2 + \overline{Q_0}^{\Delta}D_2 \\ \overline{M(z)}^{\Delta} = \overline{y_0}^{\Delta}A_3 + \overline{\varphi_0}^{\Delta}B_3 + \overline{M_0}^{\Delta}C_3 + \overline{Q_0}^{\Delta}D_3 \\ \overline{Q(z)}^{\Delta} = \overline{y_0}^{\Delta}A_4 + \overline{\varphi_0}^{\Delta}B_4 + \overline{M_0}^{\Delta}C_4 + \overline{Q_0}^{\Delta}D_4 \end{cases} \tag{17}$$

where $\overline{y_0}^{\Delta}$, $\overline{\varphi_0}^{\Delta}$, $\overline{M_0}^{\Delta}$, and $\overline{Q_0}^{\Delta}$ denote the pile top displacement, angle of rotation, external force and external moment of the virtual through-length pile, respectively. Since the through-length piles are fictional equivalents, $\overline{y_0}^{\Delta}$, $\overline{\varphi_0}^{\Delta}$, $\overline{M_0}^{\Delta}$, and $\overline{Q_0}^{\Delta}$ do not actually exist and are also equivalents, they are distinguished by the angle of rotation notation Δ.

When $L_1 \leq L \leq L_1 + L_2$, the displacements and internal forces of the through-length pile are exactly the same as the actual situation, i.e.,:

$$\begin{cases} \overline{y(z)} = \delta_{\overline{y}^{\Delta}\overline{y}_2^{\Delta}} \cdot \overline{y}_2 + \delta_{\overline{y}^{\Delta}\overline{\varphi}_2^{\Delta}} \cdot \overline{\varphi}_2 \\ \overline{\varphi(z)} = \delta_{\overline{\varphi}^{\Delta}\overline{y}_2^{\Delta}} \cdot \overline{y}_2 + \delta_{\overline{\varphi}^{\Delta}\overline{\varphi}_2^{\Delta}} \cdot \overline{\varphi}_2 \\ \overline{M(z)} = \delta_{\overline{M}^{\Delta}\overline{y}_2^{\Delta}} \cdot \overline{y}_2 + \delta_{\overline{M}^{\Delta}\overline{\varphi}_2^{\Delta}} \cdot \overline{\varphi}_2 \\ \overline{Q(z)} = \delta_{\overline{Q}^{\Delta}\overline{y}_2^{\Delta}} \cdot \overline{y}_2 + \delta_{\overline{Q}^{\Delta}\overline{\varphi}_2^{\Delta}} \cdot \overline{\varphi}_2 \end{cases}, \ L_1 \leq L \leq L_1 + L_2 \tag{18}$$

The solutions of the deflection equation for a connected pile with external loads M_0 and Q_0, upper bending stiffness E_1I_1 and lower bending stiffness E_2I_2 in the stage of double-pile cooperative work are

$$\begin{cases} \overline{y(z)} = \delta_{\overline{yy_1}}\overline{y_1} + \delta_{\overline{y\varphi_1}}\overline{\varphi_1} + \delta_{\overline{yM_0}}\overline{M_0} + \delta_{\overline{yQ_0}}\overline{Q_0} \\ \overline{\varphi(z)} = \delta_{\overline{\varphi y_1}}\overline{y_1} + \delta_{\overline{\varphi\varphi_1}}\overline{\varphi_1} + \delta_{\overline{\varphi M_0}}\overline{M_0} + \delta_{\overline{\varphi Q_0}}\overline{Q_0} \\ \overline{M(z)} = \delta_{\overline{My_1}}\overline{y_1} + \delta_{\overline{M\varphi_1}}\overline{\varphi_1} + \delta_{\overline{MM_0}}\overline{M_0} + \delta_{\overline{MQ_0}}\overline{Q_0} \\ \overline{Q(z)} = \delta_{\overline{Qy_1}}\overline{y_1} + \delta_{\overline{Q\varphi_1}}\overline{\varphi_1} + \delta_{\overline{QM_0}}\overline{M_0} + \delta_{\overline{QQ_0}}\overline{Q_0} \end{cases}, \ 0 \leq L \leq L_1 \tag{19}$$

and

$$\begin{cases} \overline{y(z)} = \delta_{\overline{y}^{\Delta}\overline{y_2}^{\Delta}} \cdot \overline{y_2} + \delta_{\overline{y}^{\Delta}\overline{\varphi_2}^{\Delta}} \cdot \overline{\varphi_2} \\ \overline{\varphi(z)} = \delta_{\overline{\varphi}^{\Delta}\overline{y_2}^{\Delta}} \cdot \overline{y_2} + \delta_{\overline{\varphi}^{\Delta}\overline{\varphi_2}^{\Delta}} \cdot \overline{\varphi_2} \\ \overline{M(z)} = \delta_{\overline{M}^{\Delta}\overline{y_2}^{\Delta}} \cdot \overline{y_2} + \delta_{\overline{M}^{\Delta}\overline{\varphi_2}^{\Delta}} \cdot \overline{\varphi_2} \\ \overline{Q(z)} = \delta_{\overline{Q}^{\Delta}\overline{y_2}^{\Delta}} \cdot \overline{y_2} + \delta_{\overline{Q}^{\Delta}\overline{\varphi_2}^{\Delta}} \cdot \overline{\varphi_2} \end{cases} \ L_1 \leq L \leq L_1 + L_2 \tag{20}$$

where $\overline{y_1}, \overline{\varphi_1}$, $\overline{y_2}$ and $\overline{\varphi_2}$ are unknown intermediate variables.

Assuming that the connectors do not deform during normal operation, it can be ensured that Sections 2 and 3 maintain the same horizontal displacement, and the difference in the angle of rotation is the structural angle of rotation φ^0. Then, the geometric conditions can be summarised as

$$\begin{cases} \overline{y_1} = \overline{y_2} \\ \overline{\varphi_1} + \overline{\varphi^0} = \overline{\varphi_2} \end{cases} \tag{21}$$

The connection is working with 100 percent transfer of internal forces to the upper and lower piles, so there is no weakening of internal forces in Sections 2 and 3, then the equilibrium conditions can be summarised as

$$\begin{cases} \overline{M_1} = \overline{M_2} \\ \overline{Q_1} = \overline{Q_2} \end{cases} \tag{22}$$

2.4. Solution for Internal Force Equations

Bringing Equations (21) and (22) into Equations (11) and (18) yields a system of equations:

$$\begin{cases} \delta_{\overline{M_1y_1}}\overline{y_1} + \delta_{\overline{M_1}\cdot\overline{\varphi_1}}\overline{\varphi_1} + \delta_{\overline{M_1}\cdot\overline{M_0}}\overline{M_0} + \delta_{\overline{M_1}\cdot\overline{Q_0}}\overline{Q_0} = \delta_{\overline{M_2}^{\Delta}\overline{y_2}^{\Delta}} \cdot \overline{y_1} + \delta_{\overline{M_2}^{\Delta}\overline{\varphi_2}^{\Delta}} \cdot \left(\overline{\varphi_1} + \overline{\varphi^0}\right) \\ \delta_{\overline{Q_1}\cdot\overline{y_1}}\overline{y_1} + \delta_{\overline{Q_1}\cdot\overline{\varphi_1}}\overline{\varphi_1} + \delta_{\overline{Q_1}\cdot\overline{M_0}}\overline{M_0} + \delta_{\overline{Q_1Q_0}}\overline{Q_0} = \delta_{\overline{Q_2}^{\Delta}\overline{y_2}^{\Delta}} \cdot \overline{y_1} + \delta_{\overline{Q_2}^{\Delta}\overline{\varphi_2}^{\Delta}} \cdot \left(\overline{\varphi_1} + \overline{\varphi^0}\right) \end{cases} \tag{23}$$

The solution for Equation (23) is

$$\begin{cases} \overline{y_1} = \delta_{\overline{y_1\varphi^0}} \cdot \overline{\varphi^0} + \delta_{\overline{y_1M_0}} \cdot \overline{M_0} + \delta_{\overline{y_1Q_0}} \cdot \overline{Q_0} \\ \overline{\varphi_1} = \delta_{\overline{\varphi_1\varphi^0}} \cdot \overline{\varphi^0} + \delta_{\overline{\varphi_1M_0}} \cdot \overline{M_0} + \delta_{\overline{\varphi_1Q_0}} \cdot \overline{Q_0} \end{cases} \tag{24}$$

where

$$\begin{cases}
\delta_{\overline{y_1}\overline{\varphi^0}} = \dfrac{\left(\delta_{\overline{Q_1}\overline{\varphi_1}}-\delta_{\overline{Q_2}\Delta\overline{\varphi_2}\Delta}\right)\delta_{\overline{M_2}\Delta\overline{\varphi_2}\Delta}-\left(\delta_{\overline{M_1}\overline{\varphi_1}}-\delta_{\overline{M_2}\Delta\overline{\varphi_2}\Delta}\right)\delta_{\overline{Q_2}\Delta\overline{\varphi_2}\Delta}}{\left(\delta_{\overline{M_1}\overline{y_1}}-\delta_{\overline{M_2}\Delta\overline{y_2}\Delta}\right)\left(\delta_{\overline{Q_1}\overline{\varphi_1}}-\delta_{\overline{Q_2}\Delta\overline{\varphi_2}\Delta}\right)-\left(\delta_{\overline{Q_1}\overline{y_1}}-\delta_{\overline{Q_2}\Delta\overline{y_2}\Delta}\right)\left(\delta_{\overline{M_1}\overline{\varphi_1}}-\delta_{\overline{M_2}\Delta\overline{\varphi_2}\Delta}\right)} \\
\delta_{\overline{y_1}\overline{M_0}} = \dfrac{-\left(\delta_{\overline{Q_1}\overline{\varphi_1}}-\delta_{\overline{Q_2}\Delta\overline{\varphi_2}\Delta}\right)\delta_{\overline{M_1}\overline{M_0}}+\left(\delta_{\overline{M_1}\overline{\varphi_1}}-\delta_{\overline{M_2}\Delta\overline{\varphi_2}\Delta}\right)\delta_{\overline{Q_1}\overline{M_0}}}{\left(\delta_{\overline{M_1}\overline{y_1}}-\delta_{\overline{M_2}\Delta\overline{y_2}\Delta}\right)\left(\delta_{\overline{Q_1}\overline{\varphi_1}}-\delta_{\overline{Q_2}\Delta\overline{\varphi_2}\Delta}\right)-\left(\delta_{\overline{Q_1}\overline{y_1}}-\delta_{\overline{Q_2}\Delta\overline{y_2}\Delta}\right)\left(\delta_{\overline{M_1}\overline{\varphi_1}}-\delta_{\overline{M_2}\Delta\overline{\varphi_2}\Delta}\right)} \\
\delta_{\overline{y_1}\overline{Q_0}} = \dfrac{-\left(\delta_{\overline{Q_1}\overline{\varphi_1}}-\delta_{\overline{Q_2}\Delta\overline{\varphi_2}\Delta}\right)\delta_{\overline{M_1}\overline{Q_0}}+\left(\delta_{\overline{M_1}\overline{\varphi_1}}-\delta_{\overline{M_2}\Delta\overline{\varphi_2}\Delta}\right)\delta_{\overline{Q_1}\overline{\varphi_0}}}{\left(\delta_{\overline{M_1}\overline{y_1}}-\delta_{\overline{M_2}\Delta\overline{\varphi_1}\Delta}\right)\left(\delta_{\overline{Q_1}\overline{\varphi_1}}-\delta_{\overline{Q_2}\Delta\overline{\varphi_2}\Delta}\right)-\left(\delta_{\overline{Q_1}\overline{y_1}}-\delta_{\overline{Q_2}\Delta\overline{y_2}\Delta}\right)\left(\delta_{\overline{M_1}\overline{\varphi_1}}-\delta_{\overline{M_2}\Delta\overline{\varphi_2}\Delta}\right)} \\
\delta_{\overline{\varphi_1}\overline{\varphi^0}} = \dfrac{-\left(\delta_{\overline{Q_1}\overline{y_1}}-\delta_{\overline{Q_2}\Delta\overline{y_2}\Delta}\right)\delta_{\overline{M_2}\Delta\overline{\varphi_2}\Delta}+\left(\delta_{\overline{M_1}\overline{y_1}}-\delta_{\overline{M_2}\Delta\overline{y_2}\Delta}\right)\delta_{\overline{Q_2}\Delta\overline{\varphi_2}\Delta}}{\left(\delta_{\overline{M_1}\overline{y_1}}-\delta_{\overline{M_2}\Delta\overline{y_2}\Delta}\right)\left(\delta_{\overline{Q_1}\overline{\varphi_1}}-\delta_{\overline{Q_2}\Delta\overline{\varphi_2}\Delta}\right)-\left(\delta_{\overline{Q_1}\overline{\varphi_1}}-\delta_{\overline{Q_2}\Delta\overline{y_2}\Delta}\right)\left(\delta_{\overline{M_1}\overline{\varphi_1}}-\delta_{\overline{M_2}\Delta\overline{\varphi_2}\Delta}\right)} \\
\delta_{\overline{\varphi_1}\overline{M_0}} = \dfrac{\left(\delta_{\overline{Q_1}\overline{y_1}}-\delta_{\overline{Q_2}\Delta\overline{y_2}\Delta}\right)\delta_{\overline{M_1}\overline{Q_0}}-\left(\delta_{\overline{M_1}\overline{y_1}}-\delta_{\overline{M_2}\Delta\overline{y_2}\Delta}\right)\delta_{\overline{Q_1}\overline{M_0}}}{\left(\delta_{\overline{M_1}\overline{y_1}}-\delta_{\overline{M_2}\Delta\overline{y_2}\Delta}\right)\left(\delta_{\overline{Q_1}\overline{\varphi_1}}-\delta_{\overline{Q_2}\Delta\overline{\varphi_2}\Delta}\right)-\left(\delta_{\overline{Q_1}\overline{y_1}}-\delta_{\overline{Q_2}\Delta\overline{y_2}\Delta}\right)\left(\delta_{\overline{M_1}\overline{\varphi_1}}-\delta_{\overline{M_2}\Delta\overline{\varphi_2}\Delta}\right)} \\
\delta_{\overline{\varphi_1}\overline{Q_0}} = \dfrac{\left(\delta_{\overline{Q_1}\overline{y_1}}-\delta_{\overline{Q_2}\Delta\overline{y_2}\Delta}\right)\delta_{\overline{M_1}\overline{Q_0}}-\left(\delta_{\overline{Q_1}\overline{y_1}}-\delta_{\overline{Q_2}\Delta\overline{y_2}\Delta}\right)\delta_{\overline{Q_1}\overline{Q_0}}}{\left(\delta_{\overline{M_1}\overline{y_1}}-\delta_{\overline{M_2}\Delta\overline{y_2}\Delta}\right)\left(\delta_{\overline{Q_1}\overline{\varphi_1}}-\delta_{\overline{Q_2}\Delta\overline{\varphi_2}\Delta}\right)-\left(\delta_{\overline{Q_1}\overline{y_1}}-\delta_{\overline{Q_2}\Delta\overline{y_2}\Delta}\right)\left(\delta_{\overline{M_1}\overline{\varphi_1}}-\delta_{\overline{M_2}\Delta\overline{\varphi_2}\Delta}\right)}
\end{cases} \tag{25}$$

When the bending stiffness of the upper and lower piles is the same, $\alpha_1 = \alpha_2$. The internal force response coefficients for mechanically connected piles with fixed constraints at the pile bottom in Section 1 are listed in Table 3 and are controlled by $\alpha_1 L_1$. The internal force response coefficients for mechanically connected piles with fixed constraints in Section 2 are listed in Table 4 and are controlled by $\alpha_2 L_1$.

Table 3. Internal force response coefficients for mechanically connected piles with fixed constraints are in Section 1.

$\alpha_1 L_1$	$\delta_{\overline{M_1}\overline{y_1}}$	$\delta_{\overline{M_1}\overline{\varphi_1}}$	$\delta_{\overline{M_1}\overline{M_0}}$	$\delta_{\overline{Q_1}\overline{M_0}}$	$\delta_{\overline{Q_1}\overline{y_1}}$	$\delta_{\overline{Q_1}\overline{\varphi_1}}$	$\delta_{\overline{Q_1}\overline{M_0}}$	$\delta_{\overline{Q_1}\overline{Q_0}}$
0.0	0.00000	0.00000	1.00000	0.00000	0.00000	0.00000	0.00000	1.00000
0.1	−0.00017	0.00001	1.00000	0.10000	−0.00500	0.00017	0.00000	1.00000
0.2	−0.00133	0.00014	0.99999	0.20000	−0.02000	0.00133	−0.00007	0.99999
0.3	−0.00450	0.00068	0.99994	0.29999	−0.04500	0.00450	−0.00034	0.99992
0.4	−0.01067	0.00214	0.99974	0.39992	−0.08000	0.01067	−0.00107	0.99966
0.5	−0.02082	0.00520	0.99922	0.49969	−0.12497	0.02082	−0.00260	0.99896
0.6	−0.03597	0.01079	0.99806	0.59909	−0.17989	0.03597	−0.00540	0.99741
0.7	−0.05704	0.01996	0.99581	0.69772	−0.24467	0.05702	−0.00996	0.99442
0.8	−0.08497	0.03398	0.99184	0.79492	−0.31917	0.08497	−0.01699	0.98914
0.9	−0.12055	0.05419	0.98538	0.88977	−0.40312	0.12056	−0.02706	0.98050
1.0	−0.16447	0.08209	0.97540	0.98087	−0.49609	0.16447	−0.04096	0.96720
1.1	−0.21717	0.11910	0.96076	1.06642	−0.59746	0.21717	−0.05937	0.94770
1.2	−0.27877	0.16652	0.94020	1.14417	−0.70631	0.27877	−0.08288	0.92032
1.3	−0.34898	0.22532	0.91246	1.21143	−0.82146	0.34897	−0.11188	0.88337
1.4	−0.42698	0.29602	0.87634	1.26524	−0.94111	0.42873	−0.14923	0.83349
1.5	−0.51144	0.37841	0.83103	1.30260	−1.06438	0.51146	−0.18659	0.77501
1.6	−0.60044	0.47147	0.77611	1.32084	−1.18843	0.60044	−0.23125	0.70203
1.7	−0.69168	0.57337	0.71181	1.31794	−1.31162	0.69173	−0.27937	0.61661
1.8	−0.78259	0.68141	0.63906	1.29292	−1.43221	0.78260	−0.32910	0.52030
1.9	−0.87067	0.79242	0.55953	1.24615	−1.54875	0.87067	−0.37866	0.41524
2.0	−0.95371	0.90297	0.47545	1.17925	−1.66040	0.95371	−0.42584	0.30446
2.2	−1.09885	1.11028	0.30389	0.99772	−1.86835	1.09885	−0.50559	0.08003
2.4	−1.21330	1.28484	0.14416	0.77944	−2.05964	1.21332	−0.55690	−0.12590

Table 3. *Cont.*

$\alpha_1 L_1$	$\delta_{\overline{M_1}\overline{y_1}}$	$\delta_{\overline{M_1}\overline{\varphi_1}}$	$\delta_{\overline{M_1 M_0}}$	$\delta_{\overline{Q_1 M_0}}$	$\delta_{\overline{Q_1}\overline{y_1}}$	$\delta_{\overline{Q_1}\overline{\varphi_1}}$	$\delta_{\overline{Q_1 M_0}}$	$\delta_{\overline{Q_1 Q_0}}$
2.6	−1.30183	1.42061	0.01039	0.55624	−2.24232	1.30187	−0.57544	−0.29352
2.8	−1.37369	1.52161	−0.09126	0.35182	−2.42376	1.37374	−0.56387	−0.41420
3.0	−1.43774	1.59667	−0.16074	0.17894	−2.60815	1.43780	−0.52806	−0.48816
3.5	−1.59636	1.72536	−0.22055	−0.09593	−3.08241	1.59642	−0.37477	−0.50400
4.0	−1.76223	1.82594	−0.18144	−0.18388	−3.54605	1.76217	−0.20311	−0.37654

Table 4. Internal force response coefficients for mechanically connected piles with fixed constraints are in Section 2.

$\alpha_2 L_1$	$\delta_{\overline{y_2}^{\Delta}\overline{y_2}^{\Delta}}$	$\delta_{\overline{y_2}^{\Delta}\overline{\varphi_2}^{\Delta}}$	$\delta_{\overline{\varphi_2}^{\Delta}\overline{y_2}^{\Delta}}$	$\delta_{\overline{\varphi_2}^{\Delta}\overline{\varphi_2}^{\Delta}}$	$\delta_{\overline{M_2}^{\Delta}\overline{y_2}^{\Delta}}$	$\delta_{\overline{M_2}^{\Delta}\overline{\varphi_2}^{\Delta}}$	$\delta_{\overline{Q_2}^{\Delta}\overline{y_2}^{\Delta}}$	$\delta_{\overline{Q_2}^{\Delta}\overline{\varphi_2}^{\Delta}}$
0.00	1.00000	0.00000	0.00000	1.00000	−1.00043	−1.50127	1.08320	1.00040
0.10	1.00000	0.00000	0.00000	1.00000	−1.04172	−1.52620	1.18248	1.04169
0.20	1.00000	0.00000	0.00000	1.00000	−1.08195	−1.55048	1.28027	1.08191
0.30	1.00000	0.00000	0.00000	1.00000	−1.12123	−1.57428	1.37664	1.12118
0.40	1.00000	0.00000	0.00000	1.00000	−1.15968	−1.59770	1.47172	1.15963
0.50	1.00000	0.00000	0.00000	1.00000	−1.19756	−1.62108	1.56568	1.19748
0.60	1.00000	0.00000	0.00000	1.00000	−1.23490	−1.64438	1.65859	1.23483
0.70	1.00000	0.00000	0.00000	1.00000	−1.27199	−1.66797	1.75067	1.27189
0.80	1.00000	0.00000	0.00000	1.00000	−1.30922	−1.69208	1.84234	1.30915
0.90	1.00000	0.00000	0.00000	1.00000	−1.34678	−1.71709	1.93364	1.34669
1.00	1.00000	0.00000	0.00000	1.00000	−1.38510	−1.74316	2.02515	1.38496
1.10	1.00000	0.00000	0.00000	1.00000	−1.42472	−1.77088	2.11736	1.42457
1.20	1.00000	0.00000	0.00000	1.00000	−1.46640	−1.80061	2.21136	1.46623
1.30	1.00000	0.00000	0.00000	1.00000	−1.51100	−1.83307	2.30797	1.51072
1.40	1.00000	0.00000	0.00000	1.00000	−1.55913	−1.86853	2.41837	1.56881
1.50	1.00000	0.00000	0.00000	1.00000	−1.61246	−1.90808	2.51488	1.61224
1.60	1.00000	0.00000	0.00000	1.00000	−1.67204	−1.95206	2.62911	1.67169
1.70	1.00000	0.00000	0.00000	1.00000	−1.74021	−2.00182	2.75555	1.74010
1.80	1.00000	0.00000	0.00000	1.00000	−1.81851	−2.05791	2.89636	1.81822
1.90	1.00000	0.00000	0.00000	1.00000	−1.91035	−2.12190	3.05858	1.91000
2.00	1.00000	0.00000	0.00000	1.00000	−2.01870	−2.19469	3.24874	2.01813
2.20	1.00000	0.00000	0.00000	1.00000	−2.30529	−2.37501	3.75800	2.30421
2.40	1.00000	0.00000	0.00000	1.00000	−2.72633	−2.61420	4.55036	2.72609
2.60	1.00000	0.00000	0.00000	1.00000	−3.37807	−2.93947	5.89616	3.37619
2.80	1.00000	0.00000	0.00000	1.00000	−4.41634	−3.38854	8.34326	4.41593
3.00	1.00000	0.00000	0.00000	1.00000	−6.19144	−4.03441	13.26690	6.19016
3.50	1.00000	0.00000	0.00000	1.00000	−24.13768	−7.97444	97.18629	23.94553
4.00	1.00000	0.00000	0.00000	1.00000	——	——	——	——

Substituting the values of the 16 coefficients from Tables 3 and 4 into Equations (24) and (25), it can obtain the intermediate variables $\overline{y_1}$ and $\overline{\varphi_1}$ in Section 1. Bringing $\overline{y_1}$ and $\overline{\varphi_1}$ into the geometric condition, Equation (21) yields intermediate variables $\overline{y_2}$ and $\overline{\varphi_2}$ at Section 2. Then, bringing the above intermediate variables $\overline{y_1}$ and $\overline{\varphi_1}$ into Equations (19) and (20), respectively, the pile displacement and internal force response of the upper and lower piles can be obtained, which is the expression of the overall pile response of the mechanically connected pile with fixed constraints at the pile bottom.

3. Finite Element Model Calculation

There are two commonly used finite element simulation modelling approaches for pile-soil interaction. One is to build a pile-soil model based on the actual situation of pile-soil interaction after simplification and assumptions. The other is to model pile springs from elastic foundation beams, using springs and damping to replace the soil around the pile. The two models have their own focus and are widely used in engineering practice. Using ABAQUS software, the pile-soil model is chosen to simulate the actual situation in the project in order to better simulate the constitutive model of the soil and the interaction between the soil and the pile body.

3.1. Basic Parameters

The relevant design parameters for the test pile are selected from the reference [21]. The numerical analysis model is established based on the test results in the reference [21]. The length of the test pile in is 18 m, of which the pile in-ground length is 17 m, and the free section length is 1 m. The pile diameter is 0.6 m. The top of the pile can be rotated freely. Nine HRB400 steel bars are used for longitudinal reinforcement. The stirrup is an HPB235 steel bar with a diameter of 12 mm and a spacing of 200 mm. The concrete strength of the pile is C30. The cover concrete thickness of the pile is 50 mm. In the area of the test pile, there are three layers of groundwater, including the stagnant water layer, upper pressurised water layer and lower pressurised water layer, and the groundwater level is located in the range of 1–2.5 m below the ground surface. The test adopts a unidirectional multi-cycle loading mode, and the maximum load of the test is 150 kN, which is divided into 10 levels for loading. The loading load of each level is 15 kN. The load loading point was 1 m from the ground, and the test pile was buried in a non-rocky foundation. According to the test result, the total displacement of the top of the pile under the maximum load of the test was 5.06 mm, and the angle of rotation was 0.18°. The physical-mechanical parameters of the soil layers are listed in Table 5. The mechanical parameters of the piles and reinforcement are listed in Table 6.

Table 5. Physical-mechanical parameters of the soil layer.

Soil Layer	T (m)	D (kg/m^3)	E (MPa)	v	Mohr-Coulomb Model Parameter	
					φ (°)	C (kPa)
Fill soil	1.90	2000.00	16.00	0.30	24.00	7.00
Pulverized soil	9.20	2040.00	18.00	0.23	26.00	16.00
Sandy soil1	2.20	2060.00	50.00	0.20	34.00	0.00
Clay 1	8.50	1970.00	20.00	0.25	22.00	16.00
Sandy soil 2	3.70	2100.00	50.00	0.20	36.00	0.00
Gravel	2.40	2150.00	80.00	0.20	38.00	0.00
Clay 2	14.7	1990	20	0.25	22	16

Note: T is the thickness of the soil layer; D is the density; E is the modulus of elasticity; v is the Poisson's ratio.

Table 6. Finite element simulation of pile and reinforcement physico-mechanical parameters.

Material	D (kg/m^3)	E (MPa)	v	Diameter (cm)	Length (m)
Concrete pile	2400.00	30,000.00	0.20	60.00	18.00
Longitudinal reinforcement	7850.00	200,000.00	0.30	2.50	18.00
Stirrup	7850.00	200,000.00	0.30	1.00	1.53

3.2. Verification of the Correctness of the m-Method

Based on the relevant information in reference [21], ordinary piles and mechanically connected piles with different boundary conditions are established. The finite element model is established using the mechanical parameters of the test piles and reinforcement [21]. The connector is set up as a hinged support with a limit. The rotation angle limit is taken as $\varphi^0 = 1000 \times 10^{-5}$. The material constitutive models for numerical simulations are shown in Table 7. The bearing capacity problem of the pile involves pile-soil contact. There are two parts of the pile-soil contact: pile perimeter and pile bottom. The master-slave contact algorithm in face-to-face contact is chosen. The surface of the pile serves as the master face, and the face in the soil body where mutual contact with the pile occurs serves as the slave face. The face-to-face discretisation method is used for the face discretisation. The mechanical behaviour between the contacting faces usually consists of two components, i.e., the normal and tangential action of the contact. According to the characteristics of the model, a hard contact model is chosen for the normal action, and a penalty stiffness algorithm for the friction model is used for the tangential action. The boundary conditions at the bottom of the pile are embedded for ordinary through-length piles and mechanically connected piles.

Table 7. Finite element simulation for each material principal structure selection.

Material	Constitutive Model	Math Expression
Soil	Mohr-Coulomb model	$\tau = c - \sigma \tan \varphi$
Concrete	Ideal elastic model	E_c = 16 MPa
Reinforcement	Ideal elastic-plastic model	$\sigma_s = \begin{cases} E_s\varepsilon_s & \varepsilon_s < \varepsilon_y \\ f_y & \varepsilon_s \geq \varepsilon_y \end{cases}$
Mechanically connection	Connector	——

Note: τ is the shear strength of the soil, σ is the positive stress at a point of the soil, c is the cohesion of the soil, and φ is the angle of internal friction; E_c is the concrete modulus of elasticity; E_s is the reinforcement modulus of elasticity; σ_s is the reinforcement modulus of elasticity; f_y is the reinforcement yield strength.

The established pile-soil model is shown in Figure 5. The soil part of the model is a cylinder. The diameter of the circle is 30 m (50 times the diameter of the pile). The height of the cylinder is 38 m (2.1 times the length of the pile). Reserve pile position in the middle of the soil body. The size of the pile position is the same as the pile, and the depth of the pile position is 17 m. The dimensions and boundaries of the established pile model are the same as those of the test pile.

(a) (b) (c)

Figure 5. Mesh diagram for the finite element model. (**a**) Soil. (**b**) Foundation pile. (**c**) Reinforcing steel.

The finite element results of the test pile are shown in Figure 6. It can be seen that the simulated displacements and angles of rotation at the top of the pile are basically consistent with the test results, with the maximum displacements and angles of rotation at the pile top. Table 8 shows a comparison of the simulated and tested horizontal displacements and angles of rotation at the pile top. The relative errors of the displacements and angles of rotation are both less than 10%, indicating that the established model has good accuracy.

Table 8. A comparison of the simulated and tested horizontal displacements and angles of rotation at the pile top.

Result	Displacement (mm)		Angle of Rotation (10^{-5} r)	
	Value	Relative Error (%)	Value	Relative Error (%)
Numerical simulation	5.23	3.25	347.40	9.56
Test	5.06	0.00	314.20	0.00

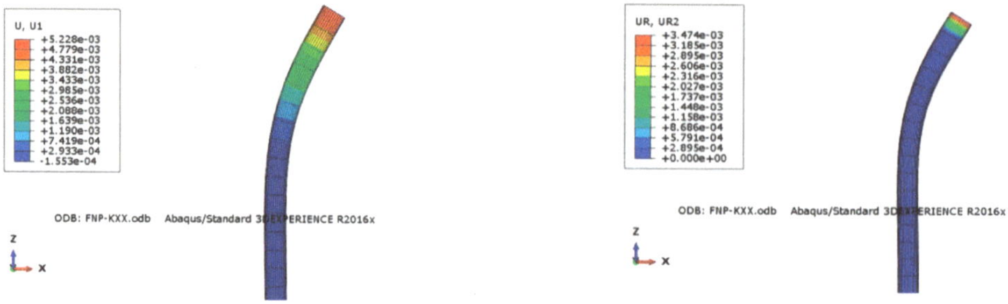

(**a**) (**b**)

Figure 6. Finite element results of the test pile. (**a**) Displacements. (**b**) Angles of rotation.

3.3. Comparison of Theoretical and Numerical Simulation Results

According to the deformation coefficient of the pile $\alpha = \sqrt[5]{\frac{mb_1}{EI}} = \sqrt[5]{\frac{12.68\times10^6\times1.26}{1.5268\times10^8}} = 0.6367\ \mathrm{m}^{-1}$, it can obtain $\alpha L = 0.6367 \times 18 = 11.4606 \geq 4.0$. Combining the pile response coefficients and external load effects, the displacement and internal force responses of the embedded ordinary pile body can be obtained in Table 9.

Table 9. Displacement and internal force response of embedded ordinary pile.

z (m)z (m)	αz	y (mm)	φ (10^5r)	M (kN·mm)	Q (kN)
0.000	0.00	24.37	−639.93	0.00	400.00
0.157	0.10	22.75	−637.93	62.58	395.41
0.314	0.20	21.13	−632.00	123.77	382.50
0.471	0.30	19.54	−622.23	182.36	362.51
0.628	0.40	17.98	−608.85	237.33	336.69
0.785	0.50	16.45	−592.10	287.88	306.22
0.942	0.60	14.98	−572.29	333.34	272.21
1.099	0.70	13.55	−549.77	373.26	235.75
1.256	0.80	12.19	−524.90	407.28	197.75
1.414	0.90	10.89	−498.03	435.34	159.16
1.571	1.00	9.66	−469.60	457.31	120.75
1.728	1.10	8.51	−439.92	473.31	83.22
1.885	1.20	7.43	−409.43	483.52	47.18
2.042	1.30	6.43	−378.47	488.23	13.10
2.199	1.40	5.51	−347.38	487.76	−19.88
2.356	1.50	4.67	−316.45	482.52	−47.62
2.513	1.60	3.90	−286.02	472.96	−73.73
2.670	1.70	3.22	−256.31	459.52	−96.85
2.827	1.80	2.60	−227.57	442.70	−116.81
2.984	1.90	2.06	−199.99	422.98	−133.74
3.141	2.00	1.59	−173.76	400.84	−147.66
3.455	2.20	0.83	−125.81	351.09	−167.22
3.769	2.40	0.30	−84.51	296.80	−177.044
4.084	2.60	−0.04	−50.26	240.62	−179.22
4.398	2.80	−0.22	−23.18	184.68	−176.11
4.712	3.00	−0.29	−3.11	130.28	−170.09
5.497	3.50	−0.15	17.90	3.29	−154.53
6.282	4.00	0.00	0.00	0.00	−100.00

Figure 7 compares the theoretical and numerical simulation displacements and internal force responses of the embedded ordinary pile. It can be seen from Figure 7 that:

(1) The theoretical value of pile displacement calculated by the m-method agrees well with the numerical simulation results. The maximum displacement occurs at the

pile top. The theoretical value calculated by the m-method is 24.37 mm, while the numerical simulation value is 24.96 mm, with a relative error of 2.42%.

(2) The maximum angle of rotation of the pile occurs at the top of the pile. The theoretical maximum angle of rotation is -639.93×10^{-5} r, while the numerical simulation maximum angle of rotation is -674.67×10^{-5} r with a relative error of 5.43%.

(3) The maximum bending moment of the pile occurs at 2–4 m from the top position of the pile. The theoretical maximum value of the bending moment is 488.23 kN·m at 2.042 m depth, and the simulated maximum value of the bending moment is 460.01 kN·m at 2.50 m depth, with a relative error of 6.13%. The theoretical point of contra-flexure is at 6.28 m, and the numerical simulation point of contra-flexure is at 7.00 m.

(4) The positive shear is 400 kN at the location of the top of the pile. The theoretical maximum value of the negative shear is −179.22 kN at the depth of 4.084 m. The simulated maximum negative shear is −190.35 kN at a depth of 4.5 m, with a relative error of 6.21%. The m-method calculates the shear as zero at 2.199 m and the numerical simulation shear as zero at 2.5 m.

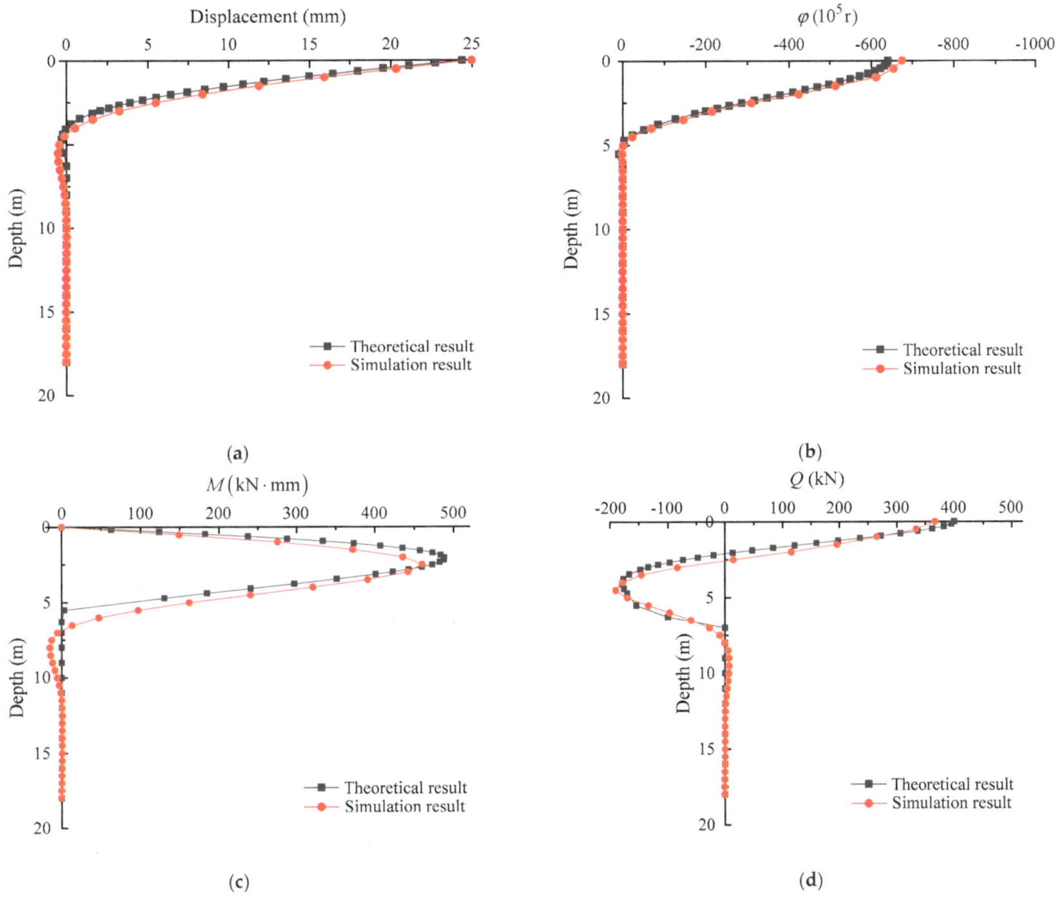

Figure 7. A comparison of theoretical and numerical simulation displacements and internal force responses of embedded ordinary pile. (**a**) Variation curve of displacement with depth. (**b**) Variation curve of the angle of rotation with depth. (**c**) Variation of shear with depth. (**d**) Variation of bending moment with depth.

According to the above analysis, it can be found that the theoretical and numerical simulation displacements and internal force responses of the embedded ordinary pile are basically consistent, which can prove the correctness of the theoretical calculation by the m-method.

Segmental calculation of embedded mechanically connected piles considering upper pile length $L_1 = 3$ m and lower pile length $L_2 = 15$ m. The deformation coefficient of the pile $\alpha_1 = \alpha_2 = \sqrt[5]{\frac{mb_1}{EI}} = \sqrt[5]{\frac{12.68\times10^6\times1.26}{1.5268\times10^8}} = 0.6367\ \mathrm{m}^{-1}$ and $\alpha_1 L_1 = 0.6367 \times 3 = 1.91$.

When, $\alpha_2 L_1 = 1.91$, the coefficients in Section 2 are $\begin{cases} \delta_{\overline{M_2}^{\Delta}\overline{y_2}^{\Delta}} = -1.91038 \\ \delta_{\overline{M_2}^{\Delta}\overline{\varphi_2}^{\Delta}} = -2.12188 \\ \delta_{\overline{Q_2}^{\Delta}\overline{y_2}^{\Delta}} = 3.05859 \\ \delta_{\overline{Q_2}^{\Delta}\overline{\varphi_2}^{\Delta}} = 1.90997 \end{cases}$.

The above coefficients are brought into Equations (17)–(20), which can be obtain:

$$\begin{cases} \delta_{\overline{y_1}^{\Delta}\overline{\varphi_0}^{\Delta}} = -0.27222 \\ \delta_{\overline{y_1}^{\Delta}\overline{M_0}^{\Delta}} = -0.04228 \\ \delta_{\overline{y_1}^{\Delta}\overline{Q_0}^{\Delta}} = 0.20291 \\ \delta_{\overline{y_1}^{\Delta}\overline{\varphi_0}^{\Delta}} = -0.63098 \\ \delta_{\overline{\varphi_1}^{\Delta}\overline{M_0}^{\Delta}} = -0.17691 \\ \delta_{\overline{\varphi_1}^{\Delta}\overline{Q_0}^{\Delta}} = -0.49999 \end{cases}.$$

When $\overline{M} = 0$, $\overline{Q} = 646.26129 \times 10^{-5}$ and $\bar{\varphi} = 1000 \times 10^{-5}$r, $\overline{y_1} = -141.08315$ and $\overline{\varphi_1} = -954.10114$. From the geometric relation, it can obtain $\overline{y_2} = \overline{y_1} = -141.08315$ and $\overline{\varphi_1} = \overline{\varphi_1} + 1000 = 45.89886 \times 10^{-5}$r. Then, the displacement and internal force response of the mechanically connected pile are obtained, as shown in Table 10.

Table 10. Displacement and internal force response of the embedded pile using mechanical connections.

z (m)z (m)	αz	y (mm)	φ (10^{-5}r)	M (kN·mm)	Q (kN)
0.0000	0.00	34.94	−1490.05	0.00	400.00
0.1570	0.10	32.60	−1486.82	62.47	393.42
0.3140	0.20	30.26	−1477.28	122.96	374.92
0.4710	0.30	27.97	−1461.67	179.71	346.30
0.6280	0.40	25.69	−1440.48	231.29	309.37
0.7850	0.50	23.449	−1414.29	276.56	265.90
0.9420	0.60	21.25	−1383.819	314.57	217.54
1.099	0.70	19.10	−1349.839	344.72	165.97
1.2560	0.80	17.01	−1313.19	366.57	112.67
1.4140	0.90	14.98	−1274.709	380.09	59.18
1.5710	1.00	13.00	−1235.30	385.26	6.89
1.7280	1.10	11.10	−1195.70	382.40	−42.87
1.8850	1.20	9.25	−1156.86	371.98	−88.88
2.0420	1.30	7.46	−1119.42	354.73	−129.95
2.1990	1.40	5.73	−1084.08	331.46	−166.81
2.3560	1.50	4.05	−1051.37	303.27	−192.86
2.5130	1.60	2.43	−1021.80	271.32	−212.56
2.6700	1.70	0.84	−995.63	236.98	−223.17
2.8270	1.80	−0.70	−973.07	201.75	−223.54
2.9840	1.90	−2.22	−954.10	167.33	−212.82
2.9840	1.90	−2.22	45.90	167.33	−212.82
3.1410	2.00	−2.13	61.43	135.22	−196.11
3.4550	2.20	−1.90	83.30	78.88	−162.69
3.7690	2.40	−1.62	94.60	32.86	−130.77
4.0840	2.60	−1.32	97.49	−3.58	−101.88

Table 10. *Cont.*

z (m)z (m)	αz	y (mm)	φ (10^{-5}r)	M (kN·mm)	Q (kN)
4.3980	2.80	−1.01	93.75	−31.64	−77.166
4.7120	3.00	−0.73	85.02	−52.62	−57.38
5.4970	3.50	−0.20	48.88	−84.74	−29.67
6.2820	4.00	0.00	0.00	−60.00	−25.04

Figure 8 compares the theoretical and numerical simulation and internal force responses of the embedded pile using mechanical connections. It can be seen from Figure 8 that:

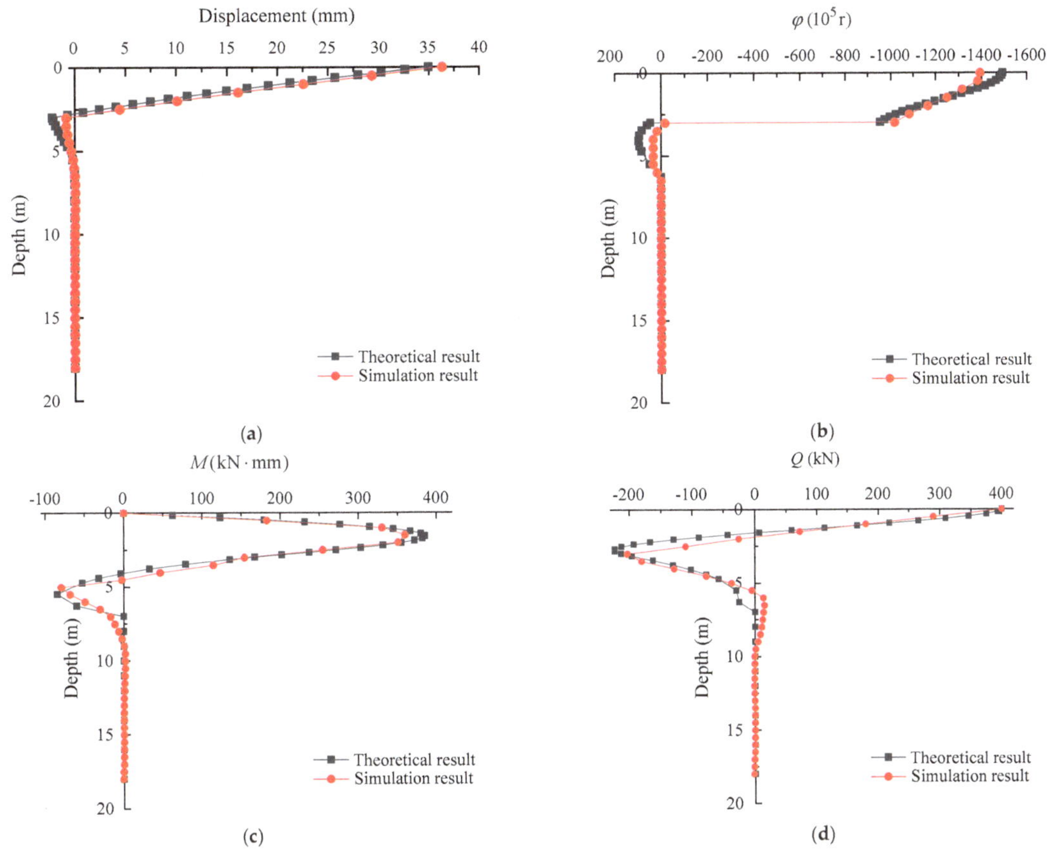

Figure 8. Comparison of theoretical and numerical simulation displacements and internal force responses of the embedded pile using mechanical connections. (**a**) Variation curve of displacement with depth. (**b**) Variation curve of the angle of rotation with depth. (**c**) Variation of shear with depth. (**d**) Variation of bending moment with depth.

(1) The theoretical and numerical simulation displacements of the mechanically connected pile fit well. The maximum displacement occurs at the top position of the pile. The theoretical maximum displacement is 34.94 mm, and the numerical simulation displacement is 36.33 mm, with a relative error of 3.83%.

(2) The maximum angle of rotation of the mechanically connected pile occurs at the top of the pile. The theoretical maximum angle of rotation is -1490.05×10^{-5} r, while the

numerical simulation maximum angle of rotation is -1393.33×10^{-5} r, with a relative error of 6.94%.

(3) The maximum positive bending moment of the mechanically connected pile occurs at 1–3 m from the top position of the pile. The theoretical maximum positive bending moment is 385.26 kN·m at 1.57 m depth, and the simulated maximum positive bending moment is 359.51 kN·m at 1.50 m depth, with a relative error of 7.16%. The theoretical maximum negative bending moment is −84.74 kN·m at 5.50 m depth, and the simulated maximum negative bending moment is −79.53 kN·m at 5.50 m depth, with a relative error of 6.55%. The theoretical point of contra-flexure is at 4.08 m, and the numerical simulation point of contra-flexure is at 4.23 m.

(4) The positive shear is 400 kN at the location of the top of the pile. The theoretical maximum value of the negative shear is −223.54 kN at a depth of 2.83 m. The simulated maximum negative shear is −203.84 kN at a depth of 4.5 m, with a relative error of 9.66%. The m-method calculates the shear as zero at 1.57 m and the numerical simulation shear as zero at 1.8 m.

According to the above analysis, it can be found that the theoretical and numerical simulation displacements and internal force responses of the embedded pile using mechanical connections are basically consistent, which can prove the correctness of the theoretical calculation by the m-method.

4. Conclusions

(1) Based on the m-method, the displacement curve equations of mechanically connected piles are established. Setting reasonable boundary conditions, the pile displacements and internal force responses of the upper and lower piles under horizontal loading are obtained, respectively, which are the expressions of the overall pile response of the mechanically connected pile.

(2) The solution method and process are expressed in the form of a table of sub-coefficients of pile internal forces, which is helpful for rapid calculation and study of the mechanically connected pile displacement and internal force response under different pile-soil conditions.

(3) The simulated displacements and angles of rotation of the pile-soil model established using ABAQUS software are basically consistent with the test results. The relative errors of the displacements and angles of rotation are both less than 10%, indicating that the established pile-soil model has good accuracy.

(4) The theoretical and numerical simulation displacements and internal forces of the mechanically connected pile fit well. The relative errors of the displacements, angles of rotation, positive and negative bending moments, and positive and negative shear forces are all less than 10%, proving the correctness of the theoretical calculation by m-method.

In summary, this study can provide effective theoretical support and methodological guidance for the displacement and internal force response of discontinuous piles.

Author Contributions: L.G. contributed to conceptualization, formal analysis, investigation and writing. M.-L.Z. contributed to methodology, funding acquisition, supervision, project administration and writing-review and editing., writing and writing-review and editing. Q.Z. contributed to formal analysis, software and data curation. G.B., X.Y. and J.D. contributed to data curation, investigation, and visualization. S.Z. and M.W. contributed to conceptualization, funding acquisition, supervision, project administration. All authors have read and agreed to the published version of the manuscript.

Funding: This research has been supported by the Natural Science Research Project of Jiangsu Province Colleges and Universities (21KJD560002 and 23KJA560007), China; Research and Innovation Team Project of Suqian College (2021TD04), China; Suqian Sci & Tech Program (H202313), China; Jiangsu Civil Architecture Society project ((2023) No. 4 Item 9), China; the Youth Fund Project of Suqian College (2023XQNA03), China; and the Fifth Provincial Research Funding Project of "333 High-level Talent Training" in 2020 (BRA2020241), China.

Data Availability Statement: The data presented in this study are available in the article here.

Conflicts of Interest: The authors declare no conflicts of interest.

List of Notations

b_0	Calculated width of the pile
EI	Flexural stiffness of the pile
E_1I_1	Flexural stiffness of the upper pile
E_2I_2	Flexural stiffness of the lower pile
L_1	Length of upper pile
L_2	Length of lower pile
M_0	External moment at the top of the upper pile
m	Proportionality coefficient of horizontal resistance factor of foundation soil with depth
Q_0	External force at the top of the upper pile
y_1	Intermediate variable
z	Pile length integral point
α_1	Horizontal deformation coefficient of the upper pile
φ_0	Displacement coefficient
φ_1	Angle of rotation at the bottom of the upper pile
δ	Displacement coefficient
$\delta_{\overline{\varphi_1}\overline{M_0}}$	Correlation coefficient between φ_1 and M_0
$\delta_{\overline{\varphi_1}\overline{Q_0}}$	Correlation coefficient between φ_1 and Q_0
$\overline{M_0}^{\Delta}$	External moment at the virtual through-length pile top
$\overline{Q_0}^{\Delta}$	External force at the virtual through-length pile top
$\overline{y_0}^{\Delta}$	Displacement at the virtual through-length pile top
$\overline{\varphi_0}^{\Delta}$	Angle of rotation at the virtual through-length pile top

References

1. Huded, P.M.; Dash, S.R. Pile foundation in alternate layered liquefiable and non-liquefiable soil deposits subjected to earthquake loading. *Earthq. Eng. Eng. Vib.* **2024**, *23*, 359–376. [CrossRef]
2. Milane, R.; Briançon, L.; Grange, S. Experimental investigation of the behaviour of pile foundations of a high-rise building. *Eur. J. Environ. Civ. Eng.* **2024**, *28*, 1394–1414. [CrossRef]
3. Lua, T.H.; Xi, X.; Tatsunori, M. Effect of pile arrangement on long-term settlement and load distribution in piled raft foundation models supported by jacked-in piles in saturated clay. *Soils Found.* **2024**, *64*, 101426.
4. Wang, J.; Jin, T.; Qin, W.; Zhang, F. Performance evaluation of post-grouting for bored piles installed in inhomogeneous gravel pebble stratum overlaid by deep and soft soil layers. *Geomech. Geoengin.* **2024**, *19*, 77–95. [CrossRef]
5. Oztürk, B. Analysis and Design of Prestressed Concrete Sheet Pile Walls. Master's Thesis, Middle East Technical University, Ankara, Turkey, 1996.
6. Chen, K.; Gan, G. Flexural Test and Numerical Simulation of a New Combination Splice for Prestressed, Precast Concrete Piles Using High-Strength Steel Strands. *Buildings* **2022**, *12*, 1371. [CrossRef]
7. Sun, C.Z.; Li, A.Q.; Miao, C.Q.; Qiao, Y. Research on cumulated damages of prestressed concrete solid pile using snap-in mechanical connection. *Earthq. Eng. Eng. Dyn.* **2018**, *38*, 211–222.
8. Sun, C.Z.; Li, A.Q.; Miao, C.Q.; Qiao, Y. Finite element analysis of mechanical properties of pile-cap connections for prestressed concrete solid pile using snap-in mechanical connection. *J. Disaster Prev. Mitig. Eng.* **2019**, *39*, 37–44.
9. Sun, C.Z.; Zhuang, M.L.; Gao, L.; Yan, J.; Yu, T.; Liu, T. Experimental study on seismic performance improvement of pile-cap joints in a prestressed concrete solid square pile using a snap-in mechanical connection. *J. Earthq. Eng.* **2023**, *27*, 3180–3201. [CrossRef]
10. Karayannis, C.G.; Chalioris, C.E. Strength of prestressed concrete beams in torsion. *Struct. Eng. Mech.* **2000**, *10*, 165–180. [CrossRef]
11. Qi, J.L.; Zhou, P.H.; Yang, X.L.; Zhou, Z.D. Application of mechanical-connection bamboo joint piles in the coastal soft soil foundation. *Build. Struct.* **2013**, *44*, 73–76.
12. Liu, L.; Xiao, L.; Li, X.Y.; Zhang, G.Z.; Wang, N.B.; Zhao, L.T. Mechanical behaviours of large-diameter rock-socketed piles under horizontal loads. *China Civ. Eng. J.* **2023**, *56*, 44–49.
13. Li, S.; Yin, J.; Zhou, J.; Yang, W. Investigation of pile-soil interaction mechanism of end-bearing foundation piles. *J. Rock Mech. Eng.* **2004**, *23*, 996–1000.
14. Yu, W.; Wu, D. Study on the joist stress of pile-supported retaining wall based on Winkler's elastic foundation beam model. *J. Zhejiang Sci-Tech Univ. Nat. Sci. Ed.* **2020**, *43*, 109–114.

15. Liu, S.; Hou, S.; Cai, Z. Lateral load test on single pile lateral bearing capacity of single pile m-method. *Chin. J. Geotech. Eng.* **2013**, *35*, 721–724.
16. Jiang, X.; Peng, R. The program development of pile foundation analysis based on m-method. *J. China Foreign Highw.* **2016**, *36*, 173–175.
17. Zhu, J.; Sun, L. Verification of m-method for Determination of Load on Piles in Soft-clay Ground. *J. Tongji Univ.* **2003**, *31*, 902–905.
18. Mao, J.; Xu, J.; Yang, L. An improved method for stabilising pile by using m-method model to the whole pile. *Rock Soil Mech.* **2018**, *39*, 1197–1202.
19. *JTG 3363–2019*; Specifications for Design of Foundation of Highway Bridges and Culverts. People's Transportation Press: Beijing, China, 2019.
20. García, F.J.A.; da Cunha, R.P.; de Albuquerque, P.J.R.; de Farias, M.M.; Bernardes, H.C. Experimental and Numerical Behavior of Horizontally Loaded Piled Rafts with a Defective Pile. *Geotech. Geol. Eng.* **2023**, *41*, 429–439. [CrossRef]
21. Chen, X.; Sun, J.; Cai, X. Horizontal static loading test and analyses of internal force and distortion on single pile. *Rock Soil Mech.* **2010**, *31*, 753–759.

buildings

MDPI

Article

Classification and Prediction of Rock Mass Boreability Based on Daily Advancement during TBM Tunneling

Zhiqiang Li [2], Yufan Tao [1,2,3,*], Yuchao Du [2,*] and Xinjie Wang [4]

1 School of Rail Transportation, Soochow University, Suzhou 215100, China
2 Geotechnical and Structural Engineering Research Center, Shandong University, Jinan 250061, China; zhiqiangli@sdu.edu.cn
3 Intelligent Urban Rail Engineering Research Center of Jiangsu Province, Suzhou 215100, China
4 Guizhou Provincial Highway Development Group Co., Ltd., Guiyang 550001, China; wxj_20242024@163.com
* Correspondence: yftao@suda.edu.cn (Y.T.); ycdu@sdu.edu.cn (Y.D.)

Abstract: The rock classification system was initially applied to drill-and-blast tunnels and subsequently adapted for TBM tunnels; however, the majority of these systems primarily focused on rock stability while neglecting considerations of boreability. Compared with conventional tunnels, TBM tunnels are characterized by their rapid tunneling speed and excellent self-stabilization of the rock mass. Therefore, it is imperative to develop a novel rock mass classification system that considers both the tunneling efficiency of TBMs and the required support measures for tunnel construction. This paper introduces a novel rock classification system for TBM tunnels that accurately predicts the construction rate by evaluating the penetration rate and daily utilization, enabling a more precise assessment of daily advancement in tunneling. Firstly, the penetration rate and construction utilization in different rock strata are analyzed based on comprehensive statistics of existing construction data. Consequently, a discriminant matrix for classifying the boreability of rock is derived. Subsequently, employing the Ensemble Classifier method, a regression prediction model for rock boreability classification is established by incorporating input parameters such as thrust, torque, rotational speed, field penetration index, and the uniaxial compressive strength of rock. The validity of the proposed model is verified by comparing predicted machine performance with actual data sets. The proposed method presents a novel approach for predicting the performance of TBM construction.

Keywords: TBM tunneling; advancements prediction; rock boreability classification; tunneling parameters; machine learning

1. Introduction

A tunnel boring machine or TBM has evolved to be a rather high-tech piece of equipment that integrates multiple components to allow for near-continuous excavation of the rock. Hard rock TBMs have played an important role in the advancement of tunnel construction in the last few decades and have set many world records in tunneling, unmatched by conventional tunneling or shield tunneling in soft ground applications. When tunneling through complex geological conditions, TBMs often encounter mixed soft and hard strata, which can significantly impact the penetration rate and working hours, making it difficult to accurately predict advancements [1,2].

There have been many studies on the performance prediction of TBMs since the 1980s [3–5]. There are several performance prediction models for TBMs that have attempted to include rock conditions by incorporating the existing rock mass classification systems or their input parameters [6–11]. The main obstacle in improving TBM performance prediction in complex geological conditions is to figure out the relationship between machine operational parameters and geological parameters [12–20]. Rostami [21] maintained that the accuracy of estimating machine performance in complex ground is still very low. Therefore, it is necessary to identify the geological conditions and predict the behavior of the surrounding rock in front of the cutterhead in TBM operations [22]. The relationship between the machine parameters and penetration rate can offer some information about the rock mass at the face and identify potential issues [1,23–26]. Analysis of TBM operational parameters can offer some insight into the behavior of the surrounding rocks during the tunneling process.

A common method for TBM performance prediction is combining TBM performance with the classification of surrounding rock mass [27]. Innaurato [28] improved Cassinelli's [4] TBM prediction model by adding uniaxial compressive strength. One of the most famous TBM prediction models based on rock mass classification is the Q_{TBM} proposed by Barton [29]. This classification method is based on an extended version of the Q system and takes all rock and machine parameters that affect the advance speed into account. Alber presented RMR_{TBM}, which is a modified RMR system for use in TBM tunneling projects [3,11]. Sapigni presented an improved RMR system for TBM construction, which indicated TBM performance reaches a maximum in the rock mass rating (RMR) range of 40–70 [30]. Bieniawski proposed the concept of the rock mass excavability (RME) index, which the authors found very effective in predicting the excavability of rock masses by TBMs using quantification of TBM performance [31]. By employing multiple and polynomial regression analyses on field data, Khademi attempted to provide a functional predictor equation of the TBM field penetration index (FPI) by using the RMR rock mass classification system [32]. Frough [33] examined the relationship between the rock mass conditions, essentially described by the rock mass rating system (RMR) and TBM performance parameters using field data. Salimi studied the possibility of developing a new rock mass classification system for TBM application by using site-specific models and Genetic Programming (GP) for the analysis of a TBM performance relative to the ground condition data [34]. Although the aforementioned studies have conducted targeted investigations on TBMs and considered the correlation between surrounding rock and driving parameters, they have not accounted for the influence of nonconstruction factors, thus impeding accurate prediction of actual construction advancements. As illustrated in Figure 1, there is no obvious relationship between traditional classification levels (II, IIIa, IIIb, IV, V) and daily advancements.

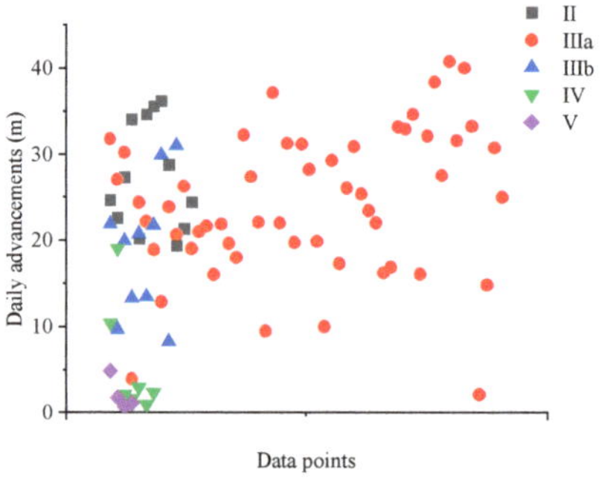

Figure 1. Relationship between traditional classification levels and daily advancements.

This paper introduces a new method of classifying and predicting rock mass for assessing rock boreability based on the analysis of driving parameters and the corresponding rock parameters, as shown in Figure 2. Firstly, the classification criteria for rock boreability based on penetration rate and utilization are established, leading to the development of a classification matrix as presented in Section 2 of this paper. Subsequently, machine learning methods are employed to establish a predictive model for new boreability classifications by correlating geological and machine operational parameters, as presented in Section 3 of this paper. In Section 4, the accuracy of this novel prediction model is validated through case studies, providing valuable insights for construction departments and facilitating better preparation.

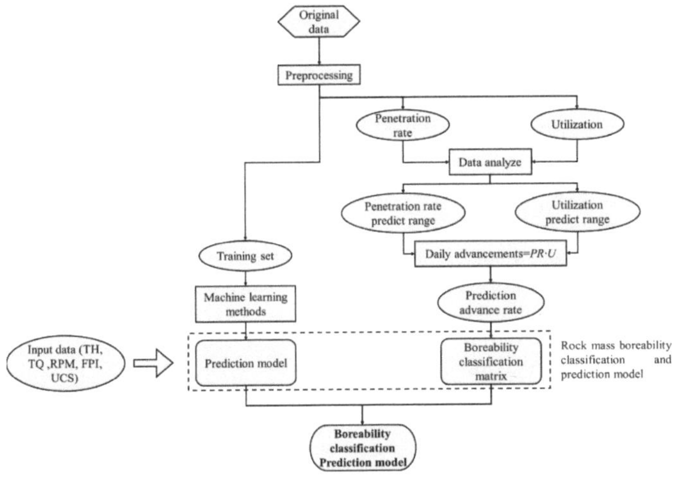

Figure 2. Flow chart of the research.

The research content of this paper can be divided into two parts:

(1) According to the TBM's surrounding rock boreability classification matrix, the surrounding rock boreability classification of the construction tunnel can be classified based on penetration rate and utilization.

(2) By adopting the machine parameters, rock parameters, and comprehensive parameters, i.e., TQ, RPM, TH, UCS, and FPI above, as the predictive input parameters, the boreability level can be used as the output result, and the machine learning methods can be used to establish the surrounding rock boreability prediction model.

2. TBM Boreability Classification System

2.1. Project Overview

This research project is a component of the water conservancy initiative in western China. It entails the excavation of a 26.041 km long tunnel with a diameter of 7800 mm, making it the longest hydraulic tunnel ever excavated by a single TBM in China. The project site ranges in altitude from 650 m to 800 m, with a general elevation difference between 10 m and 20 m, and certain areas experience height differences of up to 70 m. Along its alignment, the tunnel traverses Hualixi period granite, Carboniferous tuffaceous sandstone, and black cloud quartzite formations. The depth of the tunnel varies from 40 m to 160 m. Table 1 presents key technical parameters for the TBM employed in this endeavor.

Table 1. Key technical parameters of the TBM employed for the project.

Technical Parameters	Value
Cutterhead diameter (mm)	7830
Number of cutters	53
Rotation speed (r/min)	0~8.1
Total thrust (kN)	15,257~22,450
Specified torque (kN·m)	8800@3.8 rpm/5573@6 rpm
Instantaneous torque (kN·m)	13,200@0.3 rpm
Effective support force (kN)	45,371@345 bar
Penetration rate (mm/min)	0~120
Penetration (mm/r)	0~15

According to the geological investigation of the ground and analysis of borehole data along the tunnel alignment, the exposed strata lithology mainly consists of Devonian Middle Altay Formation (D2a) monzonite gneiss and monzonite granulite; Carboniferous upper system CzK group (C3K) tuffaceous sandstone, biotite oblique gneiss, and black cloud quartzite; and late Hualixi (γ33b) granite, biotite hornblende plagioclase. This project is situated between the trough zone and compression zone in the middle of a fold system. These two secondary tectonic units are bounded by faults: a geotrough fold belt in the north and a compression belt in the south. Regional features include 54 faults and shear zones such as f43, f55, f58~f61, f73~f75, and f106~f127. The fault zones generally have a width ranging from 5 to 30 m consisting mainly of mylonites and fractured rock.

The project area primarily lies within low hilly areas with scarce surface water resources. Only rivers adjacent to reservoirs carry notable amounts of water near the project area. The water supply in this region mainly comes from ice and snow meltwater in mountainous areas followed by seasonal precipitation. Based on differences in water storage medium and groundwater burial conditions, groundwater types within this project area mainly include Quaternary loose deposit pore diving, clastic rock pore-fracture water, and bedrock fissure water without a unified groundwater level or confrontation. Testing samples from this area revealed that groundwater exhibits medium-to-strong corrosive behavior, including weak-to-moderately corrosive steel bars within reinforced concrete lining, as well as moderately corrosive steel structures directly exposed to incoming fluids.

2.2. Penetration Rate and Utilization

Typical rock mass classification systems focus on the analysis of face and surrounding conditions to assess tunnel stability. However, the proper selection of the driving parameters to match the geological conditions plays a decisive role in the efficiency of tunneling. It is necessary to establish a new rock classification method from the perspective of tunneling efficiency to distinguish the boreability of the ground, where the higher the boreability, the higher the tunneling efficiency, indicating more efficient, faster, and easier tunneling. Simultaneously, the stability of surrounding rock is also influenced by the strength of the rock stratum, with accidents such as deformation and collapse commonly occurring in soft rock formations. These incidents lead to reduced utilization and daily advancements. The focus of this paper lies in the pivotal role of daily advancements in assessing rock boreability, which is contingent upon both penetration rate and utilization.

The penetration rate (PR) refers to the distance tunneled by a machine during a unit operation time while it is cutting through the rock. Typically, the penetration rate falls within the range of 25 to 50 ($mm \cdot min^{-1}$). It is determined by both the cutter's penetration (p) and its rotation speed. Consequently, PR is closely linked to operational parameters and rock mass properties. Figure 2 presents the research data on penetration rates, which vary from 0 to 60 mm/min with an average value of approximately 35 mm/min.

Utilization is defined as the ratio of tunneling time to total time, where the total time is the sum of TBM downtime and boring time [35], and the typical utilization ranges from 20% to 60%. The downtime primarily encompasses the time required for parameter adjustments, equipment maintenance and overhaul, and surrounding rock support operations, as well as any unplanned interruptions caused by accidents. The utilization can reflect whether the site setup is suitable, and the operation is streamlined. The utilization reflects the adaptability of TBMs to geological conditions.

As we can see from Figure 3, a normal distribution can describe the variability in penetration rate and utilization. The combination of penetration rate and utilization can be used to estimate the daily advancements under different working conditions. The project's statistical data are categorized into four groups based on daily advancements of 0~10, 10~20, 20~30, and 30~40 m per day. The average value x and standard deviation s of penetration rate and utilization are calculated for each group to determine the fluctuation range $\{(x - s) \sim (x + s)\}$, shown in Tables 2 and 3. Subsequently, the information is utilized to generate a table predicting daily advancements.

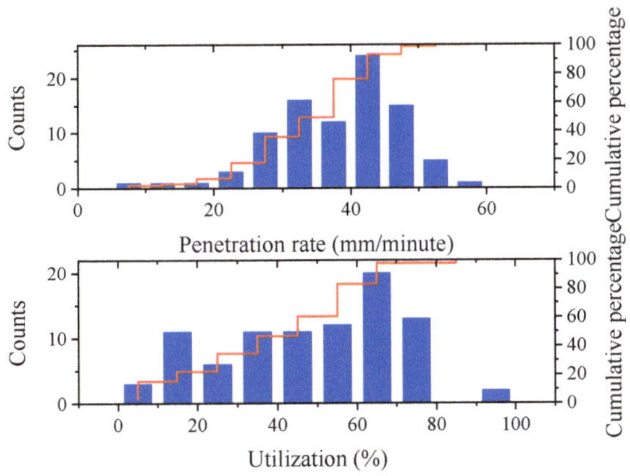

Figure 3. Statistics of penetration rate and utilization used in this research.

Table 2. Calculation and prediction table of penetration rate based on daily advancements.

Levels	Daily Advancements	Penetration Rate		
		Average Value x	Standard Deviation s	Fluctuation Range
I	30~40	41.277032	0.54	>41
II	20~30	39.268676	1.40	38~41
III	10~20	36.796099	1.11	36~38
IV	0~10	34.237046	1.61	<36

The unit of PR is mm/min; the unit of daily advancements is meters.

Table 3. Calculation and prediction table of utilization based on daily advancements.

Levels	Daily Advancements	Utilization		
		Average Value x	Standard Deviation s	Fluctuation Range
I	30~40	62%	0.06	60~80
II	20~30	55%	0.08	50~65
III	10~20	46%	0.07	40~50
IV	0~10	22%	0.14	10~40

The unit of utilization is %; the unit of daily advancements is meters.

2.3. TBM Surrounding Rock Boreability Classification

The reflection of TBM surrounding rock classification is the construction velocity. The daily advancements are reflected by the average construction rate. The construction rate is the ratio of the excavation distance to the total construction time, including excavation, overhaul, tool change and support, change step, etc., where the expression for average construction rate v is

$$v = PR \cdot U \tag{1}$$

According to Formula (1) and Tables 2 and 3, the average construction rate v interval is calculated for different combinations of penetration rate and utilization levels, and the median value of each prediction interval is used as the prediction value of the average construction rate classification. The fluctuation ranges of the penetration rate for level II and the utilization for level II are 38~41 mm/min and 50~65%, respectively. Consequently, the calculated average construction rate falls within a range of 19~26.65 mm/min. Subsequently, the average construction rate v is determined as the median value from this predicted range, which amounts to 32.868 m/day. The prediction matrix is shown in Table 4.

Table 4. Prediction value of the average construction rate v.

Utilization Level	Penetration Rate Level			
	I	**II**	**III**	**IV**
I	42.192	40.032	36.432	34.848
II	36.632	32.868	29.916	28.584
III	27.072	25.704	23.4	22.32
IV	15.48	14.544	13.176	13.104

The unit of PR is mm/min and v is m/day.

According to the TBM construction practice, in conjunction with the specific project, the surrounding rock encountered by the TBM is categorized into four levels based on the daily construction rate v: Level I—easy boreability rock: $v \geq 35$ m/day, indicating optimal TBM construction conditions; Level II—moderate boreability rock: 25 m/day $< v \leq 35$ m/day, suggesting favorable TBM construction conditions; Level III—average boreability rock: 15 m/day $< v \leq 25$ m/day, denoting typical TBM construction conditions; and Level IV—poor boreability rock: $v \leq 15$ m/day, representing unfavorable TBM construction conditions. This classification is presented in Table 5.

Table 5. TBM surrounding rock boreability classification matrix.

U \ PR	≥41	38~41	36~38	≤36
60~100	I	I	I	II
50~60	I	II	II	II
40~50	II	II	III	IV
0~40	III	IV	IV	IV

The unit of PR is mm/min; the unit of U is %.

According to the qualitative analysis of the measured data, the penetration rate and utilization during tunneling follow a nearly normal distribution (Figure 4). When the penetration rate reaches approximately 40 mm/minute, the daily utilization reaches its peak. This indicates that machine stoppages are rare during tunneling, resulting in smooth excavation. The statistical chart reveals that cutterhead utilization is somewhat reduced at higher or lower penetration rates. In cases of low penetration rates, hard rock layers pose significant geological constraints on excavation. The high thrust required for excavating such formations often leads to cutterhead damage and subsequent lengthy replacement processes, thereby reducing overall utilization. Conversely, high penetration rates encounter soft rock conditions where excessive mud in the cutterhead frequently hampers excavation progress. Consequently, TBM efficiency becomes restricted by machine-specific working conditions, leading to decreased utilization rates.

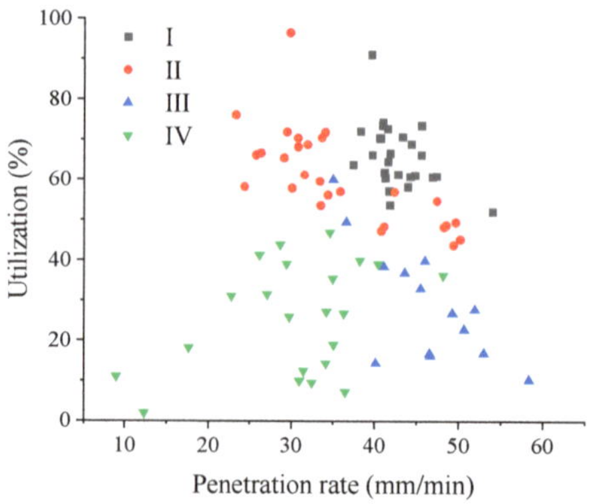

Figure 4. Data distribution scatter plot of four levels of new boreability classifications.

It can be seen from Figure 5 that the boreability classification category is quite different from the traditional one. The classification matrix presented above can be utilized to determine the rock mass classification for excavation purposes; however, obtaining real-time data on penetration rate and utilization during the process is challenging. These parameters can only be indirectly obtained through their relationships with other complex variables that cannot be directly calculated using mathematical models. Therefore, developing a multiparameter prediction model has become the preferred approach for predicting rock mass boreability classification.

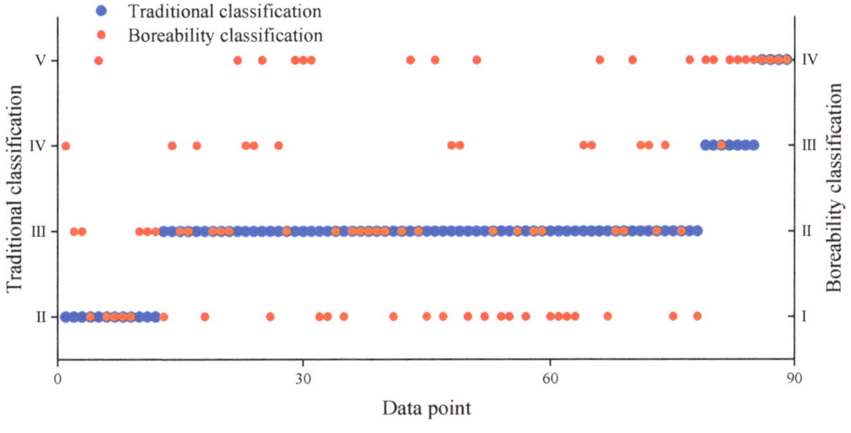

Figure 5. Comparison between traditional surrounding rock classification and boreability classification.

3. Boreability Prediction Model

3.1. Data Preparation

The factors of TBM machines encompass thrust, torque, rotation speed, cutter size, type and number, cutter spacing, support type, and more. Among these mechanical parameters that affect tunneling efficiency directly are the thrust, torque, and cutterhead of the TBM. This paper selects the thrust (TH), rotation speed (RPM), and torque (TQ) of the cutterhead as indicators for predicting rock mass boreability.

According to the engineering data studied in this paper, the thrust and the rotation speed have a relatively consistent trend with the classification of surrounding rock in traditional underground engineering, but the torque of TBMs is not completely consistent with the traditional classification (Figure 6). According to the above analysis, the three TBM mechanical parameters are selected to predict the boreability of surrounding rock.

The rock's strength is typically quantified by its uniaxial compressive strength (UCS), which serves as a key indicator for the tunneling speed of TBMs. Previous studies have consistently demonstrated a strong correlation between UCS and tunneling speed, where lower UCS values result in faster penetration rates and higher efficiency, while higher UCS values lead to slower penetration rates and reduced efficiency. However, excessively low UCS can compromise the stability of the surrounding rock, increasing surrounding rock support time, prolonging TBM downtime, and significantly impeding tunneling progress.

Boreability classification is closely related to the mechanical parameters and rock parameters. Since the TBM construction tunnel is different from the Drilling–Blasting Tunnel, the tunnel face of the TBM is closed, and it is difficult to obtain the relevant parameters of the surrounding rock during the tunneling process. Therefore, we introduced the Field Penetration Index (*FPI*) into the prediction model. The expression of the FPI is as follows:

$$FPI = TH/p \tag{2}$$

where TH is the single-cutter thrust (kN) and p is the penetration (mm/r).

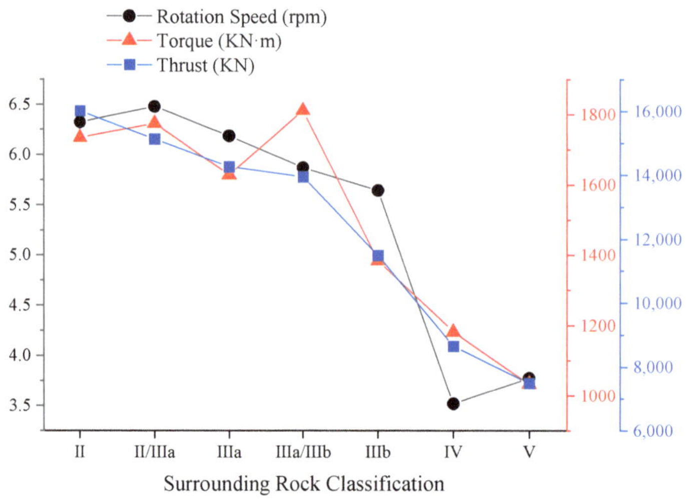

Figure 6. Machine parameter trend based on traditional rock mass classification.

The FPI represents the thrust required for penetrating the unit depth of the rock and can be used as a measure of the boreability of the rock itself. The larger the FPI value, the greater the thrust required to produce the same depth of cut, indicating that the rock mass is less likely to excavate. The FPI has a good correlation with geological parameters and tunneling parameters and can be used as a link.

3.2. Boreability Classification Prediction Method

As mentioned above, TBM field performance data used in the prediction model establishment include the rotation speed (RPM), the torque (TQ), the thrust (TH), and the Field Penetration Index (FPI), as well as uniaxial compressive strength (UCS). All data suffice for three basic conditions, Data Information Content, Data Quality, and Data Objectivity [36], and they are output from the TBM control system directly.

Tenfold cross-validation is a widely adopted approach for model evaluation, wherein the fundamental procedure is as follows. Data set partitioning: Initially, the original data set is randomly divided into approximately equal subsets of size 10. Training and validation: Subsequently, ten iterations of training and validation are performed. In each iteration, nine subsets are merged to form the training set, while one remaining subset serves as the validation set. Consequently, every subset has an opportunity to be employed as the validation set once. Model assessment: The trained model is utilized to predict outcomes on the validation set and compute corresponding evaluation metrics (e.g., accuracy, loss). Result aggregation: Ultimately, by averaging these evaluation metrics obtained from ten experiments, we evaluate the generalization capability of our model. This method possesses an advantage in that it maximizes utilization of the entire data set for both training and validation purposes while mitigating instability in evaluation results arising from diverse data partitioning methods. We adopted tenfold cross-validation in the establishment of the prediction model to prevent overfitting.

The BP neural network recognition model is established using the newff function:

$$net = newff(P, Q, S, TF, BTF) \quad (3)$$

where P is the input sample, Q is the output sample, S is the number of neurons in the hidden layer, TF is the transfer function, and BTF is the training function.

A double hidden layer BP neural network is selected with 11 and 1 hidden layer nodes, respectively. According to the recognition accuracy of the neural network, the TF between the layers is selected as tansig and purein. We compare the effects of each training function and select trainbfg. The input samples are the above five main parameters, namely RPM, TQ, TH, FPI, and UCS. The output sample is the boreability category. We set the network internal parameter learning rate to 0.1, error precision to 0.001, training step size to 50,000, and the remaining parameters to the default values. Table 5 lists part of the data used for BP neural network learning.

The following Figure 7 shows the part of data used for model training.

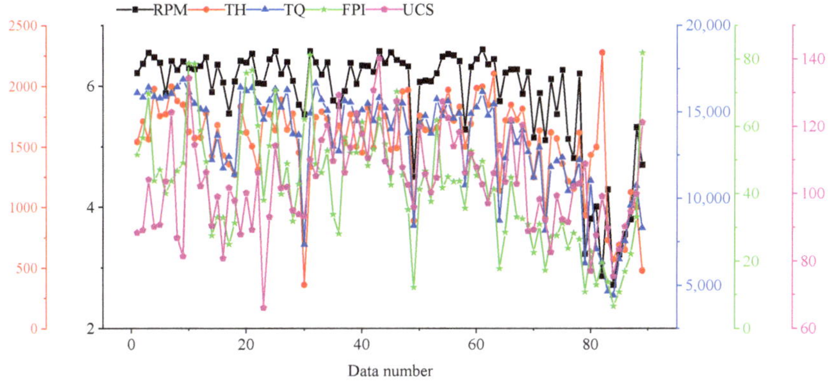

Figure 7. The data chart used for BP neural network training.

4. Results and Discussion

4.1. Case Validation

We selected 15 excavated sections from the site and determined the actual boreability level based on the penetration rate and utilization of these sections. Subsequently, a comparative analysis was conducted between the predicted results and the observed data to validate the accuracy of the model. Table 6 shows the data for the validation case and the comparison of the actual level with the prediction value.

Table 6. The validation case data and the comparison of the actual level with the prediction value.

	RPM	TQ	TH	FPI	UCS	Actual Level	Prediction Level
1	6.41	1996.71	16,058.42	43.83	124.22	I	I
2	6.60	1995.22	16,158.52	49.64	102.74	I	I
3	6.06	1427.80	11,744.40	32.91	80.86	II	II
4	5.68	1672.33	12,753.83	28.13	129.22	II	II
5	6.41	1828.05	14,912.41	43.74	118.30	I	II
6	6.22	1541.11	16,119.00	51.64	88.40	III	III
7	5.29	1502.68	10,765.68	35.72	106.18	II	II
8	5.53	360.67	7336.00	62.10	93.23	IV	IV
9	2.72	572.00	4407.80	6.61	75.55	IV	IV
10	3.23	688.75	6502.00	10.84	83.48	IV	IV
11	6.11	1617.68	11,813.59	27.13	82.64	II	II
12	5.54	1567.36	12,154.21	27.60	99.42	III	III
13	6.27	1357.3	12,370.66	31.23	92.4	I	I
14	5.13	1213.03	10,438.33	23.84	91.56	II	II
15	4.81	1152.69	10,932.04	28.33	102.21	IV	IV

The unit of RPM is r/min; the unit of TQ is kN×m; the unit of TH is kN; the unit of UCS is MPa.

Based on the results above, it can be concluded that the prediction results of the prediction model established in this paper are highly consistent with the actual level.

4.2. Discussions

The prediction result of the No. 5 section is inconsistent with the result revealed by the actual excavation. There are two possible reasons: (a) due to the small number of learning samples, there are more samples of level II and level III, and the level I samples are not enough with other samples; (b) the impact of operators in the tunneling process is large. Actually, the adjustment of tunneling parameters during the tunneling relies heavily on experience, and there is a deviation between the boreability of rock and the matching degree of tunneling parameters.

From Table 5, the specific arrangement of the boreability classification matrix can be analyzed. There is a lot of research on the prediction of TBM performance [6,14,15,19,32], which is the penetration rate in different rock masses. However, we could conclude from this research that the daily construction progress is not linearly related to the penetration rate, it is also influenced by utilization. As Khetwal's research suggested [35], the utilization of TBMs was influenced by many factors, and the utilization accounts for a large proportion of the boreability classification. This shows that during the TBM tunneling process, the construction progress is largely determined by the operation of the mechanical equipment. The key to improving mechanical construction speed is to improve the utilization of the equipment.

Different from the previous research focus on the relationship between the rock mass and machine [12,34,37], in this research, the classification method is established based on the daily progress, i.e., the penetration rate and the utilization, and the rock mass can be classified according to the boreability classification matrix. Meanwhile, the improvement of the boreability classification matrix needs plenty of tunneling data. The surrounding rock face cannot be directly observed in the TBM construction tunnel, and the actual situation of the surrounding rock cannot be accurately obtained in real time. The research content of this paper can provide some reference for TBM construction. However, the areas of prediction of faults and other unfavorable geological bodies in general need further research.

5. Conclusions

Based on the TBM-related tunneling parameters and the corresponding rock mass parameters, this paper establishes the boreability classification method for the surrounding rock of TBM tunnels. Machine learning methods are used to establish the prediction model of surrounding rock. The classification method and prediction model were applied to a TBM for case validation.

The project's statistical data are classified into four groups based on daily advancements of 0~10, 10~20, 20~30, and 30~40 m per day. For each group, the average value x and standard deviation s of penetration rate and utilization are calculated to determine the fluctuation range $\{(x - s)\sim(x + s)\}$. The average construction rate interval v is computed for different combinations of penetration rate and utilization levels, with the median value within each prediction interval serving as the predicted average construction rate classification. Based on TBM construction practices and specific project considerations, the surrounding rock encountered by the TBM is categorized into four levels according to the daily construction rate v.

The challenge of this method lies in acquiring real-time data on penetration rate and utilization. Therefore, this research aims to develop a multiparameter prediction model utilizing machine parameters, rock parameters, and comprehensive parameters (TH, RPM, TQ, UCS, and FPI) as predictive input variables. The boreability level is used as the output result for establishing the boreability prediction model using machine learning methods. Fifteen sections were selected for prediction purposes by inputting their respective parameters into the established model to obtain predicted results, which were then compared with actual levels. The findings indicate that while the predicted results of 14 sections align with the actual levels, there was a deviation observed in the predicted results of the 5th section.

In summary, the feasibility of the TBM surrounding rock boreability classification method and the prediction model is verified by engineering examples. The research provides a reference for construction departments and gives them a more visualized sense of the advancements, which helps them to make better preparations.

Author Contributions: Conceptualization, Y.T. and Z.L.; methodology, Y.T.; software, Y.T.; validation, Z.L. and Y.D.; formal analysis, Z.L.; investigation, Y.D.; resources, Y.D. and X.W.; data curation, Y.T.; writing—original draft preparation, Y.T.; writing—review and editing, Z.L. All authors have read and agreed to the published version of the manuscript.

Funding: This work is supported by Shandong Provincial Natural Science Foundation (No. ZR2022QD014) and the China Scholarship Council (201906220135).

Data Availability Statement: Data is contained within the article.

Conflicts of Interest: Author Xinjie Wang was employed by the company Guizhou Provincial Highway Development Group Co., Ltd. The remaining authors declare that the research was conducted in the absence of any commercial or financial relationships that could be construed as a potential conflict of interest.

Nomenclature

Nomenclature	Paraphrase
PR	Penetration rate
U	Utilization
TH	Thrust
TQ	Torque
RPM	Round per minute
FPI	Field penetration index
UCS	Uniaxial Compressive Strength
v	average construction rate

References

1. Farrokh, E.; Rostami, J. Effect of adverse geological condition on TBM operation in Ghomroud tunnel conveyance project. *Tunn. Undergr. Space Technol.* **2009**, *24*, 436–446. [CrossRef]
2. Yu, P.; Hu, R.; Zhang, J.; Yang, Q.; Zhao, J.; Cao, L.; Zhu, C. Numerical study on local scour characteristics around submarine pipelines in the Yellow River Delta silty sandy soil under waves and currents. *Deep Undergr. Sci. Eng.* **2024**, *3*, 182–196. [CrossRef]
3. Alber, M. Prediction of Penetration and Utilization For Hard Rock TBMs. In Proceedings of the ISRM International Symposium, Turin, Italy, 2–5 September 1996.
4. Cassinelli, F.; Cina, S.; Innaurato, N. Power consumption and metal wear in tunnel-boring machines: Analysis of tunnel-boring operation in hard rock. *Int. J. Rock Mech. Min. Sci. Geomech. Abstr.* **1983**, *20*, A25. [CrossRef]
5. Nelson, P.P.; Ingraffea, A.R.; O'Rourke, T.D.; Nelson, P.P.; Ingraffea, A.R.; O'Rourke, T.D. TBM performance prediction using rock fracture parameters. *Int. J. Rock Mech. Min. Sci. Geomech. Abstr.* **1985**, *22*, 189–192. [CrossRef]
6. Hassanpour, J.; Rostami, J.; Khamehchiyan, M.; Bruland, A. Developing new equations for TBM performance prediction in carbonate-argillaceous rocks: A case history of Nowsood water conveyance tunnel. *Geomech. Geoengin.* **2009**, *4*, 287–297. [CrossRef]
7. Hassanpour, J.; Rostami, J.; Khamehchiyan, M.; Bruland, A.; Tavakoli, H.R. TBM Performance Analysis in Pyroclastic Rocks: A Case History of Karaj Water Conveyance Tunnel. *Rock Mech. Rock Eng.* **2009**, *43*, 427–445. [CrossRef]
8. Hassanpour, J.; Rostami, J.; Zhao, J. A new hard rock TBM performance prediction model for project planning. *Tunn. Undergr. Space Technol.* **2011**, *26*, 595–603. [CrossRef]
9. Ribacchi, R.; Fazio, A.L. Influence of Rock Mass Parameters on the Performance of a TBM in a Gneissic Formation (Varzo Tunnel). *Rock Mech. Rock Eng.* **2004**, *38*, 105–127. [CrossRef]
10. Yagiz, S. Utilizing rock mass properties for predicting TBM performance in hard rock condition. *Tunn. Undergr. Space Technol.* **2008**, *23*, 326–339. [CrossRef]
11. Alber, M. Advance Rates of Hard Rock TBM's and Their Effects on Project Economics. *Tunn. Undergr. Space Technol.* **2000**, *15*, 55–64. [CrossRef]
12. Gong, Q.M.; Zhao, J. Development of a rock mass characteristics model for TBM penetration rate prediction. *Int. J. Rock Mech. Min. Sci.* **2009**, *46*, 8–18. [CrossRef]

13. Mikaeil, R.; Naghadehi, M.Z.; Sereshki, F. Multifactorial fuzzy approach to the penetrability classification of TBM in hard rock conditions. *Tunn. Undergr. Space Technol.* **2009**, *24*, 500–505. [CrossRef]
14. Ghasemi, E.; Yagiz, S.; Ataei, M. Predicting penetration rate of hard rock tunnel boring machine using fuzzy logic. *Bull. Eng. Geol. Environ.* **2013**, *73*, 23–35. [CrossRef]
15. Benato, A.; Oreste, P. Prediction of penetration per revolution in TBM tunneling as a function of intact rock and rock mass characteristics. *Int. J. Rock Mech. Min. Sci.* **2015**, *74*, 119–127. [CrossRef]
16. Fatemi, S.A.; Ahmadi, M.; Rostami, J. Evaluation of TBM performance prediction models and sensitivity analysis of input parameters. *Bull. Eng. Geol. Environ.* **2016**, *77*, 501–513. [CrossRef]
17. Liu, B.; Wang, R.; Guan, Z.; Li, J.; Xu, Z.; Guo, X.; Wang, Y. Improved support vector regression models for predicting rock mass parameters using tunnel boring machine driving data. *Tunn. Undergr. Space Technol.* **2019**, *91*, 102958. [CrossRef]
18. Liu, Q.; Jiang, Y.; Wu, Z.; Xu, X.; Liu, Q. Investigation of the Rock Fragmentation Process by a Single TBM Cutter Using a Voronoi Element-Based Numerical Manifold Method. *Rock Mech. Rock Eng.* **2017**, *51*, 1137–1152. [CrossRef]
19. Zhang, Q.; Hu, W.; Liu, Z.; Tan, J. TBM performance prediction with Bayesian optimization and automated machine learning. *Tunn. Undergr. Space Technol.* **2020**, *103*, 103493. [CrossRef]
20. Fang, Y.; Li, X.; Liu, H.; Hao, S.; Yi, Y.; Guo, Y.; Li, H. Intelligent real-time identification technology of stratum characteristics during slurry TBM tunneling. *Tunn. Undergr. Space Technol.* **2023**, *139*, 105216. [CrossRef]
21. Rostami, J. Performance prediction of hard rock Tunnel Boring Machines (TBMs) in difficult ground. *Tunn. Undergr. Space Technol.* **2016**, *57*, 173–182. [CrossRef]
22. Nilsen, B.; Ozdemir, L. Hard rock tunnel boring prediction and field performance. In Proceedings of the Rapid Excavation and Tunneling Conference, Boston, MA, USA, 13–17 June 1993; pp. 833–852.
23. Ren, T.; Zhang, H.; Guo, Y.; Tang, Y.; Okubo, S. Real-Time Estimation Method for Tunnel-Excavated Rock Strength Based on TBM Parameters. *Int. J. Geomech.* **2023**, *23*, 4023092.1–4023092.10. [CrossRef]
24. Zhang, K.; Zhao, X.; Zhang, Z. Influences of tunnelling parameters in tunnel boring machine on stress and displacement characteristics of surrounding rocks. *Tunn. Undergr. Space Technol.* **2023**, *137*, 105129. [CrossRef]
25. Zhang, Q.; Yang, B.; Zhu, Y.; Guo, C.; Jiao, C.; Cai, A. Prediction Method of TBM Tunneling Parameters Based on Bi-GRU-ATT Model. *Adv. Civ. Eng.* **2022**, *2022*, 3743472. [CrossRef]
26. Zhang, Q.; Zhu, Y.; Ma, R.; Du, C.; Du, S.; Shao, K.; Li, Q. Prediction Method of TBM Tunneling Parameters Based on PSO-Bi-LSTM Model. *Front. Earth Sci.* **2022**, *10*, 854807. [CrossRef]
27. Qiu, D.; Fu, K.; Xue, Y.; Tao, Y.; Kong, F.; Bai, C. TBM Tunnel Surrounding Rock Classification Method and Real-Time Identification Model Based on Tunneling Performance. *Int. J. Geomech.* **2022**, *22*, 4022070.1–4022070.17. [CrossRef]
28. Innaurato, N.; Mancini, A.; Rondena, E.; Zaninetti, A. Forecasting and Effective TBM Performances in a Rapid Excavation of a Tunnel in Italy. In Proceedings of the 7th ISRM Congress, Aachen, Germany, 16–20 September 1991.
29. Barton, N.R. TBM performance estimation in rock using Q(TBM). *Tunn. Tunn. Int.* **1999**, *31*, 30–34.
30. Sapigni, M.; Berti, M.; Bethaz, E.; Busillo, A.; Cardone, G. TBM performance estimation using rock mass classifications. *Int. J. Rock Mech. Min. Sci.* **2002**, *39*, 771–788. [CrossRef]
31. Bieniawski von Preinl, Z.T.; Celada Tamames, B.; Galera Fernández, J.M.; Álvarez Hernández, M. Rock mass excavability indicator: New way to selecting the optimum tunnel construction method. *Tunn. Undergr. Space Technol.* **2006**, *21*, 237. [CrossRef]
32. Khademi Hamidi, J.; Shahriar, K.; Rezai, B.; Rostami, J. Performance prediction of hard rock TBM using Rock Mass Rating (RMR) system. *Tunn. Undergr. Space Technol.* **2010**, *25*, 333–345. [CrossRef]
33. Frough, O.; Torabi, S.R.; Yagiz, S. Application of RMR for Estimating Rock-Mass–Related TBM Utilization and Performance Parameters: A Case Study. *Rock Mech. Rock Eng.* **2014**, *48*, 1305–1312. [CrossRef]
34. Salimi, A.; Rostami, J.; Moormann, C.; Hassanpour, J. Examining Feasibility of Developing a Rock Mass Classification for Hard Rock TBM Application Using Non-linear Regression, Regression Tree and Generic Programming. *Geotech. Geol. Eng.* **2017**, *36*, 1145–1159. [CrossRef]
35. Khetwal, A.; Rostami, J.; Nelson, P.P. Investigating the impact of TBM downtimes on utilization factor based on sensitivity analysis. *Tunn. Undergr. Space Technol.* **2020**, *106*, 103586. [CrossRef]
36. Erharter, G.H.; Marcher, T. MSAC: Towards data driven system behavior classification for TBM tunneling. *Tunn. Undergr. Space Technol.* **2020**, *103*, 103466. [CrossRef]
37. Salimi, A.; Rostami, J.; Moormann, C. Application of rock mass classification systems for performance estimation of rock TBMs using regression tree and artificial intelligence algorithms. *Tunn. Undergr. Space Technol.* **2019**, *92*, 103046. [CrossRef]

buildings

MDPI

Article

Effect of Moisture Content and Wet–Dry Cycles on the Strength Properties of Unsaturated Clayey Sand

Chuan Wang [1], Weimin Yang [2,*], Ning Zhang [1], Senwei Wang [2], Chuanyi Ma [1], Meixia Wang [2,*] and Zhiyuan Zhang [2]

[1] Shandong Hi-Speed Group, Jinan 250101, China; sxsdu2021@163.com (C.W.); m18553139625@163.com (N.Z.); machuanyi2006@163.com (C.M.)
[2] Department of Qilu Transportation, Shandong University, Jinan 250012, China; wangsenwei@mail.sdu.edu.cn (S.W.); zhiyuanzhang@mail.sdu.edu.cn (Z.Z.)
* Correspondence: weimin.yang@sdu.edu.cn (W.Y.); 15066231361@163.com (M.W.)

Abstract: Based on the actual situation of the project on the Weihai–Yanhai Expressway section of Rongwu Expressway, the effects of water content change and the dry–wet cycle on the mechanical behavior of unsaturated clayey sandy soil were analyzed in this study. In this study, ventilated undrained triaxial shear tests were carried out on unsaturated clayey sandy soils with different water contents (6%, 8%, 10%, 12%, 14% and 16%). Concurrently, the soil samples were subjected to three distinct wet and dry cycle pathways (2~22%, 2~12%, and 12~22%) to gain an understanding of how the mechanical features of the soil changed under the different conditions. The test findings demonstrate that when the water content increases, the unsaturated clayey sandy soil's cohesiveness and shear strength diminish. The strength of shear decline exhibits a pattern of first being quick, followed by sluggish. The strength of shear and cohesiveness of clayey sandy soil declined under the influence of the dry and wet cycles, with the first cycle primarily affecting variations in cohesiveness and strength of shear. Furthermore, the strength of shear and cohesiveness of clayey sandy soil diminish more with increasing wet and dry cycle amplitude and upper water content limits. Lastly, the drying shrinkage and hygroscopic expansion of clay particles in clayey sandy soils during wet and dry cycles are not significant, resulting in less structural damage and deterioration of the mechanical properties of the soils. The study's findings have a significant impact on the durability of roadbeds made of unsaturated clayey sandy soil in both wet and dry situations.

Keywords: clayey sand; wet–dry cycle; shear strength; moisture content

1. Introduction

Clayey sand is widely distributed in regions such as Southeast China, South China, and the Middle East. It is a common roadbed filler [1]. Since the clayey sand contains a certain amount of clay particles, the soil possesses some properties of both sand and clay [2]. In actuality, clayey sandy soil subsoils found beneath roads are frequently unsaturated soils. As a result of atmospheric rainfall and fluctuations in the water table, certain clayey sandy soil subsoils undergo extended periods of wet–dry cycling. Soils tend to deteriorate when subjected to wet-dry cycles or changes in moisture content. This has a direct impact on the roadbed pavement structure's strength, stiffness, and stability, which lowers the road's quality and lifespan [3]. The engineering properties of significant variations existed between the sand soil and clay and clayey sand [4]. However, according to the current engineering design standards, clayey sand is often regarded as sand for design, and some characteristics of the clay particles in the clayey sand are ignored, which will have a certain impact on the rationality, safety, and economy of road design [3].

Scholars have conducted numerous investigations to comprehend the impact of wet and dry cycles and moisture content on the engineering qualities of soils. And as the amount of moisture content changes, the mechanical properties such as shear strength, constitutive relation, cohesiveness, and the soil's frictional angle will alter [5–9]. For clayey sandy soils, Naser A et al. [4] investigated the impact of clay content as well as water percentage on soil cohesion through experiments and found that the cohesion of clayey sand increases and then decreases with the increase of water content. Tang et al. [10] carried out plate loading tests on sand–kaolinite mixtures with different water tables and found that the bearing capacity of clay containing sand increased with decreasing water content. Soils with a high content of clay-grained minerals are usually degraded to varying degrees under the action of wet and dry cycles [11–14]. In recent years, academics have conducted a number of studies on how wet–dry processes and the amount of moisture affect the soil's engineering qualities. Factors affected during the wet–dry cycle include soil shear strength, expansion and contraction deformation, microstructure, and the changing pattern of soil–water characteristic curves [15–17]. Studies such as those carried out by Hu et al. [18] and Ye et al. [19] have shown that different wet and dry cycling path parameters, such as cycling amplitude and upper limit moisture content, affect the degree of deterioration. The above research results have clarified the impact of wet and dry cycles and water content on the mechanical characteristics of soils, which are of great theoretical and engineering significance for the deformation and stability of roadbeds. However, most of the research objects are clay, sandy soil, or a mixture of clay and sand as clayey sand samples, and there are fewer studies on natural clayey sand. Meanwhile, the behaviors and properties of clayey sand under the action of different paths of wet and dry cycling are not clear, so the study of this topic is of great significance to accurately analyze the stability of the road subgrade under the shift in the water content and the action of wet and dry cycles.

Remodeled clayey sand is used as the object of study, wet and dry cycles and tests for triaxial shear were conducted to investigate the stress–strain curves, shear strength, cohesion, and the internal friction angle of the clayey sandy soil under the action of different water content and the different paths of the wet and dry cycles, so as to give the engineering application a reference for unsaturated clayey sandy soil.

2. Material and Methodology

2.1. Examine the Soil

The roadbed filler of the Weihai–Yanhai segment of the Rongwu Expressway provided the soil specimens used in this investigation, which possess a particular density of 2.64 g/cm^3, an ideal level of moisture of 12.5%, and a maximum dry weight of 1.79 g/cm^3. Its distribution of particle sizes is displayed in Table 1. From the perspective of particle composition, the content of clay particles in the soil samples used was about 14.46%, so it was classified as clayey sand [20], Figure 1 displays example photos of the dried and sieved soil samples and the untreated soil specimens.

Figure 1. Untreated soil specimens and the expansion site of the Rongwu Expressway's Weihai–Yanhai section.

Table 1. Particle grading of clayey sand.

>2 mm	1 mm~2 mm	0.5 mm~1 mm	0.25 mm~0.5 mm	0.075 mm~0.25 mm	<0.075 mm
13.41	12.31	18.90	21.98	18.94	14.46

2.2. Test Program and Methodology

The following test protocol was used to determine how different wet–dry cycling pathways and the amount of moisture affected the strength properties of clayey sand:

1. Preparation of a sample

According to past tests, clay-containing sandy soil has a strong cohesive force in its unsaturated state, and after testing, the clay-containing sandy soil in this test has a cohesive force of 42.8 kPa at 12% water content, which can be used to prepare the disturbed soil specimen by the compaction method. According to the Highway Geotechnical Test Regulations (GB/T 50123-2019) [21], its specific process is as follows: (1) The clayey sandy soil should be air-dried, crushed, and sieved. Soil samples should be prepared at 12.5% amount of moisture. After a day of sealed resting, the moisture content should be measured, and any errors in moisture content should be less than 0.5%. (2) The soil was compacted in 5 layers in the sample-making mold using a static pressure device, and the dry density was controlled to be 1.71 g/cm^3 (95% compaction), taking into account the compaction requirement of the highway subgrade. The soil samples measured 39.1 mm in diameter and 80 mm in height. (3) The soil specimens were removed from the sample-making mold and carefully placed on the three-flap mold to secure the two ends. The samples' initial dry density and pore space ratio were then calculated, and the height as well as the mass were measured. To ensure that we could examine the impact of moisture percentage on unsaturated clayey sandy soils, twenty-four specimens were created, and forty-five specimens underwent wet and dry cycle testing.

2. Test how the moisture amount affects the strength properties of clayey sand

Based on the results of the field tests, the moisture content of the subgrade at different positions was mostly between 6% and 14%. Therefore, six groups of tests with different moisture content were set up, which were 6%, 8%, 10%, 12%, 14%, and 16%, respectively. The initial amount of moisture and the initial specimen's drying density were 12.5% and 1.71 g/cm^3, respectively. The soil samples were sealed for 24 h after slowly air-drying or adding distilled water to the permeable stone to reach the target moisture content, followed by strength testing [5].

3. Experiment on different wet and dry cycle paths, and how they affect the strength characteristics of clayey sand

To look into how wet–dry paths for cycling affect the clayey sand's strength properties, three groups of paths were set up for wet–dry cycle tests with the following design scheme: the wet–dry cycle amplitudes of paths 1 and 2 were different, with the same lower limit for the amount of moisture; the wet–dry cycle amplitudes of paths 1 and 3 were different, with the same upper limit for the amount of moisture; paths 2 and 3 have the same amplitude of wet and dry cycles and different upper and lower moisture content moisture content. Through comparing distinct routes for the wet–dry cycle, the influence of the magnitude of the wet–dry cycling and the upper as well as lower limits of moisture on wet–dry cycling could be analyzed. Generally, the number of cycles of wet and dry clay is mostly within the scope of 3~12 times [13–19] so in this test, the most wet–dry cycles possible was set at seven. The strength test was carried out after the 0th, 1st, 3rd, 5th, and 7th wet–dry cycles, respectively. As stated by the design scheme, the cycle of wet and dry test design appears in Table 2, where the 2% lower limit for the amount of moisture is the residual amount of moisture of the sandy soil dried at low temperature, and the 22% upper limit for the amount of moisture is the amount of saturating moisture of the clayey sand.

Table 2. Wet–dry cycles test design.

Path	Initial Moisture Content/%	Lower Limit Moisture Content/%	Upper Limit Moisture Content/%	Amplitude%	Number of Cycles/Times
1	12.5	2	22	20	0, 1, 3, 5, 7
2	12.5	2	12	10	0, 1, 3, 5, 7
3	12.5	12	22	10	0, 1, 3, 5, 7

There are 45 specimens in total in the wet and dry cycle test, divided into three wet and dry cycle paths, each with 15 specimens. Three specimens per group were exposed to 0, 1, 3, 5, and 7 cycles of wet and dry cycling over the corresponding pathways. The saturated humidification of the specimen adopted the immersion method. The specimen was saturated by immersing it in distilled water for 1 day. The unsaturated humidification of the specimen was carried out by adding distilled water to the permeable stone through the test tube. The soil sample's increased amount of moisture was less than 2% each time, and the soil sample was left to stand for 2 h after each humidification process. Once the soil sample's amount of moisture reached the desired level, it was sealed so that the moisture could migrate completely and disperse evenly for one day. During the procedure of drying, the oven temperature was adjusted to 40 °C, and the weight of the specimen was measured every hour until the specimen reached the predetermined moisture content.

4. Triaxial shear test

To investigate the qualities of the strength of unsaturated clayey sandy soils, unconsolidated ventilated undrained shear tests (zero air pressure throughout the test, no measurement of pore water pressure) were conducted on remolded clayey sandy soils with a constant controlled net perimeter pressure. The soil specimens' confining pressure (σ_3) for the moisture content test was 50, 100, 200, and 300 kPa, respectively. The perimeter pressures of the specimens after the wet–dry cycle tests were taken as 50, 100, and 200 kPa, and 8% was selected as the final moisture content which was based on the information from the site investigation. The specimen's level of moisture was gauged again prior to the test to ensure that the error of the amount of moisture for the sample was less than 0.3% [13]. The triaxial test was conducted in compliance with the Standard for Unsaturated Soil Testing Method (T/CECS 1337-2023) [22], accounting for the roadbed's actual working soil conditions. The test is stopped when the specimen is entirely destroyed or the axial strain reaches 15%.

3. The Results

3.1. Impact of Moisture Content on Influencing Characteristics of Clayey Sand

To investigate the impact of moisture content on the properties of the strength of unsaturated clayey sand, the stress–strain curves of unsaturated clayey sand with different moisture content were obtained by triaxial test in this paper. As shown in Figure 2, the strength of shear of clayey sand is highly dependent on the amount of moisture present, which decreases significantly as the moisture level rises. Under 50 kPa and 300 kPa confining pressure, the strength of 16% moisture content specimens decreased by 63% and 37.3%, respectively, compared with 6% moisture content. This was mainly because of the rise in the amount of moisture, and more free water on the particle surface which had a lubricating effect, so the friction was reduced, which made the strength of soil samples lower. In addition, the rise in the amount of moisture lead to the expansion of the volume of the specimen and an increase in the pore ratio, making the structure looser, which was also one of the reasons for the reduction of shear strength.

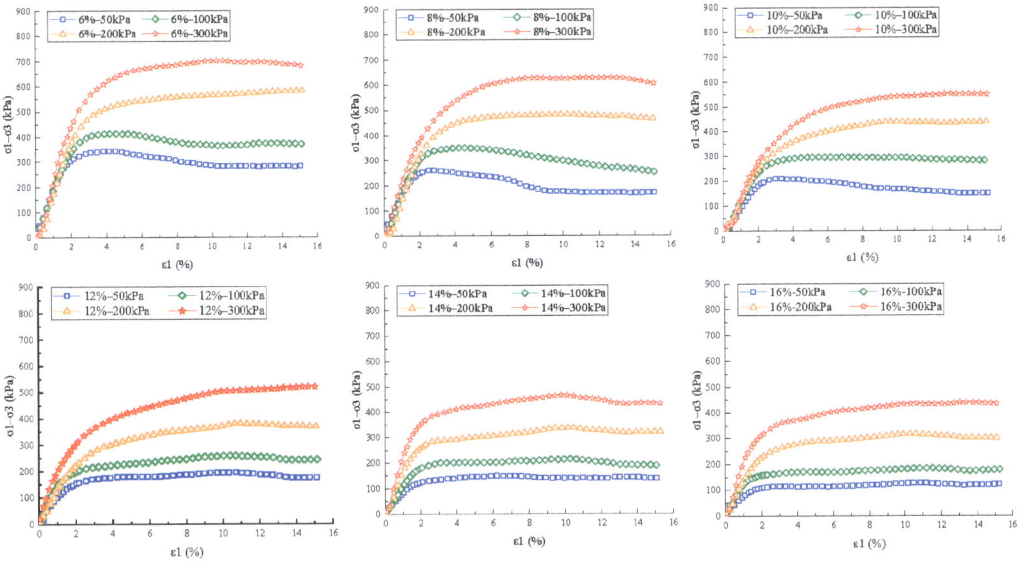

Figure 2. Curves of stress and strain for clayey sand with different moisture contents.

Meanwhile, the figure illustrates that some of the specimens of soil show brittle damage when the moisture content is small (<12%), while the soil samples basically show plastic damage when there is a lot of moisture present (≥12%). When the moisture content is less than 12%, clayey sand under low confining pressure (50–100 kPa) shows a pattern of stress softening. As the pressure to confine increases, the specimens of soil are transformed from stress softening to stress hardening, and the soil specimens' maximum shear strength appears at larger axial strains. When the moisture content is 12 percent or higher, the clayey sand specimens with different confining pressures all show a trend of stress hardening. From a microscopic perspective, the microstructure of the soil changes under high confining pressure, which results in increased intergranular embedding and a denser arrangement of soil particles. At the macro level, it was shown that the soil specimen's shear strength has a higher and stronger resistance to external deformation [6,23]. Whereas, when the confining pressure was identical, the rise in the amount of moisture makes the skeleton of the soil soften, and its shear strength decreases while showing stress hardening [7].

Figure 3 shows the test curves for the unsaturated clayey sandy soil's peak shear strength at various water contents. The difference in primary stress at 15% of the axial strain is used to determine the peak strength when the specimen lacks a discernible peak. Evidently, the strength of shear and confining pressure under each amount of moisture show a strong linear connection, which may be described by the Moore Coulomb strength relationship equation. At the same time, with the increase in moisture content, the strength of the shear of clayey sand declines. The strength of clayey sand with different enclosing pressures decreased by 12~23.6% when the 6% to 8% amount of moisture was raised, but decreased by 6~15.3% when the 14% to 16% moisture content was increased. This shows that, under similar pressure around the circumference, the peak shear strength of clayey sandy soil decreases non-linearly with a rise in the specimen's water content. In other words, the specimen's peak strength is progressively less sensitive to a rise in the amount of water, but it still decreases significantly with an up to a 16% rise in the amount of water.

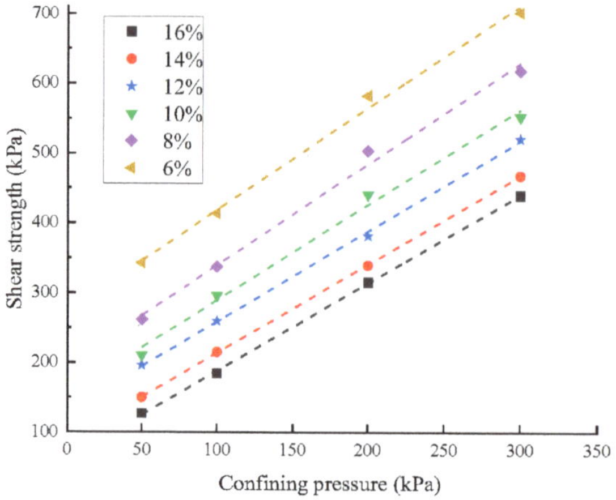

Figure 3. Shear strength of samples of soil with varying levels of moisture.

The result of the moisture percentage change on the cohesiveness and angle of internal friction of unsaturated clayey sand is shown in Figure 4. It appears that with the rise in the amount of moisture, the cohesion of clayey sandy soils shows a decreasing trend. Among them, the cohesion of 16% moisture content clayey sand decreased by 76.2% compared with that of 6% moisture content. For unsaturated clayey sand, the cohesion came from the soil's water's capillary movement [8] and the cementation body formed between clay particles and water [9]. When the amount of moisture rose, the matrix suction of soil became smaller and the water's lubricating properties reduced the soil particles' interaction with one another, both of which reduced the cohesion [6].

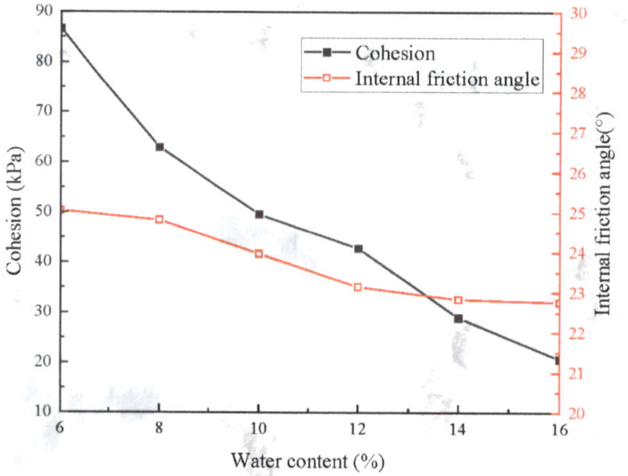

Figure 4. Impact of the amount of moisture on clayey sand's shear strength index.

Meanwhile, as the amount of moisture percentage rises, the angle of internal friction shows a decreasing trend. This is because there is more free water available between particles in the soil as the moisture percentage increases, and the free water leads to increased lubrication between the granules of soil, so the angle of internal friction decreases [9]. However, it is worth noting that the result of moisture content regarding the clayey sands' internal angle of friction is small, and the internal angle of friction of the 16% moisture percentage specimen is only reduced by 9.1% compared to that of the 6% moisture content specimen.

Through experimentation, Naser A [4] discovered that as the water content increases, the cohesiveness of clayey sandy soils first increases and subsequently declines. This could be because the unsaturated soils' shear strength increases because the matrix suction of the soil gradually increases as the amount of water drops. The presence of a peak in cohesive strength is a macroscopically manifested indicator that the area of matrix suction reduces when the water content is further reduced, and this factor has a greater influence than the increase in matrix suction. The gradation of the clayey sandy soil, however, has a significant impact on this peak; as a result, in this test, the moisture content of the clayey sandy soil varies from 6% to 16%, and the cohesiveness peak is not reached. Since this test's clayey sandy soil water content range comes from a field test, it is reasonable to infer that when the water content decreases, the roadbed's clayey sandy soil strength parameter will rise noticeably. Therefore, the water content needs to be controlled by as many anti-drainage measures as possible during the project to avoid the roadbeds being negatively affected by rainfall or rising water tables. The strength parameters of clayey sandy soils increase significantly as the water content decreases. This implies that, rather than calculating strength using the soil moisture content or optimal moisture content at the time of filling, the design strength of the roadbed should be taken into account as much as possible when it reaches equilibrium moisture content throughout the operation period.

3.2. Impact of Wet–Dry Cycling on Clayey Sand's Strength Characteristics

The strain–stress diagrams of clayey sand with distinct routes for the wet–dry cycling and different quantity of cycles between wet and dry were obtained by triaxial test, as shown in Figure 5. In the figure, 1-1-50 kPa in the legend indicates that the wet–dry cycle path is path 1, the quantity of cycles between wet and dry is 1, and the confining pressure is 50 kPa. The stress–strain curves before wet and dry cycling are shown in Figure 2 for the test results of the 8% moisture content specimens.

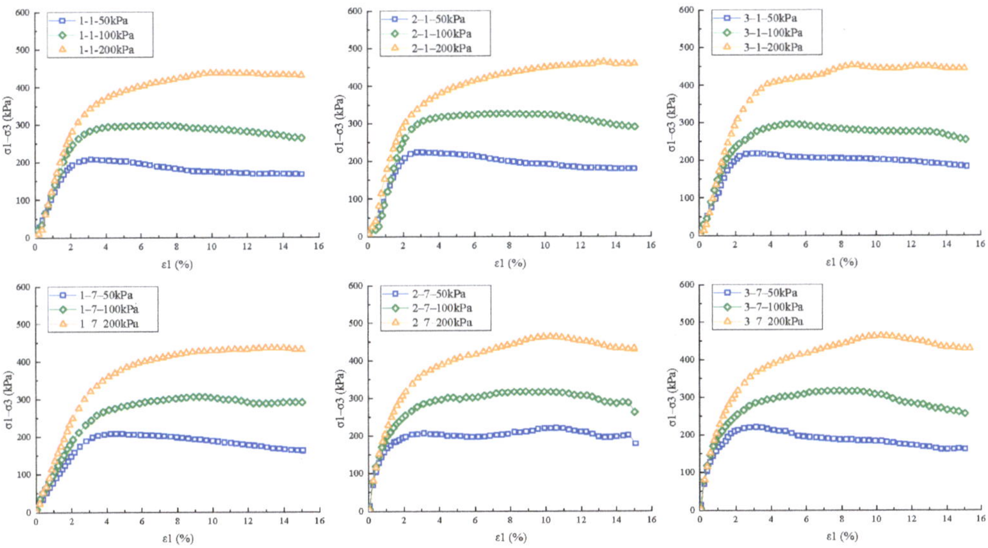

Figure 5. Soil sample stresses and strains under various conditions and the quantity of wet–dry cycles.

For wet–dry cycle action, the clayey sand had good water stability. Under different paths of wet–dry cycling, the shear strength exhibited a downward trend, and the decreasing rate ranged from 1.6% to 24.1%, which was much smaller than that of ordinary clay under wet–dry cycles. Specifically, when the confining pressure was 200 kPa, the decreasing rate was within 10%. Analyzing the reasons, it is believed that the result of wet–dry cycling on soil is represented in two aspects. First, the enlargement and reduction of clay particles brought as a result of wet–dry cycles results in cracks in soil and an increase in the soil pore ratio. Secondly, after the repeated infiltration and dissipation of pore water, the corners of soil particles are effectively washed away and abraded, and the cementing material between aggregates is dissolved [13]. After the wet–dry cycle, no significant pore and crack development was found in the cohesive sand. Based on the decrease in the shear strength of cohesive sand, it is inferred that due to the small number of cohesive sand particles, the expansion and contraction of cohesive sand are not significant under wet–dry cycling. And the alteration of soil structure caused by the repeated infiltration and dissipation of pore water is the primary cause of the strength of shear decline in clayey sand.

Meanwhile, the decrease in shear strength for clayey sand declines noticeably when confining pressure increases. When the confining pressure is 50 kPa, the strength of shear in the soil sample decreases by 15.8~24.1% and 24.1% at most. While the confining force is 100, 200 kPa, the decreases are 6.8~18.3% and 1.6~7%, respectively.

It is important to note that where the strength qualities are impacted by wet and dry cycles of clayey sandy soil differs significantly from those of clayey soil. First off, as Figure 5 illustrates, the stress–strain relationship of clayey sandy soil is less impacted by wet and dry cycles. Low confining pressures tend to soften under wet and dry cycles, whereas high confining pressures tend to harden under such cycles. On the other hand, following wet and dry cycling, the structural characteristics of clayey soils are harmed, and their stress–strain relationship tends to change in favor of stress hardening [24]. Second, the clayey sandy soil in Figure 6a did not exhibit visible cracking during wet cycling, whereas the clayey soil in Figure 6b demonstrated visible crack development following the mixing cycle. Ultimately, as Figure 7 illustrates, although clayey sandy soil's shear strength declined as a result of wet and dry cycles, the strength of shear did not exhibit a trend of gradual decline during the increased number of wet and dry cycles, and after the first rounds of wet and dry, the strength of shear of clayey sandy soil essentially remained stable. Although clayey soil fissures continuously develop as a result of wet–dry cycle, which is the primary source of clayey soil strength decay, clayey soils often exhibit a pattern of strength stabilization after three to twelve cycles [24,25]. These phenomena demonstrate that less structural damage is caused by the expansion and contraction of clay particles in clayey sand during wet and dry cycling.

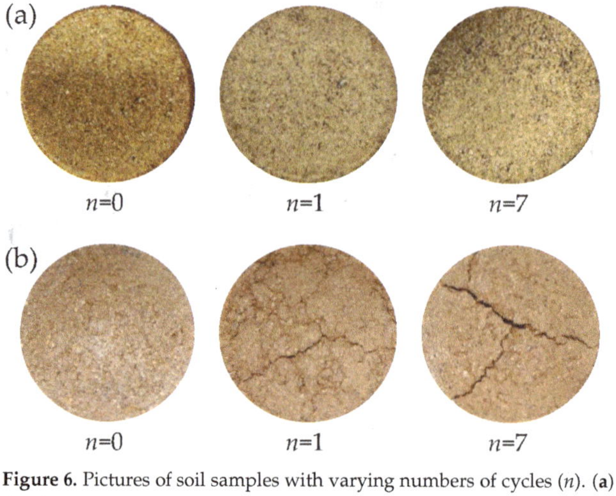

Figure 6. Pictures of soil samples with varying numbers of cycles (n). (**a**) sandy clay soil; (**b**) clay soil that is powdery.

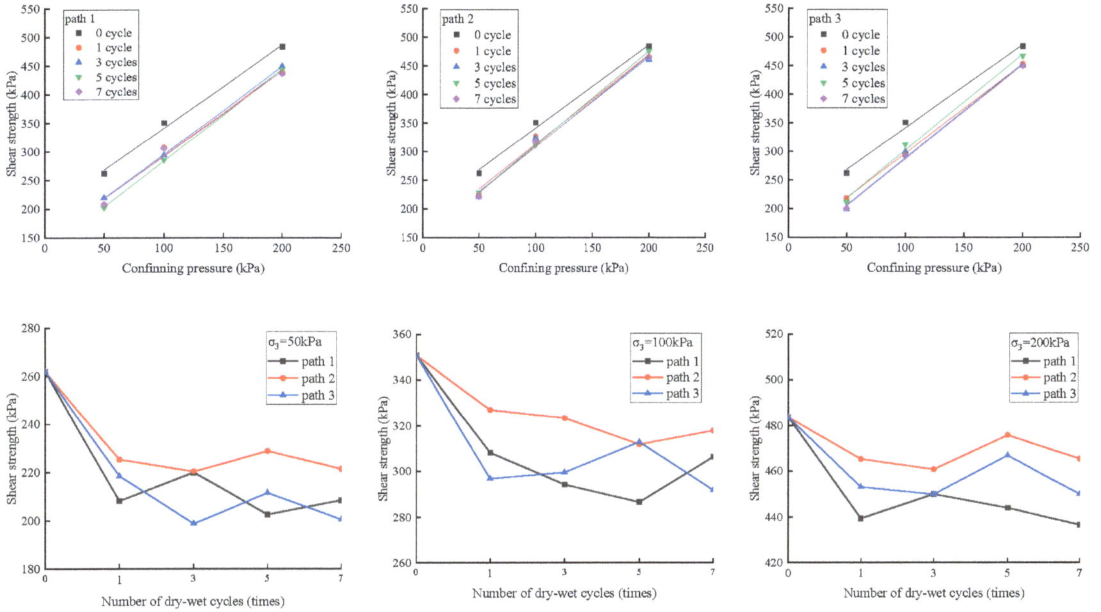

Figure 7. Peak shear strength of soil samples under different wet and dry cycle paths.

Meanwhile, the decrease of the strength of shear of clayey sand decreases significantly with the rise in confining pressure. When the stifling force is 50 kPa, the strength of shear in the soil sample decreases by 15.8~24.1% and 24.1% at most after different number of dry–wet cycles. When the confining pressures are 100 kPa and 200 kPa, the decreases are 6.8~18.3% and 1.6~7%, respectively.

According to Figure 7, compared with the soil samples without wet–dry cycling, the soil's shear strength following the wet–dry cycling of paths 1, 2, and 3 decreased by 7~22.7%, 1.6~15.8%, and 3.5~24.1%, respectively. Among them, the decrease in shear strength of soil samples under path 2 is the lowest, while the decrease in shear strength under paths 1 and 3 is close. While paths 1 and 3 have similar magnitudes of intensity decay, paths 1 and 3 have the same upper limit moisture content, different cyclic amplitudes, and similar strength attenuation. Path 2 has a small cycling amplitude and a lower upper and lower cycling limit for water content, with the smallest shear strength decay. Therefore, comparing the different cycling paths, it can be seen that a larger upper limit water content and cycling amplitude both lead to a higher degree of strength deterioration.

The connection between the specimen's cohesiveness and angle of internal friction under different paths and times of wet–dry cycle is displayed in Figure 8. It is evident from Figure 8a that the cohesion of clayey sand decreases after wet–dry cycling, and following the initial cycle, the cohesiveness attenuation amplitude is at its maximum, and then the attenuation amplitude decreases greatly and seems to remain steady. The deterioration of soil cohesion under paths 1, 2 and 3 is about 30%, 25% and 35%, respectively. Comparing different cyclic paths, it appears that the larger upper limit for the amount of moisture and cyclic amplitude will lead to a higher degree of deterioration of cohesion, which is consistent with the impact of wet–dry cycle paths regarding strength of shear. However, it is worth noting that under different paths the difference in the degree of deterioration of cohesion of clayey sand is small.

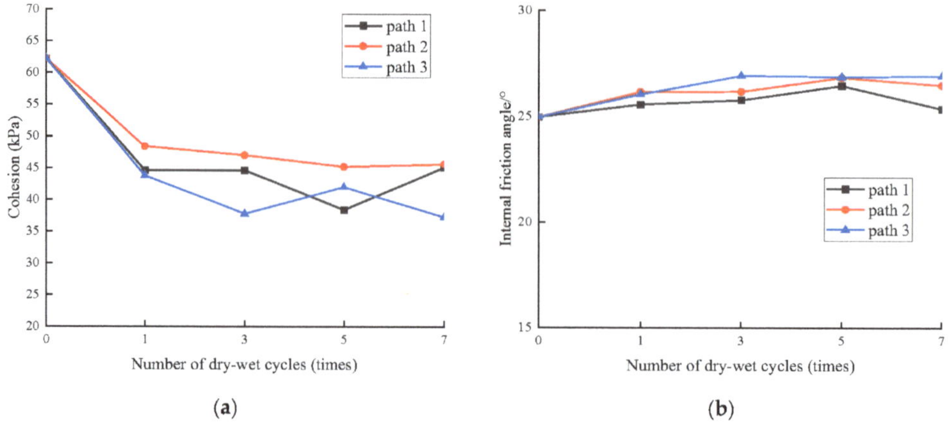

Figure 8. The impact of wet–dry cycles concerning the cohesion and angle of internal friction. (**a**) Cohesion; (**b**) Angle of internal friction.

In Figure 8b, clayey sand's angle of internal friction fluctuates within 2° following wet–dry cycling, and the spectrum of changes is small, which is much smaller than the attenuation range of cohesion. Therefore, the angle of internal friction of clayey sand is not attentive to wet–dry cycles. The analysis of the causes suggests that the angle of internal friction is relatively stable, but this does not mean that it is unaffected by wet–dry cycles, but rather that two influences lead to this phenomenon. On the one hand, the wet–dry cycles decrease the angle of internal friction and destroy the soil formation. Conversely, though the matrix suction during wet–dry cycles makes the soil compacted and increases the occlusal friction, the combined effect of these two factors causes the clayey sandy soils' internal friction angle to change somewhat, but the overall stability is maintained.

The deterioration of clayey sandy soils caused by wet and dry cycles is shown in the decrease of peak strength and cohesiveness. The project should aim to prevent wet and dry cycles from occurring or manage their amplitude in order to improve the stability of the roadbed. The more pronounced the deterioration of the soil body, the greater the amplitude of wet and dry cycles. On the other hand, clayey sandy soils degrade less under the influence of wet and dry cycles and have greater water stability than clayey soils.

4. Conclusions

In this study, the effects of moisture percentage and wet–dry cycles regarding the strength of shear characteristics of clayey sand were investigated by indoor triaxial examinations, according to which the ensuing deductions can be drawn:

1. The peak strength of unsaturated clayey sandy soil drops first swiftly and then gradually as its moisture content rises, but the strength of shear and cohesiveness of the soil decline continually. Simultaneously, the increase of water content makes the stress–strain relationship of low confining pressure clayey sandy soil change from stress softening to stress hardening. Water content has a large impact on the mechanical properties of unsaturated clayey sandy soil, and the project should try to control the increase in water content of the soil body, and in the design of the strength discount caused by the change in water content.
2. Wet and dry cycles reduced the strength of shear and cohesiveness of clayey sandy soil; the more the amplitude of wet and dry cycles and the higher the limit of moisture content, the more the soil's shear strength and cohesiveness dropped. Furthermore, the changes in shear strength and angle of internal friction were mainly concentrated in the first cycle.

3. The drying shrinkage and hygroscopic expansion of clay particles in clayey sandy soils during wet and dry cycling are not significant, resulting in less structural damage and deterioration of the mechanical properties of the soils. However, the experiments in this paper have not been carried out from a microscopic perspective to corroborate this result, and the conduct of electron microscope scanning experiments will help to further elucidate the mechanism of this phenomenon.
4. The angle of internal friction of unsaturated clayey sandy soils is increased by both decreasing the water content and wet–dry cycling. However, these effects are negligible. The variation in the angle of internal friction of soil samples under various conditions is less than 10 percent.
5. For roads that are subjected to wet and dry cycling over extended periods of time, clayey sandy soils are more suited than clayey soils with a greater clay mineral content.

Author Contributions: Conceptualization, C.W. and M.W.; Methodology, C.W. and M.W.; Formal analysis, C.W., W.Y. and Z.Z.; Investigation, W.Y. and Z.Z.; Data curation, C.W., W.Y., S.W. and C.M.; Writing—original draft, C.W. and N.Z.; Writing—review & editing, N.Z. and S.W.; Supervision, N.Z.; Project administration, C.M. All authors have read and agreed to the published version of the manuscript.

Funding: This research received no external funding.

Data Availability Statement: The datasets in the current study are available from the corresponding author upon reasonable request.

Conflicts of Interest: The research was conducted in the absence of any commercial or financial relationships that could be construed as a potential conflict of interest.

References

1. Arabani, M.; Karami, M. Geomechanical properties of lime stabilized clayey sands. *Arab. J. Sci. Eng.* **2007**, *32*, 11–25.
2. Ma, H.Y.; Yang, H.W.; Wang, X.Y.; Sun, C.S. Research into soil classification and testing of certain soil in North China. *Rock. Soil. Mech.* **2009**, *30*, 391–393.
3. Akbarimehr, D.; Rahai, A.; Eslami, A.; Karakouzian, M. Deformation characteristics of rubber waste powder-clay mixtures. *Sustainability* **2023**, *15*, 12384. [CrossRef]
4. Al-Shayea, N.A. The combined effect of clay and moisture content on the behavior of remolded unsaturated soils. *Eng. Geol.* **2001**, *62*, 319–342. [CrossRef]
5. Jiang, Y.; Chen, W.W.; Wang, G.H.; Sun, G.P.; Zhang, F.Y. Influence of initial dry density and water content on the soil–water characteristic curve and suction stress of a reconstituted loess soil. *Bull. Eng. Geol. Environ.* **2017**, *76*, 1085–1095. [CrossRef]
6. Liang, Y.; Li, W.; Wang, X. Influence of water content on mechanical properties of improved clayey soil using steel slag. *Geotech. Geol. Eng.* **2013**, *31*, 83–91. [CrossRef]
7. Kang, X.; Wang, S.; Yu, Z. Effects of soil–water interaction on the mechanical behaviors of shear-zone soils. *Int. J. Geomech.* **2022**, *22*, 06022028. [CrossRef]
8. Rasool, A.M.; Kuwano, J. Instability of unsaturated soils under constant deviatoric stress in drained conditions. *Iran. J. Sci. Technol.-Tran. Civ. Eng.* **2021**, *46*, 419–434. [CrossRef]
9. Matsumura, S.; Tatsuoka, F. Effect of compaction conditions and fines content on cyclic undrained strength of saturated soils. *Soil. Dyn. Earthq. Eng.* **2018**, *112*, 152–161. [CrossRef]
10. Tang, Y.; Chen, C.H.; Qian, B.; Ren, J.; Guan, Y.F. Plate load tests on an unsaturated sand-kaolin mixture with varying water table. *Sensors* **2022**, *22*, 2161. [CrossRef]
11. Kang, Q.R.; Xia, Y.D.; Li, X.S.; Zhang, W.Z.; Feng, C.R. Study on the effect of moisture content and dry density on shear strength of silty clay based on direct shear test. *Adv. Civ. Eng.* **2022**, *2022*, 2213363. [CrossRef]
12. Zhao, X.S.; Fu, Z.T.; Yang, Q.J.; Yao, K.; Geng, D.X.; Li, K. Subgrade fill strength and bearing characteristics of weathered phyllite blended with red clay. *Road. Mater. Pavement. Des.* **2020**, *22*, 2571–2590. [CrossRef]
13. Goh, S.G.; Rahardjo, H.; Leong, E.C. Shear strength of unsaturated soils under multiple drying-wetting cycles. *J. Geotech. Geoenviron. Eng.* **2013**, *140*, 06013001. [CrossRef]
14. Yuan, K.; Ni, W.; Lü, X.; Vecchia, G.D.; Wang, H.; Li, L.; Nie, Y. Influence of wetting–drying cycles on the compression behavior of a compacted loess from microstructure analysis. *Bull. Eng. Geol. Environ.* **2022**, *81*, 348. [CrossRef]
15. Huang, Z.; Zhang, H.; Liu, B.; Wei, B.X.; Wang, H. Using CT to test the damage characteristics of the internal structure of expansive soil induced by dry–wet cycles. *AIP. Adv.* **2021**, *11*, 075305. [CrossRef]

16. Zhou, Z.H.; Bai, Y.; Wu, Y.T.; Chen, Y.Q.; Guo, Z.; Cheng, W.K. Multiscale study on the microstructural evolution and macromechanical deterioration of expansive soil under dry-wet cycles. *J. Mech.* **2022**, *38*, 610–620. [CrossRef]
17. Zhai, J.Y.; Cai, X.Y. Strength characteristics and slope stability of expansive soil from Pingdingshan, China. *Adv. Mater. Sci. Eng.* **2018**, *2018*, 3293619.
18. Hu, C.M.; Yuan, Y.L.; Mei, Y.; Wang, X.Y.; Liu, Z. Comprehensive strength deterioration model of compacted loess exposed to drying-wetting cycles. *Bull. Eng. Geol. Environ.* **2020**, *79*, 383–398. [CrossRef]
19. Ye, W.J.; Bai, Y.; Cui, C.Y.; Duan, X. Deterioration of the internal structure of loess under dry-wet cycles. *Adv. Civil. Eng.* **2020**, *2020*, 8881423. [CrossRef]
20. Chang, W.J.; Hong, M.L. Effects of clay content on liquefaction characteristics of gap-graded clayey sands. *Soils. Found.* **2008**, *48*, 101–114. [CrossRef]
21. *GB 50123-2019*; Highway Geotechnical Test Regulations. Standard Press of China: Beijing, China, 2019.
22. *T/CECS 1337-2023*; Standard for Unsaturated Soil Testing Method. Construction Industry Press of China: Beijing, China, 2023.
23. Sadeghi, M.M.; Beigi, F.H. Dynamic behavior of reinforced clayey sand under cyclic loading. *Geotext. Geomembr.* **2014**, *42*, 564–572. [CrossRef]
24. Fang, J.J.; Yang, X.L.; Feng, Y.X.; Gong, J.; Niu, H.C. True triaxial experimental study on mechanical properties of expansive soils after drying and wetting cycles. *Chin. J. Rock Mech. Eng.* **2021**, *40*, 1043–1055.
25. Zhou, C.M.; Zhu, R.; He, Q.; Sun, D. Study on the variation law of sliding soil strength under different dry-wet cycles. *J. Disaster Prev. Mitig. Eng.* **2022**, *38*, 1–8.

buildings

MDPI

Article

Numerical Investigation on the Seismic Behavior of Novel Precast Beam–Column Joints with Mechanical Connections

Mei-Ling Zhuang [1,2,3], Chuanzhi Sun [4], Zhen Yang [2,*], Ran An [3], Liutao Bai [5], Yixiang Han [1] and Guangdong Bao [1]

1 School of Transportation and Civil Engineering, Nantong University, Nantong 226019, China; ml_zhuang99@163.com (M.-L.Z.); 18952292290@163.com (Y.H.); 18051133608@163.com (G.B.)
2 Water Resources Research Institute of Shandong Province, Jinan 250013, China
3 School of Civil Engineering, Shandong University, Jinan 250061, China; ar_ran@126.com
4 School of Civil Engineering and Architecture, Suqian College, Suqian 223800, China; schzh_xzh@163.com
5 School of Civil Engineering, Southeast University, Nanjing 211189, China; bailiutao99@163.com
* Correspondence: skyyangzhen@shandong.cn

Abstract: Traditional cast-in-place beam–column joints have the defects of high complexity and high construction difficulty, which seriously affect the efficiency and safety of the building construction line, and precast beam–column joints (PBCJs) can greatly improve the construction efficiency and quality. At present, the investigations on the seismic behavior of precast reinforced concrete structures are still mainly focused on experiments, while the numerical simulations for their own characteristics are still relatively lacking. In the present study, the seismic behavior of novel precast beam–column joints with mechanical connections (PBCJs-MCs) is investigated numerically. Based on the available experimental data, fiber models for four PBCJs-MCs are developed. Then, the simulated and experimental seismic behaviors of the prefabricated BCJs are compared and discussed. Finally, the factors influencing the seismic behavior of the PBCJs-MCs are further investigated numerically. The numerical results indicate that the fiber models can consider the effect of the bond–slip relationship of concrete and reinforcement under reciprocating loads. The relative errors of the simulated seismic behavior indexes are about 15%. The bearing capacity and displacement ductility coefficients of the PBCJs-MCs decrease rapidly as the shear-to-span ratio (λ) increases. It is recommended that the optimum λ for PBCJs-MCs is 2.0–2.5. The effect of the axial load ratio on the seismic behavior of PBCJs-MCs can be negligible in the case of the PBCJs-MCs with a moderate value of λ.

Keywords: precast beam–column joints; mechanical connection; seismic behavior; fiber models; bond–slip; load–displacement curves

1. Introduction

Compared with cast-in-place concrete structures, the most prominent feature of precast concrete structures is that the main concrete elements are prefabricated in the factory, a feature that allows them to show many advantages during construction. Each prefabricated component is equivalent to further discretizing the overall concrete structure. The composition of its basic units is more adapted to standardization and parameterization compared to the overall structure, which can enable the standardized design and factory flow production upstream and downstream of the whole industry [1]. With the main components prefabricated in the factory, the environmental pollution of the on-site construction can be effectively controlled, while the construction waste such as waste water and waste gas can be significantly reduced under the flow-through production method, and the turnover rate of auxiliary materials such as formwork can be improved to achieve energy conservation and environmental protection [2]. The standardized design is in line with the development trend of construction informationization, and, combined with BIM technology, big data technology, and information and communication technology, it can effectively improve the degree of informationization in the construction industry and realize modernization [3].

Precast reinforced concrete (PRC) structures have been widely used in civil engineering [3,4]. Beam–column joints (BCJs) in PRC structures are the core force-bearing parts, which not only affect the seismic behavior of PRC structures but also directly relate to the convenience and economy of construction. According to the presence or absence of post-cast sections at the joints, the precast concrete BCJs are divided into wet and dry connection forms [2,5]. At present, the investigations on the seismic behavior of PRC structures are still mainly focused on experiments [6–9], while numerical simulations for their own characteristics are still relatively lacking [10]. The experimental investigations on precast frame joint specimens are focused on the influence of parameters such as the connection methods on the seismic performance of BCJs [5]. However, most of the experimental studies on the seismic behavior of specimens are limited by the test conditions, time, and funding.

With the continuous development of elastic–plastic finite element theory and the rapid improvement of computer operation and processing capability, numerical simulation methods with good accuracy [10–12] have been widely applied in civil engineering. On one hand, numerical simulation can carry out a wide range of parameterized analyses on the basis of experiments and obtain richer analysis data while reducing the test input. On the other hand, numerical simulation can be used for structural optimization analysis, which can play an important role in guiding the experimental design and actual engineering design. Combined with experiment investigations, numerical methods are important for predicting the structural response of buildings. So far, many finite element models (FEMs) [13] have been developed to simulate the seismic behavior of reinforced concrete (RC) members. Precast BCJs are subjected to complex stresses and are prone to the formation of structural defects that lead to stress concentrations. In practical engineering, while the entire structure is often in an elastic phase, the joints may have transitioned to a plastic phase and suffered severe damage. This can eventually lead to structural failure. Therefore, it is crucial to focus on BCJs analysis for RC frame structures in numerical modeling. How to use the numerical simulation analysis to effectively reflect the seismic behavior of precast BCJs is of great significance to promote their development. Kremmyda et al. [14] simulated the hysteretic properties of precast joints using the ABAQUS software (ABAQUS 2011). In FEMs, a reasonable contact was set at the connection interface to simulate the shear damage of splice joints under reciprocating loads. Zoubek et al. [15] simulated the hysteresis performance of precast pin connection joints using the ABAQUS software. In Zoubek's model, solid elements were used for the beam–column members and concealed pins, which can simulate the slip effect of concealed pin connections under seismic action more accurately. Cao et al. [16] carried out an in-depth study on the numerical simulation method of PBCs using the OpenSees software, proposing a more refined analysis model applicable to both types of joints with wet and dry connections. In the model, the influence of energy dissipation elements such as prestressing and angles on the structure was considered. The seismic behavior of ten different types of precast joints was simulated to validate the accuracy of the FEMs. Most of the existing numerical simulation methods for precast concrete structures can be divided into two categories according to their modeling ideas: numerical simulation methods based on beam–column link elements and 3D solid elements [17]. The two types of numerical simulation methods either pursue the convenience of use or the accuracy of the mechanism; it is difficult to achieve unity in efficiency and precision, and each has its own advantages and disadvantages. Therefore, combined with the characteristics of the assembled concrete structures, the development of fine and efficient numerical simulation methods is still worthy of in-depth study.

The existing numerical simulation methods can approximate the specific force characteristics of precast structures such as bond–slip, shear behavior at the joint, etc. The accuracy of FEMs is closely related to the selected material constitutive model [18]. Currently, the OpenSees software has been widely adopted in various countries to conduct numerous simulations regarding practical engineering and tests. The accuracy and efficiency of the simulation results using the software have been verified [13,18]. In the OpenSees software, there are three main types of models used for the simulation of reinforced concrete members with link elements, namely member models based on test data, section models based on section stress–strain, and fiber models at the material level. Among them, the fiber models are computationally inexpensive, easy to model, and have better accuracy [17]. Paulay [19] pointed out that the deformation of BCJs consisted mainly of the shear deformation of joint shear blocks and corner deformation of the beam–column intersection through. Pantazopoulou and Bonacci [20] pointed out that the slip of reinforcement would lead to blocked load transfer at the intersection and further lead to damage of the joint shear blocks. The beam–column joint element proposed by Lowes and Altoontash [21] and improved by N-Mitra [22] consisted of three components to simulate different damage behaviors at the BCJs. The shear panel component in the middle of the beam–column joint element is used to simulate the shear behavior of the stiffness and strength degradation of the joint core under shear damage. Under low-cycle reciprocating loads, bond–slip occurs between the reinforcement and the concrete, which in turn leads to hysteresis loop pinching. A reinforcement bond–slip model, Bar–Slip, is developed in the OpenSees software using the reinforcement stress–slip relationship proposed by Eligehausen and Hawkins [23,24]. It can take into account the effects of the material properties of concrete strength, reinforcement, and the degree of anchorage, and thus analyzes the effect of the slip on the overall joint performance.

In a previous study [2], novel precast BCJs using mechanical connections (PBCJs-MCs) were proposed to improve the reliability and construction efficiency of PRC structures (see Figure 1). In Figure 1, the steel bars in the columns are bolted to the nuts [2]. Currently, there are some investigations on the seismic behaviors of PRC BCJs. Paul and Tanapornraweekit [25] evaluated the seismic performance improvement of composite BCJs using the LS-DYNA finite element software (Version 11 R 11.0.0). Yang et al. [26] tested and simulated the seismic performance of precast BCJs and found that the accuracy of the simulated results obtained from the ABAQUS software was good. Bohara et al. [27] evaluated the seismic behavior of composite wide BCJs and found that the simulated result using the LS-DYNA finite element software was in agreement with the experimental result. However, there are few numerical simulation studies on the seismic behavior of PBCJs-MCs.

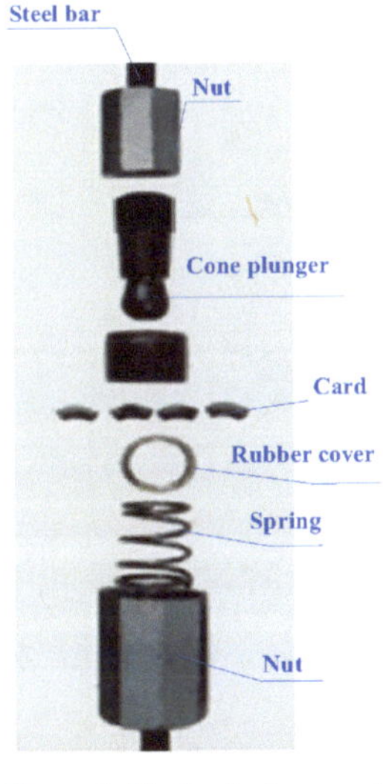

Figure 1. Mechanical connection of reinforcements.

Whether the existing constitutive models of concrete or reinforcement can accurately simulate the seismic behavior of PBCJs-MCs should be verified. Moreover, the parameters influencing the seismic behavior of PBCJs-MCs need to be analyzed using the numerical simulation method. In this present study, selecting appropriate element types and material constitutive models, the FEMs of PBCJs-MCs are established to analyze the seismic behavior of novel precast BCJs with mechanical connections. The simulation results are compared with the experimental results. Based on this, the factors influencing the seismic behavior of PBCJs-MCs are further investigated numerically.

2. Numerical Models of PBCJs-MCs

2.1. *Overall Design of the Quasi-Static Test Program*

Four PBCJs-MCs (specimens J-B_1C_2-F_1, J-B_2C_2-D_1, J-B_2C_2-F_1, and J-B_2C_2-D_2) from Ref. [2] are selected for the numerical simulation. The axial load ratios (n) of the four PBCJs-MCs are all 0.15. The shear-to-spans (λ) of the four PBCJs-MCs are all 2.0. The design drawing of the four specimens is described in Figure 2. In Figure 2, C denotes HPB300 steel bars and D denotes HRB400 steel bars. All precast columns are specified by mechanical connections. Other design details of the four PBCJs-MCs are listed in Table 1. The concrete grade of precast beams and columns are composed of C35. From Ref. [2], the material properties of concrete and steel bars can be obtained.

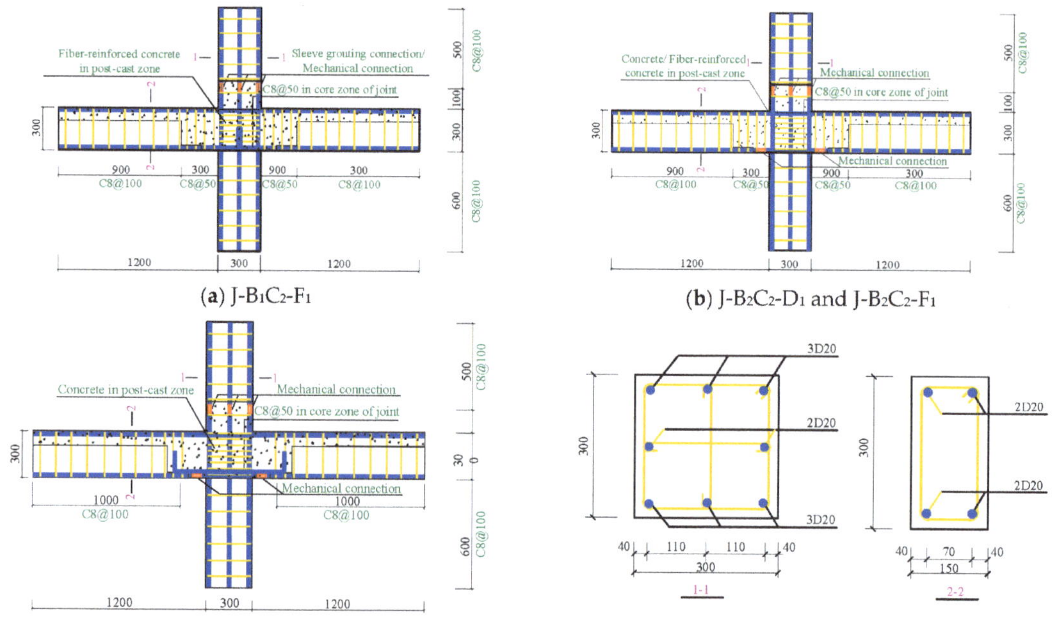

(a) J-B_1C_2-F_1 (b) J-B_2C_2-D_1 and J-B_2C_2-F_1

(c) J-B_2C_2-D_2 (d) Section drawing

Figure 2. Design details of the four PBCJs-MCs.

Table 1. Design parameter details of the four PBCJs-MCs.

Specimen	Precast Beam	Differences in the Joint Zone
J-B_1C_2-F_1	Anchor connection	PCFRC
J-B_2C_2-D_1	Mechanical connection	PCC
J-B_2C_2-F_1	Mechanical connection	PCFRC
J-B_2C_2-D_2	Mechanical connection	PCC

Note: PCFRC represents post-cast fiber reinforced concrete; PCC represents post-cast concrete.

2.2. *Establishment of FEMs*

The fiber model is used for the FEMs, as shown in Figure 3. S1 area represents the core concrete, i.e., the confined concrete area. S2, S3, S4, and S5 areas outside the confined concrete area represent the unconfined concrete area. The fiber cross-sections can be mainly discretized into confined concrete fibers, unconfined concrete fibers, and reinforcing fibers. Fiber cross-section model with the number of segments and the number of integration points of the number of segments increases, and the distribution of sectional curvature in the direction of the extended height of the member is more reasonable, but the amount of calculation also increases. Gauss–Legendre formula is used to set 4 integration points for the segmentation basis of the finite element of the rod system [28]. The fiber cross-section is divided uniformly with the number of 30–50 [29], which can ensure accuracy while significantly reducing the computational effort.

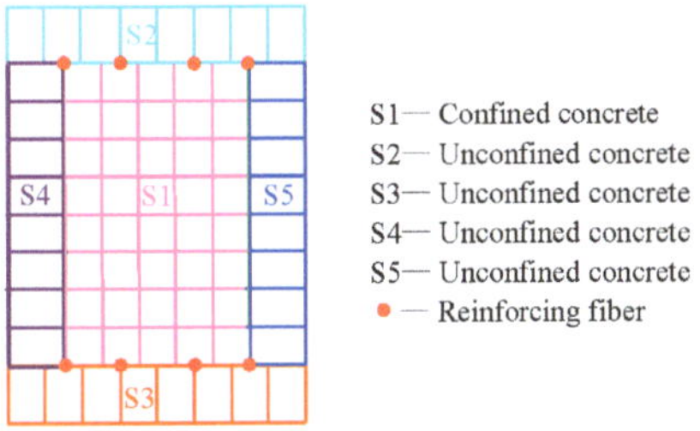

Figure 3. Schematic diagram of the fiber model.

The beam or column of the finite element model is simulated using the displacement-based nonlinear beam–column element. The action of the longitudinal reinforcement and stirrups is considered using the fiber cross-section in Figure 3. Concrete02 model is used for concrete fibers, and Reinforcing Steel model is used for steel fibers. Figure 4 shows a schematic diagram of the finite element model. Five fiber column elements are established (nodes 1–6). Six fiber beam elements are established (nodes 8–15). Beam–column joint elements are established between nodes 3, 4, 10, and 11 to simulate the bond–slip behavior of the reinforcement concrete at the joint and the shear behavior at the core of the joint. The loading mode of the FEMs (see Table 2) is the same as that of the test in Ref. [2], being drift ratio-controlled loading modes.

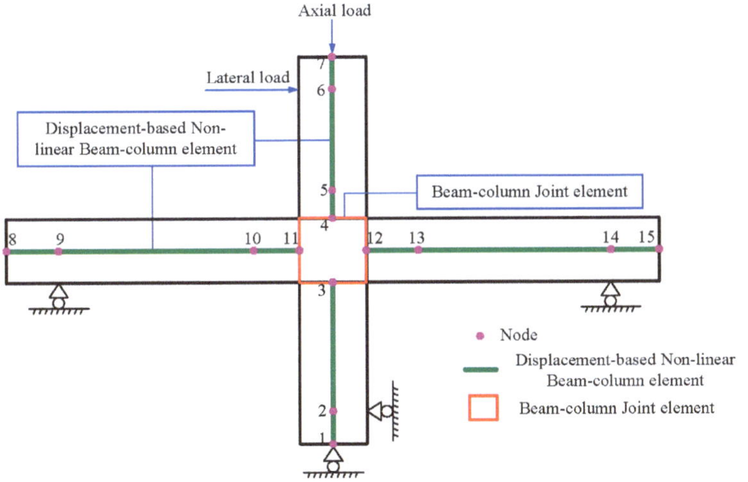

Figure 4. Schematic diagram of the finite element model for PBJs-MCs.

Table 2. Lateral load loading mode of PBCJs-MCs.

Loading Level	Drift Ratio (%)	Displacement (mm)	Number of Cycle
1	0.10	1.50	3
2	0.30	4.50	3
3	0.50	7.50	3
4	0.75	11.25	3
5	1.00	15.00	3
6	1.50	22.50	3
7	2.00	30.00	3
8	2.75	41.25	3
9	3.50	52.50	3

2.3. Element

Displacement-based nonlinear beam–column element is used to simulate beam and column elements, together with the fiber model (see Figure 4). Its advantage is that the internal force distribution of the element is more stable. The mechanical behavior of the concrete member can be simulated using fewer elements, but it is prone to the phenomenon of computational non-convergence. Beam–column joint element improved by N-Mitra [22] (see Figure 5) is used to simulate different damage behaviors of BCJs.

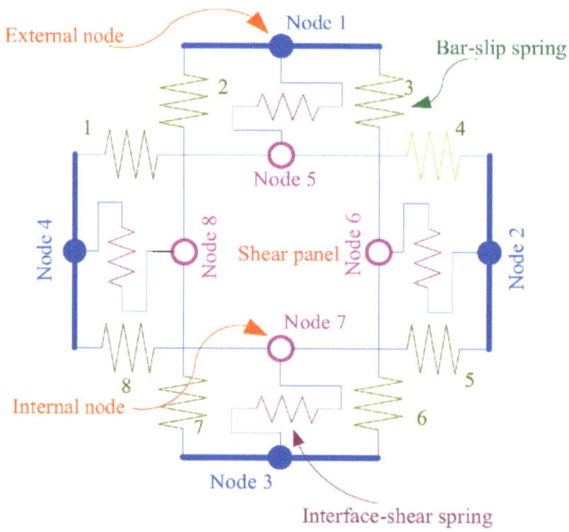

Figure 5. Beam–column joint element model.

The shear panel component is used to simulate the shear behavior of joints in terms of stiffness and strength degradation under shear damage. Eight zero-length springs (Bar–Slip springs) are used to simulate the bond–slip mechanism of the reinforcement and concrete. Four zero-width shear springs (interface-shear springs) connecting external nodes 1–4 and internal nodes 5–8 are applied to simulate the beam–column interfacial shear transfer failure mechanism. Four interface-shear springs are used to simulate the degradation of shear transfer capacity at the joint interface. Because plastic hinge rotation is at the beam–column joint during the test without significant vertical sliding, interface-shear spring element is defined as an elastic material with high elasticity to weaken the shear transfer failure mechanism in the fiber model.

2.4. Constitutive Models

2.4.1. Concrete

In nonlinear fiber beam–column elements, it is necessary to assign corresponding constitutive relationship to concrete fibers [15,16]. The Concrete02 model is used. Based on the improved Kent–Park concrete model, Concrete02 constitutive model (see Figure 6) is proposed to effectively consider the tensile properties of concrete [30], thus being able to simulate the hysteretic properties of confined concrete in tension and compression. Introducing the confinement factor K, the confining effect of stirrup reinforcement on the strength and ductility of concrete in the core zone can be considered in the FEMs, which can more accurately simulate the concrete constitutive relationship.

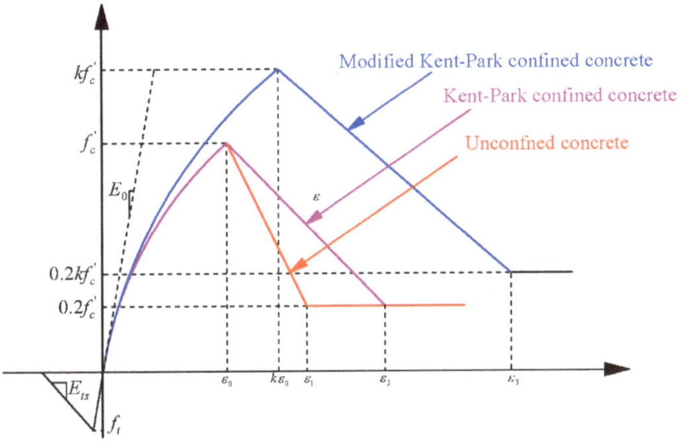

Figure 6. Concrete02 model.

2.4.2. Reinforcement

The hysteresis curve of reinforced concrete members is more closely related to the uniaxial constitutive model of reinforcement with the increase in the loading displacement. The selection of a reasonable and accurate constitutive model of reinforcement is the key to the accuracy of numerical simulation. There are currently eight constitutive models of reinforcement available in the OpenSees Version 3.0.0 software, including Steel01, Steel02, and Reinforcing Steel. Table 3 lists the different constitutive models of reinforcement; thus, Reinforcing Steel model (see Figure 7) is selected. In Figure 7, f_y, f_u are the reinforcing yield strength and ultimate strength, respectively; ε_{sh} is the reinforcing strain hardening point strain and ε_u is the reinforcing strain corresponding to f_u; E_s is the initial modulus of elasticity of reinforcement; E_{sh} is the reinforcing starting point modulus. The above parameters can be obtained from the material property tests of reinforcement in Ref. [2]. Dhakal–Maekawa buckling model [17] is used to simulate the buckling of compressed reinforcement [17]. The Coffin–Manson model is applied to simulate the adverse effects of the low-cycle fatigue accumulation damage of reinforcing steel [17]. It has three parameters, namely the strength degradation parameter C_d, the fatigue damage parameter C_f, and the fatigue damage index α_2. According to the previous study in Ref. [17], the three parameters in the Coffin–Manson model can be taken as 0.140, 0.379, and 0.379, respectively.

Table 3. Comparison of different constitutive models of reinforcement.

Constitutive Model	Stress–Strain Curve Shape	Bauschinger Effect Is Considered?	Compression Flexure Effect Is Considered?	Fatigue Damage Effect Is Considered?
Steel 01	Bending Lines	No	No	No
Steel 02	Smooth curves	Yes	No	No
Reinforcing Steel	Smooth curves	Yes	Yes	Yes

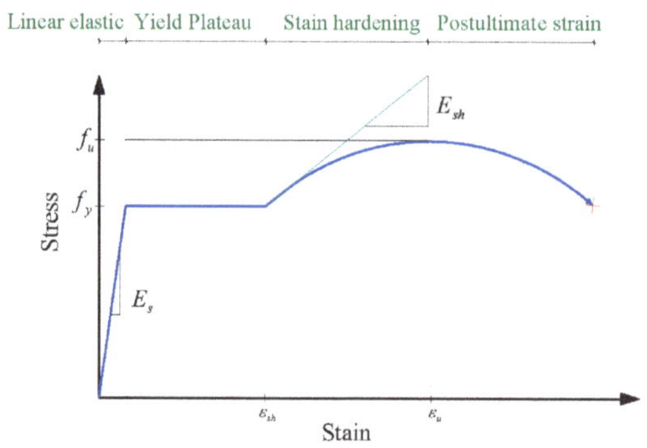

Figure 7. Reinforcing Steel model.

2.4.3. Shear Panel

The constitutive model of the shear panel is Pinching4 model (see Figure 8). It can synthesize the characteristics of strength degradation, stiffness degradation, and pinch shrinkage effect of joints. In this model, the skeleton curve envelope is multilinear and the unloading–reloading path is trilinear. The key to defining this material model is to define 16 parameters for the 8 characteristic points of the skeleton curve envelope in positive and negative directions, 6 key parameters for the unloading–reloading path, and stiffness and strength degradation criteria.

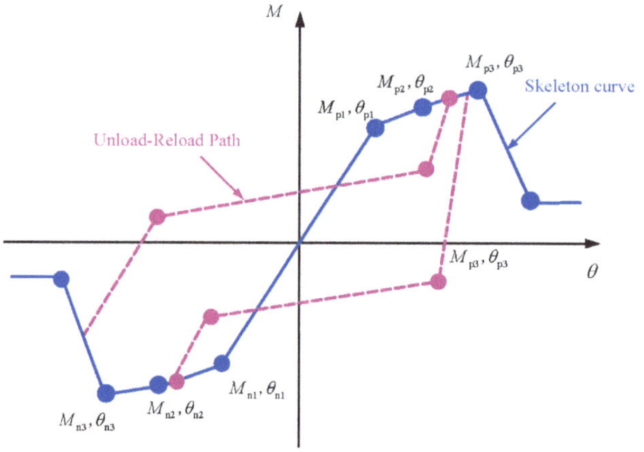

Figure 8. Pinching4 model.

Pinching4 model skeleton curves reflect the shear capacity of BCJs. The modified mild compression field theory (MCFT) assumes that the shear force is uniformly distributed and that the shear force is transmitted only through the diagonal compression bar, considers the tensile stress effect after concrete cracking, establishes the deformation compatibility condition and stress balance relationship between reinforcement and concrete, and is widely used in concrete shear calculation. According to the modified MCFT, the Membrane-2000 Version 1.0 software proposed by Bentz and Collins was used to calculate the parameters related to the characteristic points of the skeleton curve [31].

2.4.4. Constitutive Model of Bar–Slip Springs

Under low-cycle reciprocal loading, bond–slip between the reinforcing steel and concrete occurs. It leads to hysteresis loop pinching and has a great influence on the load–displacement curves. A reinforcement bond–slip model, Bar–Slip, was developed in the OpenSees Version 3.0.0. software using the stress–slip relationship for reinforcing steel proposed by Eligehausen and Hawkins [23,24]. The Bar–Slip model can consider the effects of concrete strength, reinforcement material properties, and the degree of anchorage, and thus analyze the effect of the slip on reinforcement on the overall performance of BCJs.

It should be noted that the simulation of the reinforcement bond–slip effect can be achieved more accurately with the Pinching4 constitutive model by setting two co-coordinate points and assigning fiber sections to the ends of beams and columns considering the reinforcement stress–slip relationship. After that, the bending moment–angle relationship curves were obtained from the proposed fiber section analysis and converted to the skeleton curve parameters in Pinching4 model. Considering that the mechanical connection of the specimen is at a certain distance from the end of the beam and that the stress–slip relationship cannot be accurately expressed, the Bar–Slip springs (uniaxial material constitutive) are used in the establishment of the beam–column joint element model.

2.4.5. Constitutive Model of Reinforcement Bond–Slip

There are differences in construction quality, curing conditions, and methods between wet precast concrete joints and post-cast zone concrete, and the adhesion between reinforcement and concrete is susceptible to the effects of greater reinforcement slippage in the specimens, as shown by the experimental hysteresis curves [2]. Therefore, the reinforcement at the end of the beam and column is provided a uniaxial constitutive model, Bond_SP01 model [17] (see Figure 9), to simulate the bond–slip effect of the reinforcement at the end of the beam and column, while the voids inside the mechanical joints are also equated to this reinforcement bond–slip effect. It has a total of 6 parameters, namely the yield strength f_y, slip at yield S_y, slip at failure S_u, intensification factor at the initial intensification phase b, and hysteresis factor R. Zhao and Sritharan [32] calculated and analyzed a large amount of pull-out test data to obtain a fitting formula for S_y, as well as the recommended range of parameters S_u, b, and R.

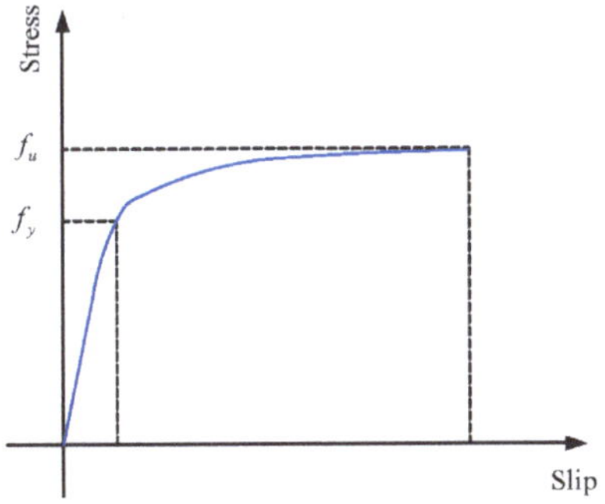

Figure 9. Bond_SP01 model.

Figure 10 shows the stress–slip curves of HRB400 steel bars-reinforced C35 concrete and PVA FRC in uniaxial tension (S is the slip value). Before yielding, the slip of the reinforcement increases very slowly and the slip value is small. After yielding, the slip of the reinforcement increases rapidly with essentially no increase in stress and the slip phenomenon is very obvious. This is also consistent with the phenomenon that the degradation rate of the bond between the steel bars and concrete (or FRC) gradually increases after the concrete or FRC has gone through the stages of cracking and crushing in the test. The slip value of the reinforcement with a greater diameter is large under the same stress conditions and its bond–slip effect is more significant, which is consistent with the findings in Ref. [33].

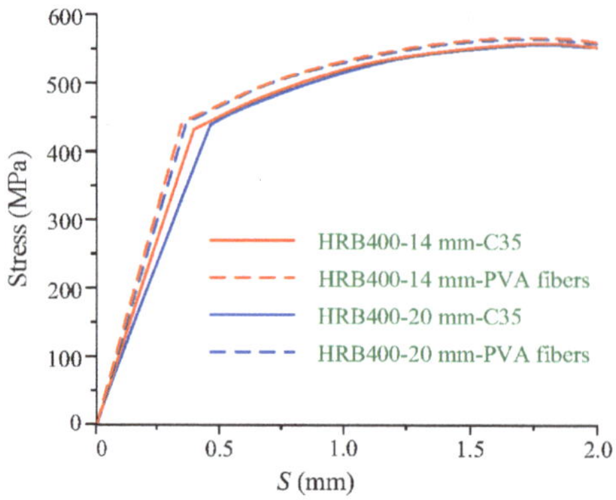

Figure 10. Stress–slip curves of HRB400 reinforcement.

3. Numerical Simulation Results and Discussion

Due to the experimental hysteresis curves in Ref. [2] being affected by various factors, the measured hysteresis curves in the positive direction (+) and negative direction (−) demonstrate asymmetric characteristics, while the numerical models are established without distinguishing the loading directions. The Pinching4 model is based on the principle that the envelopes of the skeleton curves are identical in the positive and negative directions. The specific hysteresis rules and loading and unloading stiffness in the positive and negative directions of the numerical model are adjusted according to the experimental results so that the difference between the simulated hysteresis curves in the positive and negative loading directions is minimal. Therefore, the experimental hysteresis curves are moderately shifted to compare the accuracy of the FEMs, which has no effect on the calculation of the relevant seismic behavior indexes. The comparison of the experimental and simulated peak loads of the PBCJs-MCs is shown in Table 4. The errors of the peak loads are below 7.0%, reflecting a better simulation effect. It is worth mentioning that the peak loads obtained from the simulations are generally small compared to the experimental results. Because there are differences between the strengths of new-to-old concrete in the specimens (J-B_2C_2-D_1 and J-B_2C_2-D_2), the Concrete 02 model does not reflect the mechanical properties of FRC in an integrated manner (J-B_1C_2-F_1 and J-B_2C_2-F_1), and the numerical model does not effectively consider the strength provided by the additional U-shaped reinforcement (J-B_2C_2-D_2).

Table 4. Comparison of simulation and experimental peak loads.

Specimen No.	Test Result *T* (kN)		Simulated Result *S* (kN)		E_r (%)	
	+	−	+	−	+	−
J-B_1C_2-F_1	94.3	−93.9	90.1	−90.2	−4.5	−3.9
J-B_2C_2-D_1	96.7	−89.2	91.8	−91.3	−5.1	2.4
J-B_2C_2-F_1	100.8	−99.7	99.8	−99.6	−1.0	−0.1
J-B_2C_2-D_2	97.9	−101.2	94.4	−94.3	−3.6	−6.8

3.1. Hysteresis Curves

The simulated and experimental hysteresis curves of the PBCJs-MCs are compared in Figure 11. The simulation results basically match the experimental results, reflecting the loading–unloading paths and directions of the experimental hysteresis curves, and the selected element types and material constitutive models can better reflect the shear effect and reinforcement bond–slip effect at the core zone of the joint. For specimen J-B_1C_2-F_1, the simulation is able to simulate the differences between the positive and negative hysteresis curves by adjusting the relevant hysteresis rules and degradation parameters in the Pinching4 model. In general, the numerical simulation model, element types, and material constitutive relationships selected reflect the force characteristics of the PBCJs-MCs more accurately and have a good simulation effect on the reinforcement slip effect reflected by the hysteresis curve.

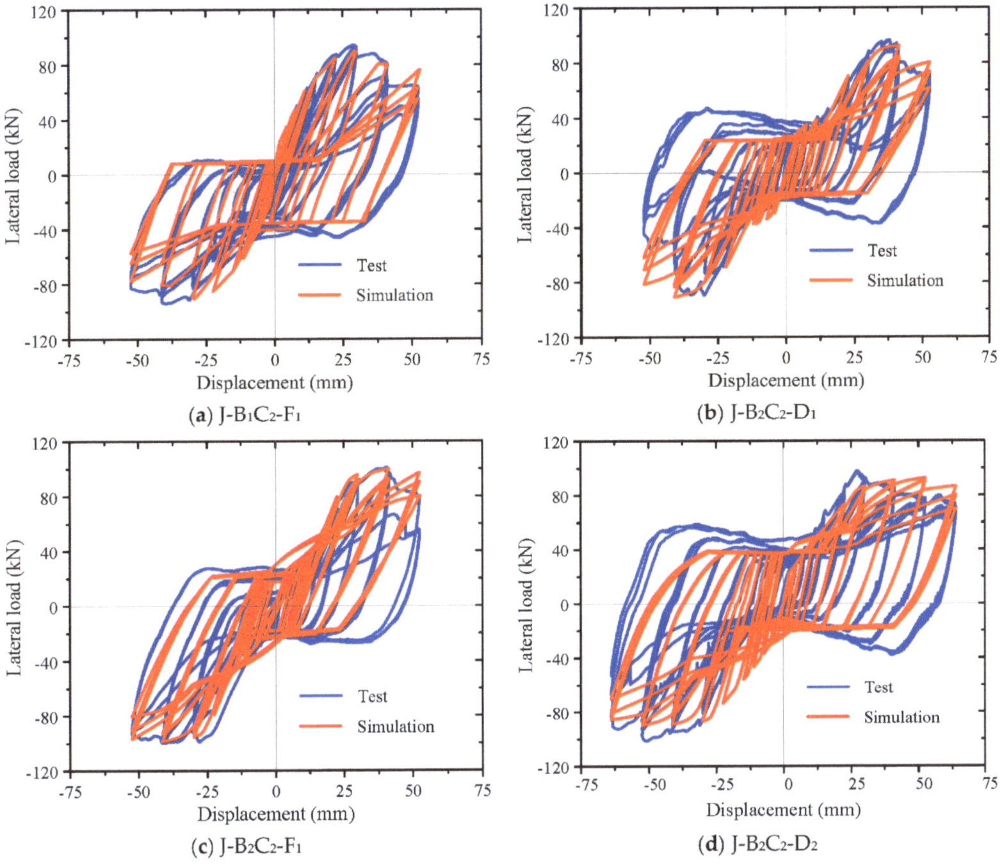

Figure 11. Comparison of experimental and simulated hysteresis curves for each specimen.

3.2. Skeleton Curves

The simulated and experimental skeleton curves of the PBCJs-MCs are compared in Figure 12. The simulated skeleton curves are closer to the experimental ones during positive loading, while the experimental results during negative loading are influenced by the larger loading device, with very slow growth initially, but the two are also closer in the later stages of loading. The rising section of the simulated skeleton curve of specimen J-B_1C_2-F_1 is better simulated. The simulated skeleton curve also reflects the process of decreasing the bearing capacity of specimen J-B_1C_2-F_1 after the damage of the column end. The falling section of the simulated skeleton curve is similar to the experimental results. For specimens J-B_2C_2-D_1, J-B_2C_2-F_1, and J-B_2C_2-D_2, the slow growth of the skeleton curve during the loading displacement amplitudes of 7.5–15 mm occurs due to the existence of internal voids in the mechanical joints, causing the reinforcing steel to slip. The rising sections of the simulated skeleton curves are rather consistent after the loading displacement amplitude of 7.5 mm.

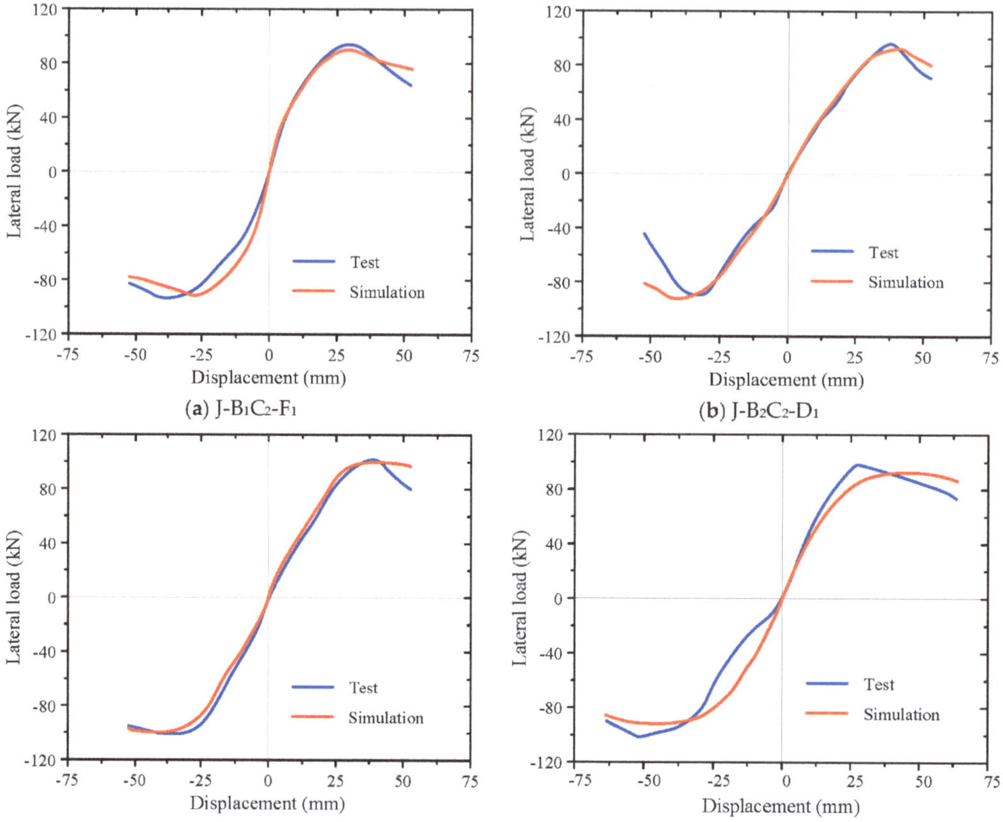

(a) J-B1C2-F1 (b) J-B2C2-D1 (c) J-B2C2-F1 (d) J-B2C2-D2

Figure 12. Comparison of experimental and simulated skeleton curves for each specimen.

3.3. Energy Dissipation Capacity

The normalized cumulative hysteretic energy coefficient E_N is calculated using the same method in Ref. [17]. The total cumulative hysteretic energy dissipation curves of the numerical simulations and experiments are calculated in Table 5. The normalized cumulative hysteretic energy coefficients of the numerical models and experimental specimens are shown in Figure 13. The total cumulative energy dissipation of the FEMs is relatively close to the experimental results. The maximum relative error of the simulated results is less than 15%. At the last loading displacement, the increase in the simulated total cumulative energy dissipation of each specimen is smaller than the experimental result. For specimen J-B_2C_2-F_1, the obvious differences between the energy dissipation–displacement curves of the test and simulation results are due to the insufficient accuracy of the model data for the bond–slip relationship, which needs to be improved by supplementing the bond–slip tests in the subsequent study. In addition, the unloading stiffness of the model at large displacements has a decreasing process with the unloading process. The Pinching4 model cannot effectively simulate this phenomenon, which needs to be corrected in further research.

Table 5. Numerical simulation results and accuracy of the total cumulative hysteretic energy dissipation.

Specimen No.	Test Value T (kN·m)	Numerical Simulation Value S (kN·m)	Relative Error ((S-T)/T)
J-B_1C_2-F_1	40.9	37.7	−7.8%
J-B_2C_2-D_1	41.9	37.0	−11.7%
J-B_2C_2-F_1	41.9	36.1	−13.8%
J-B_2C_2-D_2	69.5	69.1	−0.6%

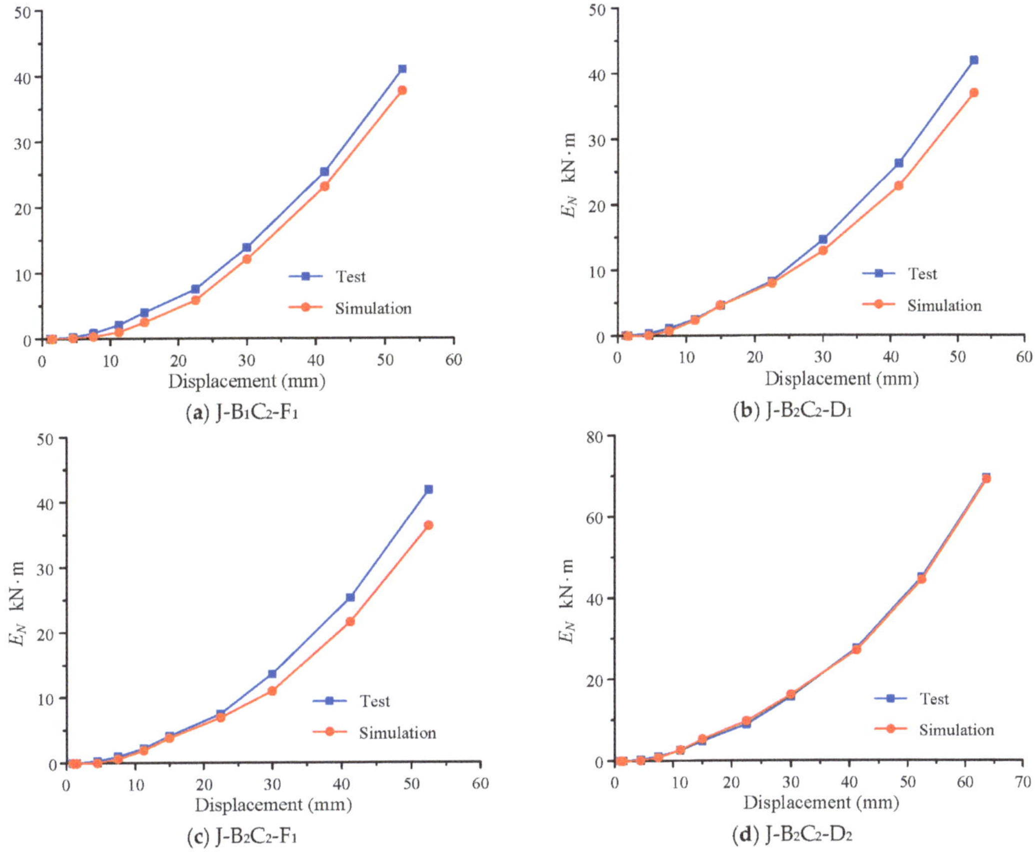

Figure 13. Comparison of experimental and simulated energy consumption capacity for each specimen.

3.4. Stiffness Degradation

The average loop stiffness K [17] is used to measure the stiffness degradation of the PBCJs-MCs. The stiffness degradation curves of the numerical simulation and experiment are compared as described in Figure 14. Some distinct branches are detectable in the stiffness degradation of the PBCJs-MCs. This is first due to the individual differences of the test specimens. Secondly, there is a slight increase in stiffness in the range of 15–20 mm, which is due to the over-limit of bond stress in the longitudinal reinforcement in the plastic hinge zone of the beam–column joints, leading to a surge in sliding displacement and resulting in the above phenomenon. The average loop stiffness of each specimen at a loading displacement of 4.5 mm differs from the experimental data to a certain extent, probably because the experimental data at this stage are susceptible to interference from external factors. After 4.5 mm, the trend of the simulated stiffness degradation curves is almost the same as that of the experiment, and the average loop stiffness values of the numerical simulation and experiment are very close to each other.

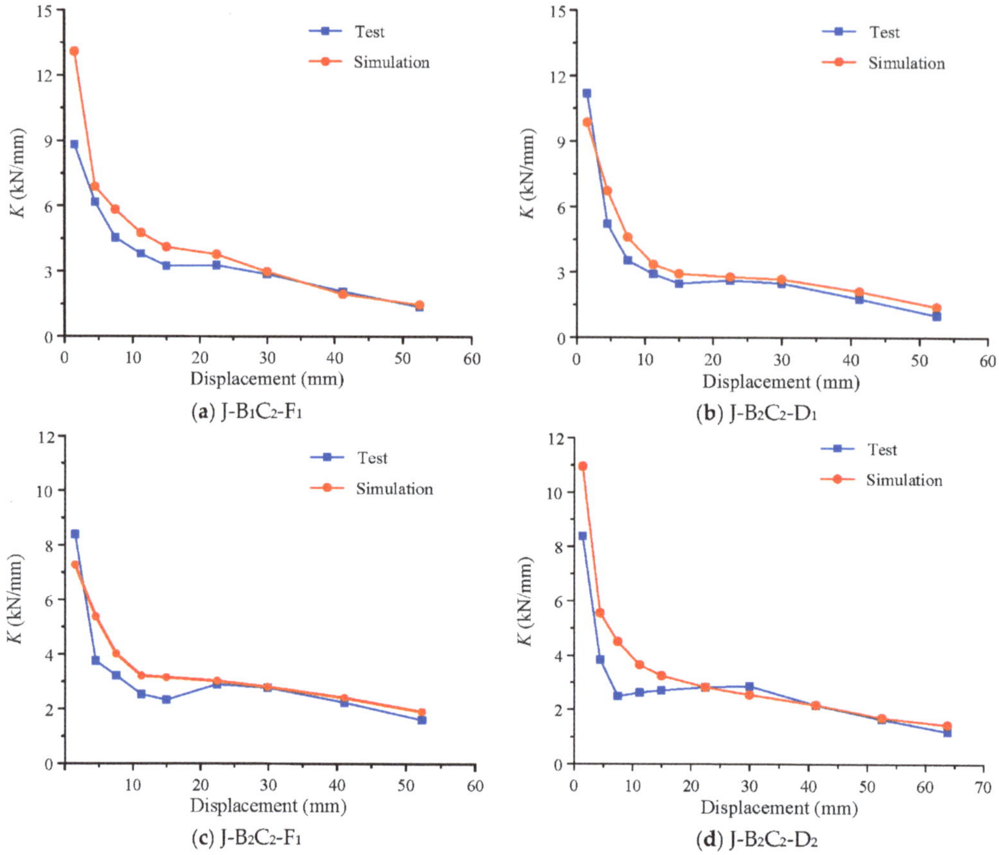

Figure 14. Comparison of experimental and simulated average loop stiffness degradation curves for each specimen.

4. Parametric Analysis of Effect Factors

4.1. Effect of the Shear-to-Span Ratio

The difference in the shear span ratios of the generally recognized members affects the difference in the damage form of the concrete member. The damage of long columns with $\lambda > 2$, is calculated, which is mainly ductile damage. The normal shear span ratio of columns in ordinary RC structures is between 2 and 3, so it is not urgent to consider the shear span ratios greater 4. The effect of the shear-to-span ratio (λ) on the seismic behavior of PBCJs-MCs is analyzed. λ is set to 2.0, 2.5, 3.0, and 3.5 to study the seismic behavior of the PBCJs-MCs with the other parameters of specimen J-B_2C_2-D_2 ($n = 0.15$) unchanged. The simulated skeleton curves of the PBCJs-MCs are shown in Figure 15. The simulated bearing capacity and displacement ductility coefficients of the PBCJs-MCs are shown in Table 6. The bearing capacity of the PBCJs-MCs gradually decreases as λ increases. Displacement ductility is the ability of a structure or member to withstand displacement before failure. With the increase in λ, the ultimate displacement of the PBCJs-MCs under the damage state because of the limit value of the PBCJs-MCs with the change in λ is not much different, but the yield displacement with the increase in λ regarding the PBCJs-MCs occurs with an increase in the yield displacement; the coefficient of ductility ratio decreases accordingly. The coefficient of ductility is a criterion for evaluating the ductility of the member. It can be concluded that the displacement ductility of the PBCJs-MCs gradually decreases as λ increases. When λ is greater than 2.5, the ductility coefficient of the PBCJs-MCs is greater than 3.0. Therefore, it is recommended that the optimum range of λ for PBCJs-MCs is 2.0–2.5.

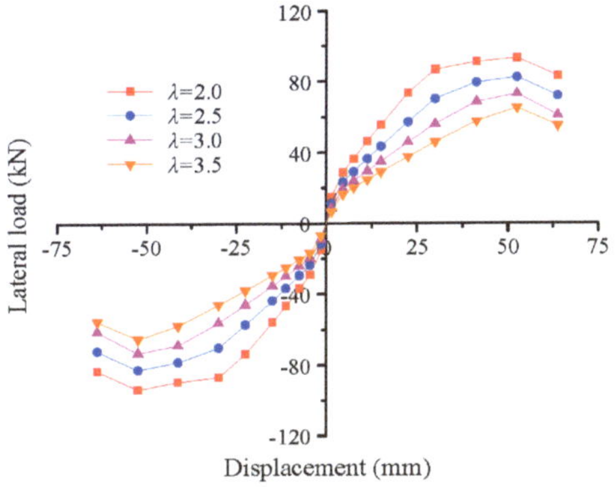

Figure 15. Skeleton curves of PBCJs-MCs with different values of λ.

Table 6. Bearing capacity and displacement ductility of PBCJs-MCs with different shear-to-span ratios.

λ	F_{yc} (kN)	Δ_{yc} (mm)	F_{mc} (kN)	Δ_{uc} (mm)	μ_c
2.0	68.7	19.6	93.4	63.8	3.3
2.5	59.2	21.1	82.5	63.8	3.0
3.0	52.8	25.9	73.1	63.8	2.5
3.5	45.9	29.9	65.2	63.8	2.2

Note: Δ_{yc} and Δ_{uc} indicate the numerical displacements at the numerical yield load F_{yc} and ultimate load F_{uc}, respectively; F_{mc} is the numerical maximum load; μ_c is the numerical ductility coefficient, and $\mu_c = \Delta_{uc}/\Delta_{yc}$.

If common building structures (e.g., classrooms, houses, hospitals, etc.) with columns have a shear span ratio greater than 3, it is not necessary to add seismic measures for the use of such RC structures. The use of columns with $\lambda > 3$ is limited in the buildings with seismic classifications of Class I and Class II. Because of the high energy dissipation capacity and ductility requirements of the columns in regions prone to both seismic and high wind events, it is prudent to use columns with $\lambda > 3$.

4.2. Effect of the Axial Load Ratio

The effect of the axial load ratio (n) on the seismic behavior of the PBCJs-MCs is also analyzed. Indeed, n is set to 0.1, 0.15, 0.3, 0.4, 0.5, and 0.7 to study the seismic behavior of the PBCJs-MCs with the other parameters of specimen J-B_2C_2-D_2 (λ = 2.0) unchanged. The simulated skeleton curves of the PBCJs-MCs are described in Figure 16. The simulated skeleton curves basically demonstrate no differences as n increases. When n is not greater than 0.4, the bearing capacity of the PBCJs-MCs increases slightly; when n is greater than 0.4, the bearing capacity of the PBCJs-MCs is almost unchanged. From the parametric analysis of λ, it can be concluded that the effect of n on the seismic behavior of the PBCJs-MCs can be negligible in the case of the PBCJs-MCs with a moderate value of λ.

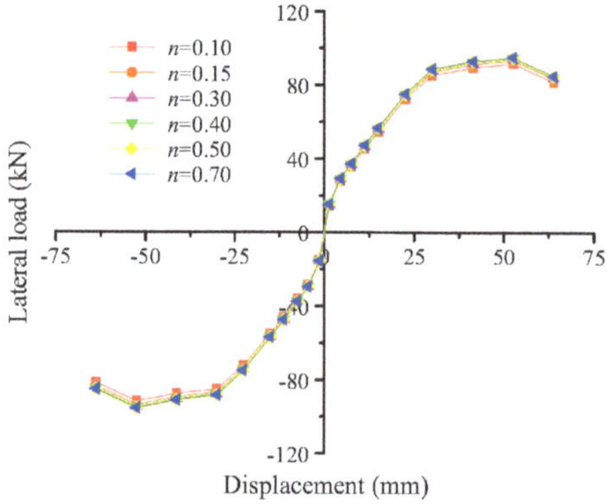

Figure 16. Skeleton curves of PBCJs-MCs with different values of n.

When the shear-to-span ratio is between 2 and 3, the axial load has less influence on the bearing capacity and ductility of the column, so, for such columns, the axial load ratio limit is appropriately relaxed, and, for the reinforced concrete frame structures of seismic Class I and Class II, the axial load ratio of the center column can be increased from 0.65 and 0.75, which are stipulated in the specification, to 0.85, which can reduce the column section under the premise of ensuring the seismic performance of the structure, save the amount of reinforcing steel and concrete, and increase the structural internal space.

5. Conclusions and Measurements

In the present study, the seismic behavior of PBCJs-MCs is investigated numerically. The following conclusions are obtained.

(1) The Concrete02 model and Reinforcing Steel model can accurately simulate the constitutive relationship of concrete and reinforcement, respectively. The beam–column joint elements can accurately simulate the different damage behaviors of the joint zone. The Bond_SP01model can accurately simulate the bond–slip between the reinforcing steel, concrete, and mechanical connections.

(2) The simulated hysteresis curves and skeleton curves of the PBCJs-MCs are similar to the experimental results. The simulated seismic behavior indexes, such as bearing capacity, energy dissipation capacity, and stiffness degradation, are not much different from the experimental results, with a relative error of about 15%.

(3) The bearing capacity and displacement ductility coefficients of the PBCJs-MCs decrease rapidly as λ increases. It is recommended that the optimum λ range for PBCJs-MCs is 2.0–2.5. Regarding the high energy dissipation capacity and ductility requirements of the members in regions prone to both seismic and high wind events, it is prudent to use columns with $\lambda > 3$.

(4) The axial load ratio has a very small influence on the seismic behavior of the PBCJs-MCs. The effect of the axial load ratio on the seismic behavior of the PBCJs-MCs can be negligible in the case of the PBCJs-MCs with a moderate value of shear-to-span ratio.

In reflecting on the broader implications of the study's findings for seismic design standards and building codes, the following measures can be taken to incorporate these insights into regulatory frameworks in a manner that both promotes innovation in precast construction techniques and ensures the safety and resilience of structures in earthquake-prone areas.

(1) Researchers should be down-to-earth, strictly abide by academic ethics, rigorously conduct research to ensure the validity of research results and the reliability of research conclusions, and actively maintain cooperation with enterprises so that the benign development of the assembly building industry can be ensured.

(2) For prefabricated assembly building construction technology, the government should take assembly building as a key research support area and provide multi-party support in terms of research funding and research conditions, strengthen the industry's supervision and review mechanism to ensure the safety of the technology, and leave its beneficial attributes to the market to decide.

The numerical model and related measures obtained in the present study help to provide an effective theoretical basis and technical support for the application of PBCJs-MCs in assembled building structures.

Author Contributions: Conceptualization, M.-L.Z. and C.S.; Methodology, L.B.; Formal analysis, M.-L.Z. and L.B.; Investigation, C.S.; Data curation, R.A., Y.H. and G.B.; Writing—original draft, M.-L.Z. and Y.H.; Writing—review & editing, Z.Y. and R.A.; Visualization, C.S., Z.Y. and G.B.; Supervision, Z.Y. and L.B.; Project administration, Z.Y. and R.A.; Funding acquisition, Z.Y. and R.A. All authors have read and agreed to the published version of the manuscript.

Funding: This research received no external funding.

Data Availability Statement: The data and materials used to support the findings of this study are available from the corresponding author upon request.

Conflicts of Interest: The authors declare no conflicts of interest.

References

1. Liu, Y.; Li, X.P.; Zheng, X.H.; Song, Z. Experimental study on seismic response of precast bridge piers with double-grouted sleeve connections. *Eng. Struct.* **2020**, *221*, 111023. [CrossRef]
2. Sun, C.Z.; Zhuang, M.L.; Bai, L.T. Experimental study on seismic behavior of novel precast concrete beam-column joints using mechanical connections. *Bull. Earthq. Eng.* **2023**, *21*, 4429–4448. [CrossRef]
3. Qiao, D.H.; Xu, Y.Q.; Zhang, X.; Gao, L.; Yu, C.R.; Yang, J.; Zhu, H.B.; Qiao, Y.; Shao, Y.F.; Zhang, W.H. Seismic behaviour and size effect of column base joints with inverted exposed grouted sleeves. *J. Build. Eng.* **2022**, *51*, 104333. [CrossRef]
4. Hadade, M.; Ferreira, M.; Carvalho, R.C.; Catoia, B. Moment-rotation response of beam-column connections in precast concrete structures. *Solid State Phenom.* **2017**, *259*, 269–274. [CrossRef]
5. Sun, C.Z.; Zhuang, M.L.; Wang, Z.B.; Chen, B.D.; Gao, L.; Qiao, Y.; Zhu, H.B.; Zhang, W.H.; Yang, J.; Yu, C.R. Experimental investigation on the seismic behavior of prefabricated fiber reinforced concrete beam-column joints using grouted sleeve connections. *Struc. Concr.* **2023**, *24*, 4702–4718. [CrossRef]
6. Chen, W.H.; Feng, K.; Wang, Y. Seismic Performance of a Novel Precast Beam-Column Joint Using Shape Memory Alloy Fibers-Reinforced Engineered Cementitious Composites. *Buildings* **2022**, *12*, 1404. [CrossRef]
7. Wang, Z.; Feng, D.C.; Wu, G. Experimental Study on Seismic Behavior of Precast Bolt-Connected Steel-Members End-Embedded Concrete (PBSEC) Beam-Column Connections. *Buildings* **2022**, *12*, 1652. [CrossRef]
8. Zheng, J.H.; Pan, Z.Z.; Zhen, H.; Deng, X.H.; Zheng, C.M.; Xie, L.P.; Xiong, Z.; Li, L.J.; Liu, F. Experimental Investigation on the Seismic Behavior of Precast Concrete Beam-Column Joints with Five-Spiral Stirrups. *Buildings* **2023**, *13*, 2357. [CrossRef]
9. Peng, W.; Lu, W.; Liu, S.; Liu, Y.; Xu, L.; Li, F. Experimental and Numerical Study on the Seismic Performances of Reinforcement-Embedded RC Column-to-Precast Cap Beams with Socket Connections. *Buildings* **2023**, *13*, 2367. [CrossRef]
10. Bert, C.W.; Malik, M. Differential quadrature: A powerful new technique for analysis of composite structures. *Compos. Struct.* **1997**, *39*, 179–189. [CrossRef]
11. Wang, Y.Y.; Gu, Y.; Liu, J.L. A domain-decomposition generalized finite difference method for stress analysis in three-dimensional composite materials. *Appl. Math. Lett.* **2020**, *104*, 106226. [CrossRef]
12. Kabir, H.; Aghdam, M.M. A generalized 2D Bézier-based solution for stress analysis of notched epoxy resin plates reinforced with graphene nanoplatelets. *Thin Wall. Struct.* **2021**, *169*, 108484. [CrossRef]
13. Ni, X.Y.; Qiang, Z.; Li, Y.Z. Cyclic test and numerical analysis of the seismic behavior of concrete columns reinforced by HRB600 steel bars. *J. Build. Eng.* **2022**, *50*, 104211. [CrossRef]
14. Kremmyda, G.D.; Fahjan, Y.M.; Tsoukantas, S.G. Nonlinear FE analysis of precast RC pinned beam-to-column connections under monotonic and cyclic shear loading. *Bull. Earthq. Eng.* **2014**, *12*, 1615–1638. [CrossRef]
15. Zoubek, B.; Isakovic, T.; Fahjan, Y.; Fischinger, M. Cyclic failure analysis of the beam-to-column dowel connections in precast industrial buildings. *Eng. Struct.* **2013**, *52*, 179–191. [CrossRef]
16. Cao, X.Y.; Feng, D.C.; Wang, Z.; Gang, W. Investigation on modelling approaches for prefabricated concrete beam-to-column connections using openSEEs. *China Civil Eng.* **2019**, *52*, 13–27.
17. Zhuang, M.L.; Sun, C.Z.; Dong, B. Experimental and numerical investigations on seismic behavior of HTRB630 high-strength steel bars reinforced concrete columns. *Case Stud. Constr. Mater.* **2022**, *17*, e01185.

18. Mazzoni, S.; McKenna, F.; Scott, M.H.; Fenves, G.L. *Open System for Earthquake Engineering Simulation User Commandlanguage Manual*; University of California: Berkeley, CA, USA, 2009.
19. Paulay, T. Equilibrium criteria for reinforced concrete beam-column joints. *ACI Struct. J.* **1989**, *86*, 635–643.
20. Pantazopoulou, S.J.; Bonacci, J.F. On earthquake-resistant reinforced concrete frame connections. *Can. J. Civil Eng.* **1994**, *21*, 307–328. [CrossRef]
21. Lowes, L.N.; Altoontash, A. Modeling Reinforced Concrete Beam-Column Joints Subjected to Cyclic Loading. *J. Struct. Eng.* **2003**, *129*, 1686–1697. [CrossRef]
22. Mitra, N. *An Analytical Study of Reinforced Concrete Beam-Column Joint Behavior under Seismic Loading*; University of Washington: Washington, DC, USA, 2007.
23. Eligehausen, R.; Popov, E.P.; Bertero, V.V. *Local Bond Stress-Slip Relationships of Deformed Bars under Generalized Excitations*; EERC Report; University of California: Berkeley, CA, USA, 1982.
24. Hawkins, N.M.; Lin, I.J.; Jeang, F.L. *Local Bond Strength of Concrete for Cyclic Reversed Loadings*; Bartos, P., Ed.; Bond in Concrete; Applied Science Publishers Ltd.: London, UK, 1982; pp. 151–161.
25. Paudel, S.G.; Tanapornraweekit, G.; Tangtermsirikul, S. Numerical study on seismic behavior improvement of composite wide beam-column interior joints. *J. Build. Eng.* **2022**, *46*, 103637. [CrossRef]
26. Yang, J.; Guo, T.; Chai, S. Experimental and numerical investigation on seismic behaviours of beam-column joints of precast prestressed concrete frame under given corrosion levels. *Structures* **2020**, *27*, 1209–1221. [CrossRef]
27. Bohara, R.P.; Tanapornraweekit, G.; Tangtermsirikul, S. Experimental and numerical investigation on seismic behavior of a composite wide beam-column interior joint. *Structures* **2021**, *30*, 560–573. [CrossRef]
28. Gharakhanloo, A. *Distributed and Concentrated Inelasticity Beam-Column Elements Used in Earthquake Engineering*; Norwegian University of Science and Technology: Tromsoe, Norway, 2014.
29. Huang, Z.M. *Finite Element Flexibility Method Based 3D Inelastic Seismic Response Analysis of Reinforced Concrete Frames*; Chongqing University: Chongqing, China, 2013.
30. Scott, B.D.; Park, R.; Priestley, M.J. Stress-strain behavior of concrete confined by overlapping hoops at low and high strain rate. *J. Am. Conc. Inst.* **1982**, *79*, 13–27.
31. Bentz, E.C. Sectional Analysis of Reinforced Concrete Members. Ph.D. Thesis, University of Toronto, Toronto, ON, Canada, 2000.
32. Zhao, J.; Sritharan, S. Modeling of Strain Penetration Effects in Fiber-Based Analysis of Reinforced Concrete Structures. *ACI Struct. J.* **2007**, *104*, 133–141.
33. Cao, W.; Liu, X.; Qiao, Q.; Tian, Z.; Lin, D. Experimental study on bond slip behavior of ordinary and high strength recycled concrete and steel bars. *J. Build. Struct.* **2016**, *37* (Suppl. S2), 127–134.

buildings

MDPI

Article

Study of Fatigue Performance of Ultra-Short Stud Connectors in Ultra-High Performance Concrete

Ran An [1], You-Zhi Wang [2], Mei-Ling Zhuang [3,*], Zhen Yang [3], Chang-Jin Tian [4], Kai Qiu [5], Meng-Ying Cheng [6] and Zhao-Yuan Lv [7]

[1] College of Civil Engineering, Qilu Institute of Technology, Jinan 250200, China; ar307309@163.com
[2] School of Civil Engineering, Shandong University, Jinan 250061, China; wyz96996@163.com
[3] Water Resources Research Institute of Shandong Province, Jinan 250013, China; skyyangzhen@shandong.cn
[4] China Construction Infrastructure Co., Ltd., Beijing 100029, China; 17862990729@163.com
[5] School of Mechanics and Civil Engineering, China University of Mining and Technology, Xuzhou 221116, China
[6] Design & Consulting Company (Shandong) of China Construction 8th Engineering Division, Jinan 250101, China; cmy451796@163.com
[7] Shimao Group Holdings Limited, Jinan 250011, China; 1365866766@163.com
* Correspondence: ml_zhuang99@163.com

Abstract: Steel–UHPC composite bridge decking made of ultra-high performance concrete (UHPC) has been progressively employed to reinforce historic steel bridges. The coordinated force and deformation between the steel deck and UHPC are therefore greatly influenced by the shear stud connectors at the shear interface. Four fatigue push-out specimens of ultra-short studs with an aspect ratio of 1.84 in UHPC were examined to investigate the fatigue properties of ultra-short studs with an aspect ratio below 2.0 utilized in UHPC reinforcing aged steel bridges. The test results indicated that three failure modes—fracture surface at stud shank, fracture surface at steel flange, and fracture surface at stud cap—were noted for ultra-short studs in UHPC under various load ranges. The fatigue life decreased from 1287.3×10^4 to 24.4×10^4 as the shear stress range of the stud increased from 88.2 MPa to 158.8 MPa. The UHPC can ensure that the failure mode of the specimens was stud shank failure. Based on the test and literature results, a fatigue strength design S–N curve for short studs in UHPC was proposed, and calculation models for stiffness degradation and plastic slip accumulation of short studs in UHPC were established. The employment of ultra-short studs in the field of UHPC reinforcing aging steel bridges can be supported by the research findings.

Keywords: shear stud connector; ultra-high performance concrete; push-out test; fatigue behavior

1. Introduction

Steel–UHPC composite bridge decks have been widely applied in bridge engineering in recent years [1–5]. The UHPC structural layer can significantly improve the overall stiffness of steel bridge decks while ensuring their lightweight and high-strength advantages, thereby effectively improving the fatigue performance of the steel bridge deck [6,7]. The total structural performance of the composite components is significantly influenced by the shear connectors that connect the UHPC and steel bridge deck. Tightly joining the two, the head of the shear connector is implanted in the UHPC structural layer and is welded to the top plate of the steel deck, as shown in Figure 1. There are several types of shear connectors; among them, shear stud connectors are the most commonly used due to their isotropic mechanical properties, high shear and tensile capacities, and convenient welding processes [8–11].

To meet anchorage requirements and prevent shear studs from being pulled out, EC4 stipulates that the length-to-diameter ratio of studs (the ratio of stud length to stud diameter) in steel–normal concrete composite structures should be greater than 4.0 [12].

However, numerous studies have shown that the shear studs embedded in UHPC may not be subject to this restriction due to the ultra-high compressive and tensile strength, as well as the elastic modulus of UHPC, effectively restricting the deformation of shear studs [13–17]. Therefore, the aspect ratio of shear studs in UHPC is usually between 2.5 and 4.0 and can be classified as short studs. Previous studies by Shi et al. [18], Chen et al. [19], and Huang et al. [20] conducted fatigue tests of steel–UHPC composite bridge decks using full-scale models, and fatigue failure of short stud connectors was observed in all tests. Therefore, it is necessary to systematically study the fatigue behavior of short stud connectors in UHPC.

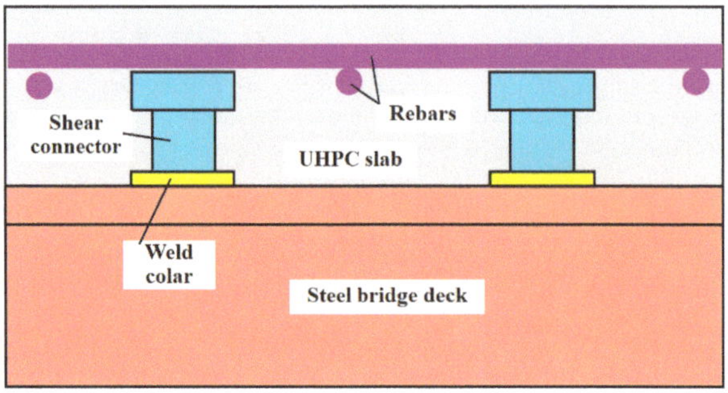

Figure 1. Steel–UHPC composite bridge deck.

There are limited studies [21–24] on the fatigue behavior of short stud connectors in UHPC, and all the aspect ratios were larger than 2.0. The fatigue failure mode in the existing studies [21–24] was shear stud failure, in which fracture surfaces were between stud shank and base steel, and the UHPC remained generally intact. Cao et al. [21] conducted four fatigue push-out tests of short studs with an aspect ratio of 2.7 in UHPC. The results showed that the fatigue strength of short studs in UHPC was higher than that of the regular studs in normal concrete. Wei et al. [22] carried out seven fatigue push-out tests of short studs in UHPC under different load ranges, with an aspect ratio of 3.12. Based on experimental and literature results, a fatigue strength design S–N curve for short studs in UHPC was provided. Huang et al. [23] established a refined finite element model for fatigue push-out tests of short studs in UHPC, with an aspect ratio of 2.69, and a formula for evaluating the fatigue life of short studs under multiple factors was developed using the fracture mechanics method. Liu et al. [24] studied the fatigue performance of short studs in engineered cementitious composites (ECCs) through fatigue push-out tests with an aspect ratio of 3.75. The research revealed that the fatigue strength of short studs in ECC was lower than that of regular studs in normal concrete, possibly due to the significantly lower elastic modulus of ECC compared to normal concrete.

Most existing studies on the fatigue performance of short studs in UHPC focus on newly built bridges, while there are several challenges for the reinforcement of aged steel bridge decks using UHPC: (1) to reduce the self-weight of the aged steel bridge decks, thin-layer UHPC is commonly used, with thickness ranging from 40 mm to 60 mm, resulting in shorter stud height; (2) to minimize the impact of welding residual stresses on aged bridge decks, shear studs should be sparsely welded. Therefore, large-diameter studs must be provided to meet the shear connection degree. These necessitate a smaller aspect ratio for short studs used in UHPC reinforcement for aged steel bridges, even below 2.0, and can be classified as ultra-short studs. Research on the fatigue performance of ultra-short studs in UHPC is still lacking.

This study aims to investigate the fatigue behavior of ultra-short studs in UHPC with an aspect ratio of 1.84 and establish prediction formulas for these behaviors. Four fatigue push-out tests under different load ranges were conducted to systematically reveal the fatigue failure modes, fatigue life, and degradation of mechanical behaviors of ultra-short studs in UHPC. Based on experimental and literature results, a fatigue strength design S–N curve for ultra-short studs in UHPC and predictive formulas for plastic slip accumulation and elastic stiffness degradation were established.

2. Fatigue Push-Out Test

2.1. Test Specimens and Fabrication

There were four identical specimens in the fatigue push-out test, as shown in Figure 2. Each specimen comprised a 400 mm long, 200 mm wide, and 270 mm high H-shaped steel column with a welded steel cover plate, and two 500 mm wide, 400 mm heigh, and 55 mm thick reinforcement UHPC slabs. The UHPC slabs were connected to the steel column by four 35 mm height and 19 mm diameter ultra-short shear stud connectors welded on each side of the 12 mm flange plate of the H-shaped steel column. The aspect ratio of the stud was 1.84. The reinforcement rebars in UHPC were 10 mm in diameter and spaced 100 mm in height and transverse direction, respectively.

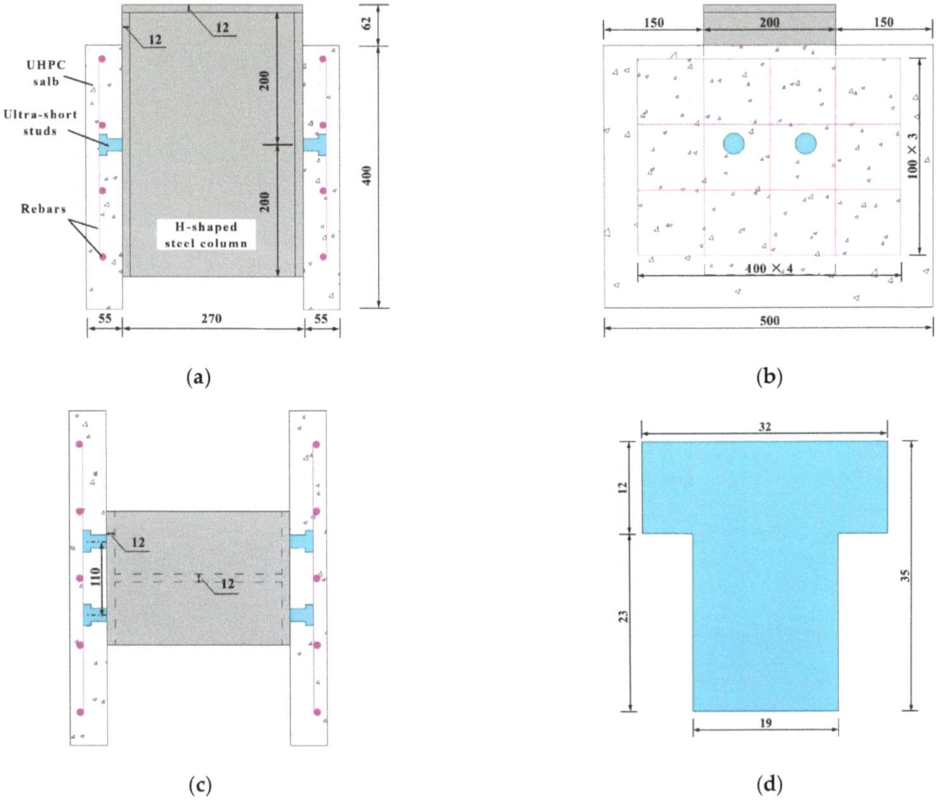

(a) (b) (c) (d)

Figure 2. Size of specimens (unit: mm): (**a**) front view; (**b**) side view; (**c**) top view; (**d**) stud dimension.

Before pouring UHPC, the steel flange plate was greased to eliminate bonding and reduce friction between the steel and UHPC. The fabrication process of the specimens is shown in Figure 3 and is as follows: ultra-short stud welding, formwork assembly, reinforcement rebar binding, UHPC pouring, steam curing, and formwork removal after 28 days.

Figure 3. Fabrication process of specimens: (a) stud welding; (b) formwork and reinforcement rebar assembly; (c) UHPC pouring; (d) formwork removal.

2.2. Material Properties

The mechanical properties of UHPC are listed in Table 1. Three 100 mm × 100 mm × 100 mm cubic specimens and three 100 mm × 100 mm × 400 mm cuboid specimens were fabricated along with test specimens to evaluate the compressive and flexural strengths of UHPC, respectively. The mix ratio of the UHPC is listed in Table 2. The H-shaped steel column was made of Q355-grade structural steel. The reinforcement bar was made of HRB400 steel with a yield strength of 400 MPa. The shear stud connectors consisted of ML15 steel. The tensile properties of steel beams, reinforcement bars, and shear studs were provided by the manufacturer, as shown in Table 3.

Table 1. Mechanical properties of UHPC (MPa).

Compressive Strength	Flexural Strength	Elastic Modulus
134	24	48,200

Table 2. Mix ratio of UHPC (unit: kg/m^3).

Cement	Fly Ash	Silica Fume	Steel Fiber	Quartz Sand (40–80 Mesh)	Quartz Sand (20–40 Mesh)	High-Range Water Reducer	Water
886	120	110	160	306	706	20	177

Table 3. Tensile properties of steel members.

Members	Yield Strength (MPa)	Ultimate Strength (MPa)
H-shaped steel column	360	470
Shear studs	350	467
Reinforcement rebars	400	574

2.3. Test Setup and Loading Protocol

The fatigue push-out test setup is shown in Figure 4. Each specimen was supported on a concrete cushion, and the bottom was poured with high-strength gypsum for leveling and bonding with the concrete cushion block to ensure the stability of the specimen during fatigue loading. Cyclic loading was applied on the specimen using a PMW800-250 electro-hydraulic pulsating fatigue testing machine produced by Jinan Lizhi Testing System Co., Ltd. (Jinan, China) with a load capacity of 500 kN. The cyclic loading was uniformly transmitted to the top surface of the specimen through the steel beam. The constant amplitude fatigue load range ($\Delta P = P_{max} - P_{min}$) was adopted during tests, and the load ratio $R = P_{min}/P_{max} = 0.20$ in accordance with a former study, where ΔP was the load range, and P_{max} and P_{min} were the upper and lower limits of fatigue load, respectively. The P_{max} did not exceed the elastic capacity of the stud, as the fatigue test was in the elastic stage. The loading frequency was set to be 3 Hz, considering testing machine properties. The UHPC plates on both sides of the H-shaped column were labeled as A and B, and the corresponding embedded studs were labeled as A1, A2, and B1, B2, respectively. Four electronic displacement sensors were arranged evenly on each side of the specimen with an accuracy of 0.001 mm to measure the relative slip between UHPC and the H-shaped column.

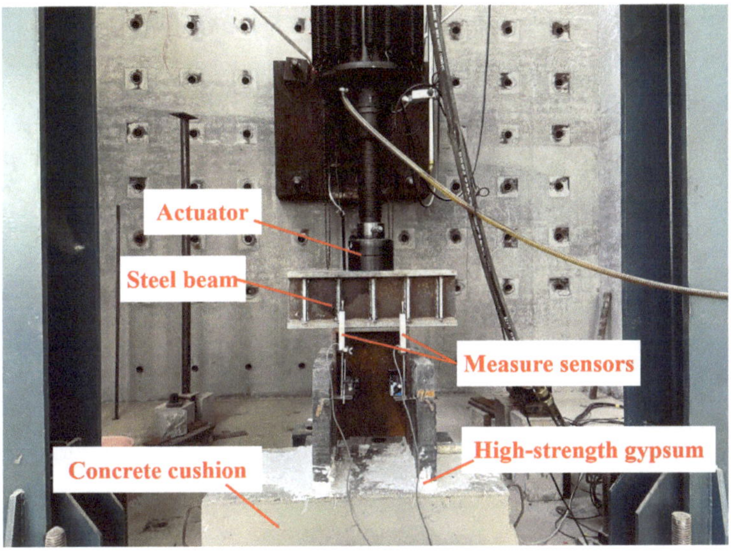

Figure 4. Fatigue push-out test.

Fatigue tests were stopped at certain intervals of fatigue loading cycles to measure the static relative slips between UHPC and the H-shaped column, and the maximum static load was taken as P_{max}. The relative slip value was the average value of the four displacement sensors on side A and side B. Sinusoidal wave cyclic loading was adopted in the fatigue tests until specimen fatigue failed to record the load–slip curve and the plastic slip accumulation.

The four specimens were labeled as FT1–FT4 due to different ΔP. Among them, FT1 was used to test the fatigue life under normal usage status based on IIW [25]. The IIW defined 90 MPa as the fatigue strength in normal usage status with corresponding 200×10^4 fatigue loading cycles. The experimental loading parameters are shown in Table 3.

The test loading parameters are shown in Table 4. The τ_{max}, τ_{min}, and $\Delta\tau$ are calculated as P_{max}/A_s, P_{min}/A_s, and $\Delta P/A_s$, respectively, and A_s denotes the sectional area of the stud shank.

Table 4. Fatigue test loading parameters.

Specimens	Loading (kN)			Shear Stress of Single Stud (MPa)		
	P_{max}	P_{min}	ΔP	τ_{max}	τ_{min}	$\Delta\tau$
FT-1	125	25	100	105.9	17.6	88.2
FT-2	150	30	120	127.1	26.5	105.9
FT-3	190	40	150	167.6	35.3	132.3
FT-4	225	45	180	198.5	39.7	158.8

3. Test Results

Before fatigue tests, static push-out tests were conducted using specimens of the same specifications to obtain the static shear strength of the studs, and the average elastic capacity and ultimate shear capacity of a single stud were 87.5 kN and 147.9 kN, respectively. The failure mode of all specimens was ultra-short stud fatigue failure, leading to a unilateral UHPC plate separated from the steel column. It indicated that UHPC can ensure stud fatigue fracture for ultra-short studs even with aspect ratios less than 2.0 and fully utilize the fatigue properties of ultra-short studs. Figure 5 shows the failure modes of the ultra-short studs, with the following three modes:

(1) Mode I: Fracture surface at stud shank. Fatigue cracks initiated at the weld toe of the stud shank-to-weld collar and propagated along the melting line between the weld collar and heat-affected zone (HAZ) in the stud shank.

(2) Mode II: Fracture surface at the steel flange. Fatigue cracks initiated at the weld toe of the weld collar-to-steel flange and propagated along the melting line between the weld collar and HAZ in the steel flange, forming a small concave surface in the steel flange after fatigue failure.

(3) Mode III: Fracture surface at stud cap. Fatigue cracks initiated at the connection between the stud shank and the stud cap and propagated along the interface between the two. Mode III has not been reported in previous studies yet and will only occur together with Mode II and not appear separately.

The fracture surface of the shear stud comprised a dull fatigue fracture zone and a bright forced fracture zone. The dull fatigue fracture zone was caused by the propagation of fatigue cracks, and the bright forced fracture zone was caused by the forced shear fracture. The fatigue cracks first propagated along the cross-section to form a dull fatigue fracture zone. When the dull fatigue fracture zone was large enough, the remaining cross-section was insufficient to bear the fatigue load, and the stud failed instantaneously, forming a bright forced fracture zone.

There were two failure modes of UHPC, as shown in Figure 6. Mode A: UHPC was intact and locally crushed below the stud. This mode is a common failure mode in the existing literature [21–24]. Mode B: UHPC split failure, and the cracks were in a central diffusion shape. In the existing literature, the UHPC failure modes were basically Mode A, while in this study, most of the UHPC cracked. Among all the observable failure sections in UHPC, Mode A only accounted for 25%. The reasons are as follows: The anchoring length of the short studs in UHPC in existing literature is relative longer than those in this study, so only local UHPC beneath the stud root was subjected to fatigue load. However, due to the ultra-low aspect ratio of the studs in this study, the entire UHPC beneath the ultra-short stud was subjected to fatigue loading. As fatigue loading cycles increased, micro-cracks in UHPC continued to propagate, weakening its constraint on the ultra-short stud. So, the deformation of the stud kept increasing, further intensifying the cracking of UHPC. When the remaining cross-section of the stud was insufficient to bear the fatigue load, the stud fractured instantly, and cracks in UHPC quickly connected, ultimately leading to the UHPC splitting.

The failure modes and fatigue life of each specimen are summarized in Table 5. As is shown, as the shear stress range of the stud increased from 88.2 MPa to 158.8 MPa, the fatigue life, N (loading cycles), decreased from 1287.3×10^4 to 24.4×10^4.

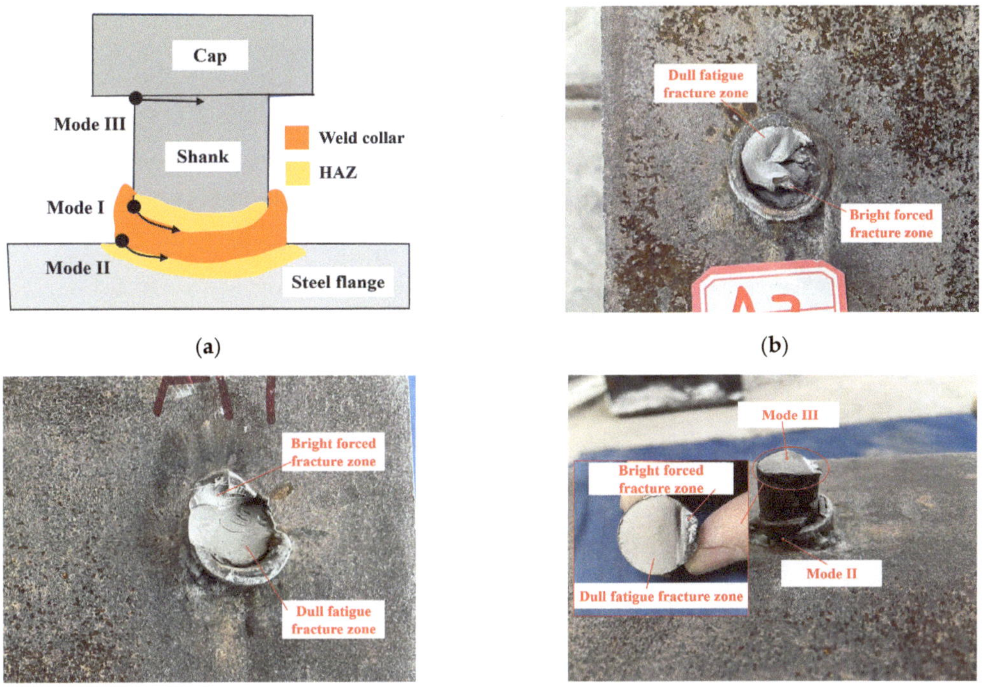

(a) (b) (c) (d)

Figure 5. Failure modes of ultra-short studs. (**a**) Stud fracture surfaces. (**b**) Mode I. (**c**) Mode II. (**d**) Mode III.

(a) (b)

Figure 6. Failure modes of UHPC. (**a**) Mode A. (**b**) Mode B.

Table 5. Fatigue push-out test results.

Specimen	ΔP (kN)	$\Delta\tau$ (MPa)	Fatigue Life, N (10^4)	Failure Modes	
				Stud	UHPC
FT-1	100	88.2	1287.3	Mode II, Mode II–III	Mode b, Mode b
FT-2	120	105.9	389.7	Mode II	Mode a, Mode b
FT-3	150	132.3	58.2	Mode I, Mode II–III	Mode b, Mode b
FT-4	180	158.8	24.4	Mode I, Mode II	Mode a, Mode b

4. Nominal Shear Fatigue Strength (S–N Curve)

Currently, the shear fatigue strength evaluation of studs in concrete mainly adopts the S–N curve of nominal shear stress range ($\Delta\tau$) versus fatigue life (N), with log($\Delta\tau$) as the independent variable and log(N) as the dependent variable, as shown in Equation (1):

$$\log N = \log C - m \log \Delta\tau \tag{1}$$

where m and logC represent the slope and linear regression constant of the S–N curve, respectively.

Table 6 lists the fatigue test results from existing literature [21,22,26,27] and from this study, totaling 17 samples, which are plotted as coordinate points on the S–N curve, as shown in Figure 7. The shear stud connectors in this literature also consisted of ML15 steel and were consistent with this study.

Linear regression analysis was conducted based on the test results using the method specified in IIW [25]. The slope of the average regression curve, denoted as m, was calculated as 7.3, taking log($\Delta\tau$) as the independent variable and log(N) as the dependent variable. That was close to the specified value of 8.0 in EC3 [28]. To align with EC3 and relevant literature calculation methods, the slope of the S–N curve was set to 8.0 during curve regression analysis. The mean regression curve was obtained as follows, as shown in Equation (2):

$$\log N = 21.354 - 8 \log \Delta\tau \tag{2}$$

By adopting the 95% survival probability curve specified in EC3 and setting the slope of the S–N curve to 8.0, the mean $\mu_{\log C}$ and standard deviation $\sigma_{\log C}$ of logC can be calculated using Equations (3) and (4), respectively:

$$\mu_{\log C} = \frac{\sum \log C_i}{n} \tag{3}$$

$$\sigma_{\log C} = \sqrt{\frac{\sum \left(\mu_{\log C} - \log C_i\right)^2}{n-1}} \tag{4}$$

where n represents the number of specimens, and $\log C_i$ represents the linear regression constant corresponding to each data point.

Substituting the experimental data from this study and the literature into Equations (3) and (4), respectively, we obtained $\mu_{\log C}$ = 22.835 and $\sigma_{\log C}$ = 0.17.

By using Equations (5) and (6), the characteristic value $\log C_k$ corresponding to a 95% survival probability can be calculated as follows:

$$\log C_k = \mu_{\log C} - k\sigma_{\log C} \tag{5}$$

$$k = 1.645\left(1 + \frac{1}{\sqrt{n}}\right) \tag{6}$$

where k represents the characteristic coefficient [29].

In this section, k was calculated as 2.04 by substituting n (n = 17) into Equation (6). Then, the fatigue constant for a 95% survival probability, $\log C_k$, was calculated as 22.483 by substituting k into Equation (5). Finally, the S–N fitting curve with a 95% survival probability was obtained, as shown in Equation (7):

$$\log N = 22.483 - 8 \log \Delta\tau \tag{7}$$

Figure 7 also presents the 95% and 50% (k = 0) survival rate S–N curve obtained based on Equation (7), and the S–N curve specified by EC3 for shear studs in normal concrete, as shown in Equation (8).

$$\log N = 21.935 - 8 \log \Delta\tau \tag{8}$$

Table 6. Fatigue test results in the existing literature and this study.

Literatures	Specimen	Short Stud (mm)		UHPC Strength (MPa)	Experimental Results	
		h × d	h/d		Δτ (MPa)	N (10^4)
Cao et al. [21]—①	FAT-1 FAT-2 FAT-3 FAT-4	35 × 13	2.69	135.9	94 117 125 135	1178.7 113.0 168.8 44.1
Wei et al. [22]—②	FT-1 FT-2 FT-3 FT-4 FT-5 FT-6 FT-7	50 × 16	3.13	120.3	170 160 150 140 130 120 110	8.7 24.7 16.3 96.0 64.0 230.0 236.0
Li et al. [26]—③	B-3	35 × 13	2.69	85.3	145	60.0
Zhang et al. [27]—④	F-1	35 × 13	2.69	150.2	112	240.5
This study—⑤	FT-1 FT-2 FT-3 FT-4	35 × 19	1.84	134.0	88.2 105.9 132.3 158.8	1287.3 389.7 58.2 24.4

From Figure 7, it can be observed that the fatigue strength of the shear stud under 200 million cycles is 116 MPa, 105 MPa, and 90 MPa for 50%, 95% survival probabilities, and the EC3 specified curve, respectively. The fitting curve for a 95% survival probability is also significantly higher than the curve specified by EC3. The S–N curve specified by EC3 is applicable for shear studs in normal concrete. This indicates that the fatigue strength of shear studs in UHPC is significantly better than that in normal concrete. As there are no specific regulations governing the fatigue strength of short studs in UHPC, it is recommended to utilize Equation (7) for the fatigue design of short studs in the UHPC composite bridge deck. It is demonstrated that Equation (7) can ensure safety and fully leverage the superior fatigue properties of short studs in UHPC.

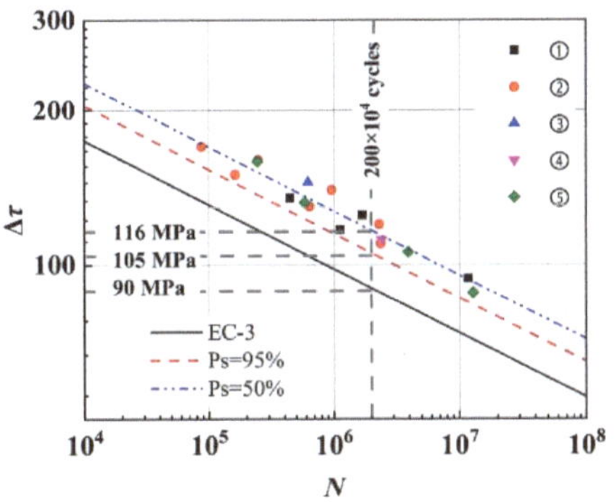

Figure 7. S–N curves of shear studs.

5. Calculation Model for Stiffness Degradation of Short Studs in UHPC

5.1. Fatigue Load–Slip Curves

Figure 8 illustrates the typical evolution of the load–slip curves of short studs after loading cycles n. It can be seen that with the increase in fatigue loading cycles, the degradation of its mechanical performance can be described by the cumulative plastic slip, $\delta_{\mathrm{pl},n}$ and the degradation of elastic stiffness, $K_{\mathrm{el},n}$.

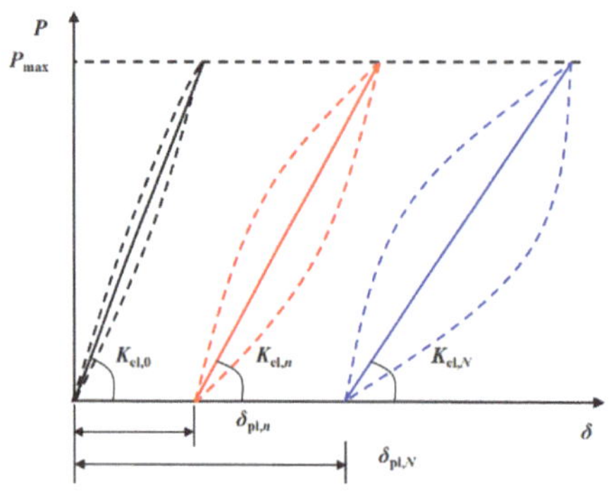

Figure 8. Evolution of the load–slip curves of short studs.

Figure 9 shows the static load–slip curves of each specimen throughout the entire fatigue life cycle. The load–slip curve comprised loading and unloading segments. The slip values were the average of four displacement sensors on side A and side B. It can be observed in Figure 9 that the relative slips increased approximately linearly with the increase in load during the initial loading stage. Although the unloading curve exhibits some nonlinear characteristics, there was no residual slip after unloading. With the fatigue loading cycles increasing, the relative slips showed more pronounced nonlinear features. The slope of the loading curve continuously decreased with loading cycles increasing, and plastic slip accumulated continuously, indicating a gradual decrease in the elastic stiffness of the studs. In the later loading stages, the slip values after unloading cannot return to the initial point, and plastic slip continued to increase with loading cycles increasing until the specimens' fatigue failure. At this stage, the area enclosed by the loading and unloading curves significantly increased, indicating a significant increase in the deformation energy produced during one loading cycle. It can also be observed that, with a higher shear stress range, the nonlinear behavior was exhibited earlier and the plastic slip accumulated more rapidly, leading to a faster decrease in elastic stiffness.

Figure 10 illustrates the evolution of plastic slip with the number of loading cycles for each specimen. During the initial loading stage, plastic slip increased rapidly due to debonding between the steel flange and UHPC, as well as the elimination of non-elastic deformations. Subsequently, plastic slip steadily increased at a constant rate, indicating that fatigue loading continuously damaged the studs. In the later stage, plastic slip increases rapidly with fatigue loading cycles increasing, ultimately leading to fatigue failure.

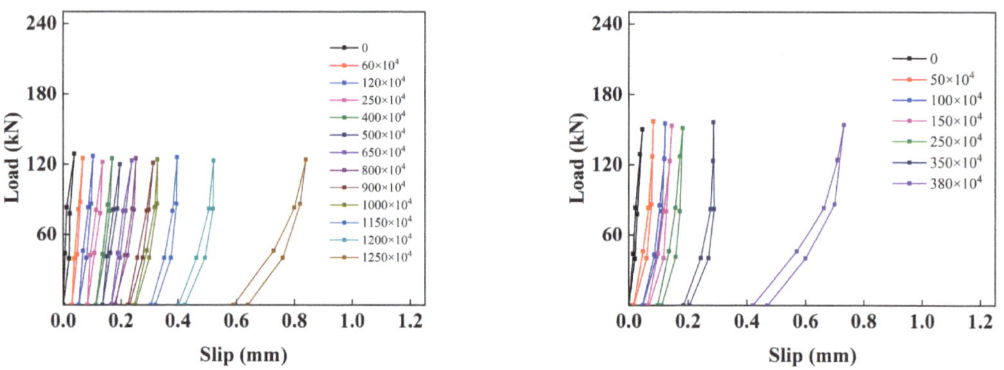

(a) (b)

Figure 9. *Cont.*

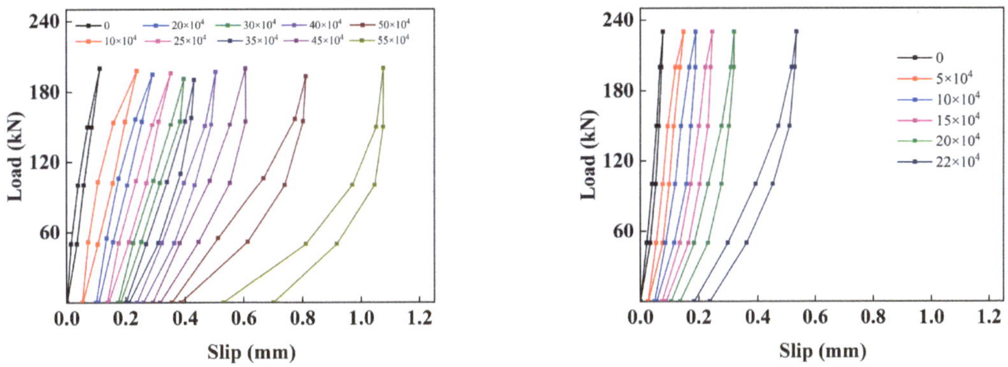

(**c**) (**d**)

Figure 9. Load–slip curves under full fatigue life cycles. (**a**) FT-1. (**b**) FT-2. (**c**) FT-3. (**d**) FT-4.

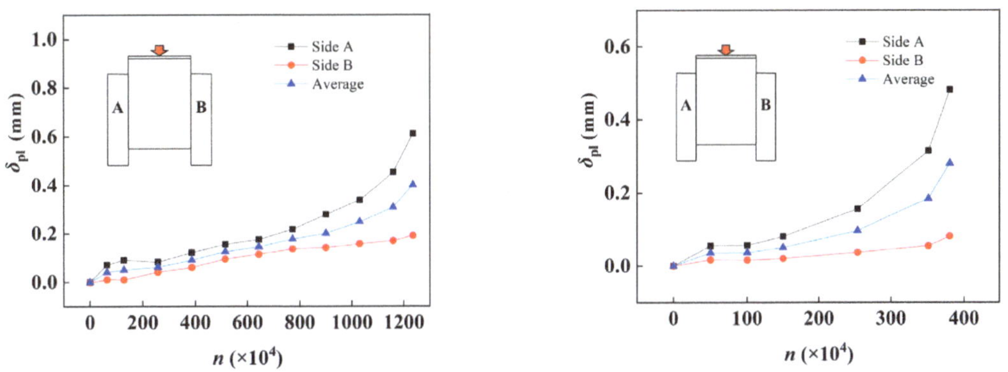

(**a**) (**b**)

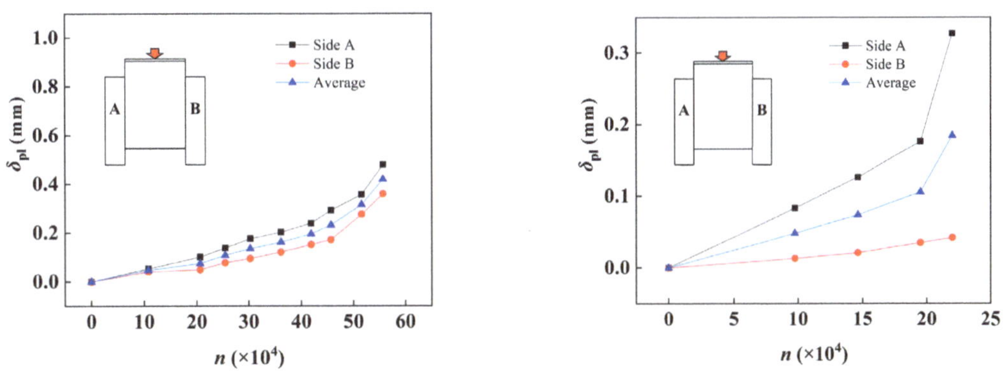

(**c**) (**d**)

Figure 10. Evolution process of plastic slip. (**a**) FT-1. (**b**) FT-2. (**c**) FT-3. (**d**) FT-4.

5.2. Stiffness Degradation Model

Based on the results shown in Figures 8 and 9, the relationship between the relative plastic slip ($\delta_{\mathrm{pl},n}/\delta_{\mathrm{pl},N}$), relative elastic stiffness ($K_{\mathrm{el},n}/K_{\mathrm{el},0}$), and the relative fatigue loading cycles (n/N) after cycles n was calculated and is plotted in Figure 11. In the figure, $\delta_{pl,N}$ represents the maximum plastic slip recorded before fatigue failure, and $K_{\mathrm{el},0}$ represents the initial elastic stiffness recorded at the beginning of the test. From Figure 11, it can be observed that the relative cumulative plastic slip followed an anti-S growth curve, while the relative elastic stiffness degradation followed an anti-*S* descending curve. The evolution of relative plastic slip and relative elastic stiffness comprised three stages: firstly, due to the elimination of bonding and non-elastic deformation between steel flange and UHPC, it entered a brief period of rapid development. Subsequently, with fatigue damage continuously accumulated in the shear studs, the plastic slip and elastic stiffness entered a linear and stable development stage, which occupied the majority of the fatigue life. Finally, before fatigue failure, plastic slip accumulated rapidly, and elastic stiffness dropped sharply, leading to the eventual fatigue failure of the specimen.

In order to quantify the evolution of relative plastic slip and relative elastic stiffness, the double reciprocal logarithmic function [22] was used to fit the experimental data, as shown in Equations (9) and (10):

$$\frac{\delta_{\mathrm{pl},n}}{\delta_{\mathrm{pl},N}} = \begin{cases} 0 & n = 0 \\ B_1 \ln[-B_2 \ln(n/N)] & 0 < n < N \end{cases} \tag{9}$$

$$\frac{K_{\mathrm{el},n}}{K_{\mathrm{el},N}} = \begin{cases} 1 & n = 0 \\ C_1 + C_2 \ln\left(\frac{1}{n/N} - 1\right) & 0 < n < N \end{cases} \tag{10}$$

where B_1, B_2, C_1, and C_2 are regression coefficients, and $B_1 = -0.1490$, $B_2 = 0.2876$, $C_1 = 0.5434$, and $C_2 = 0.0952$.

The fitting results are also presented in Figure 11, with determination coefficients R^2 of 0.916 and 0.935, respectively, indicating a high degree of fitting accuracy.

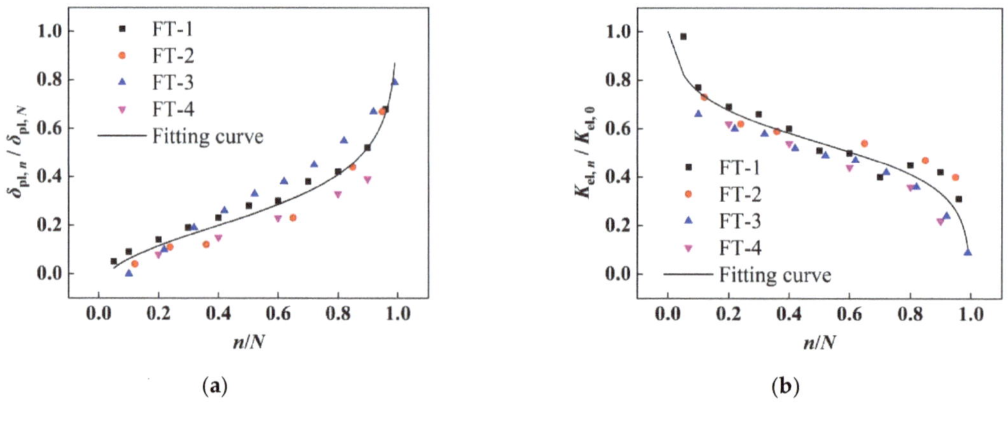

Figure 11. Degradation of mechanical behaviors. (**a**) Relative plastic slip. (**b**) Relative elastic stiffness.

6. Conclusions

This study experimentally investigated the fatigue behavior of ultra-short studs with an aspect ratio of 1.84 in UHPC. The following conclusions were drawn:

(1) The failure mode of all specimens was ultra-short stud fatigue failure, indicating that UHPC can ensure stud fatigue fracture for ultra-short stud and fully utilize the fatigue properties of ultra-short stud, which was also verified in shear studs with aspect ratios larger than 2.0 in similar studies.

(2) There were three failure modes of ultra-short studs in UHPC: Mode I, fracture surface at stud shank; Mode II, fracture surface at steel flange; and Mode III, fracture surface at stud cap. Among them, Mode III has not been reported in previous studies yet, and will only occur together with Mode II.

(3) There were two failure modes for UHPC: Mode A, UHPC was intact and locally crushed below the stud. This failure mode was in accordance with existing studies. Mode B, UHPC split failure, and the cracks were in a central diffusion shape. This failure mode of UHPC has also not been reported in previous studies yet.

(4) A fatigue strength design S–N curve for short studs in UHPC was proposed based on the fatigue test results. The fatigue strength of short studs in UHPC at 200×10^4 cycles with a 95% survival probability in the proposed curve was 105 MPa, which was significantly higher than the codified curves in EC3. The S–N curve can be used for fatigue strength design and fatigue life calculation of ultra-short studs in UHPC.

(5) Calculation models for stiffness degradation and plastic slip accumulation of short studs in UHPC were established based on the fatigue test results.

(6) Future research should focus on the fatigue failure mechanism and finite element parameter analysis of ultra-short studs in UHPC.

Author Contributions: Data curation, C.-J.T., K.Q., M.-Y.C. and Z.-Y.L.; investigation, Z.Y.; project administration, Y.-Z.W.; resources, Y.-Z.W., Z.Y., M.-Y.C. and Z.-Y.L.; writing—original draft, R.A., C.-J.T. and K.Q.; writing—review and editing, M.-L.Z. and R.A. All authors have read and agreed to the published version of the manuscript.

Funding: This research received no external funding.

Data Availability Statement: The data presented in this study are available in article here.

Conflicts of Interest: Author Chang-Jin Tian is employed by the China Construction Infrastructure Co., Ltd. Author Meng-Ying Cheng is employed by the Design & Consulting Company (Shandong) of China Construction 8th Engineering Division. Author Zhao-Yuan Lv is employed by the Shimao Group Holdings Limited. The remaining authors declare that the research was conducted in the absence of any commercial or financial relationships that could be construed as a potential conflict of interest.

References

1. Pei, B.; Li, L.; Shao, X.; Wang, L.; Zeng, Y. Field measurement and practical design of a lightweight composite bridge deck. *J. Constr. Steel Res.* **2018**, *147*, 564–574. [CrossRef]
2. Luo, J.; Shao, X.; Fan, W.; Cao, J.; Deng, S. Flexural cracking behavior and crack width predictions of composite (steel + UHPC) lightweight deck system. *Eng. Struct.* **2019**, *194*, 120–137. [CrossRef]
3. Liu, Y.; Zhang, Q.; Meng, W.; Bao, Y.; Bu, Y. Transverse fatigue behaviour of steel-UHPC composite deck with large-size U-ribs. *Eng. Struct.* **2019**, *180*, 388–399. [CrossRef]
4. Lu, K.; Du, L.; Xu, Q.; Yao, Y.; Wang, J. Fatigue performance of stud shear connectors in steel-concrete composite beam with initial damage. *Eng. Struct.* **2023**, *276*, 115381. [CrossRef]
5. Wang, Y.; Cao, J.; Shao, X.; Mao, Q.; Qu, Z. Effect of UHPC strengthening layer on static and fatigue performance of upper cutouts in orthotropic steel decks. *J. Constr. Steel Res.* **2022**, *198*, 107557. [CrossRef]
6. Luo, J.; Shao, X.; Cao, J.; Xiong, M.; Fan, W. Transverse bending behavior of the steel-UHPC lightweight composite deck: Orthogonal test and analysis. *J. Constr. Steel Res.* **2019**, *162*, 105708. [CrossRef]
7. Wang, Y.; Shao, X.; Zhang, X.; Cao, J.; Zhao, X.; Deng, S. Structural behaviors of low-profile steel plate reinforced UHPC deck panel with longitudinal ribs. *ASCE J. Bridge Eng.* **2021**, *26*, 04021043. [CrossRef]
8. Ding, F.X.; Yin, G.A.; Wang, H.B.; Wang, L.; Guo, Q. Static behavior of stud connectors in bi-direction push-off tests. *Thin-Walled Struct.* **2017**, *120*, 307–318. [CrossRef]
9. Xue, W.; Ding, M.; Wang, H.; Luo, Z. Static behavior and theoretical model of stud shear connectors. *J. Bridge Eng.* **2008**, *13*, 623–634. [CrossRef]
10. Luis, P.; Jerome, F. Headed steel stud anchors in composite structures, part I: Shear. *J. Constr. Steel Res.* **2010**, *66*, 198–212.
11. Yan, J.B.; Zhang, W.; Liew, J.R.; Li, Z.X. Numerical studies on shear resistance of headed stud connectors in different concretes under Arctic low temperature. *Mater. Des.* **2016**, *112*, 184–196. [CrossRef]
12. *EN 1994-1-1-2004*; Design of Composite Structures, Part 1-1: General Rules and Rules for Buildings. CEN—European Committee for Standardization: Brussels, Belgium, 2004.

13. Yang, H.; Zheng, Y.; Mo, S.; Lin, P. Push-out tests on studs with UHPC cover embedded in UHPC-NSC composite slab. *Constr. Build. Mater.* **2022**, *331*, 127210. [CrossRef]
14. Kim, J.S.; Kwark, J.; Joh, C.; Yoo, S.W.; Lee, K.C. Headed stud shear connector for thin ultrahigh-performance concrete bridge deck. *J. Constr. Steel Res.* **2015**, *108*, 23–30. [CrossRef]
15. Kruszewski, D.; Wille, K.; Zaghi, A.E. Push-out behavior of headed shear studs welded on thin plates and embedded in UHPC. *Eng. Struct.* **2018**, *173*, 429–441. [CrossRef]
16. Kruszewski, D.; Wille, K.; Zaghi, A.E. Design considerations for headed shear studs embedded in ultra-high performance concrete as part of a novel bridge repair method. *J. Constr. Steel Res.* **2018**, *149*, 180–194. [CrossRef]
17. Luo, Y.; Hoki, K.; Hayashi, K.; Nakashima, M. Behavior and strength of headed stud-SFRCC shear connection, II: Strength Evaluation. *J. Struct. Eng.* **2015**, *142*, 04015113. [CrossRef]
18. Shi, Z.; Su, Q.; Kavoura, F.; Veljkovic, M. Behavior of short-headed stud connectors in orthotropic steel-UHPC composite bridge deck under fatigue loading. *Int. J. Fatigue* **2022**, *160*, 106845. [CrossRef]
19. Chen, S.; Huang, Y.; Gu, P.; Wang, J.Y. Experimental study on fatigue performance of UHPC-orthotropic steel composite deck. *Thin-Walled Struct.* **2019**, *14*, 1–18. [CrossRef]
20. Huang, Y.; Chen, S.M.; Gu, P. Numerical analysis on fatigue behavior of ultrahigh-performance concrete-Orthotropic steel composite bridge deck. *Adv. Struct. Eng.* **2022**, *25*, 1151–1166. [CrossRef]
21. Cao, J.; Shao, X.; Deng, L.; Gan, Y. Static and fatigue behavior of short-headed studs embedded in a thin ultrahigh-performance concrete layer. *J. Bridge Eng.* **2017**, *22*, 04017005. [CrossRef]
22. Wei, C.; Zhang, Q.; Zhou, Y.; Cheng, Z.; Li, M.; Cui, C. Static and fatigue behaviors of short stud connectors embedded in ultra-high performance concrete. *Eng. Struct.* **2022**, *273*, 114888. [CrossRef]
23. Huang, Y.; Chen, S.M.; Gu, P. Static and fatigue behavior of shear stud connection embedded in UHPC. *Structures* **2021**, *34*, 2777–2788. [CrossRef]
24. Liu, Y.; Zhang, Q.; Bao, Y.; Bu, Y. Static and fatigue push-out tests of short headed shear studs embedded in Engineered Cementitious Composites (ECC). *Eng. Struct.* **2019**, *182*, 29–38. [CrossRef]
25. Hobbacher, A.F. The new IIW recommendations for fatigue assessment of welded joints and components-a comprehensive code recently updated. *Int. J. Fatigue* **2009**, *1*, 50–58. [CrossRef]
26. Li, J.; Yang, B.; Shao, X.D.; Li, J. Research on shear fatigue of studs for composite deck system of steel slab and thin CRRPC layer. *J. Civ. Eng.* **2016**, *49*, 67–75. (In Chinese)
27. Zhang, S.; Shao, X.D.; Huang, X.J.; Yang, B. Static and fatigue behaviors of small stud shear connector for lightweight composite bridge deck. *J. Highw. Transp. Res. Dev.* **2016**, *33*, 111–119. (In Chinese)
28. *Eurocode-3*; Design of Steel Structures. European Committee for Standardization: Brussels, Belgium, 2004.
29. Hobbacher, A.F. *Recommendations for Fatigue Design of Welded Joints and Components (XIII-2460-13/XV-1440-13)*; International Institute of Welding: Paris, France, 2016; pp. 1–143.

buildings

MDPI

Article

Minimum Error Integration Method for Quadrilateral Flat Plate Fitting in Steel Construction

Zhuoju Huang [1,2,*] and Jiemin Ding [2]

[1] Department of Structural Engineering, Tongji University, Shanghai 200092, China
[2] Tongji Architectural Design (Group) Co., Ltd., Shanghai 200092, China
* Correspondence: 52hzj@tjad.cn; Tel.: +86-15216758536

Abstract: In modern steel structures such as bridges and buildings, curved members are increasingly adopted. Current industry practices often approximate mildly curved surfaces with flat plates within tolerances, but optimizing this substitution to minimize fitting errors while reducing the quantity of curved plates remains a critical engineering challenge. While traditional approaches that rely on empirical craftsmanship or least-squares fitting lack precision, this study proposes a minimum error integration method that integrates Lagrange interpolation-based error estimation with an adaptive step-size steepest descent algorithm to reduce fitting error. Numerical experiments are performed to compare the proposed method against the least-squares method across two scenarios: (1) surfaces with typical shape and curvature and (2) a practical engineering case. Our results demonstrate at most a 75.5% reduction in fitting errors for analytical curved plates with particularly significant improvements in biconvex curvature scenarios. A practical engineering validation reveals that the method increases the proportion of planarizable plates from 27% to 45% under identical tolerance criteria, effectively reducing curved-plate fabrication demands and thus reducing cost and carbon emissions. The proposed optimization method offers a mathematically grounded alternative to experience-dependent practices. These findings validate the method's potential to enhance cost-effectiveness and manufacturing sustainability in steel structure projects, suggesting broader applicability in curvature-driven construction scenarios.

Keywords: steel structures; curve member fabrication; fitting error minimization; Lagrange interpolation

1. Introduction

In modern steel bridges and buildings, structural members such as box girders and box arches are typically composed of multiple quadrilateral steel plates (Figure 1). However, bending steel plates into curved shapes is often challenging for curved structural members [1–4]. Additionally, the bending process exhibits heightened susceptibility to crack initiation in steel plates [5]. Moreover, the metal forming process requires a significant amount of electrical energy. Formed steel consumes approximately 50% more energy compared to flat steel plates [6,7], and the bending process significantly amplifies workload in key carbon-intensive operations, including stiffener welding and flame straightening, which collectively account for 80% of carbon emissions in steel component fabrication [8–11]. Consequently, the increased utilization of curved plates introduces non-negligible carbon footprint increments that should not be overlooked in sustainable manufacturing assessments. Under current domestic fabrication techniques in China, the comprehensive unit cost of steel structural curved members remains at least 1.3 times that of planar components [12]. To mitigate production expenses, in practical engineering, for elements with small curvature, flat plates are usually substituted during fabrication [13], provided that the following criteria are met: The deviation between the flat plate and the original curved plate remains within permissible tolerances [14–16] and the unevenness of the seam between adjacent plates meets the error requirements for welding construction [17,18]. Such substitutions significantly increase fabrication efficiency, reduce project costs, and increase manufacturing sustainability [19].

(a) (b)

Figure 1. Curved members fabricated using quadrilateral steel plates: (**a**) arch and (**b**) girder.

The engineering community currently addresses such substitution problems through empirical approaches. A critical gap exists in the existing literature regarding systematic discussions of this specific issue. Traditionally, such substitutions rely on trial-and-error estimation by craftsmen [19], which lacks precision and struggles to control construction costs and quality during early project phases.

In recent years, least-squares method (LSM)-based fitting with discrete points [20] has been introduced to solve such problems. In the three-dimensional Cartesian coordinate system, a general plane can be parameterized by coefficients a,b,c expressed as Equation (1).

$$z = ax + by + c \tag{1}$$

For a set of points (x_1, y_1, z_1), (x_2, y_2, z_2), …, (x_n, y_n, z_n), a linear equation could be built to find the least-squares plane:

$$\begin{pmatrix} x_1 & y_1 & 1 \\ x_2 & y_2 & 1 \\ \vdots & \vdots & \vdots \\ x_n & y_n & 1 \end{pmatrix} \begin{pmatrix} a \\ b \\ c \end{pmatrix} = \begin{pmatrix} z_1 \\ z_2 \\ \vdots \\ z_n \end{pmatrix} \tag{2}$$

When $n \geq 3$, the parameters a,b,c of the least-squares plane for the point set can be obtained by computing the pseudo inverse of the coefficient matrix, as shown in Equation (2). When the original surface patch closely approximates a fitted plane, its boundary can be delineated by projecting it onto the fitted plane, thereby obtaining an approximate planar surface patch, which represents a flat plane in the real world.

While analytical, the least-squares method suffers from limited fitting accuracy. In recent years, some new approaches based on principal component analysis have been proposed [21,22] in a similar field of machine learning, but since they are not designed for geometric fitting, the geometric error is not considered, and this may eliminate some irregular points, which is not suitable for such an engineering problem.

To address these engineering challenges, in this paper, we propose a novel quadrilateral flat plate fitting method that achieves higher precision through the interpolation of fitting errors, offering an enhanced solution for practical applications.

2. Minimum Error Integration Method

2.1. Interpolation Method for Error Evaluation

In order to evaluate fitting error globally, sampling is necessary. To enhance computational efficiency in total error estimation, it is essential to minimize the number of error sampling points. Interpolation is a practical approach to approximate the overall error with a limited set of sample points [23]. This method achieves high accuracy while drastically reducing sampling requirements.

To sufficiently obtain the global error distribution, sampling points should ideally be uniformly distributed across the curved surface. For a bidirectionally uniform $m \times n$ grid layout, $(m+1) \times (n+1)$ sampling points are required. In typical bridge engineering practice, where the designed curvature variations within steel plates are relatively mild, a 3×3 grid division—sampling at nine key locations (edges, corners, and center of the quadrilateral)—is generally appropriate for quadratic interpolation. For rectangular integration domains, the sampling grid configuration is illustrated in Figure 2.

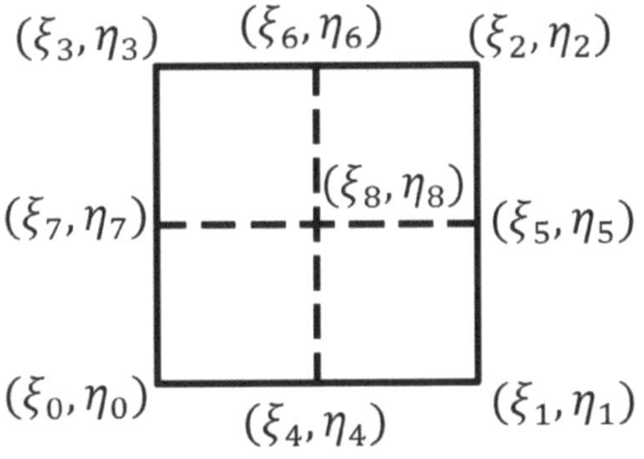

Figure 2. The sampling point grid in a rectangular domain.

Lagrange interpolation is a classical polynomial interpolation method [15,16]. The basis functions for full nine-point interpolation are expressed as Equation (3):

$$l_k(\xi) = \prod_{j=0, j\neq k}^{2} \frac{\xi - \xi_j}{\xi_k - \xi_j} \quad (3)$$

where $k = 0, 1, 2.$ and ξ_i, ξ_j, ξ_k are the ith, jth, and kth nodes in ξ direction.

Likewise, for the η direction, the basis functions are expressed as Equation (4):

$$l_k(\eta) = \prod_{j=0, j\neq k}^{2} \frac{\eta - \eta_j}{\eta_k - \eta_j} \quad (4)$$

where $k = 0, 1, 2.$ and η, η_j, η_k are the ith, jth, and kth nodes in η direction.

In a two-dimensional rectangular region, the interpolated functions are expressed as Equation (5):

$$N_{3i+j}(\xi, \eta) = l_i(\xi) l_j(\eta) \quad (5)$$

Then, the error δ in any location (ξ, η) in the two-dimensional area can be interpolated from sample points as Equation (6):

$$\delta(\xi,\eta) = \sum_{i=0}^{8} N_i(\xi,\eta)\delta_i \quad (6)$$

Since the aforementioned Lagrange interpolation applies to rectangular domains, while actual bridge plates in engineering practice often exhibit irregular geometries, isoparametric transformation, which was first used in the finite element method [24–27], is adopted to compute integrals over non-rectangular regions (Figure 3). In isoparametric transformations, the geometric configuration of arbitrary quadrilateral curved surfaces is similarly represented through interpolation functions that share identical parametric dimensions with the error interpolation scheme. This equivalence in parametric dimensions enables numerical integration of errors over any two-dimensional quadrilateral domain to be computationally evaluated via coordinate transformation to a reference rectangular domain.

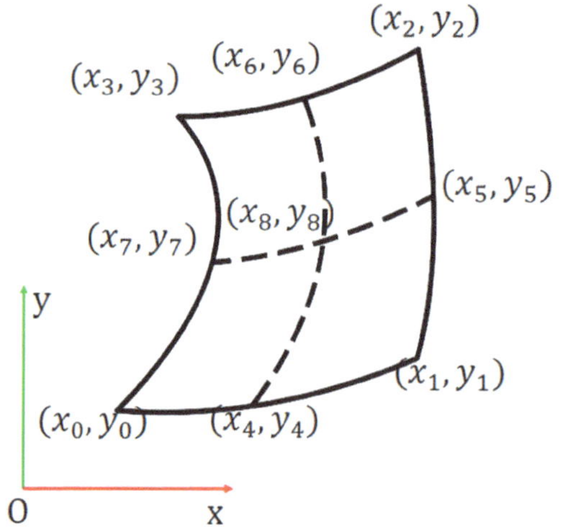

Figure 3. Isoparametric transformation for an irregular domain.

The Jacobian matrix of this transformation is defined as follows:

$$J = \begin{pmatrix} \sum\limits_{i=0}^{8} \frac{\partial N_i(\xi,\eta)}{\partial \xi} x_i & \sum\limits_{i=0}^{8} \frac{\partial N_i(\xi,\eta)}{\partial \xi} y_i \\ \sum\limits_{i=0}^{8} \frac{\partial N_i(\xi,\eta)}{\partial \eta} x_i & \sum\limits_{i=0}^{8} \frac{\partial N_i(\xi,\eta)}{\partial \eta} y_i \end{pmatrix} \quad (7)$$

This approach converts integrals over free-form regions into standard rectangular domain integrals. The absolute error integral is formulated in Equation (8):

$$\Delta(\boldsymbol{p}) = \iint \delta(\boldsymbol{p}, x, y)\mathrm{d}x\mathrm{d}y = \iint \delta(\boldsymbol{p}, \xi, \eta)\det(\boldsymbol{J})\mathrm{d}x\mathrm{d}y \quad (8)$$

where $\Delta(\boldsymbol{p})$ is the total error in the integration region given a parameter vector $\boldsymbol{p} = (a\ b\ c)^{\mathrm{T}}$, which determines the plane. And δ is relative to $\boldsymbol{p}$ in this situation, representing the fitting error, which can be taken to be the distance from the point to the plane when the difference between the two is small. And $\det(\boldsymbol{J})$ is the determinant of the Jacobi matrix.

In practice, it is the absolute magnitude of the error rather than the algebraic sum that is of interest. From a numerical computational point of view, squaring makes the local effect of a large absolute value of error magnified and is more conducive to finding the optimal approximation, so the squared error can be used in numerical integration calculations, as shown in Equation (9):

$$\Delta(\boldsymbol{p}) = \iint \delta^2(\boldsymbol{p}, x, y)\mathrm{d}x\mathrm{d}y = \iint \delta^2(\boldsymbol{p}, \xi, \eta)\det(J)\mathrm{d}\xi\mathrm{d}\eta \tag{9}$$

Hence, a standard unconstrained optimization model can be formulated:

$$\min_{\boldsymbol{p}\in\mathrm{R}^3} \Delta(\boldsymbol{p}) \tag{10}$$

2.2. Method for Optimization of the Model Solution

Generally, the convexity of this optimization problem in Equation (10) cannot be guaranteed, and the algorithm's performance depends critically on the initial guess. If the initial value lies near the optimal solution, convergence to the global optimum can be ensured. In practice, the least-squares solution may serve as the initial guess for subsequent refinement. When the state is sufficiently close to the optimum, the steepest descent method [28]—a gradient-based algorithm using adaptive optimal step sizes—becomes a suitable choice. The algorithm designed for the problem studied in this work is described as follows:

1. Initialize $\boldsymbol{p}^{<0>} \leftarrow (a\ b\ c)^{\mathrm{T}}$
2. For each time step $t = 0, 1, 2, \ldots$, do:
3. Compute gradient $g^{<t>} \leftarrow \nabla\Delta(\boldsymbol{p}^{<t>})$
4. Search for optimum step size $\alpha^* \leftarrow \mathrm{argmin}\Delta(\boldsymbol{p}^{<t>} - \alpha g^{<t>})$
5. Update: $\boldsymbol{p}^{<t+1>} \leftarrow \boldsymbol{p}^{<t>} - \alpha^* g^{<t>})$
6. Repeat step 2~5 until $\boldsymbol{p}$ converges

In the algorithm described above, the superscript denotes the iteration count for a variable. Since the fitting error is computed via numerical interpolation and integration, the one-dimensional search in Step 4 cannot resolve the analytical minimum through derivative-based methods [29]. To address this, a bisection method is employed for rapid step-size optimization, as detailed below:

1. Initialize step size $\alpha_0{}^{<0>} \leftarrow 0,\ \alpha_1{}^{<0>} \leftarrow 1$
2. For each time step $t' = 0, 1, 2, \ldots$, do:
3. $\alpha_0{}^{<t'+1>} \leftarrow \alpha_1{}^{<t'>},\ \alpha_1{}^{<t'+1>} \leftarrow 2\alpha_1{}^{<t'>}$
4. Until $\Delta\left(\boldsymbol{p}^{<t'>} - \alpha_0{}^{<t'>}g^{<t'>}\right) < \Delta\left(\boldsymbol{p}^{<t'>} - \alpha_1{}^{<t'>}g^{<t'>}\right)$, denote this time step as t^*
5. For time step $t' = t^* + 1,\ t^* + 2, \ldots$, do:
6. If $\Delta\left(\boldsymbol{p}^{<t'>} - \alpha_0{}^{<0>}g^{<t'>}\right) < \Delta\left(\boldsymbol{p}^{<t'>} - \frac{1}{2}(\alpha_0{}^{<0>} + \alpha_1{}^{<0>})g^{<t'>}\right)$
7. $\alpha_0{}^{<t'+1>} \leftarrow \alpha_1{}^{<t'>},\ \alpha_1{}^{<t'+1>} \leftarrow \frac{1}{2}\left(\alpha_0{}^{<t'>} + \alpha_1{}^{<t'>}\right)$
8. Else
9. $\alpha_0{}^{<t'+1>} \leftarrow \frac{1}{2}\left(\alpha_0{}^{<t'>} + \alpha_1{}^{<t'>}\right),\ \alpha_1{}^{<t'+1>} \leftarrow \alpha_1{}^{<t'>}$,
10. Repeat step 5~9 until α_0 converges to α_1, then take $\alpha^* = \alpha_0$

In the above process, Steps 2 to 4 delineate the interval containing the optimal step size α^*, while Steps 5 to 10 employ the bisection method to search for α^*. The entire procedure, consolidated into an intuitive workflow, is illustrated in Figure 4.

After obtaining the approximate planar surface using the steepest descent method, the boundaries of the original curved plate are projected onto this plane to delineate the boundaries of the approximated flat plate unit. Subsequent verification must confirm compliance with engineering tolerance limits. As demonstrated in the above reference, the convergence of the steepest descent method guarantees a linear convergence rate under appropriate conditions.

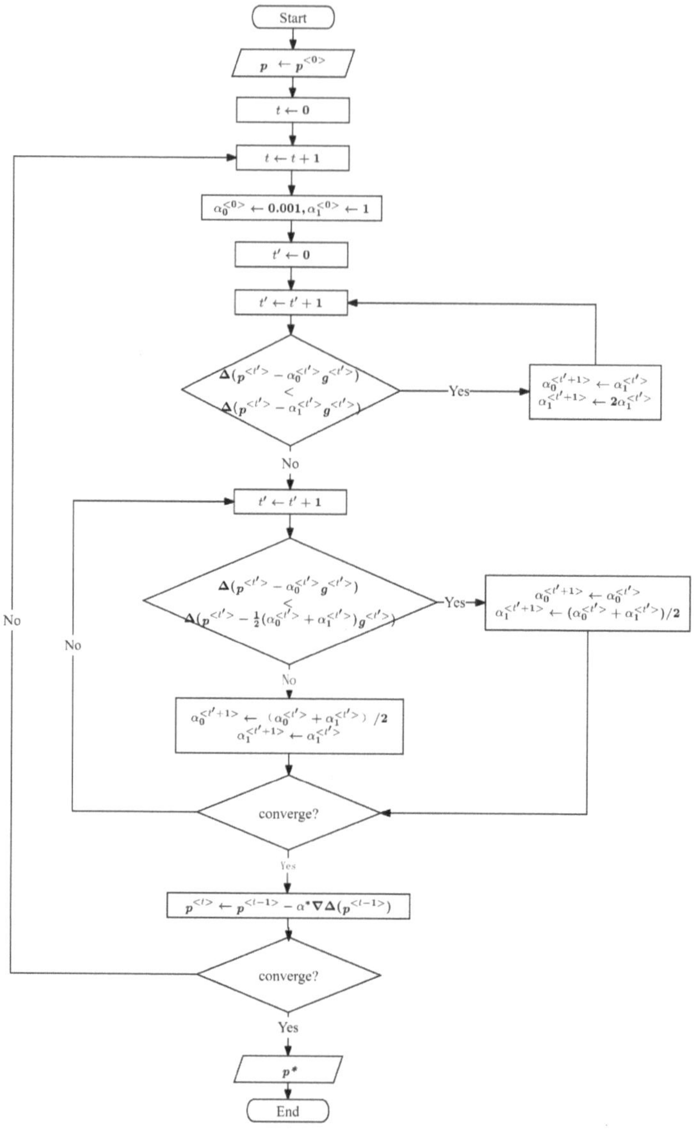

Figure 4. Workflow of the steepest descent method in this work.

3. Results

3.1. *Typical Geometry Cases*

3.1.1. Shape

Five typical planar shapes, including a square, a rectangle, a parallelogram, a trapezoid, and an arbitrary quadrilateral, were investigated, each under varying curvature conditions, including plane, ruled saddle surfaces, unidirectional/bidirectional convex surfaces, and bidirectional concave–convex surfaces. These fundamental quadrilateral geometries sufficiently represent major engineering scenarios, totaling 25 benchmark cases (Table 1).

Table 1. Surface shape of the cases.

Case	Square 1 m × 1 m	Rectangle 2 m × 1 m	Parallelogram 1 m × 1 m	Trapezoid 1 m/1.5 m × 1 m	Arbitrary 1 m on Average
Plane $\kappa_1 = 0$					
Ruled surfaces $\kappa_1\kappa_2 < 0$					
Unidirectional convex $\kappa_1 > 0$ $\kappa_2 = 0$					
Bidirectional convex $\kappa_1 > 0$ $\kappa_2 > 0$					
Bidirectional concave–convex $\kappa_1\kappa_2 < 0$					

3.1.2. Fitting Result

Using a 3 × 3 grid with nine sampling points, the fitting performance of the two methods—the basic least-squares method and the proposed minimum error integration method (MEIM)—was analyzed across surfaces with distinct geometric features, as summarized in Table 2.

Table 2. Fitting error of the nine-point LSM for different surface shapes.

Case	Square	Rectangle	Parallelogram	Trapezoid	Arbitrary
Plane	0.000	0.000	0.000	0.000	0.000
Ruled saddle	0.028	0.028	0.028	0.026	0.027
Unidirectional convex	0.200	0.200	0.200	0.200	0.200
Bidirectional convex	0.331	0.156	0.156	0.156	0.156
Bidirectional concave–convex	0.226	0.226	0.226	0.225	0.226

By taking the least-squares fitting results as initial parameters and applying the proposed minimum error integration method, the results summarized in Table 3 can be obtained.

Table 3. Fitting error of the proposed MEIM for different surface shapes.

Case	Square	Rectangle	Parallelogram	Trapezoid	Arbitrary
Plane	0.000	0.000	0.000	0.000	0.000
Ruled saddle	0.028	0.028	0.028	0.026	0.027
Unidirectional convex	0.089	0.089	0.089	0.089	0.089
Bidirectional convex	0.081	0.044	0.044	0.044	0.044
Bidirectional concave–convex	0.211	0.211	0.211	0.209	0.210

The percentage improvement of the proposed minimum error integration method over the least-squares method is tabulated in Table 4.

Table 4. Improvement of the proposed MEIM fitting.

Case	Square	Rectangle	Parallelogram	Trapezoid	ACQ
Plane	0.0%	0.0%	0.0%	0.0%	0.0%
Ruled saddle	0.0%	0.0%	0.0%	0.0%	0.0%
Unidirectional convex	55.5%	55.5%	55.5%	55.3%	55.5%
Bidirectional convex	75.5%	75.5%	75.5%	75.5%	75.5%
Bidirectional concave–convex	6.9%	6.9%	6.9%	6.9%	6.9%

Analysis of the data in Table 4 reveals the following patterns:

1. For planar and ruled surface cases, the least-squares method achieves a minimum sum of squared errors, yielding perfectly accurate fits. For ruled saddle surfaces (featuring two orthogonal linear curvature directions), planar approximations exhibit residual errors, though the least-squares method still minimizes squared errors. In both scenarios, the MEIM provides no improvement.
2. The MEIM enhances accuracy for other surface types: the improvement for bidirectional concave–concave surfaces is moderate, while the improvement for unidirectional convex and bidirectional convex surfaces is significant.
3. The MEIM is planar-shape-independent. The mean squared fitting errors remain largely unaffected by planar geometry variations.

Comparative visualizations of fitting results for distinct surface types—unidirectional convex, bidirectional convex, and bidirectional concave–convex surfaces—are depicted in Figures 5–7, with original surfaces rendered in gray and fitted planes rendered in red.

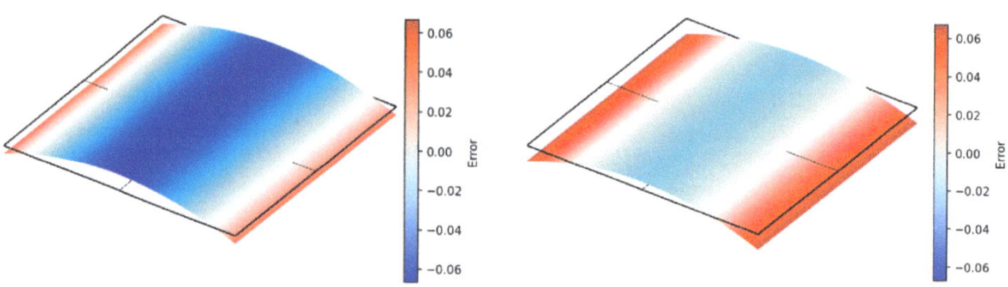

(**a**) (**b**)

Figure 5. Visualized comparison for a unidirectional convex case: (**a**) LSM and (**b**) MEIM.

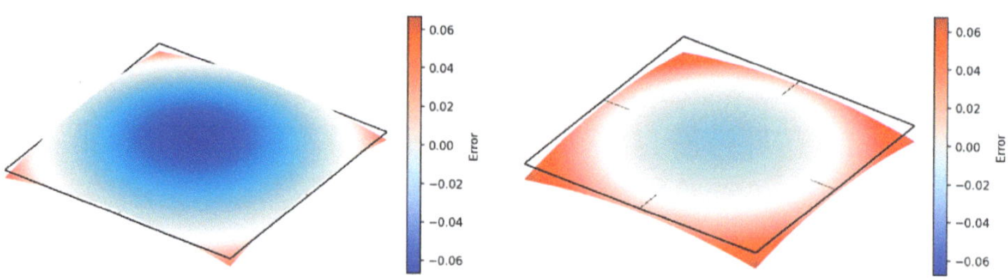

(**a**) (**b**)

Figure 6. Visualized comparison for a bidirectional convex case: (**a**) LSM and (**b**) MEIM.

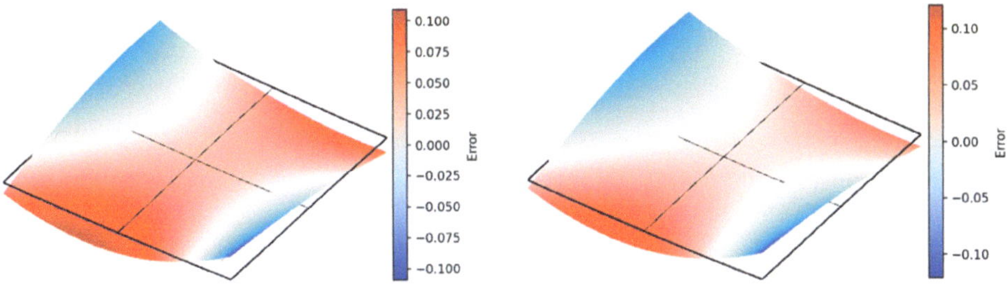

(**a**) (**b**)

Figure 7. Visualized comparison for a bidirectional convex–concave case: (**a**) LSM and (**b**) MEIM.

3.2. Practical Engineering Case

3.2.1. Overview

The Pedestrian Bridge at Shaoxing World Convention and Exhibition Center (Shaoxing World) is a curved, lightweight suspension structure spanning 120 m (Figure 8) [30]. The bridge was completed in 2023, and its curved form has significant representativeness in bridge design. Many footbridges around the world have similar curves, including Yuanshan Bridge [31] and Hemei Bridge [32] in China, Footbridge Harbor Grimberg [33] in Europe, and the Liberty Bridge [34] in the United States. Its curved main girder comprises two closed polygonal sections, forming a spatially complex geometry.

(**a**) (**b**)

Figure 8. Pedestrian Bridge at Shaoxing World: (**a**) overview and (**b**) curved girder.

The integration of multiple design factors—including a longitudinal gradient along the bridge axis, curved alignment, compound cross-sectional profiles, inclined slope rates, and pre-cambering—imparts biaxial curvature to every steel plate in the structure. However, its slender, cantilevered aesthetic demands exceptionally high fabrication precision.

The structural design strategically balances cost and constructability by optimizing plate geometries to meet economic, aesthetic, and construction quality requirements. The planar and spatial configurations of the main girder are illustrated in Figure 9.

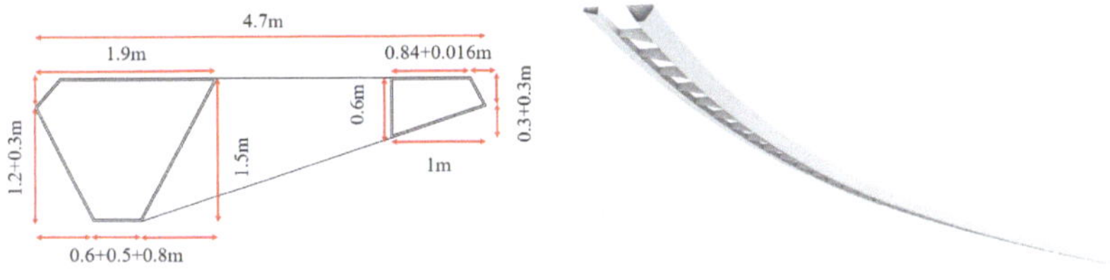

(**a**) (**b**)

Figure 9. The girder of the bridge: (**a**) cross-section and (**b**) 3D shape.

The bridge's cross-section consists of two longitudinal main girders interconnected by transverse beams. When the main girders follow curved alignments with slope variations, all plates (bottom plates, side plates, etc.) inherently develop biaxial curvature. Given that these plates are externally exposed and require seamless visual continuity, yet curved steel plate fabrication is both technically challenging and cost-prohibitive, meticulous computational verification of each plate's formed shape is essential to ensure joint compatibility and surface smoothness per design specifications.

The main girder's steel plates are subdivided into 279 quadrilateral, biaxially curved plate units based on fabrication requirements. In this study, we employ the least-squares method and the minimum error integration method to fit these geometrically complex units into flat plates, thereby streamlining the fabrication of structural components while preserving design integrity.

3.2.2. Feasibility Criteria

In practical engineering, feasibility criteria for plate fitting should align with both fabrication and aesthetic requirements. This typically involves defining maximum error thresholds $[\delta_{\mathrm{m}}]$ and average error thresholds $[\delta_{\mathrm{a}}]$ for decision-making, as shown in Equation (11):

$$\begin{cases} \delta_{\mathrm{m}} = \max\delta(x,y) < [\delta_{\mathrm{m}}] \\ \delta_{\mathrm{a}} = \frac{\iint \delta(x,y)\mathrm{d}x\mathrm{d}y}{\iint \mathrm{d}x\mathrm{d}y} < [\delta_{\mathrm{a}}] \end{cases} \quad (11)$$

For this project, the criteria are as follows:

1. Welding technical specifications mandate that the maximum deviation between fitted and target plates must not exceed 2 mm;
2. Aesthetic considerations require the average deviation across the entire plate surface to remain below 1 mm.

3.2.3. Fitting Result

Figure 10 presents the maximum and average fitting errors generated using the least-squares method and the interpolation–integration method for the main girder plates. To enhance visual clarity, the plates are sorted in ascending order based on the max fitting error results, with their data points sequentially connected to form a curve. Correspondingly, the mean fitting error results are plotted, creating the orange curve. Error threshold lines for both maximum and average deviations are annotated in the figure. Compared with the results shown in Figure 10a,b, the results of the MEIM are significantly better than those of the LSM, reducing both the maximum and mean error to about 30–50%, correspondingly.

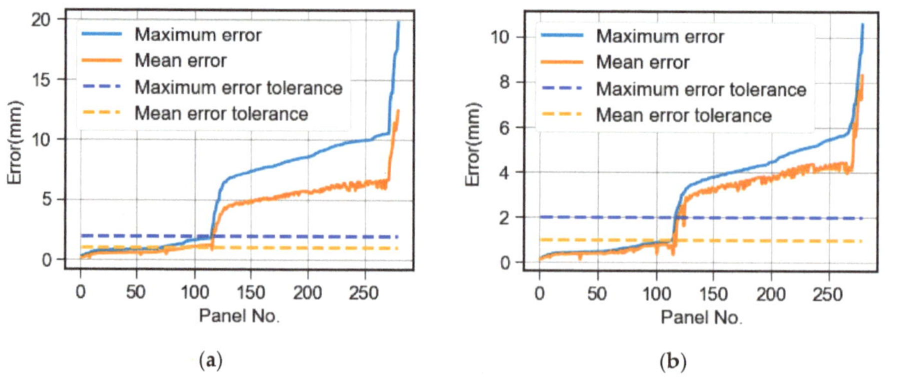

Figure 10. Fitting with $[\delta_1] = 2$ mm and $[\delta_2] = 1$ mm: (**a**) LSM and (**b**) MEIM.

Another significant observation is that the differential between maximum and average errors in the MEIM results is substantially smaller than in the LSM outcomes, indicating that for each plate segment, the fitting error distribution within the MEIM results demonstrates superior uniformity compared to the LSM. Figure 10 further highlights this contrast: under the specified error thresholds, only about 27% of plates fitted using the least-squares method meet the requirements, whereas the interpolation–integration method achieves compliance for approximately 45% of plates.

Such a result has certain benefits for both the cost and the sustainability of the practical project. For the curved beam section of this project, the steel consumption is approximately 240 tons. Based on the aforementioned optimization results combined with relevant cost data [12] estimates, the LSM demonstrates a cost reduction of approximately RMB 194,400 (equivalent to USD 26,600), while the minimum error integration method (MEIM) achieves savings of RMB 324,000 (USD 44,290). Furthermore, according to carbon emission coefficients derived from the steel structure literature [8,10], the LSM reduces carbon emissions by approximately 5205 kg CO_2, with the MEIM achieving a more substantial reduction of 8676 kg CO_2. This improvement demonstrates that the minimum error integration method enables a greater proportion of curved plates to be replaced with planar approximations without compromising technical or aesthetic standards, thereby significantly enhancing structural cost-effectiveness and manufacturing sustainability.

A direct visual comparison of the fitting results is provided in Figure 11. The target biaxially curved plates are rendered in light gray, while the least-squares and interpolation–integration fitted planar plates are shown in blue and orange, respectively. The overlapping color patterns reveal the spatial relationships between the fitted planes and the original curved plates. Notably, the interpolation–integration results exhibit more uniform alignment with the target geometry, with minimized positional deviations compared to the least-squares outcomes.

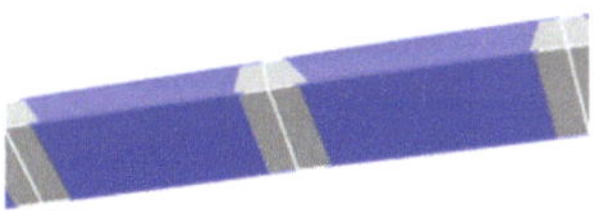

(a)

(b)

Figure 11. Visualized comparison of the fitting results: (**a**) LSM and (**b**) MEIM.

4. Discussion

In the results shown above, the interpolation–integration approach demonstrably outperforms the least-squares method, particularly for unidirectional and bidirectional convex surfaces. In these cases, the fitted plane aligns more closely with the mid-region of the plate, achieving a globally averaged optimal position. This phenomenon arises from fundamental methodological differences:

The LSM linearly sums errors, yielding exact fits for planar and ruled surfaces where errors inherently follow linear distributions. However, for curved surfaces, its linear weighting disproportionately prioritizes peripheral sampling points, degrading fit quality. But for the MEIM, the Lagrange interpolation adopted has quadratic accuracy, which better accommodates nonlinear error distributions.

Figure 12 illustrates this contrast: least-squares optimization drives the fitted plane toward error equilibrium at sampling points, minimizing local deviations δ to the state $\delta_2 = \delta_1 + \delta_3$. The minimum error integration method prioritizes error equilibrium across the integrated area A, minimizing cumulative spatial deviations to the state $A_2 = A_1 + A_3$.

Obviously, with limited sampling points, the least-squares method overweights edge regions, whereas interpolation–integration ensures holistic error balancing.

Figure 12. Geometrical explanation for the equilibrium state: (**a**) LSM and (**b**) MEIM.

For negative Gaussian surfaces, the two perpendicular curvature directions partially cancel edge-weighting biases, reducing the total error. But for non-negative Gaussian surfaces, the cumulative edge-weighting biases amplify the total error, making the MEIM's improvements particularly pronounced.

In summary, the principal innovation of the interpolation–integration method lies in quadratic error accommodation through Lagrange interpolation, and the isoparametric transformation's curvature-agnostic mapping ensures that error variations across geometries primarily stem from numerical computation limits rather than shape differences.

Although the MEIM comprises interpolation and integration, which entails greater computational complexity compared to the LSM, its computation speed remains acceptable when processing limited sample points. The computational analysis in this study was conducted on a workstation equipped with an Intel® Core™ i7-8700 CPU @ 3.20 GHz (Intel® Corporation, Santa Clara, CA, USA). The total processing time for individual panel fitting, including pre-processing and post-processing, was recorded as 63 ± 7 ms. Even in large-scale engineering structures requiring the processing of ten thousand panels, the total computation time remains approximately 10 min. This computational efficiency stems from the initial values provided by the least-squares method, which offer high-quality initial guesses that effectively reduce the number of required gradient descent iterations.

The proposed algorithm demonstrates reliable convergence when fitting continuous surfaces, with the adaptive step size calculation preventing oscillations during the optimization process. However, similar to conventional gradient descent approaches, the algorithm guarantees only a local optimum in the vicinity of the least-squares solution rather than the global optimum.

5. Conclusions

In this study, we establish a mathematical model for estimating global fitting errors through interpolation methods and employ the steepest descent algorithm to derive optimal fitting planes. The algorithm's characteristics are systematically analyzed using both typical geometries and practical engineering cases. The proposed methodology demonstrates the following advantages:

1. It achieves high-precision error estimation with minimal sampling points;
2. Compared to the LSM, it attains superior quadratic accuracy, and under particular curvature conditions, it achieves over 75% error reduction compared to the LSM;
3. While matching the least-squares method's performance for planar and negative Gaussian curvature surfaces, it significantly outperforms the least-squares method in fitting general convex surfaces;
4. Practical cases have verified the effectiveness of the method, increasing the proportion of planarizable plates from 27% to 45%, thus achieving dual benefits in construction cost reduction and carbon emission mitigation for the engineering structure.

It should be acknowledged that the proposed method exhibits certain inherent limitations. The optimization framework employed in this study demonstrates initial condition dependency. While employing least-squares solutions as initial values typically enables rapid convergence to favorable solutions, this approach does not guarantee attainment of the global optimum. Furthermore, when confronted with surfaces exhibiting excessive curvature, planar fitting may fail to maintain geometric errors within acceptable tolerance thresholds. Under such scenarios, the methodology necessitates consideration of alternative canonical surface types, such as cylindrical or conical surfaces, for satisfactory error control.

In summary, the minimum error integration method proposed in this study achieves markedly superior fitting performance for unidirectional or bidirectional convex structural plates compared to the least-squares method. These findings validate the method's potential to enhance cost-effectiveness and manufacturing sustainability in steel structure projects, suggesting broader applicability in curvature-driven construction scenarios. It is recommended to apply the method proposed in this paper during the steel structure detailing and fabrication phase and integrate it with cost analysis, carbon emission calculation, and automated fabrication systems to further leverage its value in engineering construction.

Author Contributions: Conceptualization, Z.H.; methodology, Z.H.; software, Z.H.; validation, Z.H.; formal analysis, Z.H.; investigation, Z.H.; resources, J.D.; data curation, Z.H.; writing—original draft preparation, Z.H.; writing—review and editing, J.D.; visualization, Z.H.; supervision, J.D.; project administration, J.D.; funding acquisition, J.D. All authors have read and agreed to the published version of the manuscript.

Funding: The APC was funded by Tongji Architectural Design (Group) Co., Ltd.

Data Availability Statement: The original contributions presented in this study are included in the article. Further inquiries can be directed to the corresponding author.

Conflicts of Interest: Authors Zhuoju Huang and Jiemin Ding were employed by Tongji Architectural Design (Group) Co., Ltd.

Abbreviations

The following abbreviations are used in this manuscript:

LSM	Least-squares method
MEIM	Minimum error integration method

References

1. Weng, C.; White, R. Cold-bending of thick high-strength steel plates. *J. Struct. Eng.-ASCE* **1990**, *116*, 40–54. [CrossRef]
2. Weng, C.; White, R. Residual-stresses in cold-bent thick steel plates. *J. Struct. Eng.-ASCE* **1990**, *116*, 24–39. [CrossRef]
3. Yan, R.; Lin, Y.; Xin, C.; Hu, Y. *Research on Residual Stress for Hull Plate in Cold Forming*; Zhang, X., Zhang, B., Jiang, L., Xie, M., Eds.; Scientific.Net: Schwyz, Switzerland, 2014; Volume 838–841, pp. 347–354.
4. Gerbo, E.J.; Thrall, A.P.; Smith, B.J.; Zoli, T.P. Full-Field Measurement of Residual Strains in Cold Bent Steel Plates. *J. Constr. Steel Res.* **2016**, *127*, 187–203. [CrossRef]
5. Jiang, A.; Cheng, Z.; Yu, W. Cause for Cracking of Hot-Rolled Q345 Steel Plate During Roll Bending. *Shanghai Met. Chin.* **2025**, 1–8. [CrossRef]
6. Norgate, T.E.; Jahanshahi, S.; Rankin, W.J. Assessing the Environmental Impact of Metal Production Processes. *J. Clean. Prod.* **2007**, *15*, 838–848. [CrossRef]
7. Algamal, A.; Ablat, M.A.; Ali, M.; Almotari, A.; Qattawi, A. Manufacturing Energy and Environmental Evaluation of Metallic Sheets Reshaping Based on Theoretical Models and Cradle-to-Gate Assessment. *J. Clean. Prod.* **2023**, *416*, 137795. [CrossRef]
8. Li, Q.; Chen, Z.; Yue, Q.; Feng, P. Research of Carbon Emission in Whole Manufacturing Process of Steel Structure and Carbon Emission Reduction. *Build. Struct. Chin.* **2023**, *53*, 8–13. [CrossRef]
9. Zhang, X.; Chen, H.; Li, R.; Li, Y.; Chen, Z.; Li, J.; Cao, H.; Lin, B.; Zhou, H.; Huang, Z. Research on Carbon Emissions of Production Processes in the Production Stage of Steel Structure Prefabricated Products. *Ind. Constr. Chin.* **2024**, *54*, 140–148.

10. Yang, S.; Zhou, X.; Liu, Z.; Wang, Z.; Liu, X. Research on the Calculation Method of Carbon Emission of Building Steel Structure Manufacturing and Processing Workshop. *Build. Sci. Chin.* **2023**, *39*, 19–25. [CrossRef]
11. Li, Q.; Yue, Q.; Jin, H.; Feng, P. Research Progress on Carbon Emission of Steel Structure Buildings Based on Carbon Peak and Carbon Neutral. *Build. Struct. Chin.* **2023**, *53*, 1–7+36. [CrossRef]
12. China Planning Press. *Zhejiang Province Building Construction and Decoration Engineering Budget Quota*; China Planning Press: Beijing, China, 2018. (In Chinese)
13. Hu, W.; Shi, J.; Chen, Y.; Zhu, L.; Zheng, S.; Feng, Q. Fine Modeling of Spatial Three-Dimensional Curved Steel Box Girders Based on BIM Technology. *Results Eng.* **2025**, *25*, 103729. [CrossRef]
14. *BS EN 1090-2:2018*; Execution of Steel Structures and Aluminium Structures—Part 2: Technical Requirements for Steel Structures. British Standards Institution (BSI): London, UK, 2018.
15. *GB 20205—2006*; Code for Acceptance of Construction Quality of Steel Structure. China Architecture and Building Press: Beijing, China, 2006.
16. Australian Steel Institute (AESS). *Code of Practice (for Fabricators) Architecturally Exposed Structural Steel (AESS)*; ASI AESS Document F; Australian Steel Institute (AESS): Sydney, Australia, 2012.
17. *GB 50661—2011*; Code for Welding of Steel Structures. China Architecture and Building Press: Beijing, China, 2011.
18. *ISO 13920:2023*; Welding Tolerances for Welded Constructions. ISO: Geneva, Switzerland, 2023.
19. Wang, G.; Lin, B. Detailed Design Method of Steel Plate Beam with Straight Instead of Curved. *Constr. Des. Proj. Chin.* **2023**, 102–104. [CrossRef]
20. Björck, Å. *Numerical Methods for Least Squares Problems*; Society for Industrial and Applied Mathematics: Philadelphia, PA, USA, 1996.
21. Feng, C.; Taguchi, Y.; Kamat, V.R. Fast Plane Extraction in Organized Point Clouds Using Agglomerative Hierarchical Clustering. In Proceedings of the 2014 IEEE International Conference on Robotics and Automation (ICRA), Hong Kong, China, 31 May–7 July 2014.
22. Proenca, P.F.; Gao, Y. Fast Cylinder and Plane Extraction from Depth Cameras for Visual Odometry. In Proceedings of the 2018 IEEE/RSJ International Conference on Intelligent Robots and Systems (IROS), Madrid, Spain, 1–5 October 2018; pp. 6813–6820. [CrossRef]
23. Greenbaum, A.; Chartier, T.P. *Numerical Methods*; Princeton University Press: Princeton, NJ, USA, 2012.
24. Irons, B.M. Engineering Applications of Numerical Integration in Stiffness Methods. *AIAA J.* **1966**, *4*, 2035–2037. [CrossRef]
25. Zienkiewicz, O. *The Finite Element Method in Engineering Science*; Elsevier: Amsterdam, The Netherlands, 1971.
26. Zienkiewicz, O.C.; Taylor, R.L.; Taylor, R.L. *The Finite Element Method: Basic Formulation and Linear Problems*; Elsevier: Amsterdam, The Netherlands, 1989.
27. Bathe, K.-J. *Finite Element Procedures*; Prentice Hall: Upper Saddle River, NJ, USA, 1995.
28. Curry, H.B. The Method of Steepest Descent for Non-Linear Minimization Problems. *Q. Appl. Math.* **1944**, *2*, 258–261. [CrossRef]
29. Strang, G.; Herman, E.P. *Calculus*; Wellesley-Cambridge Press: Wellesley, MA, USA, 2016.
30. Huang, Z.; Ding, J.; Xiang, S. Suspension Footbridge Form-Finding with Laplacian Smoothing Algorithm. *Int. J. Steel Struct.* **2020**, *20*, 1989–1995. [CrossRef]
31. SBP Sonne GmbH. Yuanshan Bridge (Xiamen Mountains to Sea Trail). Available online: https://www.sbp.de/en/project/yuanshan-bridge-xiamen-mountains-to-sea-trail/ (accessed on 19 April 2025).
32. Yang, Y. Hemei Bridge, Xiamen, China. *World Archit. Chin.* **2022**, 92–95. [CrossRef]
33. Adriaenssens, S.; Boegle, A. Innovative Structural Typologies. In *Innovative Bridge Design Handbook*; Elsevier: Amsterdam, The Netherlands, 2022; pp. 215–228. ISBN 978-0-12-823550-8.
34. Warner, E. Liberty Bridge Celebrates 15th Year. *Greenville Journal*, 12 September 2019. Available online: https://greenvillejournal.com/community/liberty-bridge-celebrates-15th-year/ (accessed on 19 April 2025).

MDPI AG
Grosspeteranlage 5
4052 Basel
Switzerland
Tel.: +41 61 683 77 34

Buildings Editorial Office
E-mail: buildings@mdpi.com
www.mdpi.com/journal/buildings

www.ingramcontent.com/pod-product-compliance
Lightning Source LLC
Chambersburg PA
CBHW042014260926
51850CB00020B/3383